The Plant Sciences

Series Editors

The volumes in this series form the world's most comprehensive reference on the plant sciences. Composed of ten volumes, The Plant Sciences provides both background and essential information in plant biology, exploring such topics as genetics and genomics, molecular biology, biochemistry, growth and development, and ecology and the environment. Available through both print and online mediums, the online text will be continuously updated to enable the reference to remain a useful authoritative resource for decades to come.

With broad contributions from internationally well-respected scientists in the field, The Plant Sciences is an invaluable reference for upper-division undergraduates, graduate students, and practitioners looking for an entry into a particular topic.

Series Titles

1. Genetics and Genomics
2. Molecular Biology
3. Biochemistry
4. Cell Biology
5. Growth and Development
6. Physiology and Function
7. Biotic Interactions
8. Ecology and the Environment
9. Evolution, Systematics and Biodiversity
10. Applications

More information about this series at: http://www.springer.com/series/11785

Stephen H. Howell
Editor

Molecular Biology

With 84 Figures and 10 Tables

Springer Reference

Editor
Stephen H. Howell
Development and Cell Biology
Iowa State University
Ames, Iowa, USA

ISBN 978-1-4614-7569-9 ISBN 978-1-4614-7570-5 (eBook)
ISBN 978-1-4614-7571-2 (print and electronic bundle)
DOI 10.1007/978-1-4614-7570-5
Springer New York Heidelberg Dordrecht London

Library of Congress Control Number: 2014951434

Printed on acid-free paper

Springer is part of Springer Science+Business Media (www.springer.com)

Preface

No area of plant sciences has had more spectacular achievements in the past 40 years than plant molecular biology. Though some may argue, the area had its genesis in 1976 with the first NATO meeting on plant molecular biology. At that time, two Harvard scientists, John Bedbrook and Lawrence Bogorad, announced that they had developed a restriction enzyme map of the *Chlamydomonas* chloroplast genome. It was a major accomplishment at the time, but since then, even more exciting breakthroughs have occurred – important plant genes have been cloned, plant transformation techniques developed, metabolic and signaling pathways identified, and whole genomes sequenced. Given the enormous scope of plant molecular biology that has developed over the years, it would be impossible to cover the area in one volume. Instead, special areas have been highlighted to give you a flavor of what has been accomplished and what lies on the horizon.

In the past few years, plant genomics has captured much of the attention in plant molecular biology; therefore, several of the chapters deal with the genome. Gutierrez et al. write in chapter "Replication of the Plant Genome" that replication of the plant genome is a huge challenge for a cell because at each division cycle, the entire genome must be fully and faithfully replicated. This is so that a new daughter cell will receive a genome copy identical to that of the parent. A major conundrum in DNA synthesis is the replication of the ends of chromosomes, called telomeres. Failure to replicate the ends of chromosomes would leave them open and result in chromosome instability. In the chapter "Plant telomeres and telomerase," Nelson et al. describe how the highly conserved telomerase enzyme seals the ends of chromosomes and they discuss how telomere replication in plants compares to animals.

Although plant genomes are faithfully replicated in each cell cycle, the integrity of the genome is always at stake because many biotic and/or abiotic perturbations can damage DNA and produce chromosome breaks or mutations. In the chapter "DNA repair and recombination in plants," Schröpfer et al. describe the many different mechanisms by which DNA lesions are repaired.

One of the most exciting areas of plant molecular biology in the past decade has been the discovery of small noncoding RNAs. Small RNAs play varied roles in gene regulation – silencing genes and serving to guide the machinery for chromatin modification to target genes. In the chapter "Small RNAs in plants," Won et al. describe the various classes of small RNAs in plants and discuss how they are synthesized and processed and function.

The bottomline for gene expression in plants is to make proteins. The translational apparatus that synthesizes proteins is an incredible molecular machine. The chapter "Plant translational machinery" by Browning describes the many factors that associate with ribosomes to promote the initiation, elongation, and termination of protein synthesis.

Plant cell organelles such as chloroplasts engage in two-way communication with the nucleus to maintain their integrity and regulate their functions. Larkin in the chapter "Chloroplast signaling in plants" describes plastid-to-nucleus communication during plant development and at times when plants accommodate to changes in photosynthesis activity. Since most chloroplast proteins are encoded by the nucleus, the nucleus communicates with plastids by activating the expression of plastid genes.

Another cytoplasmic organelle, the endoplasmic reticulum, also engages in two-way communication with the nucleus. The nucleus encodes proteins that make up the ER, and the ER in turn signals its status to the nucleus. In the chapter "ER stress signaling in plants," Howell describes a condition in which a plant is subjected to stress and the ER sends special sensors/transducers to the nucleus to activate stress response genes.

Normal growth and development and other responses to environmental conditions in plants are complex processes and rely on the operation of signaling pathways involving hormones and other cell-signaling components. The most prominent and versatile hormone in plants is auxin. As Rechenmann relates in the chapter "Auxin signaling in plants," the multitude of auxin functions is brought about by different auxins, the complexity of auxin metabolism, the gradients of the hormone generated by auxin transporters, and the combinatorial action of the large families of auxin receptors and coreceptors.

Cytokinin, the hormone best known for its action as a counterpoint to that of auxin in plant regeneration, also has pleiotrophic effects. As explained by Cheng and Kieber in the chapter "Cytokinin signaling in plants," cytokinin is unique in that the hormone uses a "two-component pathway" with a receptor transducing a signal through phosphorelay systems to nuclear-localized effectors called response regulators. Another class of hormones, the brassinosteroids, are also involved in a diversity of plant cellular functions, particularly in growth and developmental processes involving cell elongation. As Clouse points out in the chapter "Brassinosteroid signaling in plants," brassinosteroid signaling involves a membrane receptor, a signaling pathway involving protein phosphorylation, and the activation of a large number of genes.

COP9, originally discovered as a gene that negatively controls photomorphogenesis in the dark, is now recognized as part of a multifunctional complex called the COP9 signalosome. In the chapter "COP9 signalosome network," Franciosini et al. disclose how the COP9 signalosome takes part in many plant development processes and environmental responses including photomorphogenesis.

Gene regulatory networks allow plants to integrate a variety of inputs and/or generate diverse outputs. As described by Malapeira et al. in the chapter "Plant circadian network," numerous plant genes are regulated by the circuitry and feedback mechanisms of the circadian clock, which orchestrates many of the diurnal rhythms in plants. Gene networks are also involved in integrating stress information. Kuromori et al. in the chapter "Drought stress signaling network" detail how the gene network that responds to drought stress insures plant survival. Part of the drought stress gene network involves responses associated with the plant hormone abscisic acid, and part of the pathway acts independently of the hormone.

Altogether, these chapters highlight the many advances in plant molecular biology and provide a foundation for the study of other areas of plant sciences.

Ames, Iowa, USA
May 2014

Stephen H. Howell

Series Preface

Plant sciences is in a particularly exciting phase, with the tools of genomics, in particular, turbo-charging advances in an unprecedented way. Furthermore, with heightened attention being paid to the need for increased production of crops for food, feed, fuel, and other needs and for this to be done both sustainably and in the face of accelerating environmental change, plant science is arguably more important and receiving more attention than ever in history. As such, the field of plant sciences is rapidly changing, and this requires new approaches for the teaching of this field and the dissemination of knowledge, particularly for students. Fortunately, there are also new technologies to facilitate this need.

In this 10-volume series, *The Plant Sciences*, we aim to develop a comprehensive online and printed reference work. This is a new type of publishing venture exploiting Wiki-like capabilities, thus creating a dynamic, exciting, cutting-edge, and living entity.

The aim of this large publishing project is to produce a comprehensive reference in plant sciences. *The Plant Sciences* will be published both in print and online; the online text can be updated to enable the reference to remain a useful authoritative resource for decades to come. The broader aim is to provide a sustainable superstructure on which can be built further volumes or even series as plant science evolves. The first edition will contain 10 volumes.

The Plant Sciences is part of SpringerReference, which contains all Springer reference works. Check out the link at http://www.springerreference.com/docs/index.html#Biomedical+and+Life+Sciences-lib1, from where you can see the volumes in this series that are already coming online.

The target audience for the initial 10 volumes is upper-division undergraduates as well as graduate students and practitioners looking for an entry on a particular topic. The aim is for *The Plant Sciences* to provide both background and essential information in plant biology. The longer-term aim is for future volumes to be built (and hyperlinked) from the initial set of volumes, particularly targeting the research frontier in specific areas.

The Plant Sciences has the important extra dynamic dimension of being continually updated. *The Plant Sciences* has a constrained Wiki-like capability, with all original authors (or their delegates) being able to modify the content.

Having satisfied an approval process, new contributors will also be registered to propose modifications to the content.

It is expected that new editions of the printed version will be published every 3–5 years. The project is proceeding volume by volume, with volumes appearing as they are completed. This also helps to keep the text fresher and the project more dynamic.

We would like to thank our host institutions, colleagues, students, and funding agencies, who have all helped us in various ways and thus facilitated the development of this series. We hope this volume is used widely and look forward to seeing it develop further in the coming years.

King Abdullah University of Science & Technology, Thuwal, Saudi Arabia — Mark Tester

School of Plant Sciences, University of Arizona, Tucson, AZ, USA — Richard Jorgensen

22 July 2014

Editor Biography

Stephen Howell is a Professor of Genetics, Development, and Cell Biology at Iowa State University in Ames, Iowa. He received his B.A. degree from Grinnell College in Grinnell, Iowa, and a Ph.D. degree from Johns Hopkins University in Baltimore, Maryland. He did postdoctoral work at the University of California, San Diego, and continued as a faculty member there for 20 years. He then became the Vice President for Research at the Boyce Thompson Institute at Cornell University in Ithaca, New York. Dr. Howell went to Iowa State in 2001 as Director of the Plant Sciences Institute. In 2009, he became Director of the Division of Molecular and Cellular Biosciences at the National Science Foundation. After a 2-year stint in Washington, Dr. Howell returned to Iowa State to resume his research and teaching roles. In 1980, Howell's lab was the first to introduce recombinant DNA in plants – a cloned form of the Cauliflower Mosaic Virus genome. A few years later in 1986, Howell, along with Marlene DeLuca and Donald Helinski, introduced the firefly luciferase gene into plants to produce an iconic image of a "glowing tobacco plant." In 1992, Howell organized a Gordon Conference on Plant Molecular Biology, and from 1998 to 2001, he was Editor-in-Chief of the journal *Plant Molecular Biology*. Most recently, Howell's lab has been involved in studies of ER stress in plants. The ER stress response system confers on plants the capacity to sense and respond to adverse environmental conditions – a major issue in the face of global climate change.

Series Editors Biography

Mark Tester is Professor of Bioscience in the Center for Desert Agriculture and the Division of Biological and Environmental Sciences and Engineering, King Abdullah University for Science and Technology (KAUST), Saudi Arabia. He was previously in Adelaide, where he was a Research Professor in the Australian Centre for Plant Functional Genomics and Director of the Australian Plant Phenomics Facility. Mark led the establishment of this facility, a $55 million organisation that develops and delivers state-of-the-art phenotyping facilities, including The Plant Accelerator, an innovative plant growth and analysis facility. In Australia, he led a research group in which forward and reverse genetic approaches were used to understand salinity tolerance and how to improve this in crops such as wheat and barley. He moved to KAUST in February 2013, where this work is continuing, expanding also into work on the salinity tolerance of tomatoes.

Mark Tester has established a research program with the aim of elucidating the molecular mechanisms that enable certain plants to thrive in sub-optimal soil conditions, in particular in soils with high salinity. The ultimate applied aim is to modify crop plants in order to increase productivity on such soils, with consequent improvement of yield in both developed and developing countries. The ultimate intellectual aim is to understand the control and co-ordination of whole plant function through processes occurring at the level of single cells, particularly through processes of long-distance communication within plants.

A particular strength of Professor Tester's research programme is the integration of genetics and genomics with a breadth of physiological approaches to enable novel gene discovery. The development and use of tools for the study and manipulation of specific cell types adds a useful dimension to the research. Professor Tester received training in cell biology and physiology, specialising in work on ion transport, particularly of cations, across the plasma membrane of plant cells. His more recent focus on salinity tolerance is driven by his desire to apply his training in fundamental plant processes to a problem of practical significance.

Professor Tester was awarded a Junior Research Fellowship from Churchill College, Cambridge, in 1988, a BBSRC (UK) Research Development Fellowship in 2001, and an Australian Research Council Federation Fellowship in 2004. Professor Tester obtained his Bachelor's degree in Botany from the University of Adelaide in 1984, and his Ph.D. in Biophysics from the University of Cambridge in 1988.

Dr. Richard Jorgensen, Professor Emeritus, School of Plant Sciences, University of Arizona, Tucson, AZ, USA

Dr. Jorgensen is a recognized international leader in the fields of epigenetics, functional genomics, and computational biology. His research accomplishments include the discovery in plants of a gene-silencing phenomenon called cosuppression, which led to the discovery in animals of RNA interference, a gene-silencing tool that has major potential implications for medicine including the treatment of diseases such as cancer, hepatitis, and AIDS. In 2007, he was awarded the Martin Gibbs Medal for this groundbreaking work in cosuppression and RNAi by the American Society of Plant Biologists (ASPB). He was elected a Fellow of the American Association for the Advancement of Sciences (AAAS) in 2005 and an Inaugural Fellow of the ASPB in 2007.

Dr. Jorgensen was the founding Director of the iPlant Collaborative, a 5 years, \$50 M NSF project to develop cyber-infrastructure for plant sciences. Dr. Jorgensen also served as the Editor in Chief of *The Plant Cell,* the leading research journal in plant biology, from 2003 to 2007. He is currently Editor in Chief of *Frontiers in Plant Science*, a cutting-edge, open-access journal allied with Nature Publishing Group. He is also Series Editor for the book series *Plant Genetics and Genomics: Crops and Models* for Springer Publishing. Dr. Jorgensen has published numerous scientific articles and is regularly invited to present his research findings at universities, research institutions, and scientific conferences nationally and internationally.

Contents

Contributors

Irene Aragüez Centro de Biología Molecular 'Severo Ochoa', Consejo Superior de Investigacion, Universidad Autónoma de Madrid (UAM), Madrid, Spain

Mark A. Beilstein The School of Plant Sciences, University of Arizona, Tucson, AZ, USA

Reyes Benlloch Molecular Genetics, Centre for Research in Agricultural Genomics (CRAG), Barcelona, Spain

Consortium CSIC–IRTA–UAB–UB, Center for Research in Agricultural Genomics (CRAG), Parc de Recerca UAB, Barcelona, Spain

Karen S. Browning Department of Molecular Biosciences and the Insitute for Cellular and Molecular Biology, The University of Texas at Austin, Austin, TX, USA

Xuemei Chen Department of Botany and Plant Sciences, Institute of Integrative Genome Biology, Center for Plant Cell Biology, University of California, Riverside, CA, USA

Howard Hughes Medical Institute, University of California, Riverside, CA, USA

Chia-Yi Cheng Department of Biology, University of North Carolina, Chapel Hill, NC, USA

Steven Clouse Department of Horticultural Science, College of Agriculture and Life Sciences, North Carolina State University, Raleigh, NC, USA

Xing-Wang Deng Department of Molecular, Cellular and Developmental Biology, Yale University, New Haven CT, USA

Anna Franciosini "Charles Darwin" Department of Biology and Biotechnology, Sapienza Università di Roma, Rome, Italy

Crisanto Gutierrez Centro de Biología Molecular 'Severo Ochoa', Consejo Superior de Investigacion, Universidad Autónoma de Madrid (UAM), Madrid, Spain

Rossana Henriques Molecular Genetics, Centre for Research in Agricultural Genomics (CRAG), Barcelona, Spain

Consortium CSIC–IRTA–UAB–UB, Center for Research in Agricultural Genomics (CRAG), Parc de Recerca UAB, Barcelona, Spain

Stephen Howell Plant Sciences Institute, Genetics, Development and Cell Biology Department, Iowa State University, Ames, IA, USA

Joseph J. Kieber Department of Biology, University of North Carolina, Chapel Hill, NC, USA

Alexander Knoll Botanical Institute II, Karlsruhe Institute of Technology, Karlsruhe, Germany

Takashi Kuromori Gene Discovery Research Group, RIKEN Center for Sustainable Resource Science, Yokohama, Japan

Robert M. Larkin Michigan State University-Department of Energy Plant Research Laboratory, Michigan State University, East Lansing, MI, USA

Jordi Malapeira Molecular Genetics, Centre for Research in Agricultural Genomics (CRAG), Barcelona, Spain

Consortium CSIC–IRTA–UAB–UB, Center for Research in Agricultural Genomics (CRAG), Parc de Recerca UAB, Barcelona, Spain

Paloma Mas Molecular Genetics, Centre for Research in Agricultural Genomics (CRAG), Barcelona, Spain

Consortium CSIC–IRTA–UAB–UB, Center for Research in Agricultural Genomics (CRAG), Parc de Recerca UAB, Barcelona, Spain

Junya Mizoi Laboratory of Plant Molecular Physiology, Graduate School of Agricultural and Life Sciences, The University of Tokyo, Tokyo, Japan

Andrew D. L. Nelson The School of Plant Sciences, University of Arizona, Tucson, AZ, USA

Catherine Perrot-Rechenmann Institut des Sciences du Végétal (ISV), UPR2355 CNRS, Saclay Plant Sciences LabEx, Gif sur Yvette Cedex, France

Holger Puchta Botanical Institute II, Karlsruhe Institute of Technology, Karlsruhe, Germany

Susan Schröpfer Botanical Institute II, Karlsruhe Institute of Technology, Karlsruhe, Germany

Joana Sequeira-Mendes Centro de Biología Molecular 'Severo Ochoa', Consejo Superior de Investigacion, Universidad Autónoma de Madrid (UAM), Madrid, Spain

Giovanna Serino "Charles Darwin" Department of Biology and Biotechnology, Sapienza Università di Roma, Rome, Italy

Kazuo Shinozaki Gene Discovery Research Group, RIKEN Center for Sustainable Resource Science, Yokohama, Japan

Dorothy E. Shippen Department of Biochemistry and Biophysics, College of Agriculture and Life Sciences, Texas A&M University, College Station, TX, USA

Oliver Trapp Botanical Institute II, Karlsruhe Institute of Technology, Karlsruhe, Germany

Taishi Umezawa Faculty of Agriculture and Graduate, School of Bio-Applications and Systems Engineering, Tokyo University of Agriculture and Technology, Tokyo, Japan

So Youn Won Department of Botany and Plant Sciences, Institute of Integrative Genome Biology, Center for Plant Cell Biology, University of California, Riverside, CA, USA

Kazuko Yamaguchi-Shinozaki Laboratory of Plant Molecular Physiology, Graduate School of Agricultural and Life Sciences, The University of Tokyo, Tokyo, Japan

Rae Eden Yumul Department of Botany and Plant Sciences, Institute of Integrative Genome Biology, Center for Plant Cell Biology, University of California, Riverside, CA, USA

Replication of the Plant Genome

1

Crisanto Gutierrez, Joana Sequeira-Mendes, and Irene Aragüez

Contents

Abstract

- All multicellular organisms, whether plants or animals, have similar requirements and challenges in replicating their genomes.
- At each cell cycle, the entire genome must be fully and faithfully duplicated, so that identical copies of the genome are delivered to the new daughter cells during mitosis.
- The first experimental work aimed at elucidating the mechanism of DNA replication was performed in plants and demonstrated unequivocally that DNA and chromosomal replication occurs through a semiconservative mechanism.

C. Gutierrez (✉) • J. Sequeira-Mendes • I. Aragüez
Centro de Biología Molecular 'Severo Ochoa', Consejo Superior de Investigacion, Universidad Autónoma de Madrid (UAM), Madrid, Spain
e-mail: cgutierrez@cbm.csic.es; ueiramendes@cbm.uam.es; irenearag@yahoo.com

S.H. Howell (ed.), *Molecular Biology, The Plant Sciences 2*,
DOI 10.1007/978-1-4614-7570-5_1

- The greatest difference between the proteins forming the pre-replication complex (pre-RC) in animal and plant cells resides in the cell division cycle10-dependent transcription 1 (CDT1)-interacting proteins in both kingdoms. In animals, CDT1 interacts with geminin, while plants lack a geminin ortholog, and encode a completely unrelated CDT1-interacting protein, *GLABRA2* (GL2) *EXPRESSION MODULATOR* (GEM).
- Although consensus sequences have not been identified for *Arabidopsis* replication origins, they tend to be rich in G+C and are similar in their basic characteristics to those of animal cells.
- Plant replication origins are largely associated with histone H2A.Z and H3.3 variants, as well as with histone modifications characteristic of active transcription.
- The specific features of several components of the replication machinery are well characterized in plants. This is the case of DNA polymerases, the accessory factors, and the enzymes involved in the maturation of Okazaki fragments.
- As was first shown in plants, the replication of the genome does not occur randomly but instead is organized temporally into early and late replicating regions.
- Endoreplication, i.e., genome duplication without mitosis, is frequent in many cell types within the adult plant body. It occurs in all multicellular eukaryotes, being very common and functionally relevant in plants, in which it is necessary for cell growth and differentiation.
- DNA replication in plants has some specific features mainly derived from the regulation of protein availability, the frequent occurrence of endoreplication, and the response to hormonal, developmental, and environmental cues.

Introduction

One of the most fundamental features of all living organisms, either unicellular or multicellular, is their ability to reproduce. This relies primarily on the premise that their genetic material, the DNA, must be preserved from one generation to the next. In unicellular organisms, a single cell division leads to two new ones. In the case of multicellular organisms, made up of trillions of cells, the single fact of becoming an adult requires that the single zygotic cell and its descendants undergo an enormous amount of cell divisions to attain the large number of cells that conform the organism. In addition, during the adult life, cell homeostasis is a highly dynamic process that involves both the death of many cells and their continuous replacement with newly formed ones. These new cells are derived from special cell types, the stem cells, which are confined in certain niches within the body. The case of the renewal of blood cells or the epithelial cells in the gut, in animals, and the formation of new roots and leaves, in plants, are good examples.

The genetic information encoded in the genomic DNA must be duplicated during each cell cycle. The cell cycle is normally divided into a G1 (or gap) phase where the cell prepares to duplicate its genome during the next synthesis phase (S phase). This is followed by another gap phase (G2), where the integrity of the genome is evaluated at a checkpoint. Subsequently mitosis and cytokinesis take place, where all cellular components are divided into the two daughter cells. During the S phase of the cell cycle, the genome is replicated fully and faithfully, so that identical copies of the genome are delivered to the new daughter cells during mitosis. This is not at all a trivial task. In fact, genome replication is a highly risky process. The first challenge is posed by the size of the genome, which in many cases can be considerably large, e.g., hundreds or thousands of millions of base pairs (bp). In addition, the genome is not a single DNA molecule, but it is normally segmented into smaller molecules or chromosomes.

As mentioned above, genomic DNA must be duplicated in every cell cycle but only once (Blow and Laskey 1988; Arias and Walter 2007). Therefore, organisms have evolved mechanisms to strictly prevent parts of the genome from replicating more than once, avoiding the genomic instability that otherwise might occur. This is achieved through the so-called "licensing" mechanism that allows cells to monitor and identify genomic regions already replicated. Equally deleterious would be regions of the genome that remained under replicated at the time of cell division.

Another major challenge, derived from the large size of genomes, is that if replication started in only a single genomic site, the duplication of the entire genome would take ~2 years for a mammalian cell, considering a genome size of 3.3×10^9 bp and an average bidirectional DNA replication rate of ~1.5 kbp/min. In the case of *Arabidopsis thaliana*, with a relatively small genome of 1.2×10^8 bp, completion of the genome replication process would take ~1 month. Remarkably, full genome replication, even in the case of these large and complex genomes, occurs within a few hours. Evolution has solved this apparent problem by using multiple sites where DNA replication starts, the replication origins (ORIs), which are scattered over several kbp across the entire genome. In turn, the presence of multiple ORIs imposed the need to evolve mechanisms that define certain genomic locations as potential replication start sites (Gilbert 2010; DePamphilis and Bell 2011; Bell et al. 2012). In addition, the activity of the many thousands of ORIs must be strictly coordinated, as we will discuss later in this chapter.

Genomic DNA is not naked within the nucleus. Instead, it is associated with proteins to form chromatin. Structurally, chromatin is made up of nucleosomes, each composed of eight histone molecules; two copies of each histone (H2A, H2B, H3, and H4) form a protein core around which ~150 bp of DNA is wrapped. This pattern is repeated across the entire genome together with histone H1 and a plethora of nonhistone proteins to form chromatin. Therefore, the replication of the genome involves duplicating DNA and replicating the patterns of proteins associated with it. This involves a dynamic process of disassembly and reassembly of chromatin components, whereby not only the several dozens of proteins that form the DNA replication machinery, but also protein complexes responsible for chromatin dynamics, need to coordinate their activities.

The challenges for genomic replication involve more than managing large genomes. Genetic material is not static. Instead, it is continuously being transcribed to produce multiple RNA molecules (tRNAs, rRNA, mRNAs, miRNAs, among others). Hence, genomic replication must be coordinated with transcriptional activity of multiple regions across the genome. Transcriptional activity depends in many cases of the presence of epigenetic marks, which are modifications in the DNA, e.g., C methylation, and in the associated histones (acetylation, methylation of lysine and arginine residues, among others). These epigenetic modifications modulate the activity of certain genomic *loci* without changing the genetic information of the underlying DNA primary sequence. Likewise, DNA replication is also affected by epigenetic changes, and, in fact, chromatin and epigenetic states seem to be at the basis of DNA replication control.

All these topics are relevant for replication of any genome, both in animal and plant cells. Genomic replication in plant cells has similar challenges to those in animal cells (Bryant 2010). Here, we will discuss plant-specific features, and the reader is referred to general reviews and textbooks to find details about the mechanistic aspects of DNA replication itself, which are largely common to both plant and animal cells (DePamphilis and Bell 2011; Bell et al. 2012). Table 1 is also provided where each protein discussed in the text is referred to its gene code and a link to The *Arabidopsis* Information Resource (TAIR) where detailed information about its genomic structure and expression data is available.

DNA Replication in Plants: Early Observations

It is now 60 years since Watson and Crick discovered the structure of DNA. Their original publication was visionary in describing the structure of DNA as it relates to DNA replication. They described DNA has having two antiparallel single-stranded (ss) DNA molecules maintained together by complementary base pairing such that replication could occur by "simply" separating the two DNA strands and using each of them as a master template to generate two new identical daughter double-stranded (ds) DNA molecules. However, given the structure of DNA as they described it, DNA replication could occur in a conservative way in which an entirely new dsDNA is produced and transferred to one daughter cell while the old molecule goes to the other, in a semiconservative way in which one original and one new DNA strand go to each daughter cell or in a dispersive manner in which newly synthesized DNA is dispersed among old non-replicated regions (Fig. 1a). Several experiments were conducted in the following years to determine which mechanism operates for DNA replication. These experiments were carried out by John H. Taylor and coworkers and published in 1957 using a cytological approach to visualize chromosomal DNA replication in the common bean, *Vicia faba* (Taylor et al. 1957). They were pioneering studies and first experimental work on DNA replication in plants and actually in all multicellular eukaryotes. In these experiments, bean roots containing proliferating cells were incubated in the presence of tritiated thymidine (^{3}H-Thd) to allow first its intracellular conversion into ^{3}H-dTTP

Table 1 List of *Arabidopsis* proteins mentioned in this article, indicating their gene code and the link to TAIR (The *Arabidopsis* Information Resource) where detailed information can be obtained

Gene symbol	Gene ID	TAIR Link
ORC1A	AT4G14700	http://www.arabidopsis.org/servlets/TairObject?id=128799&type=locus
ORC1B	AT4G12620	http://www.arabidopsis.org/servlets/TairObject?id=129550&type=locus
ORC2	AT2G37560	http://www.arabidopsis.org/servlets/TairObject?id=31602&type=locus
ORC3	AT5G16690	http://www.arabidopsis.org/servlets/TairObject?id=131234&type=locus
ORC4	AT2G01120	http://www.arabidopsis.org/servlets/TairObject?id=32459&type=locus
ORC5	AT4G29910	http://www.arabidopsis.org/servlets/TairObject?id=127869&type=locus
ORC6	AT1G26840	http://www.arabidopsis.org/servlets/TairObject?id=137403&type=locus
CDC6A	AT2G29680	http://www.arabidopsis.org/servlets/TairObject?id=34670&type=locus
CDC6B	AT1G07270	http://www.arabidopsis.org/servlets/TairObject?id=26699&type=locus
CDT1A	AT2G31270	http://www.arabidopsis.org/servlets/TairObject?id=31914&type=locus
CDT1B	AT3G54710	http://www.arabidopsis.org/servlets/TairObject?id=40497&type=locus
MCM2	AT1G44900	http://www.arabidopsis.org/servlets/TairObject?id=29628&type=locus
MCM3	AT5G46280	http://www.arabidopsis.org/servlets/TairObject?id=134123&type=locus
MCM4	AT2G16440	http://www.arabidopsis.org/servlets/TairObject?id=31938&type=locus
MCM5	AT2G07690	http://www.arabidopsis.org/servlets/TairObject?id=33726&type=locus
MCM6	AT5G44635	http://www.arabidopsis.org/servlets/TairObject?id=500229907&type=locus
MCM7/ PROLIFERA	AT4G02060	http://www.arabidopsis.org/servlets/TairObject?id=129113&type=locus
RBR	AT3G12280	http://www.arabidopsis.org/servlets/TairObject?id=36966&type=locus
GEM	AT2G22475	http://www.arabidopsis.org/servlets/TairObject?id=500231580&type=locus
ABAP1	AT5G13060	http://www.arabidopsis.org/servlets/TairObject?id=135427&type=locus
ETG1	AT2G40550	http://www.arabidopsis.org/servlets/TairObject?id=34868&type=locus
ATXR5	AT5G09790	http://www.arabidopsis.org/servlets/TairObject?id=130701&type=locus
ATXR6	AT5G24330	http://www.arabidopsis.org/servlets/TairObject?id=134049&type=locus

(*continued*)

Table 1 (continued)

Gene symbol	Gene ID	TAIR Link
PCNA1	AT1G07370	http://www.arabidopsis.org/servlets/TairObject?id=29065&type=locus
PCNA2	AT2G29570	http://www.arabidopsis.org/servlets/TairObject?id=32002&type=locus
CDC45	AT3G25100	http://www.arabidopsis.org/servlets/TairObject?id=38407&type=locus
SLD5	AT5G49010	http://www.arabidopsis.org/servlets/TairObject?id=131949&type=locus
PSF1	AT1G80190	http://www.arabidopsis.org/servlets/TairObject?id=28180&type=locus
PSF2	AT3G12530	http://www.arabidopsis.org/servlets/TairObject?id=40289&type=locus
PSF3	AT1G19080	http://www.arabidopsis.org/servlets/TairObject?id=27378&type=locus
MCM10	AT2G20980	http://www.arabidopsis.org/servlets/TairObject?id=32674&type=locus
MCM8	AT3G09660	http://www.arabidopsis.org/servlets/TairObject?id=35699&type=locus
POLA1	AT5G67100	http://www.arabidopsis.org/servlets/TairObject?id=132133&type=locus
POLA3	AT1G67320	http://www.arabidopsis.org/servlets/TairObject?id=28573&type=locus
POLA4	AT5G41880	http://www.arabidopsis.org/servlets/TairObject?id=131773&type=locus
POLA2	AT1G67630	http://www.arabidopsis.org/servlets/TairObject?id=26910&type=locus
POLD1	AT5G63960	http://www.arabidopsis.org/servlets/TairObject?id=132854&type=locus
POLD2	AT2G42120	http://www.arabidopsis.org/servlets/TairObject?id=34564&type=locus
POLD3	AT1G78650	http://www.arabidopsis.org/servlets/TairObject?id=137438&type=locus
POLD4	AT1G09815	http://www.arabidopsis.org/servlets/TairObject?id=500231427&type=locus
POLH	AT5G44740	http://www.arabidopsis.org/servlets/TairObject?id=132218&type=locus
POLL	AT1G10520	http://www.arabidopsis.org/servlets/TairObject?id=136248&type=locus
RFC1	AT5G22010	http://www.arabidopsis.org/servlets/TairObject?id=500231351&type=locus
POLE1A	AT1G08260	http://www.arabidopsis.org/servlets/TairObject?id=137012&type=locus
POLE1B	AT2G27120	http://www.arabidopsis.org/servlets/TairObject?id=34456&type=locus
POLE2	AT5G22110	http://www.arabidopsis.org/servlets/TairObject?id=500231370&type=locus

(*continued*)

Table 1 (continued)

Gene symbol	Gene ID	TAIR Link
DPB4/	Gene family	
DPB3A	Gene family	
RPA1A, RPA1B	AT5G06510	http://www.arabidopsis.org/servlets/TairObject?id=130600&type=locus
RPA1C, RPA1D	AT5G45400	http://www.arabidopsis.org/servlets/TairObject?id=133243&type=locus
RPA2A, RPA2B	AT2G24490	http://www.arabidopsis.org/servlets/TairObject?id=34738&type=locus
RPA3A, RPA3B	AT3G52630	http://www.arabidopsis.org/servlets/TairObject?id=37147&type=locus
FEN1	AT5G26680	http://www.arabidopsis.org/servlets/TairObject?id=130957&type=locus
LIG1A, LIG1B	AT1G08130	http://www.arabidopsis.org/servlets/TairObject?id=137008&type=locus
CYCA1, A2, A3	AT1G44110	http://www.arabidopsis.org/servlets/TairObject?id=137805&type=locus
CYCD3	AT4G34160	http://www.arabidopsis.org/servlets/TairObject?id=127929&type=locus
CDKA	AT3G48750	http://www.arabidopsis.org/servlets/TairObject?id=39995&type=locus
CDKB	AT3G54180	http://www.arabidopsis.org/servlets/TairObject?id=36618&type=locus
E2FE/DEL1	AT3G48160	http://www.arabidopsis.org/servlets/TairObject?id=40158&type=locus
FAS1	AT1G65470	http://www.arabidopsis.org/servlets/TairObject?id=30544&type=locus
FAS2	AT5G64630	http://www.arabidopsis.org/servlets/TairObject?id=134746&type=locus
MSI1	AT5G58230	http://www.arabidopsis.org/servlets/TairObject?id=132921&type=locus

and then its incorporation into DNA by DNA polymerases during S phase. After treatment of labeled roots with colchicine, cells accumulated in mitotic metaphase allowing for chromosomes, each composed of two sister chromatids, to be visualized. In addition, the newly synthesized DNA in which ^{3}H-Thd had been incorporated was also visualized by autoradiography, in which the particles emitted by the decay of ^{3}H activate a photographic emulsion overlying the specimen. After 2–3 weeks of exposure, silver deposits were detected as black grains. In the work of Taylor and colleagues, all chromosomes blocked in the first mitotic division after labeling showed two sister chromatids labeled with grains. When colchicine was applied not immediately after labeling but some hours afterward to allow the labeled cells to reach a second mitosis with an intervening S phase in the absence of ^{3}H-Thd, the mitotic chromosomes contained one labeled chromatid and one unlabeled sister chromatid. This pattern of labeling was consistent with

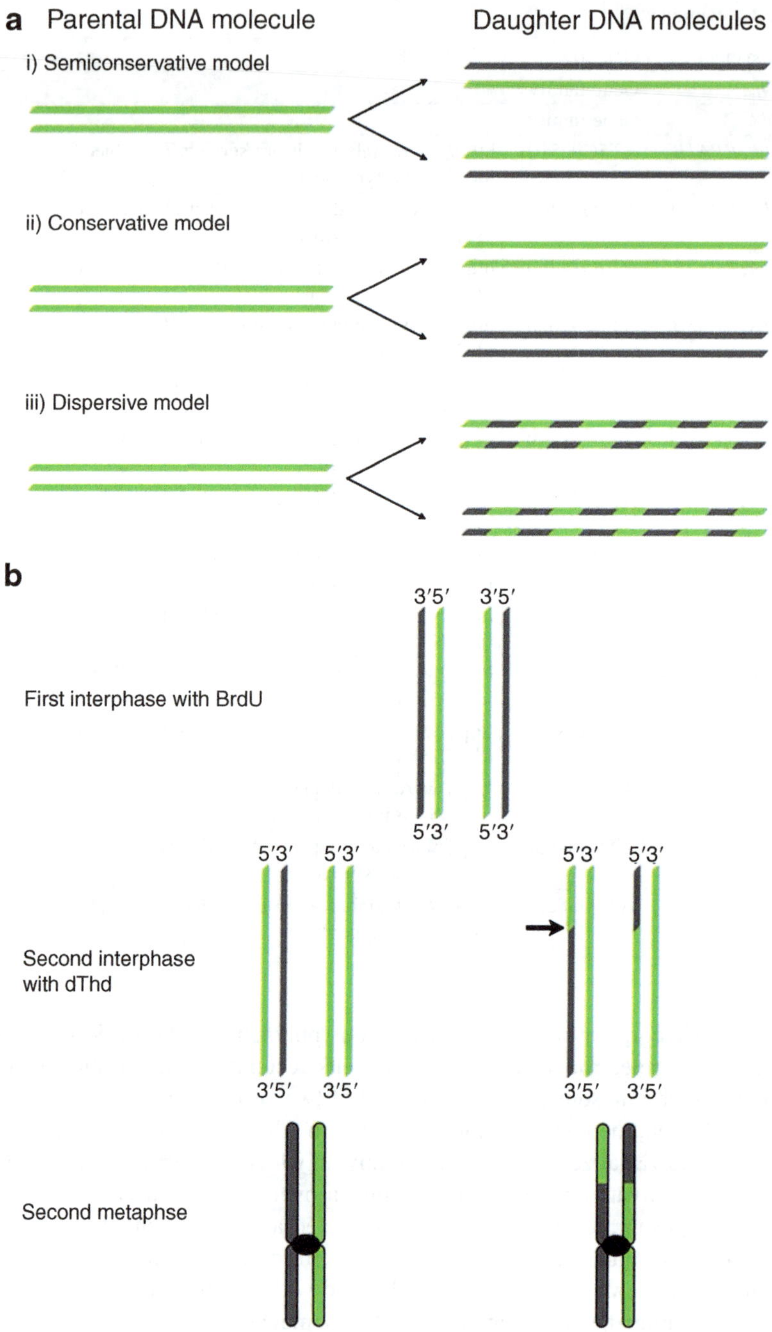

Fig. 1 Semiconservative nature of DNA replication. (**a**) The mechanism of DNA replication can be inferred depending on the pattern of labeling with a deoxynucleotide analog, e.g., tritiated

chromosomal replication occurring through a semiconservative mechanism. Had replication occurred by a conservative mode, then only one sister chromatid would have been labeled in the first mitotic division (Fig. 1a).

In addition, these experiments also served to demonstrate that processes of exchanging genetic material could occur between newly replicated DNA molecules, or chromatids, by so-called sister chromatid exchanges (SCEs; Fig. 1b), that is, reciprocal exchanges between sister chromatids that do not entitle changes of genetic information since they occur between identical DNA molecules (a major difference with exchanges occurring during meiosis). Later, the formation of SCEs was directly related to the presence of DNA damage and, in general, with altered events that take place during uncontrolled DNA replication, expanding the field considerably to connect DNA replication, DNA damage, and DNA repair.

Assembly of Pre-replication Complexes (Pre-RC)

As described above, the entire genome must be fully replicated during each cell cycle, before the cell is ready to trigger cell division, so as to deliver one completed copy of the genome to each daughter cell. However, at the same time, the genome must replicate once and only once. The importance of such restriction is evidenced by the development of multiple mechanisms in the cell to prevent over- or underreplication. Frequently, these mechanisms are species-specific. Overall, the process known as "replication licensing" allows the assembly of pre-replication complexes (pre-RCs) at potential origins of replication (ORIs) during G1 (Arias and Walter 2007). This is strictly dependent upon a suitable and accessible chromatin state. Later during S phase, a subset of the licensed ORIs will fire and start replication.

Pre-RCs are multiprotein assemblies constituted by the origin recognition complex (ORC), a complex of ORC1-6 subunits, cell division cycle 6 (CDC6), CDC10-dependent transcription 1 (CDT1), and the minichromosome maintenance (MCM) complex, formed by the six subunits MCM2-7 (Fig. 2a). Detailed mechanistic studies at the biochemical and molecular level of pre-RC assembly have been carried out in yeast and mammalian cells. In very general terms, CDC6 and CDT1 are able to recognize, by poorly understood mechanisms, some of the sites where ORC is bound to chromatin. This allows the incorporation of the MCM

Fig. 1 (continued) thymidine or 5-bromodeoxyuridine (BrdU). Newly synthesized DNA is depicted in *dark grey*. The appearance of DNA products uniformly labeled in the two daughter molecules demonstrated the DNA molecules are replicated by a semiconservative mechanism. (**b**) Diagram of the expected BrdU-labeling pattern when cells are incubated in the presence of BrdU for one cell cycle and in the absence of this analog in the following cell cycle. The application of staining procedures that allow the distinction between BrdU-containing and BrdU-free DNA molecules serves to visualized in mitotic chromosomes the occurrence of sister chromatid exchanges (SCEs; *arrow*) during the second cell cycle

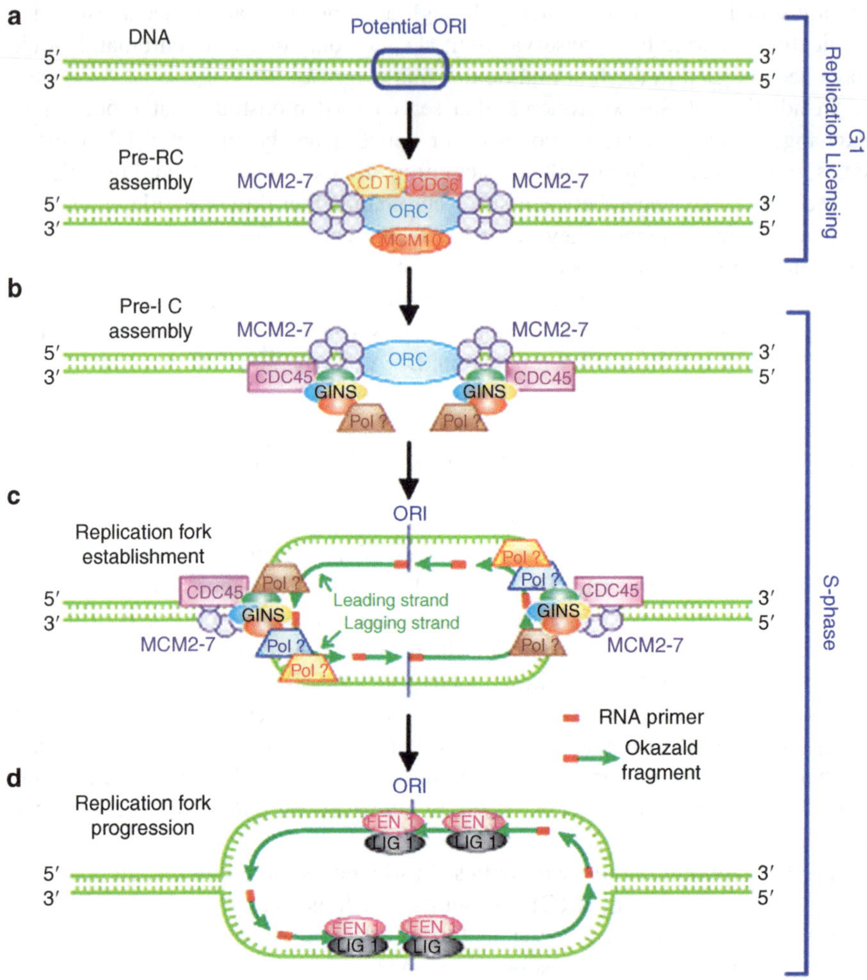

Fig. 2 Major events occurring during initiation of DNA replication in eukaryotic cells. (**a**) A potential ORI becomes licensed for replication during G1 by the assembly of the pre-RC, which is formed by ORC, CDC6, CDT1, and MCM2-7. (**b**) During S phase, DNA replication initiates by unwinding the two strands at the initiation site, which requires the participation of MCM10. This is accompanied by the disassembly of the pre-RC, the subsequent assembly of the pre-IC, and the loading of Pol epolymerases to establish two bidirectional replication forks. Loading of the factors CDC45 and GINS is essential for the establishment of the forks. (**c**) The CDC45-MCM-GINS complex, which functions as the replicative helicase, travels along with the forks. While the 3′–5′ DNA strand can be copied by the DNA polymerase in the 5′–3 direction (leading strand), the synthesis of the other DNA strand (lagging strand) proceeds in a discontinuous manner through Okazaki fragments synthesis. The DNA polymerase a-DNA primase complex synthesizes the RNA primer at each strand and extends the first nucleotides. Then, it is substituted by the DNA polymerases eand din the leading and lagging strands, respectively. In this figure, the DNA polymerase a-DNA primase is depicted in the lagging strand only, where it initiates the synthesis of each new Okazaki fragment. (**d**) The fork progresses, and, in a final step, Okazaki fragments are processed to eliminate the RNA primer at their 5′ ends and ligated. It must be noted that in order to simplify this scheme, some important proteins mentioned in the text have been omitted from this figure

complex into these sites, which will act as part of a replicative helicase once the ORI is activated. The reader is referred to other writings for details about these proteins. Due to the significant homology between yeast, animal, and plant pre-RC components, it is likely that the overall process will occur in a similar manner, although direct studies to prove this assumption are still lacking.

Genes encoding all pre-RC components have been identified in various plant species, including *Arabidopsis*, rice, and maize, among others (Table 1; Shultz et al. 2007; Gutierrez 2009). In some species, pre-RC components are encoded by two different genes, for example, *ORC1A* and *ORC1B*, *CDC6A* and *CDC6B*, and *CDT1A* and *CDT1B*, in *Arabidopsis*. In these cases the proteins are highly homologous, but their promoter regions are distinct, strongly suggesting that cell type and/or developmental stage-specific regulation of their expression is a primary mechanism controlling pre-RC availability. ORC contains a core formed by subunits ORC2, ORC3, ORC4, and ORC5, to which ORC1 and ORC6 assemble as labile subunits. The mRNAs of all ORC subunits are abundant in proliferating cells. In *Arabidopsis*, the expression of all these genes, except for *ORC5*, is controlled by the E2F family of transcription factors, whose activity is repressed early in G1 by the RETINOBLASTOMA-RELATED (RBR) protein. When cyclin-dependent kinase (CDK)/cyclin activity increases during G1, RBR is phosphorylated, and the transcriptional activity of E2Fs allows for the expression of their targets, i.e., *ORC* genes. However, some of them are also expressed in non-proliferative domains of the plant, in differentiated cells. This has been also reported for some metazoan *ORC* genes that play roles unrelated to DNA replication.

The rest of pre-RC proteins in plants are also homologous to their animal counterparts, in particular those constituting the hexameric MCM complex. Nevertheless, the amino acid sequences of other pre-RC proteins, such as CDT1, are quite divergent. The genes encoding these proteins are also E2F targets, revealing a coordinated regulation of their expression. Perhaps the most striking difference between the proteins forming the pre-RC in animal and plant cells resides in the CDT1-interacting proteins in both kingdoms. In animals, CDT1 interacts with geminin (Gmn), which in turn inhibits its activity and contributes to the regulation of the licensing mechanism. Plants lack a geminin ortholog, although they encode a completely unrelated CDT1-interacting protein called GLABRA2 (GL2) *EXPRESSION MODULATOR* (GEM; Caro et al. 2007). To date, a role for GEM in controlling CDT1 activity in the plant pre-RC has not been demonstrated, but it will be extremely important to determine the regulatory role of GEM as well as the structural relationship between plant CDT1 and GEM, in comparison to that of animal CDT1 and Gmn.

Extensive protein mechanistic studies with plant CDT1 and MCM have served to identify numerous interacting proteins such as Armadillo BTB *Arabidopsis* protein 1 (ABAP1), a CDT1 interactor, or E2F TARGET 1 (ETG1), a MCM2-7 interactor that plays roles in cell growth control and in DNA damage G2 checkpoint. These studies strongly point to a diverse and complex role of pre-RC proteins not only in DNA replication but also in other associated processes occurring during the cell cycle.

DNA Replication Origin Specification and Usage

All the genomic sites bound by pre-RC are potential ORIs, but which are the molecular features that specify only a subset of them as active ORIs? Research in this field has been driven since the early days by work with the budding yeast *Saccharomyces cerevisiae*, which uses a sequence-specific signature to define active ORIs. The first ORI identification method in *S. cerevisiae* was based on the functional autonomously replicating sequence (ARS) assay in plasmids (Brewer and Fangman 1987), where DNA elements were tested for their capacity to promote autonomous replication. Subsequent studies showed that a subset of these sequences also functioned as ORIs in their chromosomal context. Molecular dissection of one of these ORIs, the ARS1, revealed the presence of a 100–150 bp sequence, defined as the minimal replicator. The A-element comprises the ARS consensus sequence (ACS), a conserved 11 bp A/T-rich sequence that is common to all known ARSs and essential for ORI function (Fig. 3a). In fact, ORC interacts specifically with this sequence in an ATP-dependent manner. Additionally, the elements B1, B2, and B3 are key regulators of the ARS function, although they are neither necessary nor sufficient for ORI activation at the individual level. Many efforts were made subsequently to identify ORIs by sequence specificity in other eukaryotes, approaches that have all failed since DNA primary sequence, as a feature defining ORIs in eukaryotes, is the exception rather than the rule. Even so, it is worth noting that *S. cerevisiae* has ~12,000 ACSs, from which only ~400 act as true ORIs (Mechali 2010). This observation strongly suggests that the sequence-specific signature that defines active ORIs is not the only defining element in the determination of initiation sites in this model organism.

The fission yeast, *Schizosaccharomyces pombe*, does not use a particular sequence to specify ORIs, although it has been demonstrated that its initiation sites are always A + T rich (Fig. 3b; Segurado et al. 2003). In fact, almost any synthetic DNA sequence with an A + T richness greater than 80 % can work as an ORI in vivo. In multicellular eukaryotes, both animals and plants, it is only now that the molecular features that determine ORIs are being elucidated. Metazoan ORIs also lack a consensus DNA sequence. In *Drosophila*, ORIs coincide with regions of increased A/T content, which are bound by ORC. Many of these ORC binding sites localize to regions of open chromatin and coincide with a subset of RNA polymerase II binding sites, suggesting coordination between transcription and replication (Fig. 3c; Mechali 2010). In mammals, most of the ORIs initially identified mapped close to gene promoters, many of which were associated with CpG islands (Fig. 3d). In the past few years, genome-wide surveys of ORIs in distinct mammalian systems highlighted a correlation with transcriptional activity and transcriptional regulatory elements, among which CpG island-associated promoters are included. Moreover, recent findings revealed the presence of G-rich elements in the great majority of mouse and human ORIs (Gilbert 2010; Mechali 2010).

In plants, the first ORI was identified in the non-transcribed region of the ribosomal DNA (rDNA) repeats (Hernandez et al. 1988). A major characteristic

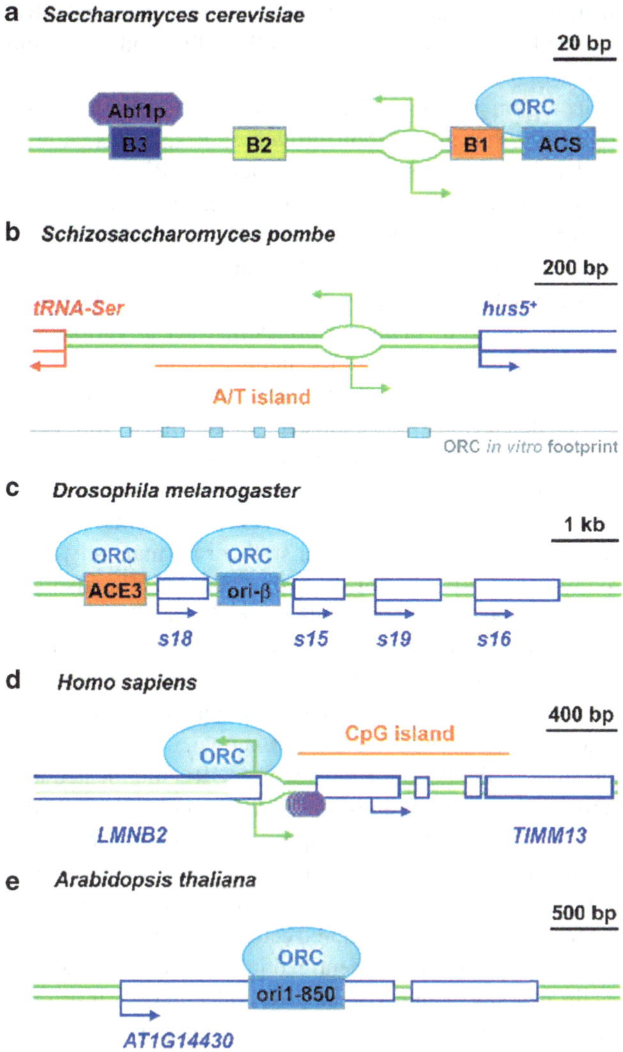

Fig. 3 Replication origins in different organisms. *Green arrows* denote bidirectional replication forks. *White-filled rectangles* indicate genes transcribed from the Watson (*blue line*) or the Crick (*red line*) strand. ORC is depicted as a *light blue oval shape*. (**a**) Schematic representation of the ARS1 replication origin of *Saccharomyces cerevisiae*. The Abf1 transcription factor binds the B3 element and is important for origin function. ORC recognizes the ARS consensus sequence (ACS) and the B1 element. (**b**) Organization of *ars1* ORI of *Schizosaccharomyces pombe* localized in the intergenic region containing the promoters of two divergently transcribed genes. The associated A/T-rich region (A/T island) is indicated in *orange*. ORC in vitro footprint is represented by the *light blue rectangles* below the *ars1* scheme. (**c**) Scheme of the chorion gene locus of *Drosophila melanogaster*. ORC complex binds to the ACE3 and ori-β elements. (**d**) Schematic representation of the human *lamin B2* replication origin at the promoter of the *TIMM13* gene. The *purple circles* represent transcription factors bound to the *TIMM13* promoter. The location of the CpG island is indicated in *orange*. (**e**) Scheme of the *Arabidopsis ori1-850* replication origin at the body of the

of rDNA replication is that due to its repetitive nature, a potential ORI is present in each rDNA repeated unit. However, not all of them fire, and only a subset is active in every S phase.

The genomic region that is synthesized from a single ORI is known as a replicon. It is known that at different growth stages, plants show variations in replicon size, that is, the distance between active ORIs within a given cell is developmentally regulated (Bryant 2010). Changes also occur in response to the presence of different nutrients and hormones. The transition from the vegetative to the reproductive stage is characterized by a decrease in replicon size, implying that more ORIs are active coinciding with this transition. This is probably the consequence of the action of plant hormones that act at various developmental stages and organs. Other cases where the number of active ORIs changes are in response to DNA damage or situations that led to fork stalling. Under these circumstances, a deleterious consequence is to finish the cell cycle with underreplicated genomic regions. To optimize replication efficiency, backup mechanisms that allow cells to recruit for replication the so-called "dormant" ORIs have evolved (Blow et al. 2011). These normally licensed ORIs are not used unless the activated start sites suffer from stalls or permanently arrested replication forks. In any case, activation of dormant ORIs must be also strictly coordinated with the temporal control of replication during S phase. Therefore, the number of active ORIs is not absolute or constant. Instead, it strongly depends on the cell type, the developmental stage, the response to internal stimuli, and the environmental conditions. Molecular details about the mechanisms controlling ORI usage under all these conditions are not known.

The identification of the genomic sites where ORIs are located as well as their genomic features is an important question, currently being addressed both in plant and animal cells. The advent of genomic approaches has contributed enormously to the advance in our knowledge of ORI specification in multicellular eukaryotes, including plants. A genome-wide map of all ORIs that are preferentially active during early S phase in *Arabidopsis*-cultured cells was recently generated, a study that revealed that ~1,500 ORIs could be identified in these cells (Costas et al. 2011a, b). The analysis of the genomic features of this ORI collection was instrumental to define common characteristics and to compare them with those specifying ORIs in animal cells. In short, *Arabidopsis* ORIs seem to be amazingly similar in their basic requirements to their metazoan counterparts (Fig. 3e). Therefore, the evolutionary acquisition of ORI specification features found early in evolution of multicellular eukaryotes must have been maintained in different branches of multicellularity. In other words, it is conceivable that these basic and evolutionary old ORI characteristics appeared concomitantly with the explosion of multicellularity and/or were already acquired in the common multicellular ancestors.

Fig. 3 (continued) AT1G14430 gene. The ORI localization was determined by massive sequencing of BrdU-labeled DNA from a synchronized *Arabidopsis* cell culture, and ORC binding was detected by ChIP-chip analysis

Origins of DNA replication in plants are mainly located toward the 5′ half of gene units, including the N-terminal part of the coding region plus the promoter sequences upstream. In general, they also tend to be associated with highly expressed genes, similar to animal ORIs (Fig. 3e). Although they do not exhibit any detectable sequence in common, they tend to be rich in G+C. This is particularly clear in the case of *Arabidopsis*, which has a relatively low C+G content over the entire genome (~37 %). However, the short C+G stretches associated with ORIs are undermethylated. These structural features of ORI DNA sequences might favor the assembly of pre-RCs, a conclusion that applies also to animal ORIs. Certain histone properties also seem to contribute to ORI specification. So far, only correlative studies have been carried out, that is, conclusions are based on results of colocalization of ORIs with particular histones and histone modifications. Specifically, plant ORIs are largely associated with the histone H2A.Z and H3.3 variants. These observations are fully consistent with the higher than average C+G content observed at ORIs and the depletion of C methylation. Furthermore, consistent with the presence of ORIs in highly expressed genes, initiation sites also colocalize with epigenetic marks typical of that chromatin state, that is, histone modifications characteristic of active transcription: they are enriched in H3K4me2, H3K4me3, and H4K5ac and depleted in H3K4me1 and H3K9me2. The overall changes in some of these markers detected by immunofluorescence studies during S phase are consistent with their association with ORIs.

Replication Origin Firing: Fork Assembly

The process of selecting the subset of ORIs that will actually fire and start replication from the total amount of pre-RC-bound genomic sites is virtually unknown. Whatever the mechanism is, DNA replication starts by unwinding the two strands at the initiation sites. This is accompanied by the disassembly of pre-RC and subsequent assembly of pre-initiation complexes (pre-IC) and loading of replicative DNA polymerases to establish two bidirectional replication forks (Fig. 2b). Disassembly of pre-RC and the concomitant inactivation of used pre-RC are the basis for mechanisms evolved to prevent re-replication. The mechanisms differ depending on the systems in which they have been studied. In some cases, some pre-RC components, e.g., ORC1 or CDC6, are phosphorylated, and this leads to changes in their subcellular localization. In other cases, some components are proteolytically degraded. Since these components will not be required until the next cell cycle, this is an excellent and secure, though expensive, mechanism to prevent pre-RC reassembly. In animal cells, geminin accumulates at active ORIs and inactivates CDT1. As already mentioned, plants lack a geminin homologue, and it remains to be determined if GEM, the plant CDT1-interacting protein, has any role in preventing re-replication.

Direct biochemical and molecular analysis of the mechanisms targeting pre-RC components has not been carried out in plants. However, it has been found that some histone modifications are crucial in controlling repeated pre-RC function. In

Arabidopsis, the H3K27me1 level is determinant for re-replication of heterochromatin regions. This mark is deposited by the action of *Arabidopsis* trithorax-related protein 5 (ATXR5) and ATXR6 monomethyltransferases. These enzymes interact with proliferating cell nuclear antigen (PCNA), a component of the replication machinery at the fork that may act as a landing pad to change the H3K27me1 status just after pre-IC assembly (Jacob et al. 2010).

Once the pre-RC is disassembled, several proteins are targeted to these sites to assemble an active fork. Prior to this, DNA needs to be fully unwound and set for DNA polymerase loading. It has been shown that purified MCM complex needs to associate with other factors before it can act as the replicative helicase that unwinds the DNA template. Two factors are crucial at this stage for establishing an active replication fork: the cell division cycle 45 (CDC45) protein and the GINS (go, ichi, ni, san; five, one, two, and three in Japanese) complex, formed by SLD5 (synthetic lethality with DPB11 5), PSF1 (partner of SLD five 1), PSF2, and PSF3. These two factors together with the MCM complex are recruited to ORIs at the onset of S phase, forming the replicative helicase (Labib and Gambus 2007; Table 1; Fig. 2b). The subunits of GINS, which were identified independently in several laboratories, form a tight complex throughout the cell cycle. However, this complex is only recruited to the ORIs during S phase. Initial unwinding of the origin region also requires the participation of MCM10, which binds to the DNA polymerase α-DNA primase complex, and perhaps to MCM8, which may favor DNA polymerase loading. Once the origin region is unwound and the replicative DNA polymerases are loaded, the CDC45 protein together with the MCM and GINS complexes travels with the fork. This macromolecular entity is also known as CMG complex, and it has been purified from various sources. It has been postulated that it constitutes the core of the replicative helicase that unwinds the DNA template ahead of the polymerization machinery. Based on this function, the CMG complex has been also named "unwindosome."

Replication Fork Progression

Plant replicative DNA polymerases have been purified and biochemically characterized from different species, and they largely conform to the properties defining the family B of DNA polymerases: DNA polymerase α(POLA), DNA polymerase δ(POLD), and DNA polymerase ε(POLE). The typical POLA holoenzyme, responsible for initiation of DNA replication at ORIs (Fig. 2c), is formed by four subunits containing the polymerase activity (POLA1) and the DNA primase activity (POLA3 and POLA4), while POLA2 is an accessory polypeptide (Table 1). Hypomorphic mutations in the *Arabidopsis POLA1* gene (*icu2*) have been identified that still allow the development of plants, although showing high pleiotropic phenotype, including abnormal leaves.

POLD also contains four subunits (POLD1-4) and is highly conserved, including the presence of a 3′–5′ exonuclease activity that confers proofreading ability to the holoenzyme. *POLD1* expression in proliferating cells correlates with the expression of PCNA, an accessory factor that confers processivity. This gene, as well as the *POLD2* gene, is regulated by nutrient availability, e.g., sucrose. PCNA is highly conserved and is encoded by two genes in many plant species. It is phosphorylated by various CDK/cyclin complexes to regulate its interaction with POLD at various stages of development, e.g., germination. It is worth noting that PCNA also participates in DNA damage response by interacting with highly specialized DNA polymerases such DNA polymerase η (POLH) and λ(POLL). These proteins belong to the DNA polymerase family Y and X, respectively, and are involved in replicating DNA regions that contain DNA damage. This complex process is called translesion synthesis and involves the recognition of the DNA damage by the DNA replication machinery, the switch to the specialized DNA polymerases, the initiation of DNA synthesis past the lesion, and the reestablishment of a normal DNA replication fork.

Assembly of the DNA POLD-PCNA complex at the 3′OH end for elongation is facilitated by the replication factor C (RFC), a heteropentameric complex, typical of highly proliferative tissues in plants (Table 1). Quite interestingly, a role of the large subunit (RFC1) in the recruitment of the DNA replication machinery to the ORI of the dsDNA forms of geminiviruses (plant DNA viruses) has been identified (Luque et al. 2002; Jeske 2009). RFC1 also plays a role in gene silencing, which is crucial for the maintenance of the repressed state of transposon elements in constitutive heterochromatin.

The POLE holoenzyme is a complex of POLE1A and POLE1B (the catalytic subunits) and POLE2, DNA polymerase B(II) subunit 4 (POLE4 or DPB4), and POLE3/DPB3A (the regulatory subunits). In addition to its participation in leading strand synthesis, POLE1A has a specific role in embryogenesis, meristem development, and floral transition, most likely through interaction with other proteins and/or through its contribution to regulate the proliferative activity at certain plant locations.

The fact that plant development is very plastic, and compatible with the presence of hypomorphic mutations in most of these essential genes, provides a unique opportunity to study their role and relevance in processes in the adult plant beyond DNA replication itself (Gutierrez 2005; Inze and De Veylder 2006).

Extensive amounts of ssDNA are produced during fork progression. This is covered by single-strand binding (SSB) protein that in all eukaryotes, including plants, is the heterotrimeric complex replication protein A (RPA1, RPA2, and RPA3). A role of RPA in the transition from pre-RC to pre-IC and beyond has been also shown in plants, where RPA subunits are encoded by a multigene family. This suggests that their expression is highly tissue-specific and/or cell-type-specific. Some of these genes, preferentially those encoding the large subunit, are required for both DNA replication and DNA repair. Furthermore, RPA2 is required for gene silencing of transposon elements, in cooperation with RFC1.

At this stage, two replication forks ready to synthesize DNA bidirectionally are assembled and associated with the CDC45-MCM-GINS complex, which will work as a replicative helicase, unwinding the DNA ahead of the fork (Fig. 2b). Mechanistically, replication fork progression has been determined in various model systems, and the reader is directed to general reviews and textbooks for details (DePamphilis and Bell 2011; Bell et al. 2012). The major steps involved in establishing a fork are as follows. Since the DNA polymerases can only function in the 5′–3′ direction and the DNA contains two antiparallel strands (5′–3′ and 3′–5′), the organization of the forks in the two strands poses a problem. Thus, the 3′–5′ DNA strand can be copied by the DNA polymerase in the 5′–3′ direction that is called the leading strand (Fig. 2c). However, the newly synthesized DNA in the other strand, the lagging strand, needs a mechanism to overcome the directional restriction of DNA synthesis.

The solution to this problem came with the discovery by R. Okazaki that lagging strand synthesis occurs in normal 5′–3′ direction but in a discontinuous manner generating short DNA pieces called Okazaki fragments. These fragments arise after repeated initiation events by the DNA primase and further elongation by its associated DNA polymerase α. Later, it was demonstrated that Okazaki fragments contained a short (5–10 nt long) RNA piece at their 5′ ends, formed as a result of each priming event. The existence of Okazaki fragments as a mechanism to synthesize the lagging strand has been demonstrated in all organisms analyzed.

It must be emphasized that at a genome scale, lagging strand synthesis represents a major endeavor since Okazaki fragments cover the entire genome and each one of them is ~150 nt long in most eukaryotes. This means that ~10^6 Okazaki fragments are needed to cover the *Arabidopsis* genome, which is relatively small (126 Mbp) compared with most plant species. This implies that the selection for each primase activity is not very strict, although it is known that when associated with DNA polymerase α, DNA primase prefers to initiate the synthesis of the small RNA primer at AT-rich stretches.

In all eukaryotes, including plants, a switch occurs from the DNA polymerase α-DNA primase initiator complex to the replicative polymerases: DNA polymerase ε in the leading strand and DNA polymerase din the lagging strand. This allows the DNA pol α-DNA primase complex to initiate a new Okazaki fragment. This switch is facilitated by a complex of the sliding clamp (the PCNA trimer) and the clamp loader (the heteropentameric RFC complex). The series of events that occur has been determined precisely and consists of the following: first, the association of the clamp loader and the sliding clamp to the primed DNA, which releases the DNA polymerase α-DNA primase complex; second, the ATP-dependent delivery of the sliding clamp to the primed template; and finally, the association of the replicative DNA polymerase to the sliding clamp. In this way, DNA synthesis proceeds in a continuous manner in one strand and through Okazaki fragment synthesis in the other. In the final step, Okazaki fragments must be processed, to eliminate the RNA part at their 5′ ends, and ligated (Fig. 2d). As in animal cells, these processes are carried out by the 5′–3′ exonuclease activity of flap endonuclease 1 (FEN1) and the DNA ligase 1 (LIG1).

While the joining of Okazaki fragments occurs as the fork progresses, the joining of neighbor replicons is delayed well until the S phase finishes. First, replicons within the same cluster are ligated to form large portions of fully replicated DNA, and, later, whole clusters are ligated to render a fully replicated chromosome. This process that does not take place right after finishing each replicon. Instead it is at the very end of the S phase, or even in early G2, when neighbor replicons join together into larger replicated units that eventually constitute the fully replicated sister chromatids (Schvartzman et al. 1981). This mechanism poses the question of whether specific termination sites are present in the genome. This has been demonstrated to be the case within the rDNA region, although it may not be a general feature across the genome.

Replication Timing

The complexity involved in thousands of ORIs firing during S phase raises the problem of coordinating their individual licensing, activation, progression, and processing at forks. The goal during S phase is to achieve full genome replication. Consequently, one theoretical possibility is that ORIs could be activated randomly while a putative mechanism would be assessing when the genome is fully replicated. However, this is not the case: early experiments in plants using DNA fiber autoradiography developed in the laboratory of Taylor and colleagues, first, and with fluorescent markers, later, together with biochemical approaches carried out in the 1970s, showed that genome replication as a whole is subjected to a strict temporal control during S phase (Bryant 2010). Thus, replication of plant genomes, as also demonstrated later for animal genomes, does not occur randomly but instead is organized into two large families of replication units: the early and the late replicating regions (Hiratani et al. 2009; Aparicio 2013). This biphasic mode of genome replication is responsible for most of euchromatin replicating early and mid in S phase and later the heterochromatin, which contains repeats and transposon elements and is frequently confined to the centromeric and pericentromeric regions of the chromosomes. Genomic approaches in *Arabidopsis* have significantly contributed to define more precisely the temporal control of genome replication. Thus, early-/mid-replicating domains during the first 2 h of S phase are associated with regions of active chromatin, enriched in highly transcribed genes and in H3K56ac marks. On the contrary, genomic regions replicating late in S phase are depleted in H3K56ac but enriched in repressive marks, such as H3K9me2 and methylated cytosines (Lee et al. 2010). However, these marks seem to be merely associated with replication, and it is not known if they are cause or effect of the replication timing control.

Endoreplication

The goal of the cell division cycle is the production of two daughter cells, each with an identical copy of the genome. However, there is an alternative cell cycle, called the endocycle (also known as endoreplication or endoreduplication cycle), where the genome is fully duplicated during S phase, but mitosis is prevented. Thus, the endoreplicating cells undergo an S phase and a G phase and then immediately start a new S phase. Consequently, since the duplicated genomes are not transferred to the daughter cells, the nuclear DNA content (or ploidy) of the endoreplicating cell increases from the normal diploid 2C content to 4C, 8C, 16C, 32C, and so forth (Gutierrez 2009; Breuer et al. 2010; De Veylder et al. 2011; Edgar et al. 2014).

Endoreplication occurs in all multicellular eukaryotes, but it is particularly common in plants, where in many species a large proportion of cells have ≥4C DNA content per cell (in some cases ~70 % of cells in the adult body). In many cases, the occurrence of endoreplication is very relevant for plant development since it is necessary for cell growth and differentiation. It is important to keep in mind that the endocycle is not similar to the abnormal re-replication by multiple reinitiation events at the same genomic locations observed when the licensing control is lost, since endoreplication involves one full genome duplication and only once per endocycle.

A large variety of pathways that lead to cell cycle arrest and trigger the endocycle have been identified in plants. These include alterations in the activity of certain DNA replication proteins, changes in the balance of cyclins by altering the mechanism that controls their degradation in mitosis, expression of some transcription factors at certain developmental stages, the initiation of organogenesis, or the presence of DNA damage or altered cell cycle checkpoints.

Regarding DNA replication proteins, the excess of some pre-RC components, e. g., CDC6 or CDT1, leads to an increase in the endoreplication process that normally occurs associated with *Arabidopsis* leaf development. This is particularly clear in the leaf epidermis, constituted by three different cell types: the guard cells that form the stomata, necessary for gas exchange, which never endoreplicate; the pavement cells, puzzle-shaped cells that constitute most of the epidermis and that endoreplicate in association with cell expansion; and the trichomes, or leaf hairs, single cells that protrude from the epidermis and contain normally three branches. Development of trichome branches is genetically controlled, and the increase in branch number depends on the occurrence of continued endocycles. An excess of CDC6 or CDT1 produces a systemic increase in endoreplication level and an increase in the number of trichomes, demonstrating the importance of endocycle progression for trichome development.

As already mentioned, the pre-RC protein genes are targets of the E2F family of transcription factors that regulate the activity of RBR. Downregulation of RBR activity also leads to an increase in the endoreplication level. This is likely due to upregulation of pre-RC proteins, although RBR may regulate the expression of additional genes with a direct or indirect effect on endocycle progression. The activity of RBR is, in turn, controlled by CDK/cyclin complexes. Plants contain multiple forms of cyclins A (CYCA) and cyclins D (CYCD) that form complexes with CDKA, negatively regulating RBR function. Thus, increases in CYCD3 lead to a strong inhibition of the endocycle, likely by a mechanism dependent on RBR. The G2/M transition is also regulated by plant-specific CDKs, the CDKBs, which are targets of E2F, and inhibition of these CDKBs also alters the normal endoreplication pattern. Other pathways triggering the endocycle depend on different cyclins, e.g., CYCA2,3, which is targeted for degradation by the anaphase-promoting complex/cyclosome (APC/C). Some activators of the APC/C are also E2F targets but in this case of the atypical E2FE/DEL1. All these data reflect that multiple interconnected pathways regulate the endocycle. However, the pathways that involve cell cycle regulators are not the only ones that impinge on endoreplication. Also, upstream cellular factors that transduce the information derived from environmental challenges, hormonal signals, and developmental cues have an effect on endoreplication. For instance, the endocycle plays a fundamental role during plant developmental processes, such as the formation of the endosperm (a reserve tissue in the seed) or fruit development.

One important aspect when multiple endoreplication cycles occur is the increased complexity of handling an increasing number of genomes that still are attached together in the nucleus, given that the new chromatids have not been separated and segregated in mitosis. Topoisomerases are enzymes involved in eliminating entanglements, catenation between DNA strands, and other kinds of tensions in DNA molecules. Interestingly, mutations in some topoisomerases prevent the normal progression of the endocycle beyond the 8C DNA content, suggesting that this is probably the uppermost limit that cells can handle in the absence of these enzymes.

A correct deposition of histones associated with DNA replication is also important for proper occurrence of the endocycle program. Dimers of histone H3.1-H4 are deposited during fork progression by the histone chaperone chromatin assembly factor 1 (CAF-1), a heterotrimeric complex, which in plants is formed by the subunits called FASCIATA1 (FAS)1, FAS2, and MULTICOPY SUPPRESSOR OF IRA 1 (MSI1). Mutants in the large subunit are lethal in metazoa but not in plants. Plants carrying the *fas1* mutation have defects in cell cycle progression and arrest the cell cycle in G2 because they have a constitutively activated DNA damage G2 checkpoint. These plants exhibit a systemic triggering of the endocycle program since very early in development, i.e., 24–48 h after germination. This probably has provided a significant evolutionary advantage to plants, because cells with a defect in chromatin assembly by CAF-1 trigger the endocycle as a way to escape from problems in cell cycle progression, instead of becoming permanently arrested.

Future Directions

A significant advance has been achieved during the past years in our knowledge of several aspects of DNA replication in plants, e.g., replication proteins, overall replication timing during the S phase, and overall features of ORIs, among others. However, this is only the beginning and the basis for future studies.

Regarding DNA replication proteins, it is probable that based on their conservation relative to yeast and animal proteins, the biochemical processes carried out by plant DNA replication proteins should be very similar to what has been demonstrated for other systems. It is likely at the level of regulation and availability of these proteins that major differences could be identified. For instance, the sessile nature of plants has forced them to evolve specific mechanisms to respond to environmental challenges and, as a consequence, to modulate the growth and proliferation properties of many cells within their body.

It is known that cell fate acquisition is accompanied by significant changes in the epigenomic and transcriptional profiles of eukaryotic cells. Since increasing evidence reveals a coupling between these features and DNA replication, particularly at the level of initiation and timing, it will be of primary relevance to analyze them in parallel. This requires the development of appropriate protocols to study DNA replication in cells within the whole plant, a challenging step due to the limited amount of proliferating cells, which are restricted to a few locations in the plant, the meristems. The availability of powerful genomic techniques will be of tremendous value in these studies. In this context, a refinement of the approach for the foreseeable future would be the identification of DNA replication features, including ORIs, in individual cell types purified from different organs. Succeeding in such approach will be one more step ahead, since it will facilitate the study of developmental factors that impinge directly on DNA replication functions.

Intense efforts should be also focused on identifying the combination of DNA and chromatin features responsible for specifying ORIs. As discussed earlier in this chapter, there are many different genomic sites where ORIs can be located, but it is clear that ORIs are not randomly distributed across the genome. Therefore, identifying the chromatin signature, or signatures, of ORIs will be a major issue in the immediate future. Likewise, finding possible variations depending on the cell type, the developmental stage, the physiological state, and the environmental conditions are extremely attractive avenues for future studies.

References

Aparicio OM. Location, location, location: it's all in the timing for replication origins. Genes Dev. 2013;27:117–28.

Arias EE, Walter JC. Strength in numbers: preventing rereplication via multiple mechanisms in eukaryotic cells. Genes Dev. 2007;21:497–518.

Bell S, Méchali M, DePamphilis ML. DNA replication. Cold Spring Harbor: Cold Spring Harbor Laboratory Press; 2012.

Blow JJ, Laskey RA. A role for the nuclear envelope in controlling DNA replication within the cell cycle. Nature. 1988;332:546–8.

Blow JJ, Ge XQ, Jackson DA. How dormant origins promote complete genome replication. Trends Biochem Sci. 2011;36:405–14.

Breuer C, Ishida T, Sugimoto K. Developmental control of endocycles and cell growth in plants. Curr Opin Plant Biol. 2010;13:654–60.

Brewer BJ, Fangman WL. The localization of replication origins on ARS plasmids in *S. cerevisiae*. Cell. 1987;6:463–71.

Bryant JA. Replication of nuclear DNA. Progr Bot. 2010;71:25–60.

Caro E, Castellano MM, Gutierrez C. A chromatin link that couples cell division to root epidermis patterning in Arabidopsis. Nature. 2007;447:213–7.

Costas C, de la Paz Sanchez M, Stroud H, Yu Y, Oliveros JC, Feng S, Benguria A, López-Vidriero I, Zhang X, Solano R, Jacobsen SE, Gutierrez C. Genome-wide mapping of

Arabidopsis thaliana origins of DNA replication and their associated epigenetic marks. Nat Struct Mol Biol. 2011a;18:395–400.

Costas C, Sanchez MP, Sequeira-Mendes J, Gutierrez C. Progress in understanding DNA replication control. Plant Sci. 2011b;181:203–9.

DePamphilis ML, Bell S. Genome duplication. Concepts, mechanisms, evolution, and disease. London: Garland Science/Taylor & Francis Group; 2011.

De Veylder L, Larkin JC, Schnittger A. Molecular control and function of endoreplication in development and physiology. Trends Plant Sci. 2011;16:624–34.

Edgar BA, Zielke N, Gutierrez C. Nat Rev Mol Cell Biol. 2014;15:197–210.

Gilbert DM. Evaluating genome-scale approaches to eukaryotic DNA replication. Nat Rev Genet. 2010;11:673–84.

Gutierrez C. Coupling cell proliferation and development in plants. Nat Cell Biol. 2005;7:535–41.

Gutierrez C. The Arabidopsis cell cycle. In The Arabidopsis Book.e0120 http://www.aspb.org/publications/arabidopsis/2009

Hernandez P, Bjerknes CA, Lamm SS, Van't Hof J. Proximity of an ARS consensus sequence to a replication origin of pea (*Pisum sativum*). Plant Mol Biol. 1988;10:413–22.

Hiratani I, Takebayashi S, Lu J, Gilbert DM. Replication timing and transcriptional control: beyond cause and effect–part II. Curr Opin Genet Dev. 2009;19:142–9.

Inze D, De Veylder L. Cell cycle regulation in plant development. Annu Rev Genet. 2006;40:77–105.

Jacob Y, Stroud H, Leblanc C, Feng S, Zhuo L, Caro E, Hassel C, Gutierrez C, Michaels SD, Jacobsen SE. Regulation of heterochromatic DNA replication by histone H3 lysine 27 methyltransferases. Nature. 2010;466:987–91.

Jeske H. Geminiviruses. Curr Top Microbiol Immunol. 2009;331:185–226.

Labib K, Gambus A. A key role for the GINS complex at DNA replication forks. Trends Cell Biol. 2007;2007(17):271–8.

Lee TJ, Pascuzzi PE, Settlage SB, Shultz RW, Tanurdzic M, Rabinowicz PD, Menges M, Zheng P, Main D, Murray JA, Sosinski B, Allen GC, Martienssen RA, Hanley-Bowdoin L, Vaughn MW, Thompson WF. Arabidopsis thaliana chromosome 4 replicates in two phases that correlate with chromatin state. PLoS Genet. 2010;6:e1000982.

Luque A, Sanz-Burgos AP, Ramirez-Parra E, Castellano MM, Gutierrez C. Interactionof geminivirus Rep protein with replication factor C ans its potential during geminivirus DNA replication. Virology 2002;302:83–94.

Mechali M. Eukaryotic DNA, replication origins: many choices for appropriate answers. Nat Rev Mol Cell Biol. 2010;11:728–38.

Ramirez-Parra E, Gutierrez C. The many faces of chromatin assembly factor 1. Trends Plant Sci. 2007;12:570–6.

Schvartzman JB, Chenet B, Bjerknes C, Van't Hof J. Nascent replicons are synchronously joined at the end of S phase or during G2 phase in peas. Biochim Biophys Acta. 1981;653:185–92.

Segurado M, de Luis A, Antequera F. Genome-wide distribution of DNA replication origins at A+T-rich islands in *Schizosaccharomyces pombe*. EMBO Rep. 2003;4:1048–53.

Shultz RW, Tatineni VM, Hanley-Bowdoin L, Thompson WF. Genome-wide analysis of the core DNA replication machinery in the higher plants Arabidopsis and rice. Plant Physiol. 2007;144:1697–714.

Taylor JH, Woods PS, Hughes WL. The organization and duplication of chromosomes as revealed by autoradiographic studies using tritium-labeled thymidine. Proc Natl Acad Sci USA. 1957;43:122–8.

Plant Telomeres and Telomerase

2

Andrew D. L. Nelson, Mark A. Beilstein, and Dorothy E. Shippen

Contents

Abstract

- The essential function of the telomere is to facilitate the complete replication of the chromosome terminus and to prevent the terminus from eliciting a DNA damage response that would cause genome instability.

A.D.L. Nelson • M.A. Beilstein
The School of Plant Sciences, University of Arizona, Tucson, AZ, USA
e-mail: andrew.d.l.nelson@gmail.com; mbeilstein@email.arizona.edu; mbeilstein@gmail.com

D.E. Shippen (✉)
Department of Biochemistry and Biophysics, College of Agriculture and Life Sciences, Texas A&M University, College Station, TX, USA
e-mail: dshippen@tamu.edu

S.H. Howell (ed.), *Molecular Biology, The Plant Sciences 2*,
DOI 10.1007/978-1-4614-7570-5_4

- Telomere failure can occur from the loss of telomere capping proteins or the prolonged absence of telomerase.
- Because of its extraordinary tolerance to genome instability, *Arabidopsis thaliana* is a powerful model for telomere biology.
- Telomerase expression is highly regulated and in both plants and animals is confined to cells with long-term proliferation capacity.
- Unlike vertebrate telomeres, plant telomeres are asymmetric with one end of the chromosome terminating in a 3′ single-stranded overhang and the other in a blunt end.
- The protein composition of plant telomeres reveals an evolutionary bridge with yeast and vertebrates; some factors are more like yeast, others more like vertebrates, and still others unique to plants.
- The identification and characterization of the CST complex in Arabidopsis paved the way for understanding the molecular basis of human stem cell disease.
- Gene duplication and diversification have had made a significant impact on the composition and regulation of Arabidopsis telomerase.
- The telomerase RNA component is evolving very rapidly in plants and is giving rise to novel regulatory mechanisms.
- De novo telomere formation by telomerase at internal double-strand breaks in the DNA causes loss of chromosomal DNA and must be strictly regulated to ensure faithful repair of DNA damage.

Introduction

In the late 1930s, while America was suffering through the great depression and Europe was appeasing Nazi Germany, two geneticists were shedding light on the remarkable segments of DNA found at the end of eukaryotic chromosomes. Barbara McClintock, a maize cytogeneticist at the University of Missouri, followed the fate of broken chromosomes. McClintock observed the phenomenon of anaphase bridges created when a chromosome with two centromeres (dicentric chromosome), connected by heterochromatin, attempted to segregate and was pulled in opposite directions during meiosis I. These chromosomes would eventually break, and the broken sister chromatids would subsequently fuse to generate a new dicentric chromosome (discussed further below). Then the cycle of breakage-fusion-bridge formation would repeat itself. Importantly, McClintock noticed that in some cell lines, rather than undergoing fusion events, the ends of broken chromosomes would "heal" themselves and generate new stable chromosomes, thereby ending the breakage-fusion-bridge cycle (McKnight et al. 2002; de Lange et al. 2006). McClintock termed the repair process "chromosome healing," since the healed chromosomes were safe from future fusion events. Around the same time, Hermann Muller (University of Edinburgh), a geneticist studying the effects of X-rays on Drosophila chromosomes, found many genome rearrangements arising

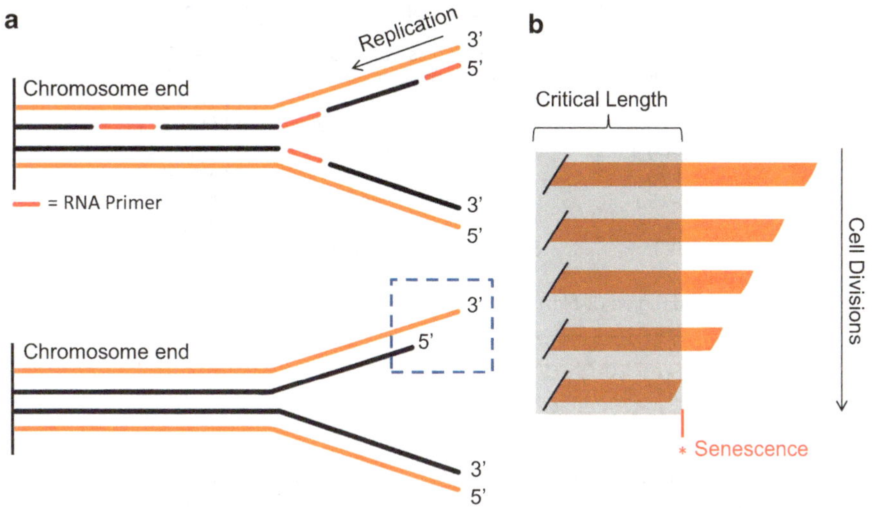

Fig. 1 The end-replication problem and replicative senescence. (**a**) Lagging strand synthesis results in incomplete replication of the daughter strand (*black*), due to the inability to fill in the 5′ end of the new daughter strand when the RNA primer is removed (*red*). See *dashed box*. (**b**) Many cells follow a replicative senescence program. After a number of divisions, telomere attrition leads to a critical length, where senescence programs are activated (*red* *). Cells that manage to bypass this initial checkpoint either find a way to extend their telomeres by telomerase activation or recombination (ALT) or undergo massive chromosomal rearrangements. See text for more details

from the formation and repair of double-strand breaks (DSBs). Muller astutely noticed that chromosomal fusion did not arise from DSBs at the ends of the chromosomes. Working independently, both Muller and McClintock recognized that the ends of chromosomes had special properties that protected them from being covalently joined end to end. Muller coined a name for these ends: telomeres.

Eventually scientists discovered that chromosomes are composed of double-stranded (ds) DNA and that dsDNA has a polarity to its structure. The helical structure of dsDNA complicates replication of the chromosome terminus, and this realization led to theories on the "end-replication problem," first laid out by Alexey Olovnikov in 1971 and later by James Watson in 1972. Due to the semiconservative mechanism of replicating linear, eukaryotic DNA, lagging strand synthesis results in an un-replicated segment of DNA at the chromosome terminus with each round of cell division. The "problem" presented by this slow erosion of genetic information is that essential genes located at or near the end of the chromosome would eventually be lost (Fig. 1). Elizabeth Blackburn provided a partial solution to the problem in the late 1970s, when she sequenced chromosome ends from the ciliated protozoan Tetrahymena and found that telomeres consist of simple G-rich repeats (de Lange et al. 2006). Blackburn's research yielded the first glimpse into the molecular nature of the telomeres that McClintock and Muller had observed 40 years before. Loss of these DNA sequences would likely not be a problem in somatic cells because the repetitive G-rich DNA at the extreme chromosome

a

Telomere length (kb)	Organism	Repeat Sequence
2-7	*Arabidopsis thaliana*	TTTAGGG
40-160	*Nicotiana tabacum*	TTTAGGG
2-40	Maize	TTTAGGG
~0.3	*Schizosaccharomyces pombe*	TTACAG$_{2-3}$
0.25-0.4	Tetrahymena	TTTTGGGG
2-30	Human	TTAGGG
20-150	*Mus musculus*	TTAGGG
-	Insecta*	TTAGG
0.3	*Saccharomyces cerevisiae*	$G_{2-3}(TG)_{1-6}$

b

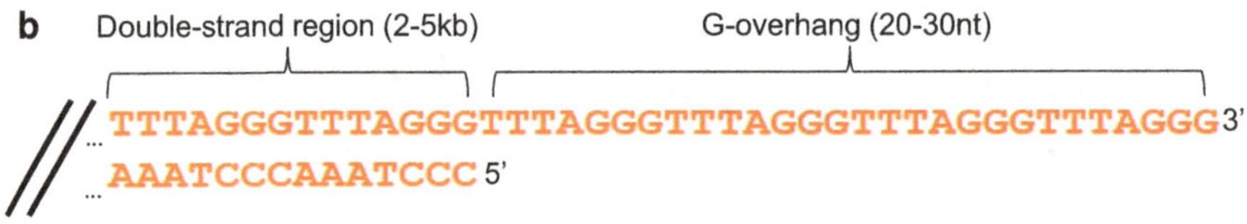

Fig. 2 **Telomere sequence conservation and general structure.** (**a**) Telomere sequence and average length from several model systems. Asterisk denotes a telomere sequence that is not conserved among all insect species, including Drosophila. (**b**) Schematic representation of the telomeric duplex region and the 3' G-overhang in *Arabidopsis thaliana*

terminus does not contain functional genes and because these cells stop replicating after ~50 cell divisions as they reach a state of senescence (Fig. 1). However, the end-replication problem was potentially more significant for germ line cells, which are responsible for passing intact genetic information to the next generation and must undergo many more rounds of replication than the soma. How can organisms with linear chromosomes faithfully transfer all of their genetic material to their offspring over many generations?

In the early 1980s, an important clue in solving the end-replication problem in long-lived cell lines, such as germ cells, was discovered. Blackburn and Jack Szostak working with the budding yeast *Saccharomyces cerevisiae* found that linear plasmids abutted by the Tetrahymena telomere sequence were not only stable, but the ends were extended by the addition of the yeast telomere repeat sequence. This observation spurred the search for the telomere repeat sequences from other organisms, quickly revealing that simple G-rich repeat arrays were a highly conserved feature of telomeres in multicellular eukaryotes (Fig. 2a). The curious exceptions are the dipteran insects, whose chromosome ends are capped by retrotransposon elements (de Lange et al. 2006). The high conservation in telomere sequence suggests that telomeres evolved as an early solution to the end-replication problem.

The experiments in yeast also revealed that telomere repeat sequences serve as a substrate for an activity that could add new repeats de novo onto the chromosome ends. Elizabeth Blackburn was at the center of another paradigm-shifting discovery, when she and her graduate student Carol Greider isolated telomerase, the enzyme responsible for adding telomere repeats onto chromosome ends. Telomerase adds long arrays of G-rich telomeric repeats to single-stranded DNA (ssDNA) primers that mimic the natural 3′ overhang on the chromosome terminus (Fig. 2b). Genetic

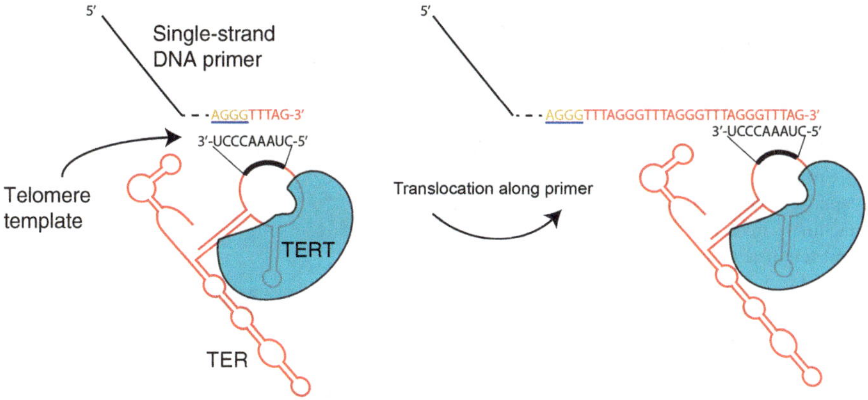

Fig. 3 Telomerase is a reverse transcriptase with repeat addition processivity. Telomerase is minimally composed of TERT (*blue*) and TER (*red*). Within TER is a template region (bold *black line*) that is complementary to the telomere repeat. TERT uses the template region within TER to reverse transcribe telomere repeats onto the G-overhang. Once telomerase finishes reverse transcribing one repeat, it translocates along the DNA, repositioning itself so that it can under-go successive rounds of telomere repeat addition in a process known as repeat addition processivity

and biochemical analysis followed, demonstrating that telomerase is indeed a very unusual enzyme. It is a ribonucleoprotein (RNP) reverse transcriptase, consisting of a catalytic subunit (TERT) and an integral RNA molecule (TER) (Fig. 3) (Autexier and Lue 2006; Egan and Collins 2012). Using a small C-rich templating sequence embedded in TER, TERT extends a ssDNA primer to generate the G-rich telomere repeat. Once the enzyme reaches the 5′ end of the TER template domain, it undergoes major conformational shift or translocation that causes the DNA to be repositioned back at the beginning of the templating domain for another round of repeat synthesis. The process of reiterative synthesis allows long arrays of telomere repeats to be synthesized and is termed repeat addition processivity (Fig. 3). It is a property of the enzyme that in vivo is modulated by accessory factors associated with the telomerase RNP and by proteins stably associated with the chromosome terminus.

The synthesis and maintenance of telomeres by telomerase is highly conserved across eukaryotes and is crucial for genome stability as well as cellular proliferation capacity. Cells lacking telomerase activity eventually submit to a process termed replicative senescence. Replicative senescence is now a well-established phenomenon in humans and is proposed to be important for tumor suppression (Artandi and DePinho 2010). Human telomerase is repressed early in embryonic development in most tissues. Replicative senescence then arises from the progressive erosion of telomeric DNA due to incomplete replication, degradation, and recombination. Critically shortened telomeres are perceived as damaged DNA and activate a DNA damage response (DDR) that can lead to end-to-end chromosome fusions and the breakage-fusion-bridge cycle described by McClintock (Fig. 4) (Murnane 2012; O'Sullivan and Karlseder 2010). Very short telomeres also force the cell to

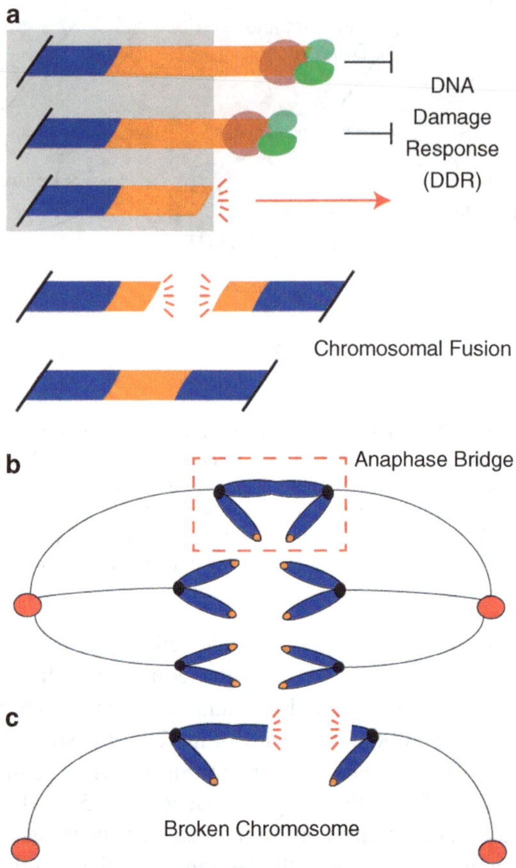

Fig. 4 Telomere dysfunction can initiate the breakage-fusion-bridge cycle. (**a**) As telomeres shorten past a critical length threshold, they are no longer able to inhibit a DNA damage response. The DNA damage machinery mistakes these short telomeres for a double-strand break (DSB) and attempts to repair them through covalent linkage to another segment of DNA, creating a dicentric chromosome. (**b**) During anaphase of the cell cycle, the dicentric chromosome (*red box*) will be pulled toward opposite ends of the cell, eventually resulting in a chromosome break (**c**). The broken chromosomes are subject to successive cycles of fusion, bridging and breakage, destabilizing the genome

exit the cell cycle and enter into the first stages of senescence (Fig. 1b). Cells can escape this fate by reactivating telomerase, which restores telomere maintenance. Immortality then gives these cells the time to accumulate mutations in tumor suppressor genes or oncogenes, giving rise to cancer.

In stem and germ line cells, telomerase prevents replicative senescence by maintaining telomere tracts. Consequently, altering the dynamics of telomere length maintenance by telomerase, or perturbing the complex protein architecture that protects the telomeric DNA, has profound effects on integrity of the entire genome (O'Sullivan and Karlseder 2010). Recently, a multitude of stem cell-related diseases in humans have been linked to deficiencies in telomerase or telomere-associated proteins, underscoring the critical role of telomeres for organismal viability (O'Sullivan and Karlseder 2010; Price et al. 2010).

Plants and mammals diverged from their last common ancestor approximately 1.4 billion years ago and likely evolved multicellularity independently. Despite this ancient divergence, many fundamental aspects of chromosome maintenance are shared between the two systems. The flowering plant

Arabidopsis thaliana is the reference species for all of plant biology, and it has been exploited as an important comparative model for telomere biology. With its wide array of genetic and molecular tools, Arabidopsis is at the forefront of several exciting discoveries in the telomere field. Plants display an amazing tolerance to genome instability, an ability to maintain stem cell niches over thousands of years and a capacity to survive for multiple generations in the absence of core telomere components shown to be crucial in other systems (Watson and Riha 2011). These attributes argue that plants have much to teach on the relationship between telomeres and genome stability. This chapter highlights what plants have revealed about the structure of telomeres and telomerase and how the telomere biology of plants compares with other major eukaryotic model systems.

Telomere Sequence and Architecture in Plants

Despite the wide variety of physiological changes that occurred as land plants colonized new habitats over the last 450 million years of evolution, the sequence of the telomere repeat array is strikingly conserved. The moss *Physcomitrella patens*, representing one of the earliest diverging clades of land plants, harbors the same telomere repeat found within the highly derived angiosperm *Arabidopsis thaliana*. What is now called the canonical plant telomere repeat, TTTAGGG, is present in a large and diverse number of early vascular plants, gymnosperms, and angiosperms as well as green algae, indicating that the evolution of this sequence predates the origin of land plants. The only known exception occurs in the monocot *Asparagus officinalis*, which has the same telomere repeat that occurs in humans (TTAGGG) (Watson and Riha 2011).

Unlike the telomere sequence, the length of the telomere tract varies widely between species and even within species (Table 1). Telomeres in the *Arabidopsis thaliana* Columbia (Col-0) ecotype span 2–5 kb, while in other *A. thaliana* ecotypes such as Niederzenz (Nd-0), telomeres are twice as long, ranging from 3.5 to 9 kb. Crosses between ecotypes with different telomere lengths produce plants with intermediate length telomeres, suggesting that telomere length is heritable and influenced by genetic factors. As in vertebrates, telomere length is not positively correlated with plant longevity. Telomeres in the perennial monocot *Othocallis siberica* reach 10 kb, and 25 kb for the bristlecone pine, *Pinus longaeva*, one of the longest living organisms, but telomeres are even longer (150 kb) in short-lived *Nicotiana tabacum* (Watson and Riha 2011).

Telomere length maintenance and prevention of a DNA damage response is influenced by the specialized architecture of the chromosome terminus. A key component of the telomere is a 3′ single-strand extension termed the G-overhang (Fig. 5a). The G-overhang, so named because it is derived from the G-rich telomeric DNA strand, is crucial for telomere length maintenance by telomerase (Nelson and Shippen 2012a). Removal of the RNA primer used for replication during lagging strand synthesis naturally creates a 3′ overhang on the daughter strand (Fig. 5b). In

Table 1 Telomere length variation and sequence conservation. Telomere sequence and average length in plants. TTAGGG is the dominant repeat sequence. Asterisk denotes a species that harbors both TTTAGGG and TTAGGG repeats, indicating a recent switch. Species designated by the superscript 1 represents a "short-lived" perennial, with an average lifespan of 100–200 years. Species designated by the superscript 2 represents a "long-lived" perennial with an average lifespan of 2,000–5,000 years

Telomere length (kb)	Organism	Repeat sequence	Lifespan
2–9	*Arabidopsis thaliana*	TTTAGGG	Annual
40–160	*Nicotiana tabacum*	TTTAGGG	Annual
10–50	*Pisum sativum*	TTTAGGG	Annual
2–40	*Zea mays*	TTTAGGG	Annual
5–11	*Oryza sativa*	TTTAGGG	Annual
>10	*Othocallis siberica*	TTAGGG*	Biennial
0.5–30	*Pinus palustris*	TTTAGGG	Perennial[1]
2–25	*Pinus longaeva*	TTTAGGG	Perennial[2]
1–5.5	*Selaginella moellendorffii*	TTTAGGG	Perennial
0.5–3.5	*Physcomitrella patens*	TTTAGGG	NA

contrast, the 3′ end of the newly synthesized leading strand is fully replicated, creating a blunt terminus that could be bound by the Ku heterodimer (see below). In most eukaryotes, a carefully controlled exonucleolytic resecting of the C-rich telomeric DNA on this terminus creates a G-overhang, making both ends of the chromosome symmetrical (Fig. 5c).

The importance of the G-overhang for telomere elongation by telomerase is reflected in the fact that G-overhangs have been detected in *P. patens* and several angiosperms, including Arabidopsis (Nelson and Shippen 2012a). G-overhang length diminishes in non-replicative tissue such as leaves, suggesting that the mechanism of telomere protection is altered in this setting, leaving the ends susceptible to nucleolytic processing. As opposed to yeast and vertebrates, recent studies indicate that plants maintain G-overhangs on only half of their chromosome ends (Fig. 5d). Blunt-ended chromosome termini have been detected in the angiosperms *Silene latifolia*, Arabidopsis, and maize and in *P. patens*. The presence of blunt-ended telomeres implies that C-strand resection observed in other eukaryotes does not take place in plants, either due to loss of the exonuclease responsible for this activity or because this step is blocked by a protective protein (discussed below) (Nelson and Shippen 2012a).

The G-overhang can assume one or more higher order configurations thought to regulate access of the 3′ terminus to replication, repair, and recombination machinery (Watson and Riha 2010). One particularly intriguing structure is the telomere loop (t-loop). T-loops form by the invasion of the G-overhang into the upstream duplex region of the telomere and can be detected by electron microscopy. Consistent with the expected symmetrical architecture of telomeres, t-loops have been observed at both ends of the minichromosomes in trypanosomes. T-loops have also been detected in the common garden pea (*Pisum sativum*), where telomeres are approximately 25 kb. T-loops must be adequately stabilized; otherwise branch

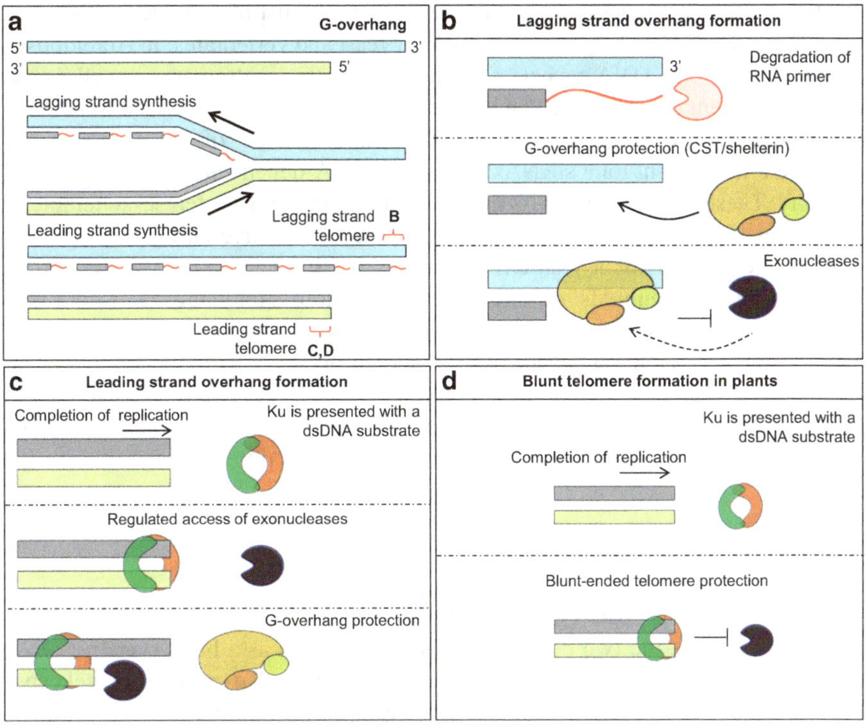

Fig. 5 Model for maintenance of chromosome ends in plants. Due to the manner in which linear DNA is replicated, G-overhangs are formed by two different mechanisms. (**a**) As replication forks move through the telomere from subtelomeric origins, multiple RNA primers (*red curvy lines*) are laid down to replicate DNA in short segments, termed Okazaki fragments. Cessation of replication results in an RNA primer remaining on the newly synthesized lagging strand telomere and a blunt-ended telomere resulting from leading strand synthesis. (**b**) Following lagging strand synthesis, degradation of the RNA primer creates a 3' overhang (G-overhang) that is bound by CST rendering it resistant to nuclease attack. (**c**) The blunt-ended telomere created by leading strand synthesis is a substrate for Ku. Ku has the capability of sliding and in vertebrates may move internally on the DNA, giving transient access to a 5' nuclease that creates a 3' overhang. (**d**) In plants, the stable association of Ku with the extreme chromosome end would prevent nuclease access, leaving the blunt-ended telomere intact (Modified from Nelson and Shippen (2012))

migration of the displacement loop formed from the G-overhang strand invasion results in a Holliday junction. Subsequent cleavage of this junction can lead to deletion of a large segment of DNA from the chromosome terminus, and the formation of extrachromosomal telomeric circles (Palm and de Lange 2008). Since G-overhangs occur on only half of the chromosome ends in Arabidopsis, t-loops are predicted to be present on only those termini. During the next round of DNA replication, the terminus with a G-overhang will be converted into a blunt end and vice versa (see below). Thus, in contrast to other eukaryotes, the chromosome terminus in Arabidopsis, and possibly other highly derived land plants, is extraordinarily dynamic.

Heterochromatic marks are a recent addition to the list of architectural features necessary for telomere length maintenance in yeast and vertebrates. In Arabidopsis, arrays of telomeric DNA repeats can be found at the centromere as well as the telomere. Recent data indicate that both regions are actively transcribed to produce telomeric repeat-containing RNAs (TERRA). A subset of these RNAs are processed into ~24 nt long siRNAs through an RNA-dependent DNA methylation pathway. These telomeric siRNAs then guide DNA methyltransferases to telomeric DNA-containing regions throughout the genome to reinforce chromatin silencing through methylation of asymmetric cytosines (O'Sullivan and Karlseder 2010). Unlike yeast and mammals, the silencing of Arabidopsis telomeres does not result in a concomitant silencing of telomere-adjacent genes. In addition, disruption of pathways critical for maintenance of the heterochromatic state in the telomeric and subtelomeric regions does not alter telomere length or the rate of recombination. More studies are required to clarify the heterochromatic state of telomeres and how this architectural feature affects chromosome end protection in plants.

Telomere-Binding Proteins

The telomere tract is coated with proteins that bind either the ss or ds region of the telomere or bridge the two domains. These proteins act as a barrier against DNA repair machinery and exonucleolytic degradation. They also serve as a gatekeeper for the replication and DNA processing enzymes that naturally act on the chromosome terminus to promote telomere integrity. Two major telomere-binding complexes have been described: shelterin and CST (Fig. 6a, b). The six-membered shelterin complex was originally discovered in vertebrates, although various components have now been described in yeast and plants. Shelterin consists of two ds telomeric DNA-binding proteins, TRF1 and TRF2, as well as RAP1, which associates with TRF2 but does not bind DNA, the heterodimer TPP1/POT1, which binds ss telomeric DNA, and TIN2, which serves as a bridging protein between the ss and ds DNA-binding components of shelterin (Palm and de Lange 2008).

Vertebrate TRF1 and TRF2 proteins show high specificity for ds telomeric DNA and through their homodimerization are capable of altering telomere architecture into a more compact, higher-order state. Both TRF1 and TRF2 are negative regulators of telomere length. TRF2 is believed to perform this role by promoting formation of t-loops, a telomeric substrate unsuitable for telomerase. TRF2 is capable of creating positive supercoils that unwind DNA and promote strand invasion. Both TRF1 and TRF2 change the secondary structure of telomeric DNA, creating loops of DNA around themselves and acting like telomere-specific histones. This TRF-mediated telomeric secondary structure serves to inhibit a DNA damage response (de Lange 2009). RAP1 and TIN2 do not directly associate with telomeric DNA but instead are critical for stabilization of the overall complex. Finally, TPP1 and POT1 form a heterodimer that functions to protect the G-overhang due to its high binding affinity for the ss 3' overhang. The high affinity of POT1/TPP1 for the G-overhang enables it to outcompete DNA damage sensing

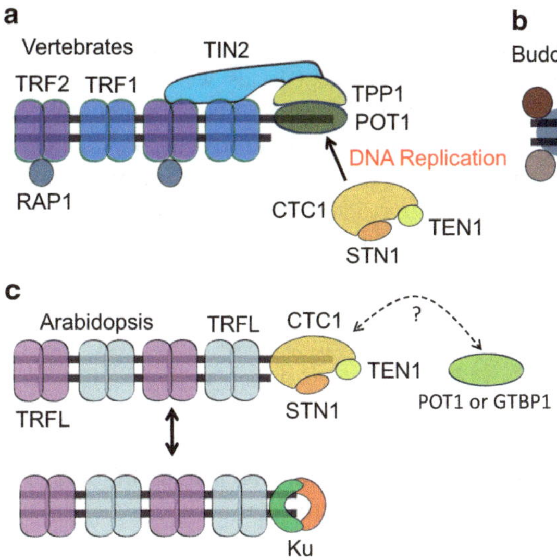

Fig. 6 Telomere protein complexes in budding yeast, vertebrates, and Arabidopsis. (**a**) Vertebrate telomeres associate with shelterin. Shelterin includes the dsDNA-binding proteins TRF1 and TRF2, which homodimerize; Rap1 which binds TRF2; and TIN2, a bridging protein that links the TRF proteins with the ssDNA-binding heterodimer, TPP1/POT1. Vertebrate CST (CTC1/STN1/TEN1) is proposed to engage telomeres only transiently during S-phase to promote DNA replication. (**b**) The telomeres of budding yeast are associated with the dsDNA-binding protein Rap1 and its associated factors: Rif1 and Rif2. G-overhangs are bound by CST (Cdc13, Stn1, Ten1). (**c**) Arabidopsis telomeres are associated with multiple TRFL proteins that bind ds telomeric DNA. The chromosome termini are asymmetrical. The G-overhang is bound by CST, while the other end is blunt and bound by Ku. Evidence from certain species suggests that POT1 and GTBP1 may also associate with the G-overhang. See text for details

machinery and, as a consequence, block a DNA damage response. In addition, interactions between TPP1 and telomerase facilitate recruitment of telomerase to the G-overhang for telomere extension. In vertebrates, loss of any of the shelterin components has a profound effect on telomere integrity and is ultimately lethal, arguing that each of these components is essential for chromosome end protection.

For many years it was surmised that budding yeast and vertebrates evolved separate telomere end-binding complexes. With the exception of RAP1, no other shelterin components could be detected in *S. cerevisiae*. The yeast RAP1 binds the telomeric duplex and is associated with two accessory factors: Rif1 and Rif2. The major telomere complex in yeast, however, is CST, a heterotrimer consisting of Cdc13/Stn1/Ten1 that bears structural similarity to RPA (Fig. 6b) (Linger and Price 2009; Price et al. 2010). CST serves as a multifunctional ss telomere-binding complex that associates with the G-overhang via contacts made by Cdc13. Cdc13 coordinates telomeric DNA replication by promoting G-strand synthesis by telomerase and C-strand synthesis through its interactions with DNA polymerase α/primase. A null mutation in any of the CST components is lethal.

Current studies have given rise to a more unified view of telomere capping components. Shelterin-like proteins have recently been uncovered in fission yeast, and Stn1 and Ten1 proteins are present in fission yeast as well as plants and vertebrates. However, bona fide Cdc13 homologs have yet to be identified outside of budding yeast, leaving open the question of whether this third component of CST exists in other eukaryotes. This conundrum was resolved by studies in Arabidopsis (discussed below) that uncovered CTC1, a large and novel protein that can form a heterotrimer with STN1 and TEN1. Genetic and biochemical studies in plants and humans indicate that the functions of CTC1 overlap with yeast Cdc13 (Nelson and Shippen 2012b).

In contrast to the highly detrimental consequences associated with loss of individual CST components in yeast or plants, human cells depleted of CST exhibit relatively mild phenotypes. Ctc1 or Stn1 knockdown results in a slight increase in G-overhangs and loss of telomeric DNA signals on some chromosome ends. The primary role of vertebrate CST appears to lie in promoting replication of telomeric DNA. Notably, CTC1 and STN1 were originally identified as accessory factors for the DNA polymerase α/primase (Price et al. 2010). Despite the mild phenotypes that accompany CST mutations in cell culture, two devastating stem cell-related human diseases have been linked to mutations in CTC1: Coats plus and Dyskeratosis congenita. Genetic data support the conclusion that shelterin and CST act in distinct pathways for telomere maintenance, and yet these two complexes must work hand in hand. How this feat is accomplished is unknown.

Although plants encode both shelterin and CST components, the functions of the complexes are more closely aligned with those of yeast than vertebrates (Nelson and Shippen 2012b). In addition, only a fraction of the shelterin proteins can be identified in plant genomes, leaving open the possibility that alternative factors have evolved to protect and maintain plant chromosome ends. In the next section, an overview of telomere-associated plant proteins is presented.

Shelterin-Like Components in Plants

Plants encode a family of ds telomere-binding proteins called TRF-like (TRFL) due to their sequence similarity with the Myb DNA-binding domains of the human TRF proteins. Unlike vertebrates where there are two TRF proteins, Arabidopsis underwent an expansion to encode 12 TRFL proteins (Watson and Riha 2010). The six members of TRFL family 1 contain a conserved region immediately adjacent to the C-terminal Myb domain that is necessary for telomeric DNA binding in vitro. The six members of family 2 do not contain this so-called Myb-extension domain and do not bind telomeric DNA in vitro. Structural data for TRFL family 1 members AtTRP1 and *Nicotiana glutinosa* TRF1 bound to telomeric DNA confirm the importance of the Myb extension in recognizing the additional nucleotide in the plant telomere repeat. Despite the expansion of the TRFL family in Arabidopsis, there are only two TRFLs in *Carica papaya*, *P. patens*, one in *N. glutinosa*, and three in *Selaginella moellendorffii*, all with the

Myb-extension domain and bearing closer sequence similarity to the telomeric DNA-binding TRFL family 1. Thus, it seems clear that the Myb extension within the TRFL proteins evolved to recognize the plant telomere repeat. It is likely that this evolutionary event already occurred by the time land plants began to diversify and quite possibly even earlier since some green algae share the plant telomere repeat.

The preponderance of TRFL genes in Arabidopsis and their apparent functional redundancy has complicated efforts to elucidate their individual contributions to telomere biology. However, some information has been gleaned. Loss of AtTBP1 (TRFL family 1) results in telomere elongation, but in contrast to vertebrates deficient in TRF1 or TRF2 does not result in any apparent developmental defects. NgTRF1 localizes to the nucleus and displays similar DNA-binding properties to AtTBP1. Moreover, alteration of NgTRF1 levels perturbs telomere length and leads to decreased cell viability in tobacco BY-2 cell culture, suggesting the TRFLs are indeed important for stability of plant telomeres.

An intensive search through plant genomes recovers only one other shelterin component besides the TRFLs: POT1. Plant POT1 proteins are structurally similar to their metazoan and yeast counterparts, containing the requisite two oligosaccharide–oligonucleotide (OB-fold) binding motifs at their N-terminus and a C-terminal protein–protein interaction domain. In the bryophyte *P. patens*, POT1 serves a very similar role to vertebrate and fission yeast POT1. It binds ss telomeric DNA and is critical for telomere length maintenance (Fig. 6c). Loss of PpPOT1 leads to increased G-overhang signals and increased chromosomal fusions, which result in developmental defects and sterility. Thus, POT1 is necessary for telomere and genome stability in this very early diverging plant lineage (Watson and Riha 2010).

Intriguingly, sometime after the divergence of vascular plants and mosses, POT1 either lost the ability to bind telomeric DNA or changed the manner in which it is bound so as to preclude in vitro detection. Evidence for this shift can be found in the lycophyte *Selaginella moellendorffii* and then throughout later diverging land plants. Interestingly, POT1 in *Asparagus officinalis*, which retains the human-type telomere repeat, as well as one of the two POT1 proteins in *Zea mays*, appears to have regained the ability to bind telomeric DNA. The Arabidopsis genome encodes two full-length POT1 paralogs and one truncated POT1 protein, none of which display telomeric DNA binding in vitro. AtPOT1b may make a contribution to chromosome end protection, as overexpression of the N-terminal portion leads to rapid telomere shortening and rampant chromosomal fusions. The major role of POT1a and POT1b, however, is connected to telomerase function and regulation (discussed below).

CST and Other Proteins that Associate with ss Telomeric DNA

Migration of POT1 from the G-overhang to telomerase and the absence of the other shelterin components left a critical vacancy at the chromosome terminus in plants. A possible replacement was identified by purifying ss telomeric DNA-binding proteins from tobacco suspension cells. Mass spectrometry identified a single protein with specificity for ss G-rich telomeric DNA, which was subsequently termed *Ng*GTBP1 (G-strand-specific telomere-binding protein) (Fig. 6c). NgGTBP1 binds ssDNA through two RNA recognition motifs, localizes to telomeres in vivo, and is homologous to the human heterogeneous nuclear ribonucleoprotein A1 (hnRNP-A1) previously found to associate with telomeric DNA. In human cells, hnRNP-A1 is postulated to regulate the switch between the nonspecific ssDNA-binding protein RPA to POT1 after DNA replication. In tobacco suspension cells, suppression of NgGTBP1 results in rampant telomere instability, highlighting the importance of this protein in plant telomere biology. However, it remains unclear whether NgGTBP1 or its orthologs in Arabidopsis have evolved a POT1-like telomere-binding function or play a role akin to hnRNP-A1 in vertebrates.

A major advance in understanding how plant telomeres are properly protected and replicated came with the identification of CST (CTC1/STN1/TEN1) in Arabidopsis (Fig. 6c). A BLAST search using *S. pombe* Stn1 as a query revealed the *A. thaliana* STN1 ortholog. STN1 proved to be a conserved telomere protein, serving many of the functions of its yeast counterparts. Plants deficient for STN1 show dramatic telomere loss, massive chromosomal fusions, and increased G-overhang signals. After STN1 discovery, the novel telomere protein CTC1 (conserved telomere maintenance component 1) was uncovered in a TILLING mutant collection. CTC1 mutants exhibit a profound telomere uncapping phenotype, similar to plants lacking STN1 (Nelson and Shippen 2012b).

In support of the idea that CST is critical for telomere maintenance and stability in plants, single copy homologs were identified in most major lineages of the plant kingdom where genome sequences are available, including mosses, lycopods, and several monocot and eudicot species. These discoveries contrast with the expansion and diversification of the POT1 and TRFL gene families.

Two Alternative Protein Complexes Cap Plant Chromosome Ends

One of the most startling discoveries in plant telomere biology is that the two ends of a chromosome are decidedly asymmetrical. Half of the chromosome ends bear a G-overhang while the other half are blunt-ended. If CST is the major cap for telomeres with a G-overhang, then an alternative capping complex must exist with a preference for blunt-ended ds telomeric DNA. A good candidate for this alternative cap is the Ku70/80 (Ku) heterodimer, a multifaceted, evolutionarily conserved protein complex with roles in DNA repair and telomere biology (Fig. 6c). At Arabidopsis telomeres, Ku serves as a negative regulator of telomere length. Loss of Ku results in a dramatic increase in telomere length in a telomerase-dependent manner, but does not significantly alter telomerase activity. Ku is also important for proper maintenance of G-overhang architecture. Ku mutants exhibit increased C-strand resectioning and longer (or more) G-overhangs (Figs. 5c, d). Furthermore, Ku prevents the formation of extrachromosomal t-circles (recombination by-products), a process that is enhanced when telomeres shorten rapidly in *tert ku* double mutants (Nelson and Shippen 2012a).

In light of the discovery that Ku is necessary for maintaining blunt-ended telomeres in Arabidopsis, the previous Ku data strongly support a role for Ku as an alternative telomere capping complex. First, loss of Ku leads to an increased G-overhang signal and a nearly complete loss of blunt-ended telomeres. Second, the absence of Ku allows exonuclease I to resect the exposed C-strand, creating a G-overhang. This newly created G-overhang would be a substrate for CST, which could in turn facilitate telomerase recruitment during the next cell division (Fig. 5d and Fig. 6c). This model not only explains end protection afforded by Ku but also provides a satisfying explanation for the increase in bulk telomere length observed in *ku* mutants (Nelson and Shippen 2012a).

Why did plants evolve two distinct telomere architectures? Although the answer is unknown, there is a potential advantage in having some blunt-ended chromosomes bound by Ku. Those termini will be inherently more resistant to nucleolytic processing and consequently more stable than chromosomes bearing a 3′ overhang. The increased stability afforded by blunt telomeres could confer greater stability to the entire genome, an advantageous outcome for sessile organisms that face hostile environmental conditions.

Telomerase

The catalytic subunit of telomerase, TERT, shares a common ancestor with the reverse transcriptase derived from Penelope-like retroelements (Autexier and Lue 2006). This observation supports the hypothesis that telomeres arose from retrotransposons. Indeed, as mentioned earlier, retrotransposition at telomeres is still at work in Drosophila. The first plant TERT was identified soon after the sequencing of the *A. thaliana* and *Oryza sativa* (rice) genomes. At roughly 130kD, AtTERT is similar in size to human TERT. As more TERT sequences were recovered, it became clear that the protein is well conserved and bears canonical RT domains as well as telomerase-specific motifs. Phylogenies constructed from aligned land plant TERT sequences reproduce the widely accepted organismal tree, confirming that a single ancestral TERT ortholog existed in the plant kingdom.

Unlike TERT, poor sequence conservation and low RNA abundance made identification of plant TERs highly problematic. The only plant TER characterized to date came from biochemical purification of Arabidopsis telomerase. Surprisingly, *A. thaliana* harbors two TER genes, which both contain the predicted telomere templating domain but are encoded by unique loci (Fig. 7a). *TER1* encodes a 748 nt RNA that contains a ~220 nt core region that is conserved in TER2. *TER2* encodes a 784 nt RNA, and intriguingly the core conserved region is separated in TER2 by a ~520 nt intervening sequence with no sequence similarity to TER1. Subsequent analysis showed that TER2 is spliced in vivo to remove the intervening sequence as well as the unique 3’ terminus producing a third TER isoform, TER2s, with strong sequence similarity to the core region of TER1

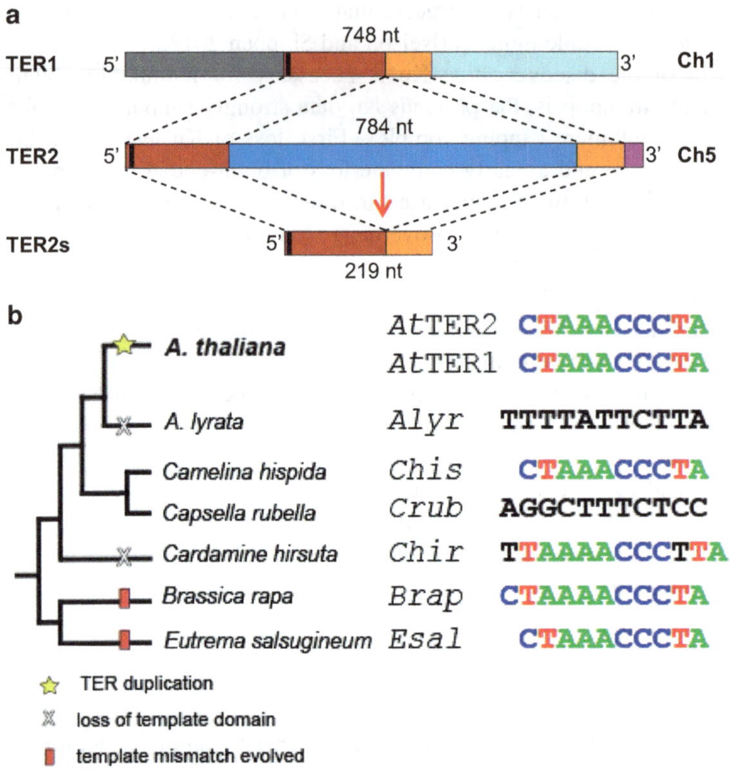

Fig. 7 **Multiple TERs in*Arabidopsis thaliana*.** (**a**) Depiction of the three *A. thaliana* TER isoforms showing regions of high similarity (>90 %) (*dashed lines*). TER1 and TER2 are encoded in the genome on chromosome 1 and 5, respectively, but TER2s is generated from TER2 (*red arrow*) by the removal of a 529 intervening sequence (*blue*) and the 3' 36 nt tail (*purple*; *red arrow*). (**b**) *Left*, partial Brassicaceae tree depicting a reconstruction of a subset of the putative TER loci known to date. *Yellow star* indicates the duplication event that produced two *TER* loci in *A. thaliana*. *Red blocks* indicate species harboring an altered TER template domain. X denotes a species in which this locus completely lacks a telomere templating sequence. *Right*, putative TER template region based on alignment of the *TAD3–RAD52* locus from the species shown to the *left*

(Fig. 7a) (Nelson and Shippen 2012b). The steady state level of TER2 is much lower than TER2s, suggesting processing is efficient. Notably, TER1 and TER2s expression coincides with telomerase activity, peaking in flowers, and other rapidly dividing tissues, such as seedlings and cell culture (Nelson and Shippen 2012b).

Despite the presence of multiple TER isoforms in Arabidopsis, only TER1 is critical for telomere length maintenance in vivo. Knockdown of TER1 leads to a decrease in telomerase activity and a concomitant telomere shortening phenotype. As expected for a canonical TER, mutation of the TER1 template results in incorporation of mutated telomere repeats in vivo. In contrast, although TER2 can serve as a template for TERT in vitro, it does not facilitate telomere repeat addition in vivo. Removal of TER2 via a T-DNA insertion within the template domain does not alter telomere length. Instead, TER2 depletion causes increased telomerase activity. The conclusion that TER2 negatively regulates telomerase is bolstered by overexpression experiments showing that increased TER2 decreases TER1-mediated telomerase enzyme activity in vivo. Thus, TER1 is the functional telomere template for telomerase, while TER2 serves as a novel negatively regulator for telomerase enzyme activity (see below) (Nelson and Shippen 2012b).

The discovery of Arabidopsis TER should have facilitated the identification of TER moieties in closely related species within Brassicaceae, the family containing *A. thaliana*. However, identification of additional plant TER homologs has proven unexpectedly difficult. TER1 and TER2 are encoded on different chromosomes in *A. thaliana* and are adjacent to different coding regions (Fig. 7b). The TER1 gene on chromosome 1 overlaps with and terminates within the *RAD52A* locus, while TER2 is encoded upstream, but in the opposite orientation of the gene *TAD3* on chromosome 5. Exhaustive analysis of syntenic loci throughout 13 Brassicaceae species revealed a single locus flanked by *TAD3* and *RAD52A*, indicating that the duplication event that gave rise to the *TER1* and *TER2* genes is restricted to *A. thaliana* (Fig. 7b). Remarkably, there is no discernible template domain at the *TAD3-RAD52* locus in *A. lyrata*, the closest relative to *A. thaliana*. Sequence divergence within the template domain is not limited to *A. lyrata*; three other Arabidopsis relatives lack a template or template-like motif entirely, while four others contain nucleotide substitutions within the template domain. Character state reconstruction of the ancestral TAD3-RAD52 locus indicated that each template loss and nucleotide substitution event occurred independently (Nelson and Shippen 2012b).

Telomerase Accessory Proteins

Aside from TERT and TER, telomerase associates with multiple accessory proteins that are necessary for RNP assembly, optimal enzymatic activity, and recruitment to the chromosome end. One such protein is the pseudouridine synthase dyskerin, which functions in vertebrate telomerase RNP maturation. Mutations in human dyskerin result in telomere shortening and culminate in the stem cell disease Dyskeratosis congenita. Dyskerin also associates with Arabidopsis telomerase, likely through a predicted H/ACA box at the 3′ end of both TER1 and TER2 (Fig. 8). As in human cells, Arabidopsis dyskerin hypomorphs display reduced telomerase activity and shorter telomeres (Egan and Collins 2012; Nelson and Shippen 2012b).

Arabidopsis POT1a and POT1b both associate with telomerase but are dispersed to two different RNP complexes (Fig. 8). Loss of POT1a decreases telomerase enzyme activity and causes gradual telomere shortening. Biochemical analysis indicates that POT1a directly binds TER1 and recruits telomerase to the chromosome end through this interaction. In contrast, POT1b associates with TER2, but the significance of this interaction is still unclear. The other land plants that have been examined encode only a single POT1 protein, which does not bind telomeric DNA

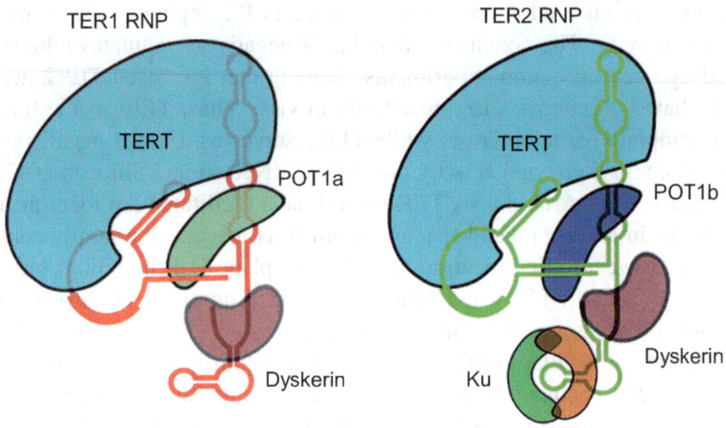

Fig. 8 **Alternative telomerase RNPs in*Arabidopsis thaliana*.** The TER1 RNP (*left*) performs the canonical role of telomere elongation whereas the TER2 complex (*right*) acts to inhibit TER1-mediated telomerase activity, particularly during DNA damage. Dyskerin associates with both RNPs and is likely critical for their proper biogenesis. POT1a physically associates with TER1 and is necessary for telomere maintenance, possibly through interactions with CST. Genetic data implicates Ku as a critical capping component and negative regulator of telomere length maintenance in *A. thaliana*. Biochemical data demonstrate that both Ku and POT1b are components of the TER2 RNP. The functions of these interactions are still unclear. TER2s (not depicted) is proposed to associate with POT1b and Ku, but not with TERT

in vitro. Since it appears that POT1 was dislodged from telomeres early in the evolutionary history of land plants, its retention in plant genomes following this event may be explained by the evolution of its role as a critical telomerase accessory protein.

In addition to protecting blunt-ended telomeres, Ku interacts with TER2 and $TER2_S$, but not TER1 (Fig. 8). Notably, Ku associates with human and budding yeast TER, but not with *S. pombe* TER (Nelson and Shippen 2012b). The interaction of Ku with $TER2/TER2_S$ in Arabidopsis is especially intriguing given that Ku negatively regulates telomere length in vivo, while TER2 negatively regulates telomerase enzyme activity. The physical interaction between Ku and TER2 raises the possibility of an inhibitory $TER2/TER2_S$-Ku complex that occludes DNA ends from an active TER1 telomerase RNP to modulate telomere length (see below). Further studies will be required to test this model.

Telomerase Regulation

In vertebrates, telomerase acts as a double-edged sword that enables cellular proliferation in both normal stem cells and in cancer cells (Artandi and DePinho 2010; Cifuentes-Rojas and Shippen 2012). While metastatic cancer is not a concern for plants, plants nevertheless modulate telomerase activity by repressing the enzyme in differentiated cells and activating it in meristems and rapidly dividing

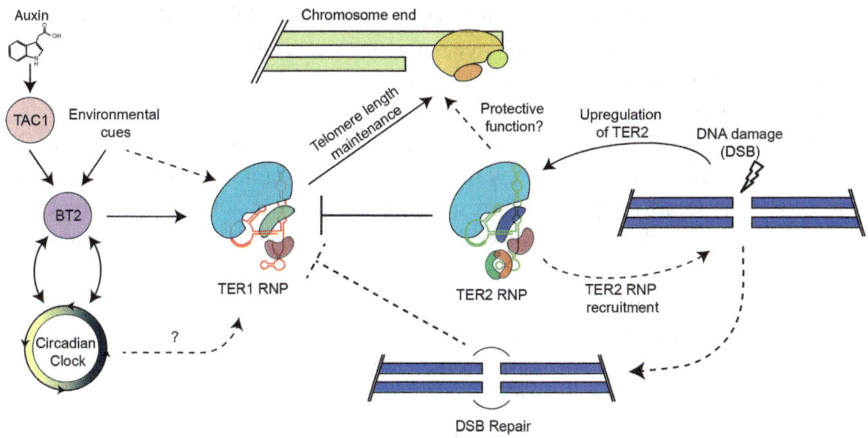

Fig. 9 **Model for telomerase regulation in plants.** Telomerase activity is regulated by many factors. Among these is the hormone auxin (*upper left*). Auxin upregulates the transcription factor TAC1, which in turns activates the calmodulin binding protein BT2, resulting in increased expression of TERT and enhanced telomerase activity. BT2 is responsive to environmental and circadian cues, suggesting that telomerase activity may also be susceptible to these stimuli. Introduction of DSBs induces TER2 RNA (*far right*), which causes a decrease in TER1-mediated telomerase activity. The mechanism of telomerase inhibition is known, but it may involve competitive inhibition as TERT has a significantly higher affinity for TER2 than for TER1. Downregulation of telomerase activity following DSBs reduces the probability of de novo telomere formation at the break site. The TER2 RNP does not direct telomere repeat incorporation itself but may play as yet undefined role in chromosome end protection

cells. A major point of control for telomerase activity in vertebrates is through TERT. As in vertebrates, plant TERT mRNA levels peak in plant settings characterized by dividing cell populations, including young seedlings, meristematic zones of the root and shoot, flowers, and suspension culture (Fig. 9). Telomerase activity follows suit, and in organs where TERT is poorly expressed such as leaves, enzyme activity is barely detectable. In addition to being limited to specific tissues, studies in tobacco show that telomerase activity is restricted to late S phase of the cell cycle.

TERT mRNA levels are modulated by auxin, a ubiquitous class of plant hormones necessary for cell division, cell expansion, and a wide range of other cellular activities. TERT is induced in an auxin-dependent manner leading to increased telomerase activity (Fig. 9). Auxin's effect on telomerase activity is potentiated in part by the auxin-responsive transcription factor TAC1, which upregulates BT2, a signaling protein capable of responding to multiple environmental cues. TERT mRNA is significantly elevated in *tac1* mutants, which may account for increased telomerase activity. Intriguingly, BT2 is circadian-regulated, suggesting that telomerase may also be affected by the circadian network (Fig. 9). Exogenous application of auxin or overexpression of either TAC1 or BT2 leads to increased telomerase activity and the presence of activity in mature tissues. In tobacco and rice this auxin-dependent increase in telomerase activity may be dependent on direct phosphorylation of the telomerase complex. In addition, the presence of multiple TERT splice variants in rice, and distinct TERT paralogs transcribed from independent loci in tobacco, may provide additional means of regulating telomerase (Watson and Riha 2011).

Finally, the recent discovery of an alternative telomerase RNA subunit in Arabidopsis (TER2) revealed a new mode of enzyme regulation (Fig. 9). Telomerase activity is rapidly repressed in response to DSBs, but not with other types of genotoxic stress such as replication fork stalling or even telomere dysfunction. Strikingly, the inhibition of telomerase coincides with the specific induction of TER2; in the absence of TER2, telomerase activity is not repressed by DSBs. DSB-induced telomerase repression provides a mechanism to dial back telomerase activity, decreasing the opportunity for the enzyme to act inappropriately at DSBs. This response would promote genome stability by favoring legitimate repair of the break (Fig. 9) (Nelson and Shippen 2012b).

De Novo Telomere Formation and Its Regulation

One interesting facet of telomerase enzymology is its highly promiscuous nature. In vitro telomerase will readily extend DNA primers lacking any 3′ complementarity to TER (Melek and Shippen 1996). How then is telomerase prevented from acting at a DSB? The addition of telomeric sequence at an interstitial break site results in the loss of the centromere-distal chromosome DNA fragment. This outcome results in partial monosomy and is the source of several human genetic diseases as well as mental retardation and cancer (Murnane 2012). In plants, McClintock observed "chromosome healing" while characterizing the recovery of fused chromosomes in maize. Dicentric chromosomes break during anaphase, and this breakage-fusion-bridge cycle continues until a stable chromosome is formed, or the cell enters senescence. McClintock discovered that broken chromosomes could be repaired in the germ line by some mechanism that prevented the chromosomes from re-fusing. Of course this mechanism requires telomerase, and thus the recruitment of telomerase to chromosome breaks is not quite so peculiar. De novo telomere formation (DNTF) by telomerase represents the ultimate in error-prone DSB repair.

Interestingly, chromosome fragmentation and DNTF are key steps in the sexual stage of the life cycle of ciliated protozoa, and in these organisms telomerase is programmed to add telomere repeats to all DNA fragments (Melek and Shippen 1996). However, this process would be highly deleterious for most genomes. Yeast and vertebrates have evolved a variety of different methods to curtail DNTF in response to DSBs, including sequestering TERT, activation of a helicase to unwind the hybrid formed between TER and the DSB, and modification of telomere-binding proteins so they cannot stabilize the terminus to recruit telomerase (Cifuentes-Rojas and Shippen 2012). *A. thaliana* TER provides a new paradigm for controlling telomerase by the induction of a noncanonical long noncoding RNA that downregulates enzyme activity (Fig. 9). Whether this type of regulatory mechanism will be found in other organisms remains to be determined.

To more directly analyze the factors that control DNTF in plants, a telomere truncation system was recently described for Arabidopsis. Small telomere repeat arrays, or telomere seed sequences, were transformed into diploid and tetraploid Arabidopsis lines. In diploid Arabidopsis the rate of insertion events leading to DNTF formation was significantly higher than that tolerated by other organisms, and the events were primarily concentrated toward the chromosome terminus. The remaining insertion events did not result in DNTF, but instead reflected stable insertions scattered randomly throughout the genome. In tetraploid Arabidopsis, where genetic redundancy diminishes the consequences of losing a chromosome arm via DNTF, successful events occurred throughout the genome and at a much higher frequency (Gaeta et al. 2012).

The tolerance to DNTF associated with tetraploid Arabidopsis has also been observed in the natural crop polyploids barley and maize. Interest in the factors controlling DNTF in plants is spurred by the possibility of engineering minichromosomes, which require telomeres for stability and retention. The properties and transmission rate of minichromosomes generated in part by DNTF has been the subject of intense study in maize and, more recently, in barley. The eventual goal of these studies is to genetically alter plants using stably transmissible minichromosomes, which could be engineered to harness a variety of selectable markers and site-specific recombination and integration systems. In addition, such organisms would present powerful systems for analyzing mechanisms that naturally suppress DNTF to promote genome integrity (Gaeta et al. 2012).

Outcomes of Telomere Dysfunction

Telomeres must be hidden from two competing forces. One of these is exonucleolytic activity, and the other is unwanted DNA repair that culminates in the end-to-end chromosome fusion and the breakage-fusion-bridge cycle. These forces are prevented by the constant vigilance of telomere capping proteins. However, the process of telomeric DNA replication provides a point of vulnerability, as the t-loop must unfold to expose the G-overhang. Following DNA replication, a new complex of proteins must assemble on the telomeres of each new chromosome (Palm and de Lange 2008; Price et al. 2010). One of the more peculiar aspects of telomere biology is how certain DNA damage repair (DDR) proteins, which should be strictly prohibited from acting on chromosome ends, have been co-opted to ensure proper chromosome end maintenance. For example, the two master DDR kinases, ATM and ATR, are critical for telomeric DNA replication and extension of telomeres (de Lange 2009; O'Sullivan and Karlseder 2010). In addition, the Ku (Ku70/Ku80) heterodimer, a core component of the nonhomologous end-joining mechanism, and the DSB processing 5'–3' exonuclease Exo1 are necessary for both protection and resecting of chromosome ends (Nelson and Shippen 2012a). Thus, distinguishing a telomere from a DSB and facilitating its complete replication presents significant challenges to eukaryotes and involves the intimate intertwining of telomere-specific factors and the DDR. Indeed, dire consequences ensue if this complex regulatory pathway is perturbed.

Telomere failure leads to activation of a powerful DDR. Cells are alerted to DNA damage primarily through ATM and ATR. ATM responds to dsDNA breaks by the break-binding complex Mre11-Rad50-Nbs1 (MRN), whereas ATR senses accumulation of ssDNA through the activities of the ssDNA-binding protein heterotrimer RPA (Riha et al. 2006; Symington and Gautier 2011). Telomere-binding proteins prohibit DDR based on their contacts with the DNA. The dsDNA-binding shelterin component TRF2 protects vertebrate telomeres from ATM-mediated DNA damage signaling, while POT1 blocks ATR (de Lange 2009). Similarly, plant telomeres are protected from an ATR-mediated DDR by CST.

Very rapid and catastrophic genome instability occurs in plants lacking core telomere-binding components such as CST. Although inactivation of telomerase leads to a slow but progressive loss of telomeric DNA, it also triggers a DDR. Remarkably, Arabidopsis bearing a null mutation in TERT can survive without telomerase for up to ten plant generations, but the plants begin to display worsening morphological defects including fasciated stems, irregular leaf and silique placement, and reduced fertility in the sixth generation. By the 8th–10th generation, all the mutants arrest at a terminal vegetative state (Watson and Riha 2011). Mutants also exhibit a high frequency of end-to-end chromosome fusion events, indicative of deprotected telomeres. Sequence analyses of telomere fusion junctions in Arabidopsis reveal that the minimal size for telomere stability is approximately 1 kb; below this length threshold chromosome ends begin to be recruited into end-joining reactions. The absolute minimal size of an Arabidopsis telomere with an intact G-overhang is approximately 300 bp, roughly the length of a yeast telomere.

Recruitment of dysfunctional telomeres into chromosome fusions is the responsibility of two complexes in the canonical nonhomologous end-joining (NHEJ) pathway: Ku and MRN (Mre11/Rad50/Nbs1) (Riha et al. 2006; Symington and Gautier 2011). In plants, loss of either MRE11 or RAD50 results in heightened DNA damage sensitivity, developmental defects, sterility, and altered telomere length dynamics. Although the absence of Ku does not lead to fertility or phenotypic abnormalities, Ku mutants are hypersensitive to DNA damage and have elongated telomeres. Sequence analysis of telomere fusion events in plants lacking Ku, MRE11, or both proteins reveals a role for Ku and MRN in telomere end-joining events and further that a non-conventional NHEJ mechanism can act on uncapped plant telomeres. These findings underscore the complex interconnectedness between DNA repair pathways and telomere length maintenance in plants.

Telomeres are maintained at a species-specific set point length primarily through the opposing forces of telomerase and the end-replication problem. Two recombination-based mechanisms can also act on telomeres to rapidly and dramatically alter telomere length. Both of these mechanisms are activated in response to telomere dysfunction. Extension of telomeres can occur via a process called ALT (alternative lengthening of telomeres), while long telomeres can be abruptly shortened by TRD (telomere rapid deletion) (Murnane 2012). Both TRD and ALT have been reported in Arabidopsis.

TRD is observed when telomeres have become deprotected, as is the case when any of the core telomere-binding components is removed. One of the hallmarks of TRD is the accumulation of extrachromosomal telomeric circles (ECTC), by-products of homologous recombination. Intriguingly, ECTCs are also associated with ALT and have been proposed to act as a template for rolling circle replication to extend the telomere tract. Although TRD can be detected at compromised telomeres, it may occur stochastically on individual telomere tracts that become hyper-elongated. In this circumstance, TRD can bring the ultra-long telomere back within the optimal size range. This mechanism appears to be regulated at least in part by Ku.

In human cells the absence of TERT is sufficient to initiate ALT. Notably, some 5 % of human tumors maintain telomeres, not through telomerase reactivation, but via ALT (Artandi and DePinho 2010; Cifuentes-Rojas and Shippen 2012). Plants exhibit evidence of ALT only after prolonged absence of telomerase. Callus derived from eighth generation (G_8) *tert* mutants can be maintained indefinitely in a laboratory setting. Examination of the chromosome termini in this tissue revealed a complete lack of telomeric DNA along with rampant genome rearrangements, fusion events, and altered ploidy, arguing that while this telomerase-independent mechanism for maintaining chromosome ends can allow undifferentiated plant cells to survive, it is unlikely to be sufficient for organismal viability.

Conclusions

Plants have provided a wealth of information on how telomeres promote long-term genome integrity in multicellular eukaryotes. By chance, or by necessity, some components of the telomere have been lost or co-opted into alternative functions in plants. Analysis of these "exceptions" has allowed for a deeper understanding of telomere dynamics and the evolving roles of telomere complexes at the chromosome terminus and in genome maintenance.

Future Directions

Future directions will leverage the vast information of plant genomes to reveal the forces driving the evolution the telomere complex and telomerase. In the near term, these studies will focus on elucidating the full complement of telomere-associated proteins and telomerase components and how these factors are regulated within the cell cycle and in response to DNA damage. The long-term outcomes of this research will not only fill in the evolutionary blanks spots for key factors that protect plant telomeres and facilitate their complete replication but also will provide important new insights into how plants exploit gene duplication and diversification to promote the integrity of the entire genome in changing environmental conditions.

References

Artandi SE, DePinho RA. Telomeres and telomerase in cancer. Carcinogenesis. 2010;31:9–18.

Autexier C, Lue NF. The structure and function of telomerase reverse transcriptase. Annu Rev Biochem. 2006;75:493–517.

Cifuentes-Rojas C, Shippen DE. Telomerase regulation. Mutat Res. 2012;730:20–7.

de Lange T. How telomeres solve the end-protection problem. Science. 2009;326:948–52.

de Lange T, Lundblad V, Blackburn EH. Telomeres. 2nd ed. Cold Spring Harbor: Cold Spring Harbor Laboratory Press; 2006.

Egan ED, Collins K. Biogenesis of telomerase ribonucleoproteins. RNA. 2012;18:1747–59.

Gaeta RT, Masonbrink RE, Krishnaswamy L, Zhao C, Birchler JA. Synthetic chromosome platforms in plants. Annu Rev Plant Biol. 2012;63:307–30.

Linger BR, Price CM. Conservation of telomere protein complexes: shuffling through evolution. Crit Rev Biochem Mol Biol. 2009;44:434–46.

McKnight TD, Riha K, Shippen DE. Telomeres, telomerase, and stability of the plant genome. Plant Mol Biol. 2002;48:331–7.

Melek M, Shippen DE. Chromosome healing: spontaneous and programmed de novo telomere formation by telomerase. Bioessays. 1996;18:301–8.

Murnane JP. Telomere dysfunction and chromosome instability. Mutat Res. 2012;730:28–36.

Nelson AD, Shippen DE. Blunt-ended telomeres: an alternative ending to the replication and end protection stories. Genes Dev. 2012a;26:1648–52.

Nelson AD, Shippen DE. Surprises from the chromosome front: lessons from Arabidopsis on telomeres and telomerase. Cold Spring Harb Symp Quant Biol. 2012b;77:7–15.

O'Sullivan RJ, Karlseder J. Telomeres: protecting chromosomes against genome instability. Nat Rev Mol Cell Biol. 2010;11:171–81.

Palm W, de Lange T. How shelterin protects mammalian telomeres. Annu Rev Genet. 2008;42:301–34.

Price CM, Boltz KA, Chaiken MF, Stewart JA, Beilstein MA, Shippen DE. Evolution of CST function in telomere maintenance. Cell Cycle. 2010;9:3157–65.

Riha K, Heacock ML, Shippen DE. The role of the nonhomologous end-joining DNA double-strand break repair pathway in telomere biology. Annu Rev Genet. 2006;40:237–77.

Symington LS, Gautier J. Double-strand break end resection and repair pathway choice. Annu Rev Genet. 2011;45:247–71.

Watson JM, Riha K. Comparative biology of telomeres: where plants stand. FEBS Lett. 2010;584:3752–9.

Watson JM, Riha K. Telomeres, aging, and plants: from weeds to Methuselah – a mini-review. Gerontology. 2011;57:129–36.

Further Reading

Amiard S, Depeiges A, Allain E, White CI, Gallego ME. Arabidopsis ATM and ATR kinases prevent propagation of genome damage caused by telomere dysfunction. Plant Cell. 2011;23:4254–65.

Cifuentes-Rojas C, Nelson AD, Boltz KA, Kannan K, She X, Shippen DE. An alternative telomerase RNA in Arabidopsis modulates enzyme activity in response to DNA damage. Genes Dev. 2012;26:2512–23.

Kazda A, Zellinger B, Rossler M, Derboven E, Kusenda B, Riha K. Chromosome end protection by blunt-ended telomeres. Genes Dev. 2012;26:1703–13.

Lee YW, Kim WT. Tobacco GTBP1, a homolog of human heterogeneous nuclear ribonucleoprotein, protects telomeres from aberrant homologous recombination. Plant Cell. 2010;22:2781–95.

Leehy KA, Lee JR, Song X, Renfrew KB, Shippen DE. MERISTEM DISORGANIZATION1 encodes TEN1, an essential telomere protein that modulates telomerase processivity in Arabidopsis. Plant Cell. 2013;25:1343–54.

Riha K, McKnight TD, Griffing LR, Shippen DE. Living with genome instability: plant responses to telomere dysfunction. Science. 2001;291:1797–800.

Surovtseva YV, Churikov D, Boltz KA, Song X, Lamb JC, Warrington R, Leehy K, Heacock M, Price CM, Shippen DE. Conserved telomere maintenance component 1 interacts with STN1 and maintains chromosome ends in higher eukaryotes. Mol Cell. 2009;36:207–18.

Vrbsky J, Akimcheva S, Watson JM, Turner TL, Daxinger L, Vyskot B, Aufsatz W, Riha K. siRNA-mediated methylation of Arabidopsis telomeres. PLoS Genet. 2010;6:e1000986.

Zellinger B, Akimcheva S, Puizina J, Schirato M, Riha K. Ku suppresses formation of telomeric circles and alternative telomere lengthening in Arabidopsis. Mol Cell. 2007;27:163–9.

DNA Repair and Recombination in Plants

3

Susan Schröpfer, Alexander Knoll, Oliver Trapp, and Holger Puchta

Contents

Abstract

- DNA damage can be caused by a large number of internal and external, biotic and abiotic sources and can affect cell viability and can lead to mutations.
- Depending on the type of damage, different evolutionarily conserved repair pathways are used.
- Some specific lesions caused by UV radiation and DNA alkylation can be repaired by direct enzymatic reversal.
- The base excision repair pathway is used for the removal of a variety of damaged bases.

S. Schröpfer • A. Knoll • O. Trapp • H. Puchta (✉)
Botanical Institute II, Karlsruhe Institute of Technology, Karlsruhe, Germany
e-mail: susan.schroepfer@kit.edu; alexander.knoll@kit.edu; oliver.trapp@kit.edu; holger.puchta@kit.edu

S.H. Howell (ed.), *Molecular Biology, The Plant Sciences 2*,
DOI 10.1007/978-1-4614-7570-5_2

- When larger modifications of nucleotides are present, the nucleotide excision repair pathway is active.
- The mismatch repair pathway can reverse the incorporation of noncomplementary nucleotides by replicative polymerases.
- DNA double-strand breaks can be repaired by the pathways of nonhomologous end-joining, single-strand annealing, and homologous recombination which lead to different outcomes.
- Different DNA damage tolerance and repair pathways can deal with DNA lesions at damaged replication forks.
- Repair of DNA has to be tightly regulated with respect to other cellular processes.
- DSB repair pathways form the basis for recently developed techniques for directed modification of genomes for research and agronomy.

Introduction

DNA is a biomolecule which represents the basis for all living organisms by encoding the information for all processes in life. The maintenance of genome stability by counteracting changes in DNA is a great challenge, which has to be achieved in every single cell.

Different kinds of mutations with diverse consequences for the cell can arise due to a multitude of factors. Point mutations are changes of a single nucleotide, for example, a substitution of one base by another base. An exchange of a pyrimidine by a pyrimidine base (e.g., T/C) or a purine by a purine base (e.g., A/G) is defined as a transition. In contrast, a substitution of a pyrimidine with a purine base (e.g., T/G or A) or a purine with a pyrimidine base (e.g., A/C or T) is called a transversion. A point mutation in an open reading frame (ORF) of a gene can lead to a substitution of an amino acid in the resulting protein, which might change the properties of the respective protein. Nonsense point mutations, e.g., by introducing a stop codon in the middle of an ORF, can result in a complete loss of function of the encoded protein. Other changes in DNA caused by insertions or deletions of one or more nucleotides might also drastically affect protein biosynthesis, e.g., by causing a frame shift. In contrast to mutations which concern a single DNA sequence, also large rearrangements of chromosomes such as inversions and translocations can arise due to aberrant recombination processes. Furthermore, huge insertions and deletions can occur as well. Insertions of many long DNA sequences into a genomic locus such as coding or regulating DNA sequences can be due to an insertion of a mobile DNA element such as a transposon. Certain bacterial species can insert huge pieces of foreign DNA into a plant genome, such as *Agrobacterium tumefaciens*, which transfers a so-called transfer DNA (T-DNA) into plant cells. T-DNA is randomly integrated into the plant genome and codes for metabolic enzymes which reprogram the metabolism of the host plant and provide nutrients specifically for the bacterium.

Mutations can arise due to a variety of DNA damage. DNA is constantly affected by DNA-damaging factors which can cause modifications in the chemical composition and structure of DNA. In contrast to the majority of other biomolecules, lesions in DNA may lead to the formation of stable mutations which will be inherited by all daughter cells and may affect their metabolism permanently. Therefore, the manifold lesions that can arise in DNA can be repaired by a complex and conserved system of DNA repair mechanisms to avoid the expression of such mutations in the genome. In plants, the repair of damaged DNA has a special relevance because plants do not have a predetermined germ line. Mutations in DNA can originate in somatic tissue which later gives rise to germ cells and can thus be passed on to the next generation.

Endogenous factors such as genotoxic by-products of the cellular metabolism or errors in DNA replication and recombination can lead to spontaneous DNA damage. Furthermore, exogenous factors coming from the abiotic and biotic environment can also affect DNA. Some examples of the various exogenous factors are energy-rich radiation such as ionizing or solar UV radiation, naturally occurring chemical genotoxins produced by some bacteria or fungi, as well as synthetic chemical agents. In the following sections the formation of different types of DNA damage by various DNA-damaging factors is discussed, followed by a description of the current state of knowledge on the diverse pathways that repair these lesions.

DNA-Damaging Factors and DNA Lesions

Replication-associated DNA damage: During the duplication of chromosomes in S phase of the cell cycle, DNA is replicated by DNA polymerases. An incorporation of a noncomplementary nucleotide opposite to the parental template DNA strand can be detected and counteracted by the proofreading activity of the replicative DNA polymerases. In such case, the misincorporated nucleotide is removed from the newly synthesized DNA strand mediated by an intrinsic exonuclease domain. Despite such proofreading activity of the replicative DNA polymerases, a small fraction of misincorporated nucleotides persist in DNA, leading to a mutation rate of about 10^{-8} per replication round. Because of nonmatching and different base pairing properties of the opposing nucleotides, a so-called DNA mismatch is formed which also affects the structure and stability of DNA. If a mismatch is not detected until the next round of DNA replication is initiated, the incorrectly incorporated nucleotide is then a component of the parental template stand, which causes a stable establishment of the mutation by the incorporation of the complementary nucleotide.

Spontaneous hydrolytic DNA damage: DNA is exposed to spontaneous modifications by hydrolysis of chemical bonds, such as deamination and depurination, leading to a change of the chemical composition of the DNA molecule. A deamination of DNA causes the removal of an amine group from DNA bases, whereas a depurination leads to a complete loss of a purine base (Fig. 1).

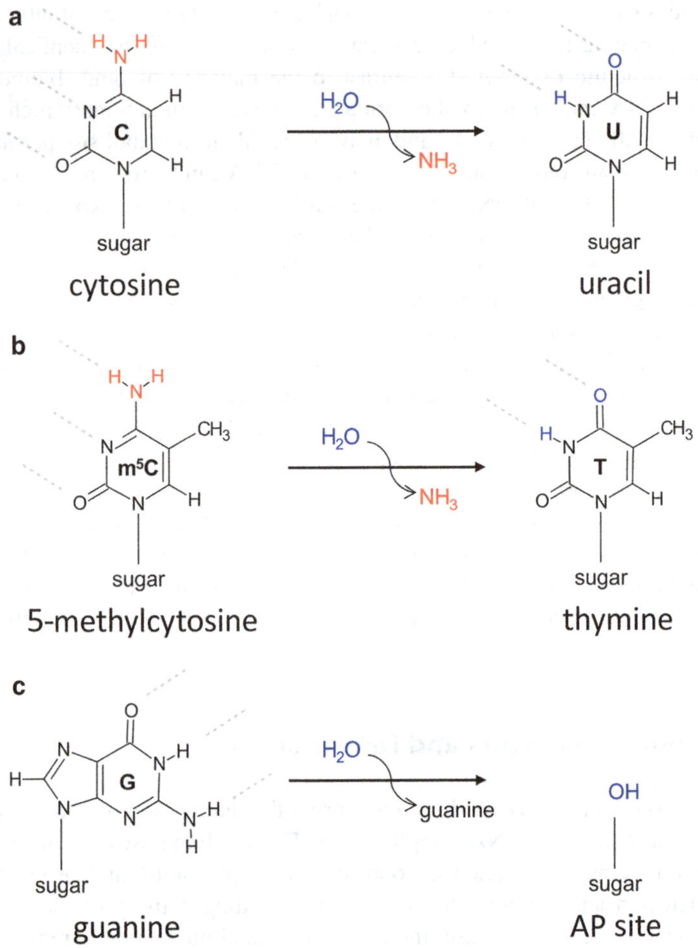

Fig. 1 Spontaneous hydrolytic DNA damage. (**a**) The deamination of the nucleobase cytosine by the hydrolytical removal of the amine group leads to the formation of uracil, which exhibits different base pairing properties than cytosine. (**b**) The deamination of 5-methylcytosine (m^5C) results in thymine. (**c**) The complete loss of the purine base guanine leading to an AP site results from a depurination

Because of its chemical instability, the pyrimidine base cytosine is the DNA base most frequently affected by deamination. The hydrolytic removal of the amine group of cytosine leads to the formation of uracil (Fig. 1a), a base which does not normally occur in DNA but rather in RNA. Thereby, a premutagenic mispairing of the resulting uracil with the opposite base guanine arises. Uracil preferentially pairs with adenine instead of guanine, leading to a point mutation by the incorporation of an incorrect nucleotide in the following replication round. Despite the deamination of cytosine, uracil can also be incorporated in DNA during the duplication of DNA, if the replicative DNA polymerase adds dUTP instead of dTTP to the newly synthetized daughter strand. In contrast to the deamination of cytosine leading to a U:G mismatch, the incorrect incorporation of dUTP during replication leads to a U:A base pair.

Other DNA bases are also affected by spontaneous deamination. The deamination of adenine results in the formation of hypoxanthine. The preferred base pairing of hypoxanthine with cytosine instead of thymine can also cause a point mutation during the next S phase. The hydrolytic removal of the amine group from guanine leads to xanthine. The base pairing features of xanthine are similar to the original, unmodified base guanine. Xanthine pairs with cytosine, but only with two instead of three hydrogen bonds, which affects the stability of the xanthine:C base pair. In contrast to the premutagenic mispair of hypoxanthine:T caused by the deamination of adenine, the formation of xanthine does not lead to an incorrect nucleotide incorporation during the following replication round.

Modified bases in DNA are not the only consequences of DNA-damaging factors. For instance, the methylation of cytosine is an epigenetic signal for the regulation of gene expression and the transcriptional state of genes. The presence of 5-methylcytosine in promoter regions of genes can be responsible for the transcriptional silencing of the respective gene. The spontaneous deamination of 5-methylcytosine leads to formation of thymine, resulting in a T:G mismatch (Fig. 1b). This mismatch poses a great challenge for the cell because thymine is naturally occurring in DNA and contains no chemical alteration to discriminate between the damaged and undamaged base in the T:G mismatch.

Further spontaneous DNA damage is caused by the depurination of purine bases. Here, the N-glycosidic bond between a purine base and the respective 2’-deoxyribose residue in the DNA backbone is hydrolyzed. This reaction leads to the release of the respective nucleic base and to the formation of a so-called abasic/apurinic site (AP site, Fig. 1c).

Oxidation of DNA: Reactive oxygen species (ROS) such as superoxide (O_2^-), hydrogen peroxide (H_2O_2), and hydroxyl radicals (•OH) have a strong oxidizing potential and are extremely reactive and unstable. They can attack various cellular components including proteins, lipids, and DNA. There are different extra- and intracellular sources which can contribute in the generation of ROS. An intracellular, endogenous source for the formation of ROS is the aerobic metabolism of a cell or, more precisely, the electron transport that occurs in mitochondria and chloroplasts. Moreover, ROS can also result from ionizing radiation and chemical substances which produce free radicals. In plants, the production of ROS can also be explained by biotic stresses such as the pathogen-activated hypersensitive response. In this defense mechanism against pathogens regulated by the phytohormone salicylic acid, ROS are generated and furthermore act as signal molecules.

DNA lesions caused by the attack of ROS are manifold. More than 50 base modifications of purines and pyrimidines caused by the ROS-mediated oxidation have been described. In addition, also single-strand breaks (SSBs) in the DNA double helix and AP sites, which are formed by the radical attack on the C-1 and C-4 atoms of the deoxyribose moiety, can be formed by ROS.

The oxidation of guanine leads to 7,8-dihydro-8-oxoguanine (8-oxoG, Fig. 2a). Compared to the other occurring DNA bases, guanine has the lowest redox potential

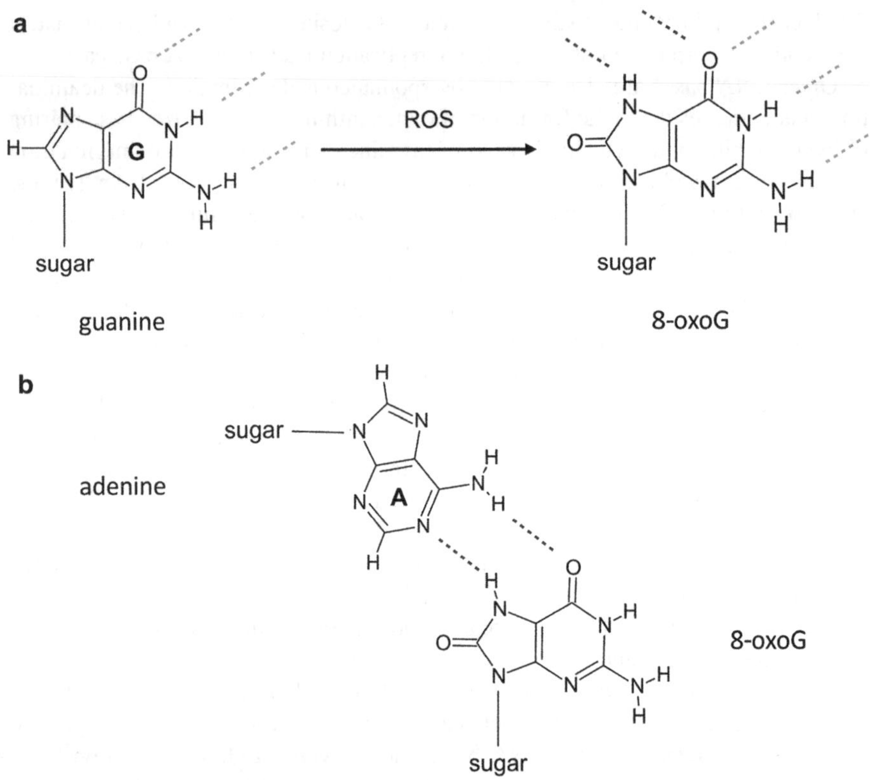

Fig. 2 **The oxidation product 8-oxoG**. (**a**) The oxidation of the nucleobase guanine by reactive oxygen species (ROS) results in 7,8-dihydro-8-oxoguanine (8-oxoG). 8-oxoG can pair with both cytosine (*grey hydrogen bonds*) and adenine (*red hydrogen bonds*). (**b**) The mispair adenine:8-oxoG pairs by two hydrogen bonds

which explains that the formation of 8-oxoG is the most abundant DNA lesion caused by oxidation. 8-oxoG can pair with both cytosine and adenine (Fig. 2b). These base pairing features lead to a strong mutagenic potential of 8-oxoG because of mispairing with adenine in the following round of replication, where a transversion from G:C to T:A is possible by the replication of an 8-oxoG-containing template strand. The impact of different oxidized bases for the cell is diverse and is not necessarily mutagenic. Thymineglycol, for instance, which results from the oxidation of the pyrimidine base thymine, is not mutagenic, but, in contrast, is cytotoxic by blocking the transcription and replication of DNA.

Alkylation of DNA: A variety of base lesions are induced by alkylating agents through the transfer of a methyl or ethyl group to reactive positions in bases or phosphate groups in the DNA backbone. Both oxygen and nitrogen atoms in DNA can be alkylated. The main targets of alkylation are the nitrogen moieties on DNA bases. The methylation of the N-7 atom of guanine and the N-3 atom of adenine is the most abundant DNA lesion. The resulting cytotoxic adducts N^7-methylguanine

(N^7-MeGua) and N^3-methyladenine (N^3-MeAde), which affect proper transcription and replication, account for more than 80 % of all methylation events. In contrast, the methylation of the oxygen bound to C-6 of guanine leading to O^6-methylguanine (O^6-MeGua) also has mutagenic potential because of changed base pairing features. O^6-MeGua predominately pairs with thymine. Accordingly, the persistence of O^6-MeGua in the parental DNA strand leads to an incorporation of thymine instead of cytosine during DNA replication.

Alkylating agents such as nitrogen mustards, nitrosoureas and alkyl sulfonates, triazines, and ethylenimines are separated into two subgroups dependent on their reaction mechanism. S_N1-type agents can alkylate oxygen as well as nitrogen atoms, whereas S_N2-type chemicals are only able to alkylate nitrogen atoms in nucleic acids. The most important environmental alkylating agent is methyl methanesulfonate (MMS), which methylates ring nitrogen residues in DNA bases, resulting in particular in N^7-MeGua and N^3-MeAde. Like another mutagenic agent ethyl methanesulfonate (EMS), used for the generation of plant mutants in research and agriculture, MMS is an S_N2-type agent and belongs to the group of alkyl sulfonates.

DNA damage induced by energy-rich radiation: Different types of energy-rich radiation types like ionizing radiation and UV radiation can lead to the formation of damaged DNA.

X-rays and γ-radiation have ionizing features and are defined as electromagnetic waves like light, but they transmit much more energy than visible light. On the one hand, ionizing radiation can have a direct effect on cellular compounds. In such a case, SSBs and double-strand beaks (DSBs) in DNA can be directly generated. Especially DSBs represent a great danger for the integrity of the genome. For instance, the presence of a single DSB during replication may lead to the complete loss of chromosome fragments and all genetic information encoded therein. Additionally, the repair of DSBs by different recombination pathways (discussed in more detail in section Mismatch Repair (MMR)) is associated with a high risk of mutations like insertions, deletions, and chromosomal rearrangements. On the other hand, ionizing radiation can also have an indirect effect on DNA by the production of ROS through the interaction with water molecules. DNA lesions caused by oxidation through ROS were described above.

Visible sunlight, on one hand, is essential for plant life to power photosynthesis. But on the other hand, the UV fraction of sunlight represents a near constant source of DNA damage. Solar UV radiation is categorized as UV-C (180–290 nm), UV-B (290–320 nm), and UV-A (320–400 nm). The energy content of radiation is inversely proportional to wavelength. For that reason, UV-C is the most and UV-A the least energetic UV radiation. Animals and plants are most affected by UV-B, the main fraction of genotoxic sunlight, penetrating and damaging their genomes. In contrast to animals, plants are not able to reduce the exposure to solar UV radiation by changing their location. The strategy of plants to minimize UV-induced DNA damage is to accumulate secondary metabolites (e.g., UV-absorbing flavonoids) in the epidermal layers which capture UV radiation and attenuate the UV dose. Despite such shielding, a portion of UV radiation attacking DNA reaches epidermal levels and tissues beyond.

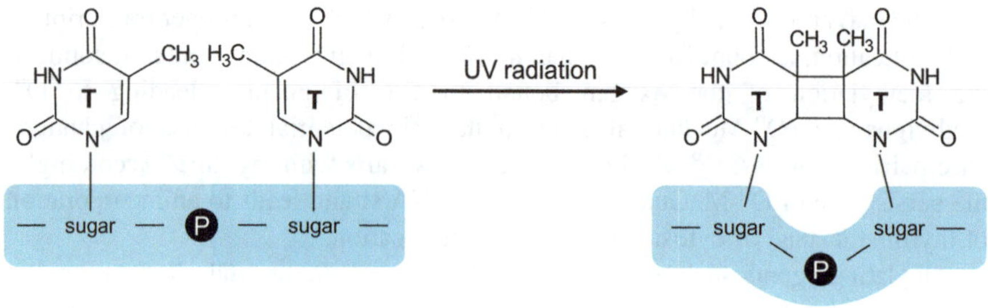

Fig. 3 **UV-induced pyrimidine dimer.** UV radiation can induce the formation cyclobutane pyrimidine dimers (CPDs). A cyclobutane ring between adjacent thymine bases is newly formed, which covalently connects the two thymines. This bulky adduct results in a bending of the DNA structure

The DNA-damaging effect of UV radiation is explained by the absorption spectrum of the DNA molecule. DNA has a maximum of absorption at 260 nm. The absorption of energy from UV radiation by DNA leads to the generation of so-called bulky DNA lesions, which are formed through the dimerization of neighboring pyrimidine bases. These bulky DNA lesions such as cyclobutane pyrimidine dimers (CPDs) and pyrimidine (6–4) pyrimidone photoproducts (6–4 PPs) are not able to base pair anymore and inhibit replication and transcription by blocking DNA and RNA polymerases. CPDs, which are more frequent than 6–4 PPs, arise through the dimerization of neighboring thymine bases (Fig. 3). Covalent bonds between the C-5 and C-6 atoms of the adjacent thymine bases are thus newly formed, resulting in a cyclobutane ring, whereas the double bonds of the thymine bases between C-5 and C-6 are converted into single bonds. The presence of a CPD results in a slight bending of the DNA double helix, in contrast to 6–4 PPs, which have a much stronger bending effect on the DNA structure. 6–4 PPs are characterized by covalent bonds between the C-6 and C-4 atoms of directly neighboring DNA bases in one DNA strand. Depending on the wavelength of UV radiation and on neighboring DNA sequences, 6–4 PPs can be formed between adjacent thymine bases (TT), cytosine bases (CC), as well as thymine and cytosine (TC).

DNA damage caused by genotoxic agents: Besides the already described DNA-damaging effects of chemical substances like alkylating agents, there are many more genotoxic agents known. These include chemicals with other modes of action, for example, base analogs, intercalating agents, cross-linking agents, and protein inhibitors.

Base analogs exhibit structural similarity to DNA bases, which are naturally occurring in the DNA molecule. Base analogs can be metabolized much like dNTPs, the substrates for DNA polymerases, and can be incorporated into DNA during replication. Depending on the respective chemical structure, the incorporation of the base analogs can lead to altered properties in base pairing in comparison to the substituted DNA base. A well-known base analog is 5-bromouracil, which is similar to thymine. The frequently occurring enol tautomer of 5-bromouracil can pair with guanine instead of adenine, which explains the mutagenic effect of 5-bromouracil.

DNA intercalating agents such as proflavine, ethidium bromide, or acridine orange are flat molecules containing a polycyclic ring system. They can intercalate into DNA by inserting between stacked base pairs, which may introduce changes in DNA winding, and subsequently lead to insertions or deletions during replication. The mutagenic potential is explained by failures in DNA synthesis caused by the introduced changes in DNA winding. For instance, DNA polymerase might incorporate an additional nucleotide if an intercalating agent is integrated in the template strand, resulting in an insertion event, or the polymerase can also jump over a nucleotide causing a deletion.

DNA cross-linkers are able to form covalent adducts with DNA and can generate cross-links (CLs) in one DNA strand (intrastrand CL) or between both complementary DNA strands (interstrand CL). The synthetic molecule cisplatin, for example, preferentially produces intrastrand CLs. In contrast, mitomycin C (MMC), which is a secondary metabolite of the bacterium *Streptomyces caespitosus*, leads predominantly to the formation of interstrand CLs. After the uptake of the cross-linker into the cell, the molecule is bioactivated, resulting in an instable and reactive intermediate. Following a first reaction step, in which the cross-linker forms a monoadduct on DNA, the second reactive center of the cross-linker can form a further covalent bond with DNA.

Genotoxic agents can also act as inhibitors of enzymes which are involved in the DNA metabolism. The alkaloid camptothecin (CPT), which is present in all organs of the Happy tree *Camptotheca acuminata*, is an inhibitor of the topoisomerase I. Topoisomerase I catalyzes the relaxation of supercoiled DNA arising during replication and transcription by the formation of a reversible SSBs in DNA, the strand passage, and the subsequent resealing of the break. CPT binds the DNA-topoisomerase I complex and inhibits the resealing of the introduced SSB, which then persists in DNA. During replication, topoisomerase I-bound SSBs can also be converted into DSBs when the replication fork meets the SSB. Caffeine, another genotoxin, comes from the secondary metabolism of different plants, most notably coffee and tea plants. Caffeine can lead to the accumulation of various types of DNA damage by inhibiting the important kinases ATM and ATR, which have a role in cell cycle progression after DNA damage (see section Tolerance and Repair Processes at Damaged Replication Forks).

The accumulation of DNA damage leads to genotoxic stress and is a risk for genome stability. The chemical modifications of DNA damage often result in structural changes in DNA, which can be detected by the DNA repair machinery. The following sections of this chapter deal with different specialized DNA repair mechanisms, which are indispensable to maintain genome stability. For more information about DNA damage and repair pathways in plants, see also other general reviews on this topic (Vonarx et al. 1998; Kunz et al. 2005; Roldan-Arjona and Ariza 2009; Tuteja et al. 2009; Spampinato and Gomez-Casati 2012).

Enzymatic Reversal of DNA Damage

The formation of various DNA lesions, such as UV photoproducts or modified bases, can be enzymatically reversed instead of being repaired through the excision pathways (see also reviews Weber 2005; Goosen and Moolenaar 2008; Dalhus et al. 2009; He et al. 2011). In this section the reversion of these lesions is discussed. Direct repair mechanisms are quite simple, as they only need a single enzyme, compared to the complex multi-protein excision repair pathways.

Most of the research in this field was done on bacterial enzymes, but further work demonstrated that there is a very high level of conservation of these proteins between the different kingdoms. Some of the information presented here was gained through research in bacteria, but the basic findings are also applicable to the plant proteins.

The direct reversal reactions of UV photoproducts, termed photoreactivation, are extremely important for plants, because they cannot avoid UV radiation like animals due to their sessile lifestyle. First, the photoreactivation will be described, followed by a short excursion to cryptochromes, which are related to photolyases but offer only limited DNA repair capability. Instead, most of them have acquired a role as light receptor and signaling component. Afterward, the direct reversal of modified bases, which works in parallel to excision repair pathways, is explained in the following sections.

Photoreactivation by photolyases: UV radiation is very toxic for the genome of a cell, as the wavelength of the energetic UV light (180–400 nm) overlaps with the absorption spectrum of DNA, which has a maximum at around 260 nm. The most energy-rich UV-C radiation (180–290 nm) is effectively filtered out by the ozone layer around the earth and therefore plays almost no role as a genotoxic factor. UV-A radiation (320–400 nm) is not energetic enough to harm DNA directly, but can be mutagenic through the production of harmful intermediates, like reactive oxygen species. The principal component of UV radiation that damages DNA is UV-B (290–320 nm), which mainly produces pyrimidine dimers. 70–80 % of all UV photoproducts are CPDs and 20–30 % are 6–4 PPs. The type of damage is in this case dependent on the DNA sequence and structure, but both exhibit genotoxic effects by blocking the polymerases during transcription and replication and are important factors for the development of skin cancer.

Plants usually accumulate shielding compounds like flavonoids that absorb UV radiation in order to minimize the potential of getting harmed, but of course this does not render them immune to UV radiation. In order to deal with the described DNA lesions induced by UV radiation, plants can utilize the nucleotide excision repair pathway to cut out the damaged DNA strand or use a direct reversion pathway, photoreactivation.

The name "photoreactivation" comes from the fact that it is a direct reversal of harmful photoproducts ("reactivation") and its need for light ("photo") energy to function. It is facilitated by specialized enzymes called "photolyases," which are basic and widespread DNA repair proteins that are conserved in most of the species living today. They act as monomers and can be classified by their specific substrates

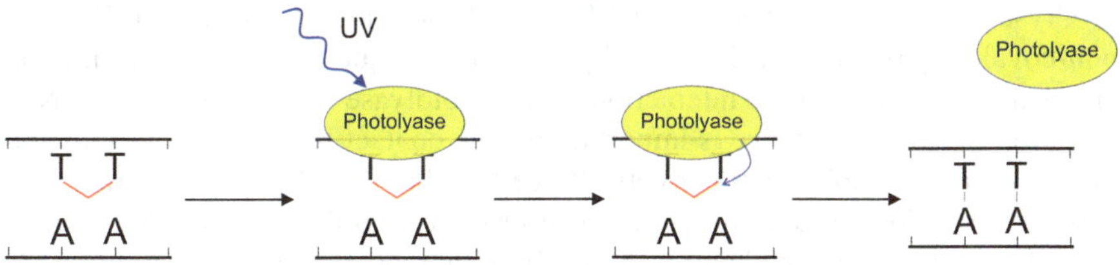

Fig. 4 Photoreactivation by photolyases. The photolyase recognizes and binds the thymine dimer. Then, the photolyase is activated by visible light and catalyzes the dissolution of covalent bonds between the neighboring thymine bases by an electron transfer onto the thymine dimer

into CPD photolyases and 6–4 PP photolyases. Both kinds of photolyases recognize and bind to a pyrimidine dimer, which is then extruded out of the DNA into the active site of the photolyase. Subsequently, light-activated photolyases transfer an electron to a pyrimidine dimer, which induces the dissolution of covalent bonds between the neighboring pyrimidine bases (see Fig. 4).

CPD and 6–4 PP photolyases each contain a flavin adenine dinucleotide (FAD) as a catalytic cofactor, which is needed for splitting pyrimidine dimers. In CPD photolyases either pterin methenyltetrahydrofolate (MTHF) or deazaflavin 8-hydroxy-5-deazariboflavin (8-HDF) is non-covalently bound as a second cofactor. These second cofactors are required to harvest light and make photoreactivation more efficient. Concerning 6–4 PP photolyases, data shows that they might only use MTHF as a second cofactor.

CPD photolyases are very well studied and can be further divided into two classes by their amino acid similarity. Class I photolyases are generally found in microbial organisms, while class II photolyases are mainly found in more complex organisms, for example, in *Arabidopsis thaliana* and the fruit fly *Drosophila melanogaster*. The human genome contains no active photolyase genes, and humans rely only on the excision pathways in order to repair UV-induced DNA damage, which could lead to mutations and skin cancer, if not repaired.

Photolyases repair DNA damage caused by the energy-rich, but for human eyes invisible, UV light. Photolyases use less energetic wavelengths in the visible spectrum in order to divide pyrimidine dimers. This is achieved through an electron transfer from a two-electron reduced $FADH^-$ to the pyrimidine dimer. The FAD cofactor in the photolyases can be directly excited by light in order to facilitate photoreactivation, but this is a very inefficient process, as FADH- shows only weak absorption of wavelengths under 400 nm and almost no absorption of longer wavelengths. In order to increase efficiency and absorbance spectra, photolyases harbor secondary chromophores that serve as a light antenna. Methenyltetrahydrofolate (MTHF)-containing photolyases exhibit their maximum catalytic activity when light between 377 and 410 nm is present, and the deazariboflavin photolyases work best when irradiated by light of wavelengths between 430 and 460 nm. The secondary antenna chromophores increase the absorption and catalytic activity of photolyases significantly. The energy absorbed by the second chromophore is transferred via resonance energy transfer onto the FAD in order to generate the fully reduced and catalytically active $FAHD^-$ and excite it.

Crystal structure analyses of CPD photolyases have shown that FAD is non-covalently bound in a C-terminal groove in a unique U-shaped conformation and that it is needed for the interaction of the photolyase with the damaged DNA. The area around this groove is important for the contact between the protein and the DNA. The pyrimidine dimer gets "flipped out" of the DNA double strand in a way that it resides in the groove in close proximity to the $FADH^-$. The excited $FADH^-$ then transfers an electron to the CPD, thereby generating a semiquinone FADH• radical and a CPD anion radical, which then undergoes monomerization and transfers the electron back to the FADH• radical. By this reaction, the CPD lesion is effectively repaired and the photolyase dissociates from the DNA double strand.

Despite CPD and 6–4 PP photolyases being very similar at the protein sequence level, much less information is available about 6–4 PP photolyases compared to CPD photolyases. But from today's point of view, although 6–4 PP photolyases are not able to repair CPDs, the molecular mechanisms with which they repair 6–4 PPs seem to be identical to the ones described for CPD photolyases.

Cryptochromes: Closely related to CPD photolyases are another group of proteins, called cryptochromes. Cryptochromes were first identified in *Arabidopsis thaliana* and are less widespread than the photolyases (see review Chaves et al. 2011). They are found in many plants and animals, but are rare in other eukaryotes and prokaryotes. The proteins are basically photolyases that have lost their ability to repair UV-induced lesions in the genome. Instead, they have acquired a role as important blue light receptors and are involved in many processes, like development, defense response, stress response, induction of flowering, and the circadian clock.

Apart from the cryptochromes involved in blue light reception and signaling, there is another class of cryptochromes, cryDASH, where DASH stands for *Drosophila*, *Arabidopsis*, *Synechocystis*, *Homo*. However, it needs to be mentioned that despite their name, they are not present in *Drosophila* and humans. cryDASH cryptochromes are still able to facilitate repair by light-induced photoreactivation of pyrimidine dimers, but only on damaged single-stranded DNA (ssDNA) substrates, whereas photolyases can efficiently repair ssDNA and double-stranded DNA (dsDNA). Research showed that this is based on the fact that cryDASH cannot flip the damaged base out of a DNA double strand. Another difference is that cryDASH proteins only use MTHF, but no deazaflavin, as second cofactor, although they are structurally similar to the photolyases. Whether cryDASH proteins also exhibit signaling activity like other cryptochromes, is, however, not yet known.

As all of the abovementioned photolyase-related protein classes are present in plants, a view of the evolution of these proteins is very interesting: All classes seem to have originated from one photolyase predecessor and are independent from each other. Even cryDASH and cryptochromes represent independent classes and have not developed from each other, although the functional characteristics of cryDASH proteins would place them in between photolyases and cryptochromes.

Enzymatic repair of base modifications: Not only UV-induced DNA damage but also potentially mutagenic base modifications can be repaired directly through specialized enzymes and are not necessarily restricted to repair through the excision repair pathways. Little is known about these proteins in plants to date, and most of the studies on the basic mechanisms of these direct reversal proteins have been conducted in bacterial or mammalian cells, leading to merely a basic understanding in plants.

Through endogenous or exogenous substances, a lot of bases are modified every day in each cell. These modified bases are potentially cytotoxic as they may block replication and transcription or be mutagenic by having different base pairing characteristics. One such example is the alkylated guanine O^6-MeGua, which predominantly pairs with thymine instead of cytosine and therefore can lead to mutations, if it is not repaired. The methylguanine-methyltransferase is a specialized repair enzyme that can directly reverse this damage by removing the methyl group from the guanine and transferring it to one of its own cysteines. The covalent binding of the methyl group is irreversible, and therefore the protein can only repair one lesion before it needs to be degraded. By definition, the methylguanine-methyltransferase is not even a proper enzyme, as it is not able to catalyze the reaction more often. Another example is the direct removal of 1-methyladenine and 3-methylcytosine, which can be reverted by the oxidoreductase ALKBH2 in *Arabidopsis thaliana*, a homolog to AlkB from *E. coli*, where initial studies were conducted.

Direct repair mechanisms are quite conserved throughout evolution and pose important ways to secure genomic stability. However, as mentioned above, such lesions can not only be repaired through enzymatic reversal but also through the excision pathways.

Base Excision Repair (BER)

Instead of a direct repair of the damaged nucleotides, DNA lesions can also be removed from the DNA by different excision repair systems such as base excision repair (BER) and nucleotide excision repair (NER), which are discussed in this and the following sections.

BER is a well-studied repair mechanism in mammals, and it is highly conserved in prokaryotes and eukaryotes (see also reviews Fortini et al. 2003; Baute and Depicker 2008; Dalhus et al. 2009; Roldan-Arjona and Ariza 2009; Wallace et al. 2012). BER is a cellular process with a variety of different enzymatic functions that occur in sequential steps, including the damage-specific recognition and then removal of the base lesion followed by cleavage of the sugar-phosphate backbone and excision of the abasic site (AP site). Subsequently, the resulting single-stranded gap is filled and resealed, using the opposite, undamaged DNA strand as template for DNA synthesis. By this repair mechanism, different types of DNA lesions can be repaired, such as modified bases originating from deamination, oxidation, and alkylation. Also AP sites resulting from the spontaneous hydrolysis of a base are repaired by BER.

Principal process of BER: The initiating step of BER is carried by a DNA glycosylase which recognizes the damaged base and hydrolyzes the N-glycosidic bond between the damaged base and the respective deoxyribose residue. By the action of the glycosylase, an AP site is generated (Fig. 5). For the further processing of the AP site by the endonucleolytic cleavage of the DNA backbone, there are two different possibilities (Fig. 5).

(1) Besides the recognition and the generation of AP sites, so-called bifunctional DNA glycosylases can further process the AP site by their intrinsic 3' AP lyase activity. Here, a 5' phosphate and a 3' blocking lesion such as a 3' α,β-unsaturated aldehyde (PUA) after a β-elimination is generated. (2) Another way to process AP sites is mediated by an AP endonuclease (such as the human APE1) after base lesion removal by a monofunctional DNA glycosylase. The activity of the AP endonuclease results in other products flanking the gap, compared to the action of bifunctional glycosylases. At the 3' terminus, a free hydroxyl (–OH) group is generated, which can be directly used by the DNA polymerase without further processing. At the opposite 5' terminus of the single-strand gap, a deoxyribose-phosphate (5' dRP) is left by the AP endonuclease.

To fill the gap via polymerase-dependent DNA synthesis and a subsequent ligation reaction, a 3' OH group and a 5' phosphate flanking the gap are required. For that reason the 3'α,β-unsaturated aldehydes and also the 5' dRPs have to be converted into conventional 3' OH and 5' phosphate ends to allow gap filling (Fig. 5). (1) 3' blocking lesions generated by bifunctional DNA glycosylases are removed by the 3' diesterase activity of the AP endonuclease, whereas (2) 5' dRPs resulting from the action of an AP endonuclease are removed by the 5' dRPase activity of polymerase β.

For gap filling and rejoining, there also exist two different sub-pathways: (A) short-patch BER and (B) long-patch BER, which differ in the size of the repair gap and also in the enzymes involved in the pathway (Fig. 5). (A) Short-patch BER, which is also called single nucleotide BER, is characterized by the incorporation of the correct single nucleotide and the processing of the 5' end by polymerase β (or bacterial polymerase I), followed by the ligation of the remaining nick by a complex consisting of DNA ligase III and XRCC1 (X-ray cross-complementing 1 or bacterial ligase I). (B) During long-patch BER, polymerase β also incorporates the initial missing nucleotide, but instead of a direct resealing of the nick, further DNA synthesis occurs by the replicative polymerases δ or ε while the DNA strand downstream of the initial damage site is displaced. Because of this differing mechanism in long-patch BER, additional protein factors are required which have also a well-known function in DNA replication. For example, the clamp protein PCNA (proliferating cell nuclear antigen) is needed for the loading of the replicative polymerases. Furthermore, the overhanging displaced DNA single strand, also called a DNA flap, is removed by FEN1 (flap endonuclease 1), which endonucleolytically cleaves at the base of the DNA flap structure. The resulting nick is sealed by ligase I.

DNA glycosylases: Organisms possess a set of different DNA glycosylases, which all exhibit several specificities for damaged DNA bases. In humans, there

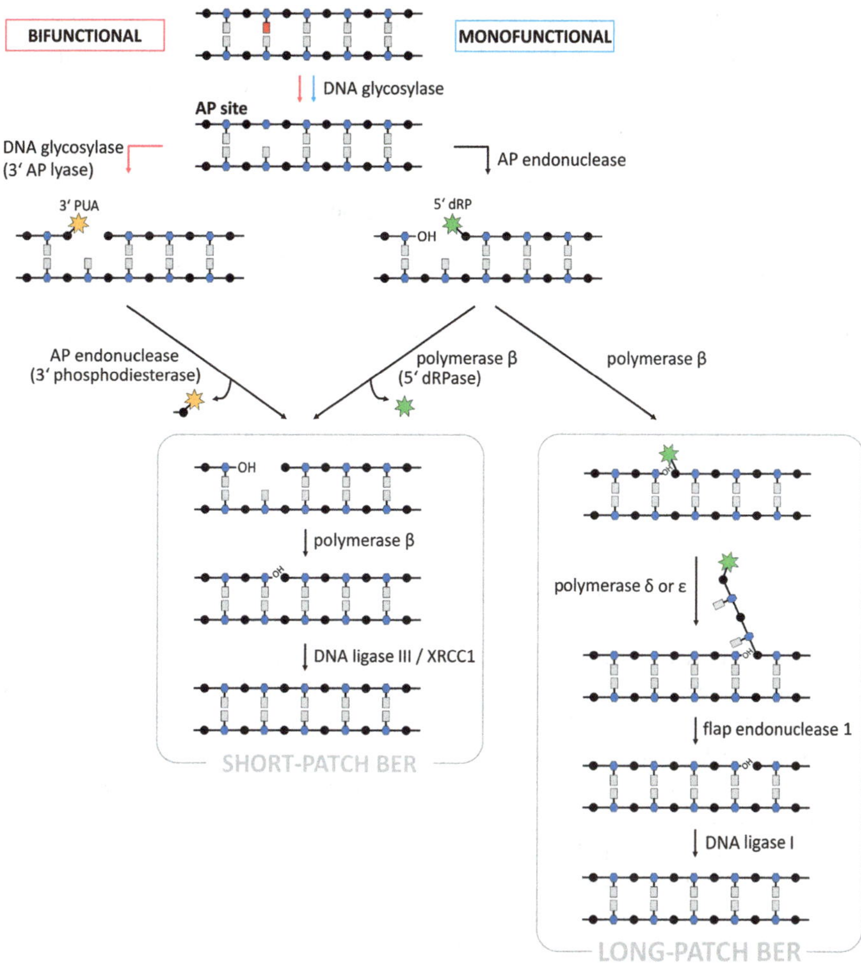

Fig. 5 **Base excision repair**. The damaged base is recognized and excised by a bifunctional (*left*) or a monofunctional (*right*) DNA glycosylase, both generating an AP site. Bifunctional DNA glycosylases can further process the AP site by their intrinsic 3' AP lyase activity, leading to a 5' phosphate and a 3' blocking lesion such as a 3' PUA (*left*). The 3' PUA is then removed by the catalytic 3' phosphodiesterase activity of the AP endonuclease, which results in a free 3' OH group. When the damaged base is excised by a monofunctional DNA glycosylase (*right*), the further processing of the AP site is done by the AP endonuclease. By the action of the AP endonuclease, a 3' OH group and 5' dRP is generated. The 5' dRP can be removed by the 5' dRPase activity of polymerase β. There are two sub-pathways for gap filling and ligation. During short-patch BER (*left*), polymerase β incorporates the missing nucleotide followed by a ligation reaction, which is fulfilled by the DNA ligase III/XRCC1 complex. During long-patch BER (*right*), the initial nucleotide is incorporated by polymerase β. The replicative polymerase δ or ε continues DNA synthesis by simultaneous displacement of the DNA strand downstream of the damaged site. The flap endonuclease 1 removes the single stranded DNA flap structure and DNA ligase I reseals the resulting nick

have been eight DNA glycosylases described at present. In most cases, there is a broad substrate spectrum of the DNA glycosylases that explain the overlapping specificities of different glycosylases in some cases. The recognition of the DNA damage by DNA glycosylases can be described as a pinch-push-plug-pull mechanism. The DNA glycosylase slides along the minor grove of the DNA helix scanning for specific modified bases, thereby bending the DNA double helix. A kind of reading head is inserted into the minor groove of DNA (pinch), which then pushes on the damaged base. This leads to a base flipping of the damaged base from the interior of the DNA double helix into the active site pocket of the glycosylase, where hydrogen bonding groups interact with the extrahelical base (plug and pull). The active site pocket of the glycosylase determines their specificity, because the extruded base has to match within. The substrate specificity depends on the shape, the hydrogen binding, and also the electrostatic potential of the base.

There are two classes of DNA glycosylases. Monofunctional DNA glycosylases only exhibit a glycosylase activity using an activated water molecule as a nucleophile for the attack to release the damaged base. Thereby, no covalent intermediates between the DNA glycosylase and the nucleotide are formed, and there is no interruption of the DNA backbone generated by the DNA glycosylase. For the breakage of the DNA backbone, a different enzyme, an AP endonuclease, is required. In contrast, bifunctional glycosylases combine base excision by the glycosylase activity with a DNA nicking activity leading to an SSB. The base excision mechanism of a bifunctional glycosylase includes the formation of a covalent Schiff base intermediate between a conserved lysine located in the active site pocket of the glycosylase and the aldehyde group of the sugar moiety of the nucleotide. A subsequent β-elimination reaction leads to the release of the damaged base.

There are different hypotheses about the scanning mechanism for damaged DNA bases. First, it was speculated that every base is flipped out of the DNA and presented into the active site pocket of the DNA glycosylase. This would imply a massive consumption of energy. A second hypothesis is founded on the breathing of DNA and the spontaneous extrusion of damaged bases. The DNA glycosylase may stabilize the open conformation during DNA breathing and recognize damaged bases. The third hypothesis considers the destabilizing effect of modified bases on the base pairing and the stability of the DNA double helix. The DNA glycosylase, which inserts an intercalating reading head into the DNA double helix, could examine very quickly the structure and the energetics of the base pairs. A damaged base would be discovered by the further destabilization of the target base pair. To support this hypothesis, fast movement of the DNA glycosylase OGG1 along normal DNA duplexes could be visualized.

OGG1 (8-Oxoguanine DNA glycosylase 1) is a bifunctional DNA glycosylase and is able to recognize and remove 8-oxoG paired with cytosine. 8-oxoG results from the oxidation of guanine and is the most abundant DNA lesion caused by oxidation. In bacteria, there is a different DNA glycosylase, MutM, which can also initiate the BER of oxidized purines including 8-oxoG. Interestingly, plants possess homologs with redundant functions for both enzymes, OGG1 and MutM. It was speculated that both glycosylases might be located in different organelles of the plant cell, for instance, in the nucleus and the chloroplast. Alternatively, both proteins might have evolved different specificities in plants during evolution.

The repair of uracil in DNA can be initiated by the monofunctional uracil DNA glycosylase (UDG). Uracil can be generated in DNA by the spontaneous deamination of cytosine, representing the most frequent product of deamination, or be wrongly incorporated during replication. UDGs are well conserved throughout evolution and present in bacteria, yeast, plants, and animals. In humans, the UDG named UNG2, located in the nucleus, is cell cycle-regulated with highest levels in the G1-to-S transition. Because of this expression pattern, it is likely that the major role of UNG2 consists in counteracting U:A base pair formation due to the misincorporation of uracil during replication. The specialized uracil glycosylase activity could also be identified in many plants such as carrot (*Daucus carota*), onion (*Allium cepa*), pea (*Pisum sativum*), and wheat (*Triticum aestivum*). In the model organisms *Arabidopsis thaliana* and rice, a homolog to UDG could be identified in silico, but is yet not characterized.

Interestingly, studies revealed that plants also possess two plant-specific DNA glycosylases for which no homolog outside of the plant kingdom is known yet to exist. These genes are called *ROS1* (repressor of silencing 1) and *DME* (DEMETER). ROS1 and DME code for bifunctional DNA glycosylases which surprisingly are not involved in the repair of damaged bases. Rather, they play a role in the regulation of gene expression by mediating the demethylation of 5-methylcytosine in an indirect manner (see review Zhu 2009). 5-methylcytosine, representing a signal for transcriptional gene silencing, is recognized and removed through the action of these glycosylases initiating BER. In this way, 5-methylcytosine is replaced by the unmethylated base cytosine. DME is specifically required to regulate the expression of the imprinted *MEDEA* gene, which is involved in proper female gametophyte and seed development. The *MEDEA* gene is generally methylated and inactivated in vegetative tissue. By the demethylating activity of the glycosylase DME, the maternal target allele *MEDEA* is specifically expressed in the central cell of the female gametophyte.

AP endonuclease: The AP endonuclease APE1 is involved in short- as well as in long-patch BER. The endonuclease activity of APE1 is characterized by nicking the phosphodiester backbone 5' to the AP site, which results in a 3' hydroxyl group and a 5' dRP flanking the nucleotide gap. APE1 also possesses further enzymatic activities such as a 3' phosphodiesterase or 3' phosphatase activity that can remove 3' terminal blocking groups formed by the 3' AP β-lyase activity of bifunctional glycosylases. Additionally, also a 3'-5' exonuclease function for 3' termini of internal nicks in DNA has been described.

Besides catalytic functions, APE1 also plays a role in a variety of interactions with several factors involved in BER. APE1 can stimulate the activity of DNA glycosylases. By direct protein-protein interaction with polymerase β, APE1 facilitates the binding of polymerase β to the AP site and stimulates the removal of 5' dRPs. APE1 is also involved in the direct recruitment of long-patch BER factors by the interaction with PCNA. Later steps in long-patch BER such as the trimming of the DNA flap as well as ligation are also stimulated by the interaction of APE1 with FEN1 and ligase I, respectively.

Repair synthesis and ligation in short-patch BER: Different polymerases are alternatively involved in the DNA repair synthesis of BER. But in short-patch BER, only the activity of polymerase β is needed. Polymerase β acts as a DNA polymerase incorporating the missing nucleotide and also as a 5' dRP lyase. Because polymerase β does not possess a proofreading activity, the polymerase β-mediated DNA synthesis is error prone and shows low fidelity (error rate of 10^{-4}). The scaffold protein XRCC1 binds to polymerase β. XRCC1 possesses no enzymatic activities but interacts with most of the factors involved in short-patch BER, which emphasizes its function in the coordination of BER. XRCC1 can stimulate the enzymatic activity of APE1. The direct protein-protein interaction with both APE1 and polymerase β may be important for the positioning of polymerase β to perform its lyase activity. Furthermore, XRCC1 provides physical linkage between the polymerase and DNA ligase IIIα, by the formation of a stable complex. DNA ligase IIIα can bind to nicked DNA via its DNA-binding domain and reseal the nick.

It is not clear whether short-patch BER occurs in plants because some of the proteins involved in this sub-pathway are missing in plants. No homologs of polymerase β have been identified in any plant genome. For that reason, it has been proposed that polymerase λ can assume the function of polymerase β in plants. As it has been shown in rice, polymerase λ exhibits a 5' dRP lyase activity like polymerase β. Furthermore, plants lack DNA ligase IIIα. However, plant DNA ligases I and/or IV may function as DNA ligase IIIα. In accordance with the lack of polymerase β and DNA ligase IIIα in plants, the plant homolog of XRCC1 does not contain the interaction domains which are responsible for the protein-protein interaction with polymerase β and ligase IIIα in humans. But XRCC1 contains the PARP1 interaction domain, which represents an additional factor involved in long-patch BER. Altogether, it seems likely that only the long-patch sub-pathway is present in plants.

Repair synthesis, flap removal, and ligation in long-patch BER: DNA repair synthesis in long-patch BER is mediated by different polymerases. Polymerase β incorporates the initial nucleotide, followed by a switch of polymerases during repair synthesis. The further synthesis of the repaired strand and the displacement of the DNA single-strand downstream of the initial damage site are mediated by polymerases δ or ε, which are involved in long-patch BER and DNA replication, but not in short-patch BER. FEN1 is responsible for the removal of the resulting DNA flap. Like the replicative polymerases, FEN1 fulfills functions in long-patch BER and also in DNA replication, processing the 5' ends of Okazaki fragments during lagging-strand synthesis. The highly conserved structure-specific endonuclease cleaves at branched DNA structures containing an overhanging single-stranded 5' flap. To accomplish this, FEN1 tracks along the ssDNA flap from the 5' end to the site of cleavage. Modifications of the 5' end like the dRP residue, left by the action of the AP endonuclease, are simultaneously removed with the trimming of the DNA flap. Thus, a dRPase activity mediated by DNA polymerases is not absolutely necessary in long-patch BER. Apart from the catalytic activity of FEN1, it is also known that FEN1 can stimulate strand displacement and DNA synthesis by polymerase β. The religation of the remaining nick is done by DNA ligase I, which is also involved in different DNA repair pathways and replication.

Besides the described factors, several additional proteins like PCNA, RFC, PARP1, RPA, and WRN are also involved in long-patch BER. RFC (replication factor C) binds to the 3' terminus, which serves as DNA synthesis primer. There, RFC helps to load PCNA onto DNA. PCNA forms a ring-shaped clamp, tracking along DNA and serves as a docking platform for the replicative DNA polymerases. Therefore, PCNA is required for the loading of the replicative polymerases onto DNA and also enhances DNA polymerase activity. Furthermore, PCNA can enhance the binding stability of FEN1 and modulate the activity of the endonuclease by protein-protein interaction. The stable association of DNA ligase I to nicked DNA duplexes also requires PCNA. RPA (replication protein A) binds to ssDNA and is needed for DNA synthesis by the replicative DNA polymerases. The poly (ADP-ribose)polymerase PARP1 binds to the SSB, which leads to the activation of poly-ADP-ribosylation of specific nuclear proteins. By its activity, PARP1 is important for the protection of the SSB and thus for preserving the substrate for BER. WRN (Werner syndrome helicase) is a RecQ helicase possessing an additional 3'-5' exonuclease activity. It was shown that WRN can stimulate the strand displacement by polymerase β dependent on its helicase activity.

SSB repair and pathway choice: As described above, BER is a highly coordinated DNA repair pathway in which DNA repair intermediates are transferred from one repair protein to the next. This transfer mechanism may avoid the occurrence of unfinished DNA repair intermediates which may have a cytotoxic effect on the cell. SSBs, for instance, can arise directly due to DNA-damaging factors or indirectly as intermediates of BER. The repair of SSBs by the SSB repair pathway (SSBR) is similar to the process of BER. In the SSBR pathway, PARP1 is involved in the recognition and the binding of the SSB, followed by a recruitment of XRCC1, which also acts in SSBR as a molecular scaffold protein. Dependent on the types of modifications, the ssDNA ends flanking the gap are processed by specific AP endonucleases for DNA synthesis and ligation. The processed ends act as substrate for short- or long-patch BER.

Different factors can influence the choice between the two BER sub-pathways, short- and long-patch BER. One factor is the type of DNA termini flanking the single-stranded gap. The occurrence of unconventional ends such as 5' dRP triggers long-patch BER, whereas the presence of conventional 5' ends leads predominantly to short-patch BER. A second parameter determining the pathway choice is the local concentration of BER components and protein-protein interactions at the repair site, which differ between short- and long-patch BER. Furthermore, the phase of the cell cycle plays a role in selection of the respective pathway. It could be shown that long-patch BER is more frequent in cells passing through S phase than in non-replicating cells.

Nucleotide Excision Repair (NER)

A multitude of diverse types of DNA damage is repaired by the nucleotide excision repair (NER) pathway including numerous of bulky adducts on DNA such as UV-induced DNA lesions like CPDs and 6–4 PPs (see also reviews de Laat et al. 1999; Costa et al. 2003; Kunz et al. 2005; Roldan-Arjona and Ariza 2009; Spampinato and Gomez-Casati 2012). NER proteins are also able to recognize structural changes in the DNA, e.g., those leading to a distortion in the double helix. In contrast to BER, not only a single base but also a 24–32-nucleotide-long oligomer containing the DNA lesion is excised, resulting in a single-stranded gap in DNA. Afterward, the undamaged DNA strand is used as a template for repair synthesis to fill the gap. The NER pathway is class-divided into two sub-pathways: global genome NER (GG-NER) and transcription-coupled NER (TC-NER). GG-NER recognizes DNA damage which is randomly distributed in the genome, whereas TC-NER repairs DNA lesions which are present in transcribed DNA strands and thus blocking transcription.

The NER proteins of eukaryotes are conserved during evolution. In humans, defects in NER can lead to the hereditary disease xeroderma pigmentosum (XP), which is characterized by an extremely high sensitivity to sunlight, in particular to UV radiation, causing various lesions in skin tissue and a predisposition for skin cancer. Mutations in seven genes involved in DNA repair are associated with XP, and for that reason they are named XPA–XPG. It was shown that XPA, XPC, XPD, XPF, and XPG are involved in NER. Like XP, the distinct recessive disorders called Cockayne syndrome (CS) and trichothiodystrophy (TTD) are also associated with defects in NER and share the common clinical feature of photosensitive skin.

Recognition and recruitment by XPC in GG-NER: The DNA damage sensor and recruitment protein XPC (xeroderma pigmentosum group C) is the first factor which is specifically involved in the GG-NER sub-pathway. XPC is able to detect DNA lesions and directly binds to damaged DNA (Fig. 6), with a high affinity for both ssDNA and dsDNA and a preference for UV-damaged DNA. Deformations in the DNA double helix are also recognized by XPC. By its binding to DNA, XPC introduces changes in the DNA conformation around the lesion which produces a local distortion in the DNA double helix structure. Furthermore, XPC is capable of recruiting further factors in the NER repair machinery to the DNA lesion such as the transcription factor TFIIH (see below). A second protein HR23B, which forms a complex with XPC, is involved in these processes by stimulating the XPC activity. The affinity of the XPC/HR23B complex for different DNA lesions as well as its localization in terms of accessibility of the respective lesion affects the DNA repair rate of GG-NER. An additional factor, CEN2 (CENTRIN2), is required for the stabilization of the complex and the stimulation of NER.

Recognition of DNA damage in transcription-coupled NER: The detection of DNA lesions and the activation of further repair processes in TC-NER are independent of XPC, which is the crucial DNA-damaging sensing and recruitment factor in GG-NER. Despite an active sensing of DNA lesions in GG-NER, the detection of DNA lesions in TC-NER is rather coincidental, depending on the

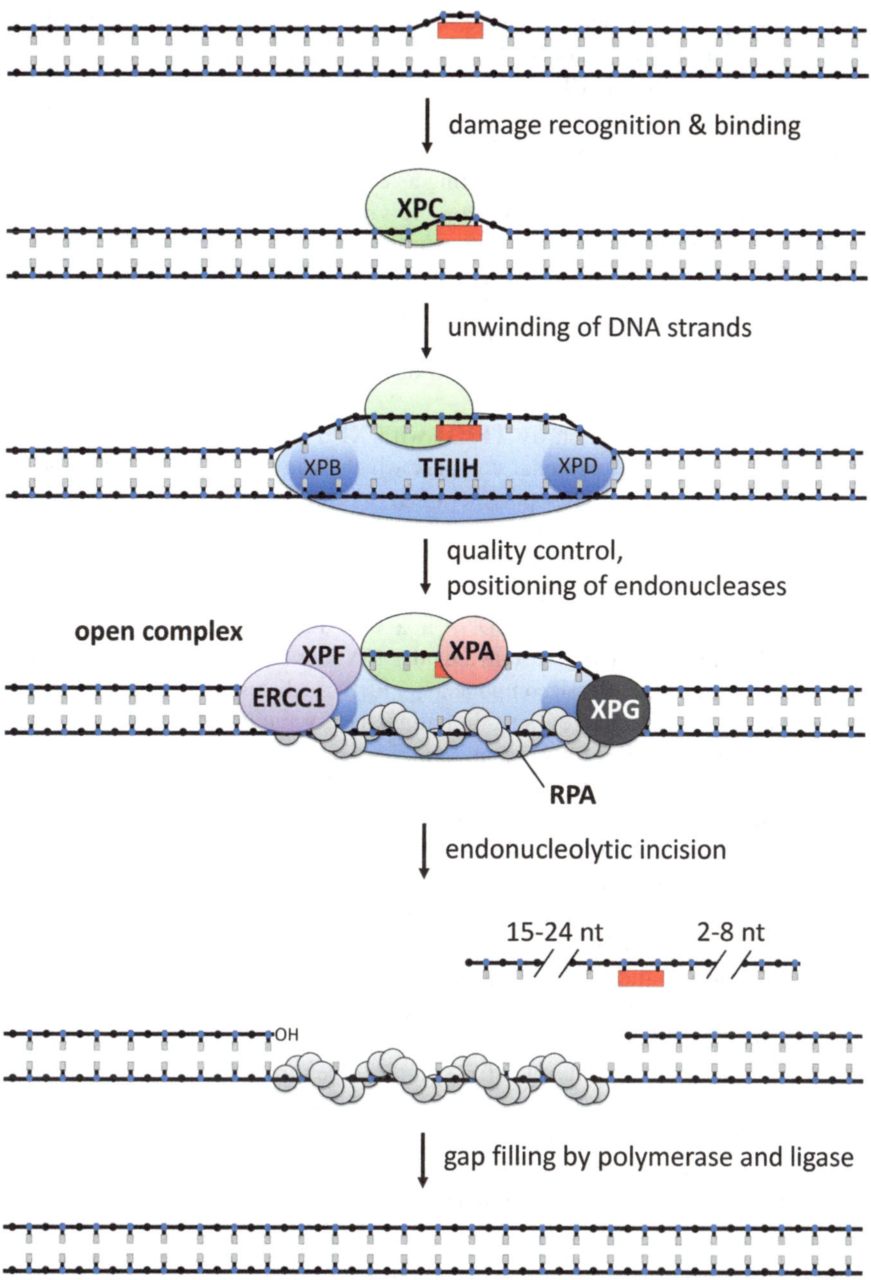

Fig. 6 **Nucleotide excision repair**. The GG-NER-specific factor XPC recognizes and binds the DNA lesion, followed by the recruitment of further NER proteins. The two helicases XPB and XPD, which are components of the transcription factor TFIIH, unwind the DNA double helix leading to strand separation in the region of the DNA lesion. The DNA damage is verified by XPA, which specifically binds to the DNA lesion. The single strand binding protein RPA binds to the

transcription process in actively expressed genes. DNA lesions in the transcribed DNA strand lead to a block of RNA polymerase II while it is moving along DNA. The transcription factor TFIIH, which is involved in both transcription and NER, is associated with the captured RNA polymerase sitting on the active, transcribed gene. From this point of view, RNA polymerases encountering a DNA lesion represent a further class of proteins acting as efficient damage sensors. Until now, it is not clear whether the RNA polymerase II is displaced or dissociated from DNA, in order to allow the accessibility for further NER factors. TC-NER requires additional repair factors which are specifically involved in this sub-pathway, such as the assembly factors CSA and CSB (for Cockayne Syndrome protein A and B, respectively). CSB, for instance, possesses nucleosome remodeling activity and therefore is able to alter the conformation of DNA and is possibly involved in the recruitment of TFIIH.

Open complex formation by TFIIH with XPB and XPD: The transcription factor TFIIH is a multifunctional complex which is involved in the initiation of transcription of DNA by RNA polymerase II, as well as in both sub-pathways of NER. The TFIIH protein complex is organized in a ringlike structure and consists of 10 subunits including the two helicases XPB (also known as ERCC3) and XPD (also known as ERCC2). The directionality of the helicase function differs between XPB and XPD. XPB can unwind dsDNA in a 3'-5' direction, whereas XPD has the opposite direction of activity (it unwinds dsDNA in a 5'-3' direction). By having the two helicases as integral components, TFIIH mediates the separation of the DNA strands at the damaged site, initiating the formation of a so-called open complex (Fig. 6). The fully open complex spans about 20–30 base pairs (bp) around the DNA lesion and contains ss- to dsDNA transition sites, which are required for the cleavage by the structure-specific endonucleases XPG and the ERCC1-XPF complex (described below). The formation of the fully open complex occurs in a two-step reaction. The initial opening spans about <10 bp and requires XPC and TFHIIH. The subsequent extension of the open complex to about 30 bp is dependent on further NER factors such as XPA, RPA, and XPG.

Quality control and positioning of the endonucleases: After the separation of the DNA strands surrounding the DNA lesion, the DNA-binding protein XPA validates the DNA damage in the open complex formation. XPA binds the damaged DNA (Fig. 6), preferentially at NER-specific types of DNA damage including CPDs and 6–4 PPs.

In principle, the endonucleases ERCC1-XPF and XPG, which catalyze the incision in the damaged DNA strand, are able to cut both DNA strands at the border of the open DNA intermediate. But in NER, the incisions by the endonucleases are

Fig. 6 (continued) undamaged DNA strand and is involved in the positioning of the endonucleases ERCC1-XPF and XPG. After the formation of the fully open complex, XPG cleaves at the 3' site 2–8 nucleotides distant from the lesion and ERCC1-XPF catalyzes the 5' incision 15–24 nucleotides away from the DNA lesion. The gap filling is done by replicative polymerases and DNA ligase I

restricted to the damaged DNA strand only. This fact is crucial for the success of DNA repair by NER and points to the important role of nuclease positioning by the NER machinery via protein-protein interactions.

Aside from binding of the damaged DNA, XPA also interacts with several factors in the NER pathway such as TFIIH, the endonuclease ERCC1-XPF1, as well as RPA. By mediating these protein-protein interactions, XPA has a role in the correct assembly and positioning of the DNA repair machinery around the DNA lesion. An additional protein, RPA, is also involved in the positioning of the endonucleases by direct protein-protein interactions. Primarily, RPA binds to the undamaged ssDNA strand by its ssDNA-binding activity to stabilize the open complex (Fig. 6). RPA has a defined DNA-binding orientation which is relevant for the coordination of the nucleases. The 3'-oriented side of RPA interacts with ERCC1-XPF, whereas the 5'-oriented side binds XPG (Fig. 6). Furthermore, RPA stimulates the endonucleolytic cleavage of the damaged DNA strand and inhibits incisions in the undamaged DNA strand. Altogether, the assembly of the fully open complex is dependent on a variety of protein-protein interactions. For example, the positioning of ERCC1-XPF requires the interaction with RPA and XPA, which facilitates and stabilizes the positioning of the endonuclease.

Incision by the endonucleases ERCC1-XPF and XPG: Two different endonucleases, ERCC1-XPF and XPG, are recruited to the pre-incision complex as described above. The activities of both endonucleases lead to an asymmetrical cleavage with respect to the DNA lesion site (Fig. 6). First, XPG cleaves at the 3' site 2–8 nucleotides distant from the lesion. Following the 3' incision, a second incision at the opposite 5' site introduced by ERCC1-XPF is 15–24 nucleotides away from the DNA lesion. The exact incision positions are dependent on the type of DNA damage and the sequence context. For this reason, the size of the replaced DNA oligonucleotide varies from 24 to 32 nucleotides.

XPG is a structure-specific endonuclease responsible for the 3' incision at the border of the open DNA intermediate. There, XPG acts with a defined cleavage polarity which is characterized by the incision in one strand of the DNA duplex at the ss to dsDNA transition site. Furthermore, XPG also has another important role during NER, as it is required for the fully open complex formation described above.

The two proteins ERCC1 (excision repair cross-complementing 1) and XPF form a stable complex which acts as a structure-specific endonuclease. Besides NER, the complex is also required in other DNA repair pathways such as interstrand cross-link repair and homologous recombination. During NER, ERCC1-XPF catalyzes the incision at the 5' site of the open complex resulting in a free hydroxyl group at the 3' end. In contrast to XPG, ERCC1-XPF does not appear to have a structural function in the open complex formation.

Resealing of the gap: The 3' end flanking the single-stranded gap, which results from the incision by ERCC1-XPF, can be directly used as a DNA primer for DNA synthesis to fill the gap (Fig. 6). Repair synthesis in NER requires several factors which are also involved in DNA replication, such as RPA, RFC, PCNA, and polymerases δ and ε (for further details see section Base Excision Repair). The role of RPA in NER is manifold. Besides the previously described function in positioning of the endonucleases, RPA is also involved in repair synthesis. The DNA binding of RPA leads to the protection of the undamaged DNA strand against nucleases, which then serves as a template for repair synthesis. Furthermore, RPA facilitates DNA synthesis because its presence stimulates the activity of the replicative DNA polymerases δ and ε. The ligation reaction of the remaining nick between the newly synthesized DNA strand and the original DNA strand is carried out by DNA ligase I.

Kinetics of NER: The recruitment and the activity of NER proteins is dependent on different parameters such as the location and the type of the DNA lesion. Therefore it was suggested that there are two different NER responses. The immediate NER response is characterized by the removal of DNA lesions which have a great effect on the DNA structure or which are detected through the transcription of genes. The remaining DNA lesions are repaired at a much slower rate in a secondary NER response. For instance, 6–4 PPs, leading to large distortions of the DNA double helix, are repaired five times faster than CPDs, which have only a slight bending effect on the DNA structure.

Evolution of NER: The basic NER mechanisms including recognition of DNA lesions, DNA incision, fragment excision, and repair synthesis have been functionally conserved during evolution. The basics of NER in eukaryotes and prokaryotes are similar, but more complex in eukaryotes because of the involvement of more than 25 factors, compared to only four in prokaryotes. These so-called Uvr proteins are required to detect and remove the DNA damage. The UvrA/UvrB complex scans the genome to find distortions of DNA. After the detection and binding of the DNA lesion by the complex, UvrA dissociates and UvrB catalyzes the local melting of the DNA double strand at the site of the DNA damage. UvrC associates with UvrB and cleaves the damaged DNA strand eight nucleotides upstream (5') and 4–5 nucleotides downstream (3') from the DNA lesion. By this endonucleolytic cleavage, a 12–13-nucleotide-long oligomer containing the DNA lesion is created. By the unwinding activity of the UvrD helicase, the damaged ssDNA strand is excised from the DNA duplex. However, despite an obvious conservation of the NER mechanism itself, the enzymes involved in this process differ between prokaryotes and eukaryotes. Because of the lack of sequence homology, when comparing the NER proteins of the both groups, it is likely that the analogous functions in the NER mechanisms evolved independently in prokaryotes and eukaryotes.

In eukaryotes, not only the NER mechanism but also the involved proteins are well conserved, suggesting a conserved repair pathway in these organisms. Most of the genes involved in NER in yeast and mammals can also be found in plant genomes, such as *XPC*, *HR23B*, *CEN2*, *XPB*, *XPD*, *RPA*, *XPG*, *ERCC1*, *XPF*, and *XPE*. Like human patients with defects in the NER pathway, plants containing mutations in NER genes display UV hypersensitivity. The same holds for different *Arabidopsis* mutants with defects in the plant homologs of XPD (*Atuvh6*), XPG (*Atuvh3*), XPF (*Atrad1*), ERCC1 (*Atuvr7*), and CEN2 (*Atcen2*).

In the case of the helicase gene *XPB*, two homologs *AtXPB1* and *AtXPB2* have been identified in the genome of *Arabidopsis thaliana*. Both duplicated homologs contain a DNA-binding domain and seven helicase motifs, which are present in human XPB, as well. Further, the other helicase in the TFIIH complex, XPD, is also present in *Arabidopsis* and shows conservation of the helicase domain. It is likely that the function of AtXPD is conserved during evolution, because it was shown that the plant XPD homolog can interact with TFIIH components in yeast. Another example of a high degree of conservation is the endonuclease complex ERCC1-XPF. Both components of the complex could be identified by sequence similarity in plants, and it was shown that XPF from *Arabidopsis* (also called AtRAD1) is able to interact with AtERCC1.

Despite conservation of most of the NER proteins in plants, there are some differences between the phylogenetic groups present. No homolog of XPA, which is involved in the quality control by binding the DNA damage and assembly of the NER machinery in mammals and yeast, could be identified in plants until now. It is possible that functions which are fulfilled by XPA in mammals are not essential for the NER mechanism in plants, or another unknown plant protein has functions similar to mammalian XPA.

Mismatch Repair (MMR)

Mismatched base pairs in dsDNA lead to distortions of the double helix that may hinder or block replication and transcription. If unrepaired, such mismatches can become fixed mutations after another round of replication. Mismatched bases may occur through chemical reactions, e.g., by spontaneous deamination or by chemical genotoxins. A further source of mismatches is replicative DNA polymerases. Even though they possess a proofreading activity to correct errors, they nonetheless place the wrong base opposite to their template strand in about one in 100,000,000 (10^{-8}) bases. To counteract this source of mutations, a specific repair pathway called mismatch repair (MMR) has evolved (see also reviews Jiricny 2006; Spampinato and Gomez-Casati 2012). It can detect and correct 99 % of mismatched bases introduced by replicative DNA polymerases, reducing the error rate to about 10^{-10}.

Although details vary in the MMR pathway between prokaryotes and eukaryotes, the basic steps are the same (see Fig. 7): At first, the mismatched bases have to be found by proteins that recognize the distortion of the double helix. Then, a stretch of the newly synthesized strand in which the wrong nucleotide has been introduced is removed. Finally, using the correct sequence information of the parental strand, the single-stranded gap is closed by a polymerase and ligase.

In *E. coli*, where MMR was initially studied, recognition of a mismatch is done by a homodimer of the protein MutS which forms a ring structure that encloses dsDNA. After it has found a mismatch, the MutS homodimer interacts with a homodimer of MutL and moves away from the mismatch along DNA until it encounters the endonuclease MutH bound to dsDNA at GATC sequences. In *E. coli*, the GATC sites are normally methylated on the adenine, but following replication, the newly synthesized daughter strand is not methylated for a short time. MutS-MutL can now activate the MutH endonuclease activity, which only cuts the unmethylated strand. This ensures that that source of the mismatch, the

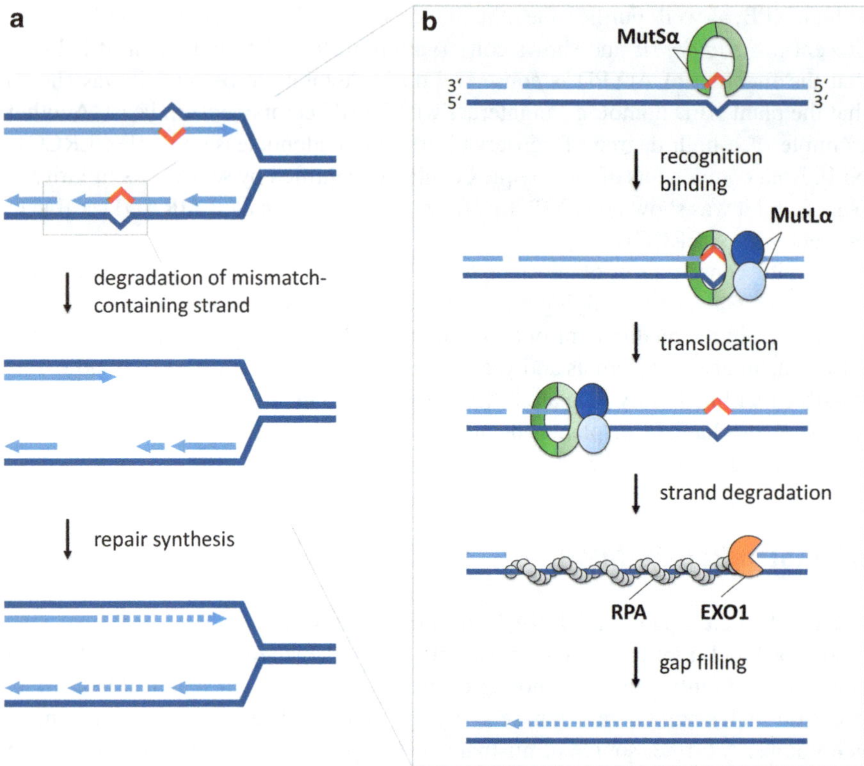

Fig. 7 Mismatch repair. (**a**) Mismatched base pairs can arise during replication due to the incorporation of false nucleotides (*red*) by the replicative polymerases. The daughter strand, which contains the mismatch, is exonucleolytically degraded up to the site of the next occurring interruption of the DNA strand. Such an interruption can be a SSB formed by the end of an Okazaki fragment in the lagging strand or the 3' end of the leading strand. The resulting gap is filled by repair synthesis. (**b**) One scenario of MMR is depicted in more detail. The heterodimer MutSα (MSH2-MSH6) recognizes and binds the mismatch. This results to a conformational change of the heterodimer, leading to the formation of a clamp structure surrounding the mismatch. A further heterodimer MutLα (MLH1-PMS2) interacts with MutSα. The ternary complex acts as a sliding clamp and translocates in n ATP-dependent manner to the next occurring SSB, where it interacts with replicative factors such as PCNA and RFC (not shown). Subsequently, the DNA strand which contains the falsely incorporated base is degraded by the exonuclease EXO1. The resulting ssDNA strand is coated by RPA and serves as a template for repair synthesis

newly synthesized strand, is repaired instead of the parental strand. Starting from this nick, the UvrD DNA helicase unwinds the duplex, thereby exposing the modified strand to exonucleases that degrade it. The gap that is formed is then filled by DNA polymerase and sealed by DNA ligase.

Eukaryotic cells contain homologs to MutS and MutL, named MSH (for MutS homolog) and MLH (for MutL homolog), respectively. In fact, several homologs to MutS and MutL can be found in plants, fungi, and animals. In mammals, the proteins MSH2 and MSH6 form a heterodimer called MutSα that is thought to act similarly to the MutS homodimer in *E. coli*. Furthermore, MSH2 and MSH3 form the heterodimer MutSβ that is involved in a more specialized repair of large DNA loops formed by insertions or deletions. A heterodimer of the MutL homologs MLH1 and PMS2, named MutLα, can interact with MutSα to form a ternary complex similar to the MutS-MutL complex in bacteria.

Comparable to MutS-MutL in bacteria, the eukaryotic MutSα-MutLα complex moves along dsDNA after mismatch recognition through MutSα. The critical step in the endonucleolytic cleavage by MutH in the daughter strand, however, is specific to gram-negative bacteria, since MutH is not conserved outside this group. So how can the degradation of the daughter strand be initiated in eukaryotes, and how is it ensured that the parental strand is not modified? In contrast to bacteria, it is thought that eukaryotic cells use preexisting gaps and nicks present during replication specifically in the newly synthesized daughter strand: the free 3' end of the leading strand or both ends of the Okazaki fragments in the lagging strand (Fig. 7). In accordance with this idea, it was shown that MSH6 from MutSα can interact with proteins usually found in the replication complex at the elongating 3' end of DNA: the sliding clamp protein PCNA and its clamp loader RFC. Such a free DNA end may now serve to initiate strand degradation in MMR. Depending on the orientation of the gap to the mismatch, a number of exonuclease activities with either 3'-5' or 5'-3' polarity are known to be involved in this degradation. While the parental strand is stabilized by the ssDNA-binding protein RPA, DNA polymerase δ closes the gap, and DNA ligase I seals the nick.

In plants, the MMR pathway seems to work similarly as described for mammals above. Counterparts of all mammalian MMR proteins are also found in plants, and mutants of MSH2 show an expected increase in point mutations and genome instability. Loss of the MMR system in plants specifically affects morphology, fertility, and seed development in a generation-dependent manner. In addition to the role of MMR proteins in the repair of mismatches during replication, there is also a further role in meiosis. In *A. thaliana*, the plant-specific MSH7 interacts with MSH2 to form another heterodimer, MutSγ, which is required for the detection of mismatches during meiotic recombination. Furthermore, the MSH4-MSH5 heterodimer and the MLH1-MLH3 heterodimer are also involved in meiotic recombination rather than classical MMR (also see section Repair of Double-Strand Breaks).

Since mismatched bases may also arise in the pairing of similar, but not identical, sequences during homologous recombination (see section Repair of Double-Strand Breaks), several proteins from the MMR pathway have been found to affect the outcomes of this, so-called homologous, recombination. Organisms containing polyploid genomes with homologous chromosomes from hybridization events, e.g., hexaploid wheat, are especially prone to pair homologous sequences during HR. Here, it is thought that MMR proteins recognize mismatches in heteroduplex DNA and abort the deleterious recombination reaction. Loss of MMR protein therefore leads to an increase in homologous recombination in addition to the elevated level in point mutations.

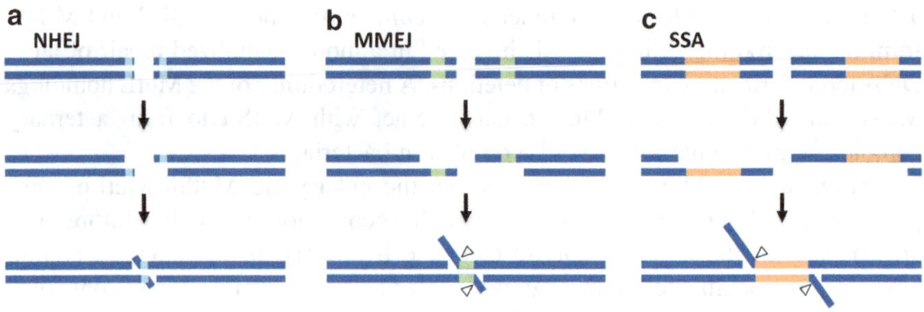

Fig. 8 **Repair of a DSB by NHEJ, MMEJ and SSA.** (**a**) In the NHEJ pathway, the DSB ends are processed and directly ligated. (**b**) In contrast to NHEJ, during MMEJ short single-stranded overhangs are formed at the ends of the break. A small complementary region of about 5–25 bp (*green*, micro homologies) are used for hybridization; through which the two ends are reconnected. (**c**) Followed by an extensive resection in the SSA pathway, the generated single-stranded overhangs can hybridize using direct sequence repeats (*orange*), which are flanking the DSB site. All three DSB repair pathways are associated with genomic changes. Non-homologous overhangs are removed by nucleases (*triangle*)

Repair of Double-Strand Breaks

DSBs, the most severe type of DNA lesion, require timely repair. A single unrepaired DSB may lead to cell death due to failure to correctly distribute chromosomes to the daughter cells. Thus it is not surprising that a number of different pathways, which at least partially can compete with each other, have evolved to cope with DSBs (see also reviews Schuermann et al. 2005; Waterworth et al. 2011). These pathways differ in the frequency they are used to repair lesions as well as in the fidelity of repair.

Repair of DSBs by nonhomologous end-joining: The predominantly used and mechanistically most simple pathway in somatic cells is called nonhomologous end-joining (NHEJ), which ligates the broken ends at DSBs (Fig. 8a). In the case of a clean break, such a ligation will result in the wild-type sequence without any mutations. In most cases, however, a break will not be blunt ended or it will be chemically modified, making a direct ligation without mutations impossible. DSB repair by NHEJ therefore is a quick, but highly mutagenic pathway.

Following the formation of a DSB, the heterodimer KU70/KU80 binds to the ends of the break. The KU heterodimer forms a circular structure with positively charged amino acids on the inside and a diameter just wide enough for dsDNA, which enables it to slide onto the broken ends, but not elsewhere along the chromosomes. Recruitment of further repair factors, such as the trimeric MRN complex (for MRE11, RAD50, NBS1), enables processing of the two ends and brings them into close proximity to each other by interaction of the MRN complexes. Finally, the ends are ligated to close the break by a heterodimer of the DNA ligase IV (LIG4) and XRCC4.

If the broken ends contain short complementary sequences, an alternative NHEJ pathway can occur that utilizes base pairing of the ends. The so-called microhomology-mediated end-joining (MMEJ) or alternative nonhomologous end-joining (aNHEJ) was described in yeast, mammals, and plants. Here, short single-stranded stretches of about 5–25 base pairs at the ends of a DSB are annealed with each other before ligation (Fig. 8b). It is not clear if the single-stranded ends are generated by specialized proteins or if they are the result of break formation itself, e.g., when two SSBs occur in two different strands in close proximity to each other. After hybridization, there might also be single-stranded flaps present that need to be cut before ligation can happen. Interestingly, MMEJ is independent of the KU70/KU80 heterodimer. It has even been shown that the KU dimer suppresses MMEJ and promotes regular NHEJ at break sites.

Repair of the DSB by annealing of direct sequence repeats: If a break occurs between two direct sequence repeats, in somatic cells a second pathway utilizing these repeats will become active. Although superficially similar to MMEJ, in single-strand annealing (SSA), other proteins are active (Fig. 8c). An important difference is the resection of single strands from the 5' ends of the breaks that will result in long single-stranded overhangs. After extensive resection over several kilobases, the direct repeats will now be located in the overhangs. These single-stranded regions will be bound by the ssDNA-binding protein RPA, protecting them from further nuclease activity. The single-stranded nature of the overhangs enables base pairing with the help of the recombination protein RAD52, closing the break. Depending on the specific break situation, there will be gaps or flaps remaining that have to be processed by polymerases and nucleases in order to make ligation of the DNA backbone possible. Since all sequences between the two repeats and also one of the repeat sequences are lost in the processing of the break, the SSA pathway is inherently mutagenic and will always lead to deletions at the break site. Interestingly, although SSA, like homologous recombination (HR), uses single-strand resection of the break ends and several recombination-associated proteins, the HR-specific recombinase RAD51 is not required.

Repair of DSBs using homologous sequences in the genome: In addition to these nonconservative pathways described above, there exists a further pathway that has the possibility to repair a DSB without changes to the DNA sequence at the break site. As explained above, when DNA damage occurs in one strand, repair pathways such as BER, NER, and MMR can utilize sequence information in the complementary DNA strand for repair. In the case of a DSB, this is not possible. There is, however, sequence information still present in the genome: in sister chromatids, homologous chromosomes and also – in the case of repeat sequences – in other loci throughout the genome, called ectopic sites. If it is possible to copy from these homologous sequences, it should be possible to repair the DSB without altering the sequence. This is how the homologous recombination (HR) pathway works, although compared to NHEJ and SSA, it is rarely used to repair DSBs in somatic cells.

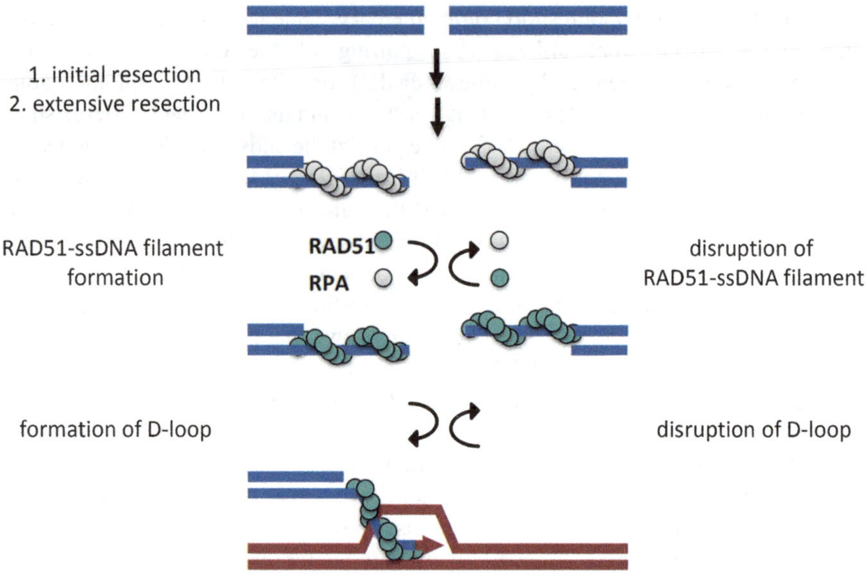

Fig. 9 **Initial steps in HR**. The DSB ends are resected in a sequential process, initial and extensive resection, to generate 3' overhangs. The ssDNA binding protein RPA binds to the resulting single-stranded 3' overhangs and is sequentially exchanged by the recombinase RAD51. The resulting RAD51-ssDNA filament can perform the search for homologous sequences in the genome and mediates the strand invasion into the homologous donor DNA molecule to form a D-loop. The generation of the RAD51-ssDNA filament and the formation of the D-loop are reversible steps, as specialized helicases can disrupt the respective repair intermediate

Homologous recombination has been most intensively studied in yeast, but the general progression of the pathway and the proteins involved seem to have been conserved throughout eukaryotes. Furthermore, HR has been studied as the pathway of meiotic recombination, and there are only minor differences in somatic cells. In meiosis, DSBs are actively induced by the cell to activate HR, while in somatic cells DSBs are the result of physical factors or chemical genotoxins. In meiosis, the preferential outcome of recombination is the creation of new genetic material, while in somatic cells HR favors the suppression of genetic change. Today's view of the progression of HR can be seen as a combination of several independently described sub-pathways. Historically, HR was studied as a black box: known genetic markers went into recombination, out came a new combination of these markers. Different models were proposed how DNA may be processed to enable the recombination outcomes in the form and also in the numbers they were observed.

Resection of the DSB: Following the formation of a DSB, the ends are resected in a 5'-3' direction by the MRN complex and the exonuclease COM1, a homolog to yeast Sae2 and human CtIP (Fig. 9). In yeast, this first resection is only a few hundred base pairs long, and the single-stranded overhangs produced by it lead to a weak HR response. To fully activate HR, further resection of several thousand base pairs is necessary. This second resection is performed by other proteins. In yeast (and homologous proteins in humans), there are two parallel and synergistic ways to resect the breaks: One way uses the exonuclease EXO1 for resection from the 5' end. Alternatively, a DNA helicase of the RecQ family unwinds the double strand, enabling a cut at the transition from ssDNA to dsDNA by the endonuclease DNA2. Loss of one of these pathways slows resection, but loss of both leads to a nearly complete cessation of HR. In plants, this resection step has not been researched in depth, but homologs of both RecQ helicases and EXO1 have been described, indicating that this step is conserved.

Searching for homologous sequences and strand invasion: Once the single-stranded overhangs have been formed, they are quickly coated by the ssDNA-binding protein RPA to protect them from degradation (Fig. 9). To initiate homologous recombination, RPA is exchanged for the recombinase RAD51. This process is facilitated by mediator proteins such as BRCA2, which is known for its human homolog in which mutations lead to a hereditary predisposition for breast cancer. In complex with ATP, RAD51 binds in a polymerized form to ssDNA, forming a helical RAD51-ssDNA nucleoprotein filament. This filament is the structure that is able to search for homologous sequences that could be used in repair. Once a homologous donor locus is found, the RAD51-ssDNA filament invades the donor double strand by displacement of the noncomplementary strand and base pairing with the other. This process forms an opened structure called displacement loop (D-loop) in which the two strands of the intact donor molecule are separated from each other (Fig. 9). The formation of the D-loop is also promoted by BRCA2. The free 3' end of the invaded single strand (the end of the break) can now be elongated by a DNA polymerase that uses the homologous donor locus as a template. In this step the D-loop migrates with the polymerase along the donor molecule, or it is enlarged by a DNA helicase.

Double-strand break repair (DSBR) model of HR: The double-strand break repair (DSBR) model describes a class of HR events that may result in different end products depending on the resolution of the recombination intermediates (Fig. 10). Recombination as described by this model mainly occurs in meiosis, but seldomly in somatic cells. According to the DSBR model for HR, the D-loop will be enlarged until the second resected end of the DSB is able to base pair with the displaced second strand of the donor molecule, a process called second end capture. Following the action of DNA polymerases and ligases, the remaining gaps in the DNA backbones of the broken molecule are closed, repairing the DSB. The resulting structure, however, now contains two interconnected DNA molecules, which are also unviable because they may lead to subsequent DNA breaks during transcription, replication, or cell division. The two molecules need to be separated to finalize repair. The connections formed at this point are two cruciform structures, called Holliday junctions (HJ) after Robin Holliday who first described them. A double HJ (dHJ), as proposed by the DSBR model, can be separated into two double strands by a process called resolution. There, specialized structure-specific endonucleases (called resolvases) recognize the HJ structure, bind to it, and cut it at the junction. A number of such resolvases have been described, e.g., RuvC in bacteria, Yen1 in yeast, and homologous GEN1 in mammals. In plants, two GEN1 homologs are present but have not yet

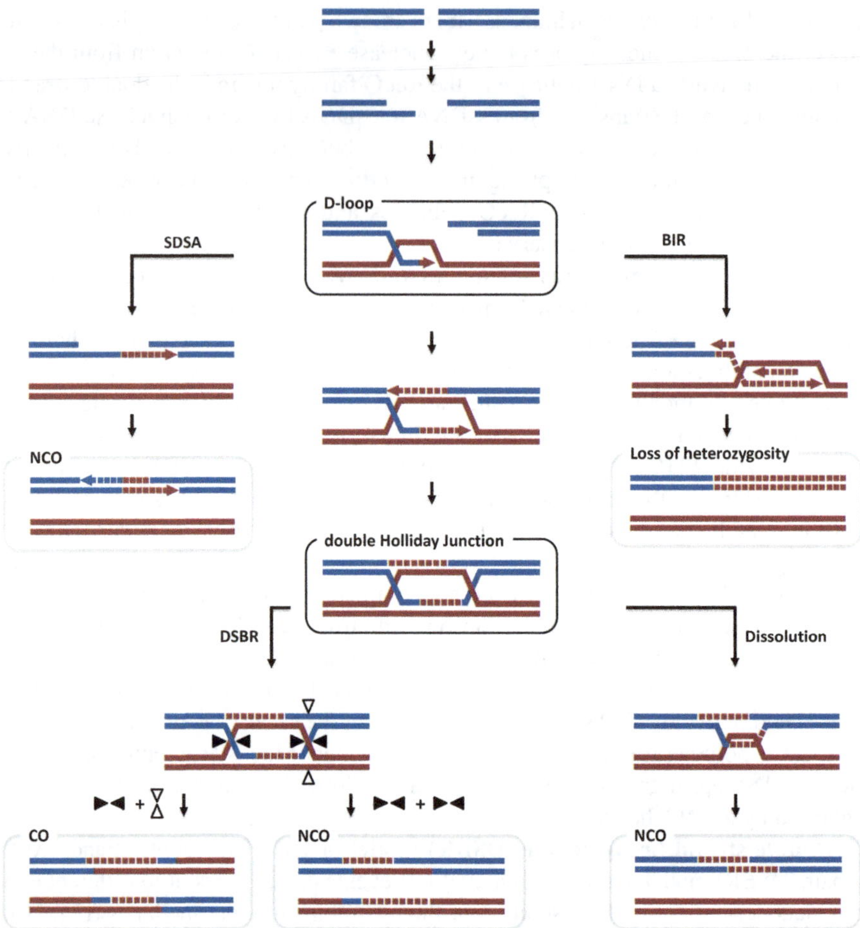

Fig. 10 Different sub-pathways and products of HR. The DSB is resected, followed by the formation of a D-loop, which represents the key intermediate of HR. According to the SDSA model, at first the D-loop is extended and subsequently, the newly synthetized DNA strand is replaced out of the D-loop and paired with the other end of the DSB. In the SDSA pathway, only NCO products are formed. In the DSBR model of HR, the D-loop is converted into a double Holliday junction after capture of the second free DSB end, which is then resolved by the endonucleolytical cleavage catalyzed by resolvases (*triangles*). Dependent on the axis of cleavage, NCO or CO products can arise via the DSBR pathway. A different pathway named dissolution can also process the double Holliday junction, but in contrast to DSBR, only NCO are generated. During the dissolution pathway, a hemicatenane is formed by a helicase, which is then dissolved by a type IA topoisomerase. Beside SDSA, DSBR and dissolution, the D-loop intermediate can expand to a replication fork like structure in the BIR pathway, if the second end of the DSB is not present

been studied in much detail. In addition, all eukaryotes including plants contain the endonuclease MUS81 that is also active on HJs but might not be a "true" resolvase since it does not cut the junctions in a direction that enables direct ligation of the products. After the resolvase activity, a DNA ligase seals the breaks and finalizes DSB repair. Depending on the orientation of the cut at the two HJs, the product of the HR reaction may be either a crossover (CO) or a noncrossover (NCO). In a NCO, which has also been called gene conversion, a nonreciprocal exchange of genetic information takes place. Information is copied from the donor to the damaged DNA double strand at the break site. When a CO happens, on the other hand, flanking sequences of the break site are exchanged between donor and damaged double strand. In meiosis COs are the basis for the mixing of parental genomes.

Synthesis-dependent strand-annealing (SDSA) model of HR: Since the resolvases acting according to the DSBR model cannot differentiate between the two possible ways to cut a HJ, one would expect a distribution of 50 % COs and 50 % NCOs. However, experiments have shown that there are far more NCOs produced. An alternative to the DSBR model is the synthesis-dependent strand-annealing (SDSA) model (Fig. 10), in which HR proceeds from a DSB as described above. In this model, single DNA strands are resected from the broken ends, and they are coated first by RPA and then RAD51, followed by homology search and strand invasion to form a D-loop. From this step onward, the SDSA model differs from the DSBR model. SDSA-like recombination occurs during meiosis as well as in somatic cells. After elongation of the invaded strand using the donor molecule as a template, the SDSA model proposes – instead of second-end capture – the rejection of the invaded strand from the D-loop, resulting in a disruption of the D-loop. This allows for the annealing of the single-stranded break ends, followed by repair synthesis to fill the gaps and religation to seal the backbone. Such a process cannot lead to a CO product, but only to NCO outcomes.

Regulation of pathway choice: Since both the DSBR and SDSA pathways of HR require the formation of a D-loop, the pathway choice is made at this crucial step. Which pathway will be used depends on the ability of the second, free end of the DSB to anneal to the displaced donor strand at the D-loop. Therefore, either elongation or stabilization of the D-loop will increase the chance of second-end capture and, therefore, the chance of the DSBR pathway, while early disruption of the D-loop due to displacement of the invaded single strand will promote SDSA. Two classes of proteins have been shown to be involved in this pathway decision process in meiosis: DNA helicases and ring- or clamp-forming proteins. In the case of DNA helicases (see review Knoll and Puchta 2011), the specific activity of the respective protein will determine which of the two pathways will be promoted. Simply by the act of strand separation, a helicase can enlarge the D-loop, forming a larger region of unpaired donor DNA and helping the capture of the second strand. Alternatively, a similar activity can separate donor strand and invaded strand, thereby promoting SDSA. A number of DNA helicases have been described that might act in such processes. The yeast Srs2 helicase as well as the animal Rtel1 helicase has been shown to disrupt model D-loops in vitro and to promote SDSA outcomes in vivo. Homologs of both proteins are conserved in plants. On the other hand, in meiotic recombination the DNA helicase ROCK-N-ROLLERS (RCK)/MER3 has been shown to promote CO formation via the DSBR pathway, most probably through D-loop elongation. Also in meiotic recombination, where CO formation is absolutely required, a heterodimer of the bacterial MutS homologous proteins MSH4 and MSH5 is thought to form a ring structure that encloses the D-loop to suppress its disruption. Loss of either of the two proteins leads to a strong reduction in the number of crossovers observed, since more D-loops will enter the SDSA pathway.

Dissolution of double Holliday junctions: In addition to the classical DSBR repair pathway, in which dHJs are processed by resolvases, it has been shown that at least one further distinct pathway exists to process dHJs. Although it was theoretically proposed about 30 years ago, it was demonstrated in vitro and in vivo only during the last decade. Like SDSA, this pathway also can only form noncrossover products, but will proceed through DSB repair in the same manner as the DSBR pathway up to the point of dHJ formation, facilitating a rather late decision for NCOs compared to the earlier decision at the D-loop in the SDSA pathway. This pathway does not require endonucleolytic resolvases and is therefore called "dissolution" to differentiate it from the resolution step in DSBR (Fig. 10). But how can the two dsDNA molecules be separated from each other without the activity of endonucleases? Dissolution requires in its most basic form a DNA helicase and a topoisomerase, although some structural proteins are necessary as well to facilitate the interaction of these proteins. Starting from a dHJ structure, a DNA helicase from the widely conserved RecQ family – in *Arabidopsis* RECQ4A specifically – can migrate the two HJs toward each other through its branch migration activity. When the two HJs meet, a new structure is formed, a so-called hemicatenane, in which one strand of each dsDNA molecule is topologically interlocked with the other. This hemicatenane represents the substrate for type IA topoisomerases, such as yeast Top3 or plant and animal TOP3α, which can process this structure to unlink the two DNA molecules. The helicase and topoisomerase-dependent dissolution reaction results exclusively in the formation of NCO products. In addition to these two enzymes, the structural proteins RMI1 and RMI2 are needed for the interaction of the helicase and the topoisomerase, and loss of RMI proteins impacts the dissolution reaction to a similar extent as the loss of the enzyme partners in the complex. In yeast, in animals, and also in plants, it has been shown that this so-called RTR (for **R**ecQ helicase, **t**opoisomerase 3/3α, **R**MI) complex suppresses the formation of CO products after DSB formation in somatic cells. Additionally, two members of the plant RTR complex are essential for the completion of meiotic recombination: mutations in *TOP3A* or *RMI1* lead to severe chromosome fragmentation during meiotic recombination and arrest at the end of meiosis I. Therefore, it seems that at least some meiotic recombination intermediates must be processed through the dissolution pathway and that topoisomerase TOP3α is the only plant protein capable of acting on hemicatenane structures.

Tolerance and Repair Processes at Damaged Replication Forks

DNA damage often occurs in nonproliferating cells. However, if DNA is damaged in proliferating cells, there are a number of checkpoints that will arrest the cell cycle until the damaged DNA has been repaired. In some cases it is still possible that cells enter S phase making DNA repair processes at the replication fork necessary.

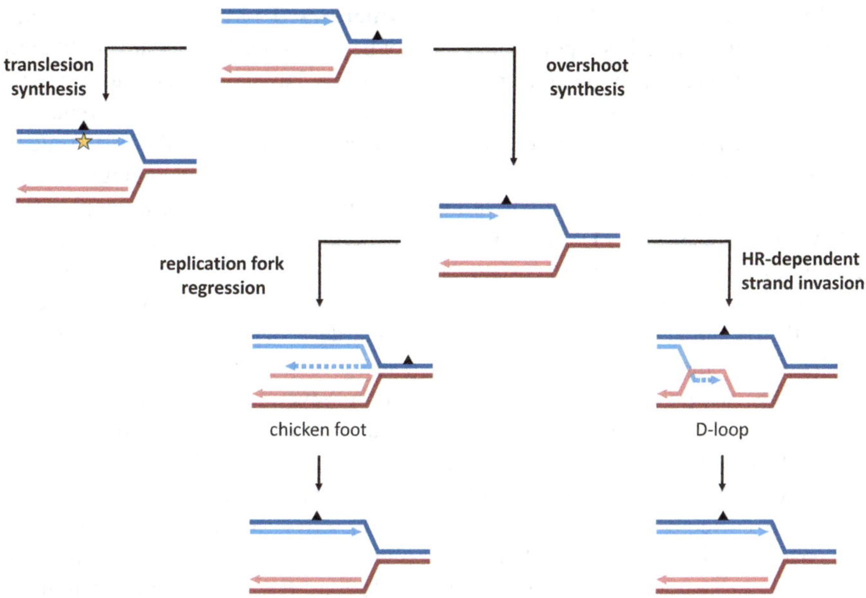

Fig. 11 **DNA damage tolerance pathways at blocked replication forks**. DNA synthesis of the leading strand of the replication fork is blocked by DNA damage in the parental template strand (*black triangle*). The DNA damage can be bypassed by the error-prone DNA synthesis of translesion polymerases (*left*), which can accept the damaged base as template. Possible mutations formed by this process are marked by a star. The other damage tolerance pathways are dependent on the uncoupled DNA synthesis of the lagging strand, called overshoot synthesis, and require the switch of the template strand for DNA synthesis. By the regression of the replication fork (*left*), a chicken foot intermediate is generated in which the newly synthesized daughter strand (lagging strand) can be used as template. After the reversal of the chicken foot, an intact replication fork is generated. Another template switch mechanism is mediated by a HR-dependent strand invasion of the shorter daughter strand (example in the figure) into the newly formed sister chromatid

DNA replication has to be tightly regulated so that each DNA molecule will be replicated only once per S phase. To ensure this, replication is initiated from replication origins, which after firing are inactivated during the ongoing S phase. Therefore, any replication fork that encounters severe DNA damage has to be processed by repair mechanisms: Replication across damaged DNA is usually not possible, but replication cannot be started again either downstream of the damaged site or upstream at its origin of replication. With replication blocked at replication forks, the chance for chromosomal rearrangements is high, which may lead to dysfunctions or even to the death of the cell, indicating the important function of the replication fork rescue.

When a replication fork is stalled due to damaged DNA that cannot be processed by replicative polymerases, cells will employ a number of different strategies. As an alternative to the repair of the lesion by one of the damage repair pathways outlined in this chapter, often DNA damage is just bypassed during replication (see also reviews Kunz et al. 2005; Garcia-Diaz and Bebenek 2007; Roldan-Arjona and Ariza 2009) and might be repaired after the completion of S phase (Fig. 11).

DNA damage bypass by translesion polymerases: To bypass damaged nucleotides, the replicative DNA polymerase cannot be used since it is not able to fit both the damaged parental strand and a newly synthesized daughter strand into its active center. To facilitate damage bypass, alternative proteins are therefore required. There are two different strategies to bypass damaged DNA, with the first strategy being error prone and the second error-free. The error-prone strategy, called translesion synthesis, uses alternative DNA polymerases instead of the replicative DNA polymerase (Fig. 11). Such translesion polymerases possess modified active centers that are able to accommodate damaged DNA to place a new nucleotide in the daughter strand. Since such polymerases most of the time are not able to correctly recognize chemically modified nucleotides, they cannot add the correct nucleotide to the growing DNA strand, leading to the incorporation of wrong nucleotides and possibly point mutations later on. Several translesion polymerases from a number of different DNA polymerase families have been described to date, but the best understood are the polymerases of the Y family. DNA polymerase η of the Y family of polymerases is conserved in fungi, animals, and plants and has the ability to correctly bypass T-T CPDs, enabling error-free translesion synthesis (TLS) of this lesion. Another Y-family polymerase, REV1, has a propensity to insert a cytosine opposite to a number of DNA lesions. Here, it is thought that this is a "best guess" strategy for the repair of abasic sites following the frequent spontaneous depurination of guanine. REV1 also interacts with another TLS polymerase, ζ of the B family, that is used to elongate the strand following REV1 activity. Polymerase ζ by itself is also able to bypass CPD lesions, but only with about 10 % fidelity. This illustrates a general rule of translesion synthesis: Whether the bypass of a lesion by a TLS polymerase will be error prone or error-free greatly depends on the type of lesion and the kind of polymerases available or involved.

DNA damage tolerance by postreplication repair: Besides the strategy of translesion synthesis that facilitates damage bypass by direct polymerase synthesis, there is also a second strategy that switches daughter strand synthesis to a new template. In the so-called postreplication repair (PRR) pathway, a feature of a stalled replication fork is used by the cell (Fig. 11).

Since damage to nucleotides usually occurs only on one strand, replication can proceed on the second, undamaged parental strand for some time, leading to replication fork uncoupling and finally stalling. So-called overshoot synthesis results in a newly synthesized daughter strand that contains exactly the sequence information missing in the damaged parental strand. To use this information, the stalled replication fork has to be rewound by a DNA helicase – in *Arabidopsis* probably the homolog of yeast Rad5, RAD5A, and the plant RecQ helicase RECQ4A – so that both daughter strands can anneal to each other instead of their respective parental strands, forming a DNA structure also known as a "chicken foot." This process is also called replication fork regression. Here, the shorter daughter strand, whose synthesis was blocked by the DNA damage, can be elongated by a DNA polymerase using the longer daughter strand as template. Once DNA synthesis has proceeded past the damaged site, the chicken foot structure can be rewound, again by a DNA helicase, to reform the replication fork. Now, however, both daughter strands lie downstream of the damaged site, allowing replication to proceed normally again.

Similar to the formation of a chicken foot by replication fork regression is also another proposed template switching mechanism, one that is reminiscent of the SDSA pathway of DSB repair by HR. At a stalled replication fork, the formation of a D-loop is also possible, e.g., by strand invasion of the short daughter strand from the damaged duplex into the undamaged and longer sister duplex. Here, elongation of the invaded strand facilitates mutation-free bypass of the damage site, so that eviction of this strand and re-annealing with its parental strand enables replication restart.

Repair processes at damaged replication forks: Besides these mechanisms to bypass damaged DNA during replication, it is also possible that repair mechanisms are initiated at a stalled replication fork (Fig. 12). Such DNA repair will become essential whenever the type of lesion will not allow a bypass. This has been shown for SSBs; if the backbone of one DNA strand is not closed, the separation of the two strands at the replication fork will lead to the formation of a so-called one-ended DSB, where only a single free end of dsDNA is present. There are many reasons why an SSB might form, but in S phase one source has been especially well researched. DNA supercoiling due to unwinding of the double strand at the replication fork will induce tension that has to be relieved by a type I topoisomerase, typically topoisomerase 1 (*TOP1*). TOP1, like other type I topoisomerases, will covalently bind a tyrosine residue in its active center to the DNA backbone, transiently creating an SSB. When a replication fork encounters this TOP1-DNA cleavage complex, a one-sided DSB can be created. Molecules such as CPT are known to stabilize the TOP1-DNA cleavage complex so that the chance of formation of a one-sided DSB is increased. Obviously, repair pathways requiring two free DNA ends, such as NHEJ and SSA, are not able to repair such a lesion. Here only HR pathways as described above will be able to repair the break by strand invasion of the free end into the sister chromatid.

Besides lesions in a single DNA strand, it is also possible that both strands in a DNA molecule may be covalently linked with each other in a so-called interstrand cross-link (CL). A number of genotoxins such as MMC and cisplatin have been described that can cross-link two DNA strands, but the cell's own metabolism might also be a source of interstrand CL-forming molecules, specifically aldehyde compounds. Whenever a replication fork encounters an interstrand CL, it cannot proceed, and obviously bypass pathways are also not feasible. Here, the covalent connection between the two strands has to be opened first, before any other repair or bypass mechanism can occur. Interstrand CL repair in animals depends on a large number of proteins, many of which belong to the Fanconi anemia (FA) pathway, since mutations in these genes are associated with this hereditary disease. FA proteins are thought to recognize interstrand CL lesions in DNA at stalled replication forks, to regulate signaling to affect cell cycle, repair, recombination, and other pathways and finally to also be directly involved in the repair of interstrand CLs themselves. Repair is initiated by the action of endonucleases like MUS81, which can "unhook" the cross-link and thereby create another type of lesion, where the two strands are not

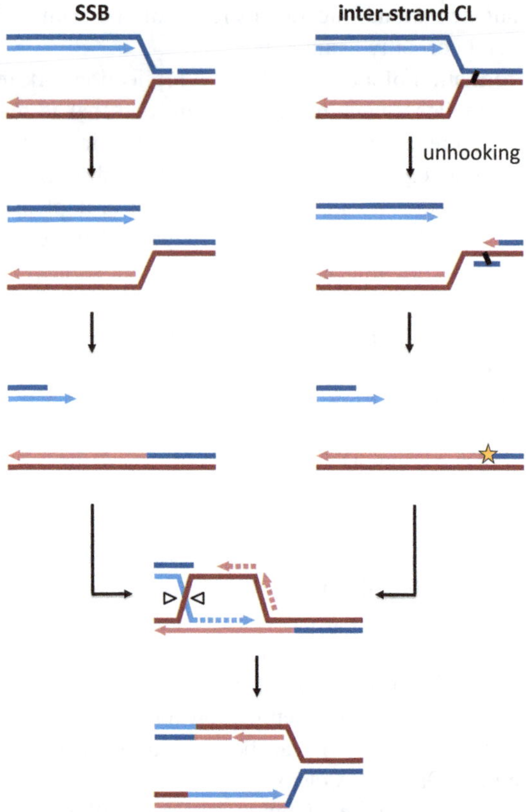

Fig. 12 **DNA damage repair pathways at damaged replication forks.** The repair of SSBs and interstrand CLs at replication forks is based on the formation of a one-sided DSB and the following initiation of HR. After the blockade of the DNA synthesis by a DNA lesion in one strand (*left*), an endonuclease can perform an incision into the ssDNA, generating a DSB. The DNA damage can further be removed e.g., by the action of an exonuclease. When the replication fork encounters a SSB which is present in the parental DNA strand (*middle*), a one-sided DSB is formed. The initial step in the repair of an interstrand CL at the replication fork (*right*) is the unhooking of the crosslink by two incisions in one DNA strand, up- and downstream of the interstrand CL. A TLS polymerase can bypass the site where the extruded crosslinked adduct is present. Subsequently, the crosslinked adduct can removed by NER proteins while the remaining one-sided DSB is repaired as described before

covalently connected anymore. The strand in which the endonucleolytic cuts occurred now contains a gap of a few bases, while in the complementary strand one base is modified by an adduct. Following this step, repair is thought to be concluded by NER proteins to remove the adduct and by HR to close the gap. In plants, many of the FA proteins are not conserved, and there has not been much research into the FA proteins that are found in plants. Interstrand CL repair, however, seems to be regulated differently in plants, than in animals. *Arabidopsis* contains at least three parallel pathways to repair cross-links, dependent on the plant homolog of

MUS81, the helicase RAD5A that is also involved in PRR (see above), and the RecQ family helicase RECQ4A. So while it seems possible that the unhooking of the interstrand CL by MUS81 is also important for the repair of this type of lesion, there seem to be alternative pathways in plants that do not depend on it.

Break-induced replication (BIR): Just as a damaged replication forks can initiate repair mechanisms, situations have been described where HR reactions initiate replication-like DNA synthesis to the end of the chromosome. When there is only a single end of dsDNA present, for example, following a cut of one fork at a stalled replication fork, initiation of an HR reaction will be possible, creating a typical D-loop. However, since there is no second end to reconnect and since replication in such a situation cannot be reinitiated after repair, the D-loop migrates with the elongating broken end toward the end of the chromosome (Fig. 10). Such a reaction, termed break-induced replication (BIR), will result in the loss of all genetic information from the broken chromosome from the breakpoint to the telomere. BIR reactions have been repeatedly described in yeast, but have not yet been found in plants.

Cellular Changes and Signaling After DNA Damage

The previous subchapters described a complex network of DNA repair pathways needed for the efficient repair of different kinds of DNA lesions. These repair pathways need to be tightly regulated in order to efficiently repair the different kinds of DNA damage and ensure cell survival. Unfortunately, the mechanisms and pathways for regulation and signaling after DNA damage are still poorly understood in plants. Many of the factors involved in DNA repair are conserved between the different kingdoms; therefore, it is tempting to speculate that the regulation of the different repair pathways might show a high level of conservation, too. One very important factor of orchestrating the different DNA damage repair pathways is a change in the expression of genes involved in the DNA damage response (see also review Mannuss et al. 2012). This is quite well studied in plants and will be the main focus of this section.

Cells do not waste resources in order to have all the repair proteins present in abundance when DNA damage occurs. It is rather a dynamic process to synthesize the enzymes that are needed to repair the particular damage. Many of the mentioned repair proteins, like RAD51 or BRCA1, which is thought to be an important mediator of DNA damage repair via HR, are present in a low number in plant cells, but if DSBs are induced in the genome, the expression of the corresponding genes is elevated several hundredfold. The Arabidopsis homolog of the breast cancer gene *BRCA1* is the gene that is most upregulated in response to DSBs in plants, which is consistent with its important function in DNA repair via HR. The large subunits of the single-strand binding protein RPA, which covers the single-stranded DNA recombination intermediates and the open complex intermediate during NER, are also among the genes that are most highly expressed in response to DNA damage. Not only are the proteins involved in the repair of DSBs through homologous recombination upregulated, but also the proteins that work in different DSB repair pathways, such as NBS1, which is part of the MRN complex, are upregulated, too. KU70 and KU80 that are involved in NHEJ are also expressed at higher levels following induction of DSBs. Not only genes involved in the repair of DSBs are upregulated but also genes involved in the BER pathway, such as *XRCC1* and *PARP1*, although there might be an overlap of functions of these proteins with the NHEJ pathway. Interestingly, most of the NER proteins are not upregulated upon DNA damage, not even through UV radiation, where it is an important repair pathway.

Apart from sophisticated repair genes, other genes – for example, the replicative DNA polymerases Polδ and Polε, which are needed for proper replication and furthermore also have a function in NER – are also upregulated after γ-irradiation.

Which factors are involved in the signaling between detection of DNA damage and the change of gene expression is an important question. Interestingly, two factors, the kinases ATM and ATR, mediate the signaling cascade following DNA damage through phosphorylation of downstream factors. ATM is mainly activated by DSBs, and ATR is primarily activated by single-stranded regions following stalled replication forks. It has to be mentioned though that there seems to be some level of redundancy between the two kinases in plants. ATM is responsible for most of the observed changes in gene expression after DSBs, whereas dependency on ATR is observed in only a few cases. It was demonstrated for Arabidopsis that especially upregulation of genes needed for the repair of DNA damage is dependent on ATM. The importance of ATM and ATR for proper signaling after DNA damage and the subsequent repair in plant cells is further strengthened by the fact that *atm/atr* double mutants show growth defects and sterility.

In the next step the signal is passed on from ATM and ATR to a transcription factor called SOG1 (suppressor of gamma response 1), which is activated by the kinases through phosphorylation. Subsequently, SOG1 is responsible for the changes in regulation of almost all genes in response to DNA damage. Since it operates downstream from both ATM and ATR, it has been proposed to be a master regulator of gene expression after DNA damage. In mammals, the most prominent tumor suppressor is a transcription factor called p53 that, among other functions, regulates gene expression after DNA damage. Although plants possess no direct homolog to p53, SOG1 might have a similar function after DNA damage.

An additional factor involved in the regulation of gene expression after DNA damage is called TEBICHI. It was postulated to regulate the expression of genes through chromatin remodeling, which is an important means for gene regulation. How the signal is transmitted to TEBICHI is not known, yet. Furthermore, some E2F transcription factors are strongly downregulated after DNA damage. E2Fs are a family of transcription factors that are highly conserved in plants and mammals. In normal plant cells, the transcription factor E2Fe suppresses the expression of a number of DNA repair factors, for example, the aforementioned *BRCA1*, *PARP1*, and *SOG1* and the photolyase gene *PHR1*. Through downregulation of E2Fe, the expression of the DNA repair genes can be increased. Therefore the E2F transcription factors seem to represent another level of regulation of gene expression after DNA damage.

The change of gene expression is a well-studied subject in plants, but of course it is not the only change a cell undergoes after DNA damage. Another important factor to increase cell survival is the regulation of the cell cycle in order to give the cell time to repair its DNA. By arresting the cell cycle at the different possible checkpoints, a cell can repair DNA damage before more severe effects occur, such as progressing into mitosis while DSBs are present. The proteins involved in signaling cell cycle arrest are also the kinases ATM and ATR and the transcription factor SOG1. It was shown that ATR is of utmost importance for the G2/M checkpoint. Additionally, a kinase called WEE1, the expression of which is also upregulated after DNA damage, is the main regulator of the intra-S phase checkpoint.

Connected to DNA damage is also an increased level of endoreduplication, in which the cells replicate their genome but skip cell division, resulting in polyploidy. It has been speculated that a higher copy number of genes can help the cells cope with the DNA damage. This step was shown to be dependent on signaling by ATM, ATR, and SOG1, too.

Future Directions

As much of the basic mechanisms of DNA repair have been elucidated, some intriguing questions remain that are especially relevant for plants.

How important Is DNA repair for plant genome evolution? In contrast to mammals, the germ line in plants is established late in development from somatic cells. Thus, genomic changes that occur during vegetative growth can become heritable. Although many changes that have adverse effects on the viability of the respective cells will be eliminated, there is ample of evidence that changes in DNA that occur during vegetative growth can contribute to plant genome evolution. Important contributors to somatic change are activated transposons and retrotransposons. Due to retrotransposition, the size of genomes of various plant species has grown over evolutionary time. A prominent example is maize, where most of the genome is composed of retroelements. Genome size can also increase by genome duplications. A number of such duplications are documented during the evolution of mono- and dicotyledonous crop plants. However, as there are also plants with surprisingly small genomes, the question arose how DNA is lost from the genome. It is now generally accepted that DSB repair processes contribute to "shrinking" of the genome. DSB repair by NHEJ is often accompanied by the loss of a few nucleotides at the break site, while DSB repair by SSA between tandemly duplicated sequences results in the loss of a repeat and all sequence information between the repeats. Bioinformatic studies indicate that both mechanisms shape plant genomes. The structure of plant retrotransposons, which consist of a multifunctional ORF flanked by homologous long terminal repeats (LTRs), reflects some important aspects of genome evolution shaped by DNA repair mechanisms. Indeed, often single LTRs instead of complete retrotransposons are found in genomes indicative of sequence loss of genomic elements due to most probably SSA. In many cases, smaller deletions have been documented by comparing two

LTRs of a single specific retroelement. Often these deletions are accompanied by the presence of microhomologies at the deletion site, which are a strong indication for nonhomologous DNA repair events such as NHEJ or MMEJ. Thus, both SSA and NHEJ contribute to genome shrinking in plants. There are also indications that species-specific variations in the efficiencies of the different DSB repair pathways might influence the evolution of genome sizes to different extents in different plant species.

DNA repair, the key for constructing the synthetic plant genome? In somatic cells the repair of DSBs is often mutagenic, leading to a change in information content at and/or around the break site. This applies to NHEJ as well as to SSA and HR. It has been demonstrated that by induction of a unique DSB at a specific site in the plant genome, different kinds of repair reactions are initiated. This could be demonstrated by cutting the genome with restriction endonucleases that cleave at rare cutting sites producing only one or a few breaks at definable sites in the genome (see review Puchta and Fauser 2013). In recent years, different types of artificial nucleases have been developed, namely, meganucleases, zinc finger nucleases (ZFNs), transcription activator-like effector nucleases (TALENs), and the CRISPR/Cas system. By programming the respective artificial DNA recognition sites of the respective nuclease, it is now possible to induce DSBs at any site of interest in plant genomes. These DSBs are then repaired either by NHEJ or, in the presence of a respective homologous template, by HR. If a DSB in an ORF is repaired by NHEJ, deletions often occur that lead to the loss of information and can also lead to frame shifts. Thus, mutations in any gene of interest can now be easily obtained by producing an endonuclease that is specific for a unique genomic site in the specific gene. Also, gene targeting to introduce predefined changes of specific amino acids or to stack transgenes at specific genomic locations by HR can now be performed with ease. But even larger rearrangements are now becoming possible. By simultaneously inducing more than one DSB within a chromatid, defined deletions can be produced. Moreover, it has already been demonstrated that by induction of two DSBs, arms between different chromosomes can be exchanged in plants. The principle behind these manipulations is that although in the majority of outcomes the correct ends of the respective breaks will be rejoined by NHEJ, once in a while ends are set free resulting in a joining of the "wrong" ends that have not been linked before. To find the combination of interest, one can screen for cells in which the "wrong" ends of the DSBs have been joined. As multiplexing, especially with the CRISPR/Cas system is becoming routine, a number of DSBs can be induced at the same time in somatic cells. Thus, the reconstruction of complex plant genomes is within reach. In the long run, plant breeding therefore might be revolutionized by the application of techniques based on DNA repair.

References

Baute J, Depicker A. Base excision repair and its role in maintaining genome stability. Crit Rev Biochem Mol Biol. 2008;43:239–76.

Chaves I, Pokorny R, Byrdin M, Hoang N, Ritz T, et al. The cryptochromes: blue light photoreceptors in plants and animals. Annu Rev Plant Biol. 2011;62:335–64.
Costa RM, Chigancas V, Galhardo Rda S, Carvalho H, Menck CF. The eukaryotic nucleotide excision repair pathway. Biochimie. 2003;85:1083–99.
Dalhus B, Laerdahl JK, Backe PH, Bjørås M. DNA base repair–recognition and initiation of catalysis. FEMS Microbiol Rev. 2009;33:1044–78.
de Laat WL, Jaspers NG, Hoeijmakers JH. Molecular mechanism of nucleotide excision repair. Genes Dev. 1999;13:768–85.
Fortini P, Pascucci B, Parlanti E, D'Errico M, Simonelli V, et al. The base excision repair: mechanisms and its relevance for cancer susceptibility. Biochimie. 2003;85:1053–71.
Garcia-Diaz M, Bebenek K. Multiple functions of DNA polymerases. CRC Crit Rev Plant Sci. 2007;26:105–22.
Goosen N, Moolenaar GF. Repair of UV damage in bacteria. DNA Repair (Amst). 2008;7:353–79.
He XJ, Chen T, Zhu JK. Regulation and function of DNA methylation in plants and animals. Cell Res. 2011;21:442–65.
Jiricny J. The multifaceted mismatch-repair system. Nat Rev Mol Cell Biol. 2006;7:335–46.
Knoll A, Puchta H. The role of DNA helicases and their interaction partners in genome stability and meiotic recombination in plants. J Exp Bot. 2011;62:1565–79.
Kunz BA, Anderson HJ, Osmond MJ, Vonarx EJ. Components of nucleotide excision repair and DNA damage tolerance in Arabidopsis thaliana. Environ Mol Mutagen. 2005;45:115–27.
Mannuss A, Trapp O, Puchta H. Gene regulation in response to DNA damage. Biochim Biophys Acta. 2012;1819:154–65.
Puchta H, Fauser F. Synthetic nucleases for genome engineering in plants: prospects for a bright future. Plant J. 2013. doi:10.1111/tpj.12338.
Roldan-Arjona T, Ariza RR. Repair and tolerance of oxidative DNA damage in plants. Mutat Res. 2009;681:169–79.
Schuermann D, Molinier J, Fritsch O, Hohn B. The dual nature of homologous recombination in plants. Trends Genet. 2005;21:172–81.
Spampinato CP, Gomez-Casati DF. Research on plants for the understanding of diseases of nuclear and mitochondrial origin. J Biomed Biotechnol 2012: 836196
Tuteja N, Ahmad P, Panda BB, Tuteja R. Genotoxic stress in plants: shedding light on DNA damage, repair and DNA repair helicases. Mutat Res. 2009;681:134–49.
Vonarx EJ, Mitchell HL, Karthikeyan R, Chatterjee I, Kunz BA. DNA repair in higher plants. Mutat Res. 1998;400:187–200.
Wallace SS, Murphy DL, Sweasy JB. Base excision repair and cancer. Cancer Lett. 2012;327:73–89.
Waterworth WM, Drury GE, Bray CM, West CE. Repairing breaks in the plant genome: the importance of keeping it together. New Phytol. 2011;192:805–22.
Weber S. Light-driven enzymatic catalysis of DNA repair: a review of recent biophysical studies on photolyase. Biochim Biophys Acta. 2005;1707:1–23.
Zhu JK. Active DNA demethylation mediated by DNA glycosylases. Annu Rev Genet. 2009;43:143–66.

Further Readings

Aguilera A, Rothstein R. Molecular genetics of recombination. Berlin/New York: Springer; 2007.
Friedberg EC, Elledge SJ, Lehmann AR. DNA repair, mutagenesis, and other responses to DNA damage: a subject collection from Cold Spring Harbor perspectives in biology. Cold Spring Harbor: Cold Spring Harbor Laboratory; 2013.
Friedberg EC, Walker GC, Siede W. DNA repair and mutagenesis. Washington, DC: ASM Press; 2006.

Small RNAs in Plants

4

So Youn Won, Rae Eden Yumul, and Xuemei Chen

Contents

S.Y. Won • R.E. Yumul
Department of Botany and Plant Sciences, Institute of Integrative Genome Biology, Center for Plant Cell Biology, University of California, Riverside, CA, USA
e-mail: swon004@student.ucr.edu; reyumul@gmail.com

X. Chen (✉)
Department of Botany and Plant Sciences, Institute of Integrative Genome Biology, Center for Plant Cell Biology, University of California, Riverside, CA, USA

Howard Hughes Medical Institute, University of California, Riverside, CA, USA
e-mail: xuemei.chen@ucr.edu

S.H. Howell (ed.), *Molecular Biology, The Plant Sciences 2*,
DOI 10.1007/978-1-4614-7570-5_3

Abstract

- Small RNAs 20–30 nucleotides in length are sequence-specific regulatory RNAs that guide nucleic acid-based processes in eukaryotic organisms.
- In plants, small RNAs are classified as microRNAs (miRNAs) or small interfering RNAs (siRNAs) based on differences in their precursors and biogenesis.
- miRNAs are generated from precursors with an imperfect intramolecular hairpin structure and silence their target genes at the posttranscriptional level through mRNA cleavage or translational inhibition.
- Endogenous siRNAs are processed from long double-stranded RNAs with perfect base-pair complementarity and are classified as heterochromatic siRNAs (hc-siRNAs) or *trans*-acting siRNAs (ta-siRNAs).
- Heterochromatic siRNAs (hc-siRNAs) are endogenous siRNAs generated from transposons and repeats and guide cytosine methylation, which induces heterochromatin formation and transcriptional gene silencing.
- *Trans*-acting siRNAs (ta-siRNAs) are endogenous siRNAs whose biogenesis is triggered by specific miRNAs. Like miRNAs, ta-siRNAs repress their targets *in trans* at the posttranscriptional level and are important for plant development.
- Exogenous siRNAs are generated in both virus-infected plants and transgenic plants.
- The steady-state levels of small RNAs are precisely regulated through their biogenesis and turnover.
- A specific modification of small RNAs protects them from uridylation and truncation, processes associated with small RNA turnover.

Introduction

Ribonucleic acid (RNA) is a major macromolecule that executes biological events in living organisms. According to the central dogma of gene expression, RNA serves as an intermediate in the flow of genetic information from DNA to protein in the form of messenger RNA (mRNA). Additionally, there are different types of non-protein-coding RNAs (ncRNAs), whose classification is based primarily on their molecular functions but also reflects differences in size and accumulation. Several classes of ncRNAs are abundant and perform housekeeping duties: ribosomal RNAs (rRNAs) and transfer RNAs (tRNAs) are involved in protein synthesis, small nuclear RNAs (snRNAs) are involved in pre-mRNA splicing, and small nucleolar RNAs (snoRNAs) guide the modification of other RNAs. The length of ncRNAs is another important criterion for their classification. Long noncoding RNAs (lncRNAs) refer to ncRNAs that are longer than the arbitrary size of 200 nucleotides (nt). While it remains unclear whether the vast number of intergenic lncRNAs detected in microarray or RNA-seq experiments are transcriptional noise or functional RNAs, there has been increasing evidence of their regulatory functions in gene expression, especially in animals. Small RNAs 20–30 nt in length, the focus of this chapter, have come to be recognized as an important class within the broad spectrum of ncRNAs because their regulatory functions are critical for biological processes.

Small RNAs are a central component of RNA-mediated silencing in all eukaryotes and regulate genes in a sequence-specific manner through the recognition of complementary nucleic acid sequences. Small RNAs are known to act via two main mechanisms. In posttranscriptional gene silencing (PTGS), small RNAs guide the cleavage or translational inhibition of their target mRNAs. In transcriptional gene silencing (TGS), small RNAs guide DNA or histone methylation, resulting in heterochromatin formation.

In plants, small RNAs are classified as microRNAs (miRNAs) or small interfering RNAs (siRNAs) based on their precursors and biogenesis. miRNAs derive from longer RNA precursors containing a stem-loop or hairpin structure with imperfect base-pairing in the stem region. While the mature miRNA is the single most abundant species generated from a precursor, the passenger strand (miRNA*) is the second most abundant species and may also be found in vivo. In contrast, siRNAs derive from longer double-stranded RNAs (dsRNAs) that exhibit nearly perfect sequence complementarity. Typically, multiple siRNA species are generated from a single precursor. Despite the differences in precursors and biogenesis that distinguish the different classes of small RNAs, however, it is important to emphasize that all small RNAs function as sequence-specific guides in target regulation.

Research over the past two decades has significantly improved the understanding of small RNAs and their regulatory mechanisms. Since the initial discovery of an miRNA, *lin-4*, from genetic screens of the nematode *Caenorhabditis elegans* in 1993, hundreds of thousands of small RNAs have been identified, particularly with the aid of next-generation sequencing technology. Along with the improved understanding of its critical regulatory functions for numerous biological processes, small RNA-mediated gene silencing is also recognized as a powerful research tool in biology. The use of small RNAs to knock down selected genes permits the dissection of the molecular functions of those genes and related pathways. Small RNA-based gene silencing has also been used for crop improvement and fighting human diseases. The awarding of the 2006 Nobel Prize in Physiology and Medicine for the discovery of RNA interference (RNAi), a homology-based gene silencing phenomenon conferred by small RNAs, further exemplifies the significance of small RNA-mediated gene regulation.

In this chapter, small RNAs in the model plant *Arabidopsis thaliana* will be discussed in terms of their biogenesis, their molecular mechanisms for target repression, and their biological functions. Major differences with animal small

RNAs will also be introduced. Finally, mechanisms that contribute to small RNA homeostasis such as degradation will be discussed.

miRNAs

Biogenesis

miRNAs are small regulatory RNAs 20–22 nt in length that act in a sequence-specific manner primarily through PTGS (reviewed in Chen 2009). Their biogenesis involves the following steps: transcription of a MIR gene to produce the miRNA precursor, cleavage to yield the mature precursor, stabilization by methylation, nuclear export, and incorporation into effector proteins (Fig. 1).

miRNA precursors that give rise to mature miRNAs are encoded by *MIR* genes, which are located in intergenic regions. *MIR* genes are individual gene units with their own promoters and terminators and are transcribed by RNA polymerase II (Pol II) (Fig. 1). Accordingly, *MIR* gene promoters harbor *cis*-acting elements for transcription by Pol II. As with protein-coding genes, the expression of *MIR* genes is subject to regulation, with Pol II transcription affected by spatiotemporal inputs specific to particular developmental stages and organs. In addition to these endogenous signals, exogenous cues from the environment, such as biotic and abiotic stresses, also affect transcription. Thus, the transcription of *MIR* genes and, ultimately, miRNA abundance are governed by regulatory frameworks that respond to various signals. Mediator, a multi-protein complex, serves as a general transcription factor and is thought to integrate various signals to promote the recruitment of Pol II to promoters. Mediator is required for the transcription of *MIR* genes in *Arabidopsis*. After and/or during transcription, miRNA precursors are capped and polyadenylated at their 5′ and 3′ ends, respectively, and introns are spliced out in a manner similar to the processing of Pol II-transcribed pre-mRNA. These *MIR* gene transcripts, or primary miRNAs (pri-miRNAs), form hairpin structures with imperfect base-pairing in the stem regions and are subsequently processed by small RNA biogenesis enzymes.

Through the successive action of Dicer-like (DCL) RNase III enzymes in the nucleus, a pri-miRNA is processed into a precursor miRNA (pre-miRNA), which is in turn processed into a mature miRNA/miRNA* duplex (Fig. 1). This duplex contains both the guide strand, the functional miRNA species that promotes PTGS, and the miRNA* passenger strand, which is eventually degraded. DCL1, an RNase III enzyme that specifically cleaves dsRNA, is responsible for the processing of most miRNAs. However, the *Arabidopsis* genome encodes four DCL genes and several evolutionarily young miRNAs are processed by DCL4 instead of DCL1. The first dicing step generates the pre-miRNA from the pri-miRNA: DCL1cleaves the stem approximately 15 nt away from the base of the stem and generates a 2-nt 3′ overhang. The second dicing step by DCL1 cleaves the newly formed pre-miRNA at a position closer to the terminal loop, generating a 20–22-nt miRNA/miRNA* duplex with 2-nt 3′ overhangs. During this process, the

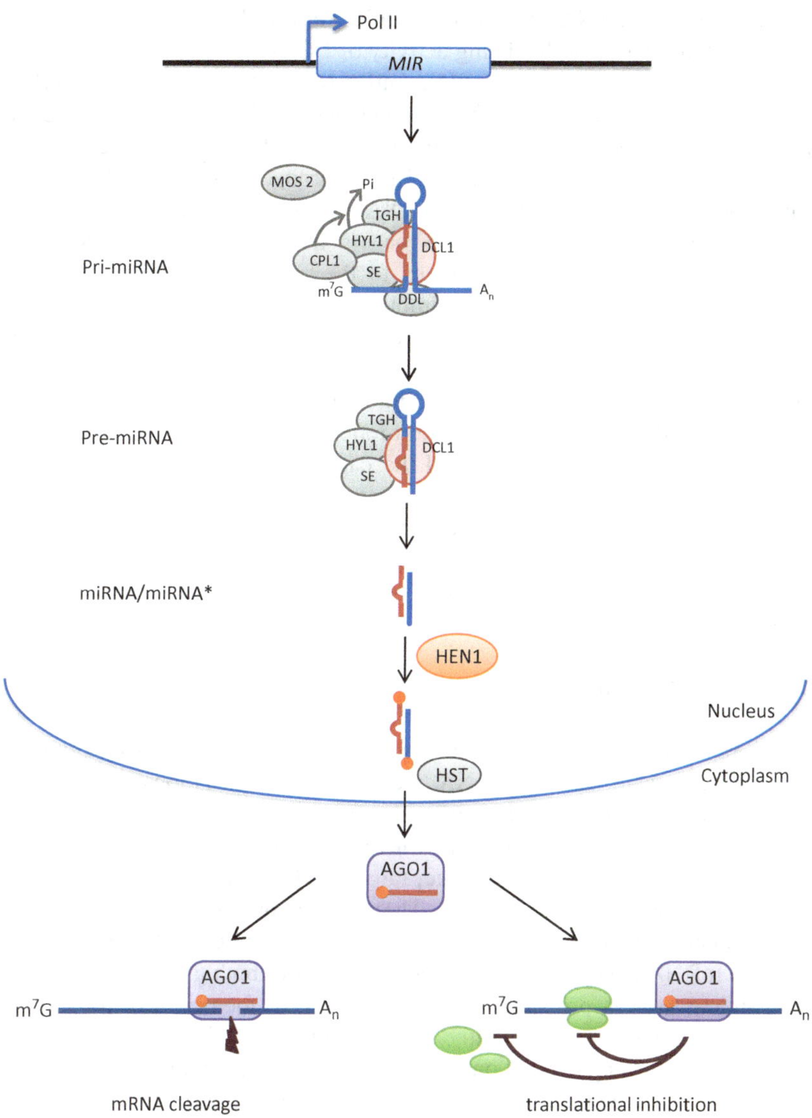

Fig. 1 miRNA biogenesis and silencing mechanism. Pri-miRNA transcripts are generated from *MIR* gene loci by RNA polymerase II and processed into pre-miRNAs by DCL1 with the assistance of HYL1, SE, DDL, CPL1, MOS2, and TGH. Pre-miRNAs are further processed into miRNA/miRNA* duplexes by DCL1. Both strands of the miRNA/miRNA* duplex are methylated by HEN1 either before or after HST-mediated transport to the cytoplasm. Mature miRNAs are loaded into AGO1 and guide target repression via mRNA cleavage or translational inhibition

DCL1 enzyme is aided by DAWDLE (DDL), a forkhead-associated (FHA) domain protein; SERRATE (SE), a zinc finger protein; and HYPONASTIC LEAVES1 (HYL1), an RNA binding protein. It has been proposed that DDL facilitates the recognition of pri-miRNAs by DCL1, while SE and HYL1 may improve the accuracy and efficiency of the dicing activity of DCL1.

Following the release of the miRNA/miRNA* duplex from the pre-miRNA, HUA ENHANCER 1 (HEN1) methylates both ends of the duplex (Fig. 1). Specifically, HEN1 deposits a single methyl group at the 2′-OH position of the 3′ terminal ribose. As discussed in section "Small RNA Turnover" below, HEN1-mediated methylation enhances the stability of miRNAs. That miRNAs are generated in the nucleus while miRNA-directed PTGS occurs in the cytoplasm indicates that miRNAs are exported from the nucleus to the cytoplasm. HASTY (HST), an Exportin-5 (Exp-5) homolog, has been implicated in the nuclear export of miRNAs, but it is unknown whether HEN1-mediated methylation precedes or follows nuclear export.

In the cytoplasm, miRNAs are loaded into ARGONAUTE (AGO) effector proteins and small RNA-mediated repression reflects the functions of these two key players: guidance by small RNAs and the catalytic activity of AGO-containing protein complexes (Fig. 1). The *Arabidopsis* genome encodes ten AGO homologs. Among these, AGO1 functions as the major effector protein for miRNA-mediated PTGS and binds most miRNAs. AGO7 specifically binds miR390, while AGO10 exhibits a binding preference for miR166/165 over other miRNA species.

Molecular Mechanism

miRNAs repress the expression of targets via PTGS, which is associated with two modes of repressive action: mRNA cleavage and translational inhibition. In miRNA-pathway-compromised mutants, these changes can be assessed through the detection of the mRNA transcript levels or protein abundance of miRNA targets. In the case of mRNA cleavage, target mRNAs are sliced at the center of the sequence bound by the miRNA. The cleaved products, particularly the 3′ fragments, can be detected in wild-type plants. When miRNAs that regulate their targets via mRNA cleavage are disrupted, the levels of both target mRNAs and the corresponding protein products increase. In contrast, there is a disproportionate accumulation of target protein relative to that of target mRNAs when miRNAs that regulate their targets by translational inhibition are disrupted. mRNA cleavage and translational inhibition may occur in parallel. For instance, a fraction of the mRNA pool targeted by a single miRNA may be repressed by cleavage while the remaining fraction is regulated by translational inhibition.

In plants, miRNA-guided cleavage has been observed for most miRNAs and is considered to be a widespread regulatory mechanism of plant miRNAs. The endonucleolytic activity of AGO1 cleaves (or slices) the phosphodiester bond linking two nucleotides in the target mRNA that correspond to the 10th and 11th nucleotides from the 5′ terminal end of the miRNA. The newly exposed 5′ and 3′ fragments are subsequently degraded by the exosome with 3′–5′ exonuclease activity and EXONUCLEASE4 (XRN4) with 5′–3′ exonuclease activity, respectively. The degradation of the 5′fragment is further accelerated by template-independent oligo-uridylation. Uridine tails are attached at the 3′ end of the 5′ fragment, which promotes decapping activity at the 5′ end. The 5′ fragment is thus rendered susceptible to 5′–3′ degradation, and translation of the cleavage fragment is prevented.

miRNA-directed translational inhibition is less commonly observed in plants than transcript cleavage, and there are two main explanations for this. While miRNAs in animals require perfect base pairing with the target mRNA only in the seed region, which corresponds to the 2nd to 7th nucleotides from the 5′ end of the miRNA, plant miRNAs require much more extensive complementarity with the target mRNA for PTGS. This difference may underlie the predominance of distinct repressive mechanisms in the two kingdoms, i.e., translational inhibition in animals and transcript cleavage in plants. Alternatively, the perceived predominance of miRNA-directed transcript cleavage in plants may reflect technical limitations. Although monitoring the effects of miRNAs on their targets is facilitated by their sequence complementarity, high-quality antibodies for the proteins corresponding to the targeted mRNAs are necessary to assess the occurrence or extent of translational inhibition. Thus, the technical challenge of producing high-quality antibodies may contribute to the less frequent observation of translational inhibition in plants.

The earliest reports of miRNA-directed translational inhibition in plants were the findings that *APETALA2* (*AP2*) and *SQUAMOSA PROMOTER BINDING PROTEIN-LIKE3* (*SPL3*) are translationally repressed by miR172 and miR156/157, respectively. Subsequent studies identified additional miRNAs that regulate their target mRNAs via translational inhibition. Moreover, several players in miRNA-directed translational inhibition have been identified from forward genetic screens in *Arabidopsis*. Mutations disrupting the microtubule-severing enzyme KATANIN1 (KTN1), the P body component VARICOSE (VCS)/Ge-1, the GW-repeat protein SUO, and the ER membrane protein ALTERED MERISTEM PROGRAM1 (AMP1) compromise miRNA-mediated translational inhibition of exogenous reporter constructs and endogenous miRNA targets. Notably, these studies show that miRNAs, such as miR156/157, miR164, miR165/166, miR172, and miR398, which are known to guide target transcript cleavage, also inhibit the translation of target mRNAs. The fact that genetic mutations (in *KTN1*, *VCS*, *SUO*, and *AMP1*) can uncouple the transcript cleavage and translational inhibition activities of these miRNAs suggests that the two repressive modes of action are independent and occur in parallel to regulate target transcripts. The molecular mechanism of translational inhibition and the events that follow remain unclear. However, two studies in zebra fish and fruit fly suggest that translational inhibition primarily affects the initiation step rather than elongation or termination and that the subsequent stimulation of mRNA deadenylation and decay occurs in an miRNA-cleavage-independent manner (Bazzini et al. 2012; Djuranovic et al. 2012).

Biological Function

Because miRNAs repress or silence their target mRNAs, studies of the targets of miRNAs have also been critical for understanding miRNA function. According to the miRBase miRNA database (www.mirbase.org, Release 19) (Kozomara and Griffiths-Jones 2011), the numbers of mature miRNAs and precursors in *Arabidopsis* are 338 and 299, respectively. Among known miRNA targets, transcription factors are the most highly represented functional group. By recognizing *cis*-acting elements at the promoters of their target genes, transcription factors can systematically activate or repress genes belonging to a particular regulatory or functional network. The regulation of the transcription of downstream genes by transcription factors is affected by both endogenous and exogenous signals. miRNAs regulating transcription factors therefore provide an additional layer of regulation for specific biological processes.

Although miRNA-mediated PTGS affects a wide variety of biological phenomena, its role in development is particularly well established, and a large number of miRNA-targeted transcription factors are implicated in developmental processes (reviewed in Chen 2009). Interactions between miRNAs and their targets have been reported in a wide range of developmental contexts, such as embryogenesis, cell differentiation, pattern formation, phase transition, and hormone signaling. Loss- or gain-of-function mutations in *MIR* genes or their targets often result in specific developmental phenotypes that are informative about their functions. Expressing miRNA-resistant targets under their endogenous promoters also affects plant morphology, demonstrating that miRNA-mediated PTGS is a critical regulatory component of developmental programs. Mutations disrupting miRNA biogenesis genes and *AGO1* consistently result in pleiotropic developmental defects. For example, the null *dcl1* allele is embryonic lethal, and the morphological phenotype of hypomorphic *dcl1* is similar to those of null *hyl1*, *hen1*, and *hst* alleles. Mutations disrupting *AGO1* also yield phenotypes similar to those of mutants with disrupted *DCL* function. Taken together, the developmental defects of miRNA pathway mutants further establish the vital functions of miRNAs in development.

Other miRNAs affect the gene regulatory networks that govern responses to environmental cues. Although less is known about miRNAs involved in stress responses compared to miRNAs involved in developmental processes, deep sequencing of small RNA populations has increased the number of identified miRNAs that are specifically expressed under certain environmental conditions, such as biotic and abiotic stresses. For example, stress-related hormones such as abscisic acid can activate or repress the expression of certain miRNAs. In turn, stress-responsive miRNAs may target genes involved in detoxification or enhancing resistance. Because stress response signals also impact the developmental network, these changes may be critical for the ability of plants to alter their developmental program under harsh external conditions and to resume the normal program when the stress condition is removed.

Autoregulation of the miRNA Pathway

Several self-feedback mechanisms are known to regulate the miRNA pathway. Two critical components of the miRNA pathway, *DCL1* and *AGO1*, are themselves targets of miRNA-mediated repression. Whereas DCL1 protein catalyzes miRNA biogenesis, there are three possible fates for *DCL1* transcripts: they may be translated into DCL1 protein, recognized as a cleavage target of miRNA-mediated PTGS, or processed as a pri-miRNA. When miRNA levels are high, miR162-directed cleavage of *DCL1* transcripts by an AGO1-containing complex is likely favored. Alternatively, the foldback RNA structure within the *DCL1* transcript may recruit the miRNA biogenesis machinery and be diced by DCL1 to generate miR838. Thus, high miRNA levels or high DCL1 protein levels lead to a decrease in *DCL1* mRNA abundance and consequently reduced *DCL1* expression and activity. *AGO1* mRNA contains a miR168 binding site, which similarly permits feedback regulation of *AGO1*. In this manner, the autoregulation of critical enzymes ensures the balanced dynamics of the miRNA pathway.

miRNAs in Animals

As in plants, miRNAs in animals represent an essential regulatory module of gene expression (reviewed in Kim et al. 2009; Krol et al. 2010). Although miRNAs and miRNA targets are not well conserved between the two kingdoms, the general principles of miRNA biogenesis and function are held in common: stem-loop-containing precursors are processed into mature miRNAs, and target repression is accomplished by miRNA-directed AGO function. Nevertheless, there are specific characteristics of animal miRNAs that are not observed in plants. First, a large fraction of miRNA genes in animals are clustered together and generate polycistronic precursors. Moreover, miRNA genes may reside within the transcriptional units of other genes and consequently depend on the transcription of these genes for their own expression. Following the transcription of miRNA genes in animals, the processing of the precursors involves two RNase III enzymes, Drosha and Dicer. These enzymes perform two independent dicing events, in contrast to DCL1 in plants, which performs both dicing steps. Animal pri-miRNAs are initially processed into pre-miRNAs by Drosha within the nucleus, and the pre-miRNAs are subsequently transported to the cytoplasm by Exp-5. Further processing of pre-miRNAs to generate the miRNA/miRNA* is performed by Dicer in the cytoplasm. In contrast to the methylation of plant miRNAs by HEN1, the miRNA/miRNA* duplex is not methylated in animals. Stable incorporation of the mature miRNA into an AGO protein subsequently directs target repression.

In animals, miRNA-directed target repression generally occurs through translational inhibition and RNA decay. Transcript cleavage is not widely observed for animal miRNAs, which may reflect the relatively low complementarity between animal miRNAs and their targets. miRNA binding sites are typically located in the 3′ untranslated region (3′ UTR) of target transcripts, with single mRNA molecules bound by multiple miRNAs for silencing. As previously described, miRNA-mediated PTGS in plants requires high sequence complementarity at target binding sites, which are generally found within coding sequences. In animals, perfect complementarity of the miRNA and its target within the seed region alone (the

2nd to 7th nucleotides from the 5′ end of the miRNA) is sufficient for target recognition. As a result, a single miRNA generally has a large number of mRNA targets. Thus, miRNA-mediated silencing in animals involves a less stringent hybridization requirement between miRNAs and their targets, and target transcript cleavage is uncommon in animals.

Evolution of *MIR* Genes

High-throughput deep sequencing of small RNA populations has led to the discovery of many miRNA species. The miRNA database (www.mirbase.org, Release 19) (Kozomara and Griffiths-Jones 2011) currently lists 338 mature miRNAs in *A. thaliana*, 2,042 in humans, and 368 in *C. elegans* in addition to the thousands of mature miRNAs from 190 other species. Despite the large number of identified miRNAs in both plants and animals, the poor conservation of miRNA sequences between the two kingdoms suggests that miRNA families in plants and animals evolved independently or that miRNAs were present in the common ancestor but the genes evolved so fast that no sequence similarities could be detected in plants and animals.

There are two major hypotheses regarding the evolution of miRNAs. The first proposes that *MIR* genes evolved from inverted duplications of miRNA targets. In this model, the duplication of protein-coding genes in a head-to-head or tail-to-tail orientation yields stem-loop structures whose stem regions exhibit extensive base-pair complementarity and are processed by the siRNA pathway rather than the miRNA pathway. Small RNAs from these young *MIR* genes regulate their homologous targets. Over evolutionary time, both *MIR* genes and their targets may undergo duplication and accumulate mutations. Some of the double-stranded precursors eventually acquire the characteristic hairpin structure of miRNA precursors and are processed by the miRNA pathway rather than the siRNA pathway. Over time, the targets come to be regulated by a limited number of specific small RNAs. According to this model, recently evolved *MIR* genes are expected to have a higher degree of sequence similarity with their targets, which has been observed for *MIR161* and *MIR163*.

Because the precursors of many young miRNAs do not match any other sequences in the genome, the duplication hypothesis cannot explain the genesis of all of the young miRNAs in plants. The random hairpin theory was proposed in part to address this shortcoming. Organisms produce a large number of hairpin structures that could potentially generate foldback precursors of small RNAs. For example, the *A. thaliana* genome contains more than 130,000 imperfect inverted repeats. The random hairpin theory proposes that *MIR* genes can evolve when the following conditions are met: a DNA segment that generates a foldback structure retains a transcriptional unit; by chance, a small RNA produced from the structure targets a protein-coding gene; and the resulting regulatory relationship confers an evolutionary advantage.

The classification of an miRNA as old or young is based on its degree of conservation among different species. While ancient miRNAs are conserved among animal or plant species of great evolutionary distance, young miRNAs are specific to a species or genus. miRNA genes may be duplicated as a result of gene duplication, whole segment duplication, or genome duplication, thereby giving rise to an miRNA family. The miRNAs produced by these families are considered old miRNAs, and their abundance is high probably because they are encoded by multiple genes. The processing of the precursors of old miRNAs by the DCL1-mediated biogenesis pathway is more precise than the processing of young miRNA precursors. Having multiple *MIR* genes within each conserved family may result in a complex relationship among the individual miRNA family members and their targets. In contrast, non-conserved miRNAs tend to be encoded by a single locus and are characterized by their short evolution times. In addition to being weakly expressed, young miRNAs typically regulate few, if any, genes, and the processing of their precursors is less precise.

Heterochromatic siRNAs

siRNAs are small RNAs 21–24 nt in length that are generated from long dsRNAs or single-stranded RNAs (ssRNAs) that produce longer and more perfect hairpin structures compared to miRNA precursors. While only one miRNA is produced from a single precursor, siRNA precursors generate multiple siRNAs. siRNAs trigger TGS or PTGS and direct the enzymatic action of their effector proteins through sequence-specific interactions with their targets.

Plants have two major families of endogenous siRNAs: heterochromatic siRNAs (hc-siRNAs) and *trans*-acting siRNAs (ta-siRNAs). siRNAs may also derive from exogenous sources, such as viruses and transgenes.

Biogenesis

Hc-siRNAs are small RNAs 21–24 nt in length that derive from heterochromatic regions, including repeats, transposons, and intergenic regions (reviewed in Law and Jacobsen 2010; Castel and Martienssen 2013). They comprise the most abundant and diverse small RNA family: approximately 80 % of all small RNAs are hc-siRNAs, with tens of thousands of unique hc-siRNAs present in wild-type *Arabidopsis* (Zhang et al. 2007; Mosher et al. 2008). Hc-siRNAs mediate TGS of heterochromatin by guiding DNA methylation and histone modification in a sequence-specific manner. In *Arabidopsis*, the biogenesis of hc-siRNAs involves the transcription of primary precursors, the conversion of the precursors to dsRNAs, further maturation by dicing, methylation, and association with AGO proteins (Fig. 2).

That the sequences of hc-siRNAs often map to repeats and transposons implies that hc-siRNA precursors are transcribed from these regions. In *Arabidopsis*, this transcription is probably performed by the plant-specific DNA-dependent RNA polymerase Pol IV (reviewed in Haag and Pikaard 2011). Although Pol IV is evolutionarily derived from Pol II, the largest subunit of Pol IV, NUCLEAR

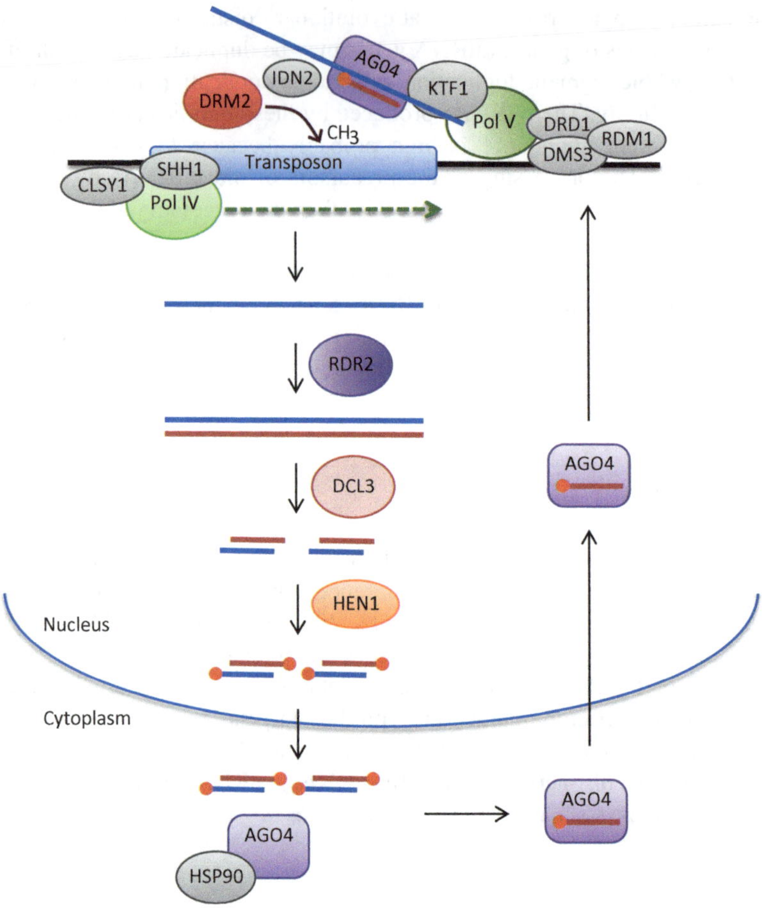

Fig. 2 Hc-siRNA biogenesis and silencing mechanism. RNA polymerase IV generates long noncoding RNAs (lncRNAs) from target regions with the aid of CLSY1 and SHH1. These lncRNAs are made double stranded by RDR2. DCL3 dices the double-stranded RNAs into 24-nt siRNAs, which are subsequently methylated by HEN1 and transported to the cytoplasm. siRNAs are loaded into AGO4 with the assistance of HSP90 then transported into the nucleus. AGO4-loaded siRNAs recognize the nascent transcripts generated by RNA polymerase V, and the DRM2 methyltransferase is recruited to the target. These interactions confer sequence-specific target methylation

RNA POLYMERASE D1 (NRPD1), is distinct from that of Pol II and confers the specific catalytic activity of Pol IV. Pol IV-dependent transcripts have not been detected experimentally, but more than 90 % of hc-siRNAs are lost in the *nrpd1* mutant. The current model suggests that Pol IV produces long ssRNAs from regions that spawn siRNAs (Fig. 2). The chromatin remodeling factor CLASSY1 (CLSY1) may promote hc-siRNA biogenesis through Pol IV, while the homeodomain protein SAWADEE HOMEODOMAIN HOMOLOG1 (SHH1) has been implicated in the recruitment of Pol IV to regions that produce hc-siRNAs.

siRNAs are typically generated from the cleavage of long dsRNAs into smaller fragments of a precise length. Hc-siRNAs are produced in this manner, following the conversion of Pol IV-dependent ssRNA transcripts into dsRNAs (Fig. 2). Of the six RNA-dependent RNA polymerase (RDR) homologs encoded in the *Arabidopsis* genome, RDR2 is responsible for this conversion. RDR2 physically interacts with Pol IV and converts Pol IV-transcribed ssRNAs into long dsRNAs with perfect sequence complementarity. Thereafter, the siRNA precursors are diced into 24-nt small RNAs with 2-nt overhangs by DCL3 and methylated at the 2′-OH group of the 3′ terminal nucleotide by HEN1. A mutation disrupting DCL3 function can be compensated by other DCL homologs: in *dcl3*, DCL2 and DCL4 generate hc-siRNAs 22 nt and 21 nt in length, respectively. As previously mentioned, the siRNA precursors processed by DCL3 yield multiple siRNAs rather than a single species from a locus.

After processing, mature hc-siRNAs are loaded into AGO effector proteins belonging to the AGO4 clade (Fig. 2). Although hc-siRNAs are synthesized in the nucleus, their abundance is ten times greater in the cytoplasm than in the nucleus. It has been reported that hc-siRNAs are incorporated into AGO4 in the cytoplasm with the assistance of HEAT-SHOCK PROTEIN90 (HSP90) and then transported to the nucleus. Four of the ten AGO homologs in *Arabidopsis*, AGO4, AGO6, AGO8, and AGO9, belong to the AGO4 clade. All of these except AGO8 function in hc-siRNA-mediated TGS, and AGO4 is the major binding partner of hc-siRNAs. AGO6 and AGO9 have been reported to function as hc-siRNA effector proteins in specific cell types and organs, while *AGO8* may be a pseudogene.

Molecular Mechanism

As the name implies, hc-siRNAs are generated from heterochromatin, and they guide cytosine methylation at the site of their transcription. In *Arabidopsis*, hc-siRNAs play a major role in determining the methylated targets of the RNA-directed DNA methylation (RdDM) pathway (reviewed in Law and Jacobsen 2010; Castel and Martienssen 2013). In RdDM, two classes of ncRNAs and a methyl transferase enzyme are required for target selection and catalytic activity, respectively. A number of subsidiary players have also been implicated in RdDM.

In addition to hc-siRNAs, another type of ncRNA is also generated from RdDM target regions and helps direct cytosine methylation (Fig. 2). The biogenesis of these long ncRNAs requires a second plant-specific RNA polymerase, Pol V (reviewed in Haag and Pikaard 2011). Like Pol IV, Pol V evolved from Pol II, and its largest subunit, NUCLEAR RNA POLYMERASE E1 (NRPE1), is distinct from those of Pol II and Pol IV. The recruitment of Pol V to its targets in the genome is facilitated by the DDR complex, whose major components are DEFECTIVE IN MERISTEM SILENCING3 (DMS3), a structural maintenance of chromosome (SMC) domain protein, DEFECTIVE IN RNA-DIRECTED DNA METHYLATION1 (DRD1), an SNF2-like chromatin remodeling protein; and REQUIRED FOR DNA METHYLATION1 (RDM1), a single-stranded methyl

DNA-binding protein. At a subset of the methylated targets of RdDM, Pol II is responsible for synthesizing the long ncRNA rather than Pol V.

AGO4-loaded hc-siRNAs are recruited to their targets through two distinct interactions: physical contact between AGO4 and WG/GW motif-containing proteins and the binding of hc-siRNAs to Pol V-dependent ncRNAs. The WG/GW motif is an AGO hook motif found in AGO4-binding proteins, and the interaction between AGO4 and WG/GW motif-containing proteins governs downstream molecular events (reviewed in Azevedo et al. 2011). It has been proposed that hc-siRNAs recognize the nascent Pol V-dependent ncRNA through base-pair complementarity and guide silencing of the target DNA in a sequence-specific manner. NRPE1 contains a WG/GW motif in its C-terminal region and is known to physically interact with AGO4. Thus, an hc-siRNA-containing AGO4 protein may be shuttled to the target region through the interaction with NRPE1 while the hc-siRNA recognizes the nascent, long ncRNA generated by Pol V. Another WG/GW motif-containing protein, SUPPRESSOR OF TY INSERTION 5-LIKE (SPT5L)/KOW DOMAIN-CONTAINING TRANSCRIPTION FACTOR1 (KTF1), interacts with both Pol V-dependent ncRNA and AGO4. Thus, SPT5L/KTF1 may stabilize the hybridization of AGO4-loaded hc-siRNAs and Pol V-dependent ncRNAs by bridging AGO4 and Pol V-dependent ncRNAs.

These complex interactions among proteins and ncRNAs must ultimately provide a stable foundation for the recruitment of a methyl transferase enzyme to the hc-siRNA targets. It has been proposed that the binding of AGO4-loaded hc-siRNAs to Pol V-dependent ncRNAs is followed by the release of AGO4. The INVOLVED IN DE NOVO2 (IDN2)/REQUIRED FOR DNA METHYLATION12 (RDM12)-containing complex is critical for the consolidation and integration of factors required for the downstream methylation event. In addition to IDN2/RDM12, the complex contains an IDN2/RDM12 paralog, either FACTOR OF DNA METHYLATION1 (FDM1)/IDN2-LIKE1 (IDNL1)/IDN2 PARALOG1 (IDP1) or FDM2/IDNL2/IDP2. The protein domains of IDN2/RDM12 and its two paralogs in *Arabidopsis* occur in the following order from the N-terminus: a zinc finger for RNA and/or DNA binding, an XS domain for dsRNA recognition, and a coiled-coil domain and XH domain for protein dimerization. The zinc finger domain may bind the methylated target DNA or function as a second RNA binding motif alongside the XS domain. XS domains bind RNAs with 5′ overhangs; in the context of RdDM, the XS domain may stabilize the duplex formed by hc-siRNAs and Pol V-dependent ncRNAs. IDN2/RDM12 dimerizes with FDM1/IDNL1/IDP1 or FDM2/IDNL2/IDP2 through the coiled-coil and XH domains in an antiparallel manner, permitting the recruitment of two distinct hc-siRNAs in a single IDN2/RDM12-containing complex. The IDN2/RDM12-containing complex may anchor the dsRNA duplex formed by hc-siRNAs and long ncRNAs to the target DNA. While the recruitment of DOMAIN REARRANGED METHYLTRANSFERASE2 (DRM2) to the target remains unclear, the RdDM effector protein RDM1 has been proposed to mediate this recruitment based on its ability to physically interact with AGO4 and DRM2. Methylation by DRM2 involves the deposition of a methyl group to the fifth carbon of cytosine residues. Considering these complex interactions in aggregate, hc-siRNAs ultimately guide DRM2, which is responsible for the methylation of the target regions.

After de novo methylation of the siRNA target regions by RdDM, cytosine methylation is maintained by different methyl transferase enzymes based on the sequence context of the methylated cytosines (reviewed in Chan et al. 2005; Law and Jacobsen 2010). The three possible sequence contexts of cytosine are the symmetric CG and CHG contexts and the asymmetric CHH context, where H stands for A, C, or T. Following DNA replication, fully methylated symmetric cytosines are hemi-methylated: the template DNA strand maintains the methylated cytosine, while the newly synthesized DNA strand is unmethylated. Using the methylated cytosine in the template as a methylation cue, CG and CHG cytosines in the nascent DNA are methylated by METHYLTRANSFERASE1 (MET1) and CHROMOMETHYLTRANSFERASE3 (CMT3), respectively. The enzymatic activity of MET1 is facilitated by the chromatin remodeling factor DECREASED IN DNA METHYLATION1(DDM1), and there is some degree of cross talk between CMT3 and another repressive chromatin modifier, SUPPRESSOR OF VARIEGATION3-9 HOMOLOGUE4/KRYPTONITE (SUVH4/KYP), which mediates histone H3K9 methylation. In contrast to CG and CHG residues, CHH residues require hc-siRNAs for methylation maintenance after DNA replication.

Biological Function

Hc-siRNAs and the RdDM pathway determine the methylation landscape in the genome. Genome-wide analyses in *Arabidopsis* have revealed a high degree of overlap among regions containing transposons and repeats, cytosine methylation, and hc-siRNAs in terms of their distribution and abundance. By guiding methylation at repeats and transposons, hc-siRNAs are critical for the maintenance of genome integrity and gene expression regulation.

Transposons and repeats occupy large portions of the *Arabidopsis* genome (reviewed in Slotkin and Martienssen 2007; Lisch 2009). Some of these elements have the ability to jump to other regions or to amplify themselves, which may disrupt functional genes or be detrimental to the organization of the host genome. However, there are several defense mechanisms that protect the genome from the movement or amplification of transposons and repeats. For example, epigenetic modifications such as cytosine methylation help silence and immobilize repeats and transposons. Reduced cytosine methylation and derepression of the expression of transposons and repeats have been observed in loss-of-function RdDM pathway mutants. Similarly, loss of MET1 and DDM1 function leads to a reduction in cytosine methylation and induces amplification and mobilization of some transposons.

Some repeats and transposons are located in intergenic regions, particularly in the promoters of protein-coding genes, and generate 24-nt small RNAs. These mobile elements are regulated by RdDM, and their methylation level affects the expression of nearby genes. As one example, two tandemly arranged repeats are found within the promoter of the gene encoding the homeobox transcription factor FLOWERING WAGENINGEN (FWA) and are highly methylated in most tissue types in wild-type *Arabidopsis*. Accordingly, *FWA* is transcriptionally silenced in these tissues. However, *FWA* is actively transcribed in the endosperm, where an active demethylation mechanism removes methylated cytosines from the repeats of the maternal *FWA* promoter. In one epigenetic *fwa* mutant, the repeats in the promoter region are hypomethylated, resulting in the transcriptional activation of *FWA* and late flowering.

Hc-siRNAs and Reproductive Growth

During the reproductive growth stage of *Arabidopsis*, the embryo and endosperm are characterized by opposing hc-siRNA-mediated DNA methylation programs. This difference in DNA methylation is also observed between the gametes and their supporting cells (reviewed in Law and Jacobsen 2010; Castel and Martienssen 2013). The female gametophyte contains the egg cell, the central cell, and five accessory cells, while the male gametophyte contains two sperm cells and one enlarged vegetative cell. The process of double fertilization in angiosperms involves the fertilization of both the egg cell and the central cell by the two sperm cells, thereby producing the embryo (zygote) and the endosperm, respectively. As companion cells, the central cell and the vegetative cell support the development of their adjacent cells, the egg cell and the sperm cells, respectively. Similarly, the endosperm supports the development of the zygote. The gametes and zygote exhibit a sharp contrast with their respective companion cells and the endosperm in terms of hc-siRNA biogenesis and DNA methylation. While the nursing cells lose CG DNA methylation and exhibit increased expression of transposons and siRNAs, the gametes and zygote maintain their CG methylation and other repressive marks at repeats and transposons. It has been proposed that the decrease in CG methylation in companion cells and the increased transcription from transposons and transcribed RNAs enlarge the siRNA pools. Subsequently, transposon-specific siRNAs may be transported from the companion cells and endosperm to the gametes and zygote to enhance transposon silencing through cytosine methylation. Whereas a germ line is established during the early stages of animal embryogenesis, the differentiation of germ cells from somatic stem cells occurs late in the plant life cycle. In effect, the changes in DNA methylation in the companion cells, whose genetic material is not transferred to the next generation, may help overcome problems resulting from the delayed establishment of the germ cells in plants and ensure the integrity of the parental genomes transferred to the offspring. DDM1 and MET1 are repressed in the vegetative cell and the central cell, respectively, resulting in a global decrease in cytosine methylation. In the companion cells of both gametes, DEMETER (DME), an active demethylase enzyme, further reduces the level of methylation through the demethylation of methylated cytosines. Transcripts are generated from the demethylated transposons and made double stranded. The resulting precursor dsRNAs are further processed into 21- and

24-nt small RNAs in the vegetative and central cells, respectively. Expression of a GFP transgene and an artificial transgene-targeting miRNA in the sperm and the vegetative cell, respectively, leads to the suppression of the GFP expressed in the sperm. A similar outcome is also observed in the egg cell and central cell, indicating that siRNAs produced in one cell can move into an adjacent cell and induce silencing during reproductive growth in *Arabidopsis*.

piRNAs in Animals

A specialized class of small RNAs known as piRNAs is enriched in animal germ line cells and is also present in somatic cells (reviewed in Ishizu et al. 2012; Castel and Martienssen 2013). piRNAs are 25–30 nt in length and are incorporated into an AGO protein belonging to the PIWI clade. These small RNAs guide heterochromatin formation in a manner similar to hc-siRNAs. Unlike small RNAs in plants, however, piRNA biogenesis is Dicer independent and entails a "ping-pong" mechanism of primary biogenesis and amplification; whether one or both of these two processes occur depends on the cell type. *Drosophila* germ cells require three members of the PIWI protein subfamily for piRNA-mediated genome protection: AGO3, Aubergine (AUB), and PIWI. Transposon fragments or relics aggregate into large clusters that generate piRNAs, and the transcription of these piRNA clusters generates long ssRNAs that are antisense to the transposons. During primary processing, these transcripts are processed into antisense piRNAs. These antisense piRNAs exhibit a uridine bias at their 5′ ends and bind AUB or PIWI. During the amplification phase, sense RNAs from transcribed transposons are recognized and cleaved by AUB-loaded antisense piRNAs, generating sense piRNAs. Transposon-specific sense piRNAs exhibit an adenosine bias at the 10th nucleotide from the 5′ terminus and are incorporated into AGO3. Finally, AGO3-loaded sense piRNAs trigger the biogenesis of antisense piRNAs from long piRNA cluster transcripts, and the ping-pong cycle is reiterated. As sense RNAs from transposons are consumed during the ping-pong cycle, the ping-pong pathway promotes the posttranscriptional silencing of targeted transposons. piRNA-guided slicing requires the endonucleolytic activity of the PIWI-family proteins. Furthermore, piRNAs are stabilized by HEN1-mediated methylation. In addition to the repression of transposons at the posttranscriptional level via transcript slicing, piRNAs also guide the deposition of repressive epigenetic marks at homologous chromatin to induce transcriptional silencing.

ta-siRNAs

siRNAs such as hc-siRNAs function at their origin or at homologous regions. Thus, hc-siRNAs act *in cis*, as their sources coincide with their targets. In contrast, *trans*-acting siRNAs (ta-siRNAs) function at loci distinct from the site of their biogenesis. As their name indicates, ta-siRNAs act *in trans*, and their regulatory mechanism is similar to that of miRNAs.

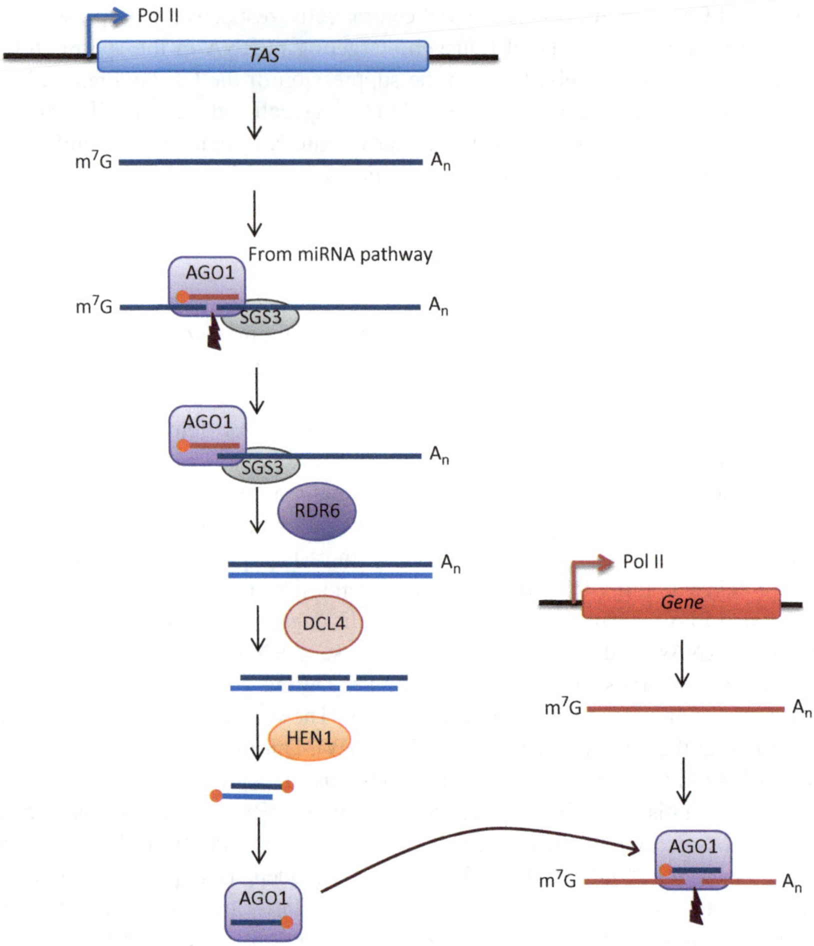

Fig. 3 **ta-siRNA biogenesis and silencing mechanism.** *TAS* transcripts are generated by Pol II and cleaved by miRNA-associated AGO1. The cleaved 3′ fragments are protected from degradation by miRNA-containing complexes, RISC and SGS3, and then made double stranded by RDR6. The double-stranded RNAs are processed into 21-nt siRNAs by DCL4 and methylated by HEN1. ta-siRNAs are loaded into AGO1 and regulate their targets in the same manner as miRNA-mediated target repression

Biogenesis

Arabidopsis ta-siRNAs are small regulatory RNAs 21 nt in length whose precursors are generated from *TAS* loci (reviewed in Allen and Howell 2010). Their biogenesis is clearly distinct from that of other small RNAs: transcripts generated from *TAS* loci are targets of miRNA-directed cleavage, and the cleavage products serve as the sources for the biogenesis of secondary siRNAs (Fig. 3). The biogenesis of the secondary siRNAs is similar to that of hc-siRNAs, following the conversion of the single-stranded cleavage products into dsRNA precursors.

There are four groups of *TAS* genes in *Arabidopsis*. *TAS1* and *TAS3* are each encoded by three isoforms: *TAS1a*, *TAS1b*, and *TAS1c* for *TAS1* and *TAS3a*, *TAS3b*, and *TAS3c* for *TAS3*. *TAS2* and *TAS4* are transcribed from single loci. *TAS* transcripts are non-protein-coding RNAs and contain RNA sequences that are recognized and regulated by miRNAs. miR173 targets both *TAS1* and *TAS2*, and miR390 and miR828 target *TAS3* and *TAS4*, respectively. *TAS* transcripts are subject to miRNA-directed cleavage carried out by miRNA-bound AGO1 or AGO7 (Fig. 3). The stabilization of the cleavage products involves SUPPRESSOR OF GENE SILENCING3 (SGS3), and the ssRNA products are made double stranded by RDR6. The resulting long dsRNA precursors are processed into 21-nt small RNAs by DCL4 in a precisely phased manner: DCL4 successively cleaves the dsRNA precursor beginning at one end and generating multiple small RNAs at 21-nt intervals. Although diverse small RNAs are generated from a single dsRNA precursor, only some of the small RNAs are stable and incorporated into AGO1. Like miRNAs, ta-siRNAs are methylated by HEN1, and their downstream function is similar to that of miRNAs.

Although hundreds of miRNAs have been identified in *Arabidopsis*, only three miRNAs (miR173, miR390, and miR828) are known to initiate the biogenesis of ta-siRNAs. Several studies have revealed specific factors that influence the initiation of secondary siRNA biogenesis: a specialized AGO protein, the length of the miRNAs, the structure of the miRNA/miRNA* duplex, the position of the miRNA binding site within the target RNA, and the degree of sequence complementarity between the miRNAs and their targets. Among the ten AGO homologs in *Arabidopsis*, AGO7 is the only family member that can generate ta-siRNAs from *TAS3*. AGO7 exclusively binds miR390, which has two target sites in the *TAS3* transcript. The 3′ miRNA-target site has nearly perfect complementarity with miR390, and AGO7-mediated cleavage at this site triggers ta-siRNA biogenesis. However, ta-siRNA biogenesis at the *TAS3* locus also requires the 5′ target site of miR390 in the *TAS3* transcript, as described by the two-hit trigger model. Although the 5′ miR390 target site is resistant to AGO7 cleavage, AGO7 must be recruited to both the 5′ and 3′ target sites to initiate ta-siRNA biogenesis from *TAS3* transcripts.

miRNA length is another determinant of ta-siRNA biogenesis. While the majority of miRNAs are 21 nt in length, ta-siRNA-generating miRNAs are 22 nt in length with the exception of miR390, which is 21 nt in length. In a transient expression study in *Nicotiana benthamiana*, artificially engineered miRNAs (miR173, miR472, and miR828) 21 or 22 nt in length were tested for their ability to trigger secondary siRNA biogenesis from a co-infiltrated target construct. Only the 22-nt forms of the miRNAs successfully triggered secondary siRNA biogenesis.

The asymmetric structure of the miRNA/miRNA* duplex has also been found to affect the initiation of ta-siRNA biogenesis. In a transient system similar to that described in the preceding paragraph, four artificial miR173/miR173* duplexes were examined: 22/21-, 21/22-, and 21/21-nt duplexes with asymmetric bulges along with a symmetric 21/21-nt duplex. The experiment showed that duplexes with asymmetric bulges could generate secondary siRNAs, regardless of their length. High-throughput sequencing techniques have identified more pairs of miRNAs and mRNAs that generate secondary siRNAs. When miRNAs were found to induce the production of secondary siRNAs from target mRNAs, the miRNA/miRNA* duplexes were found to contain a 22-nt strand and a 21-nt strand (i.e., either 21/22- or 22/21-nt duplexes). Additionally, the miRNA/miRNA* duplexes tended to be asymmetrically bulged in terms of their structure. It has been proposed that the AGO1-containing RNA-induced silencing complex (RISC) induces either target repression or secondary small RNA biogenesis based on the structure of the miRNA/miRNA* duplex.

Two additional factors that influence the initiation of ta-siRNA biogenesis are the location of the miRNA binding site in the target transcript and the degree of sequence complementarity between the miRNA and its target site. In a study in which ta-siRNAs were generated in plants from a synthetic GFP reporter by miR173, the efficiency of ta-siRNA generation was maximal when the miR173-binding site was located immediately after the stop codon. When premature stop codons were introduced further upstream of the miR173-binding site, the abundance of ta-siRNAs decreased while the abundance of longer transcripts from the transgene increased. These observations suggest a link between translation by ribosomes and *TAS* precursor processing. Finally, reduced complementarity at the 3′ end of synthetic miR173 has been shown to abolish the generation of ta-siRNAs, while reduced complementarity at the 5′ end has a less detrimental effect.

Molecular Mechanism and Biological Function

As observed for miRNAs, ta-siRNAs regulate the expression of their target genes *in trans*. Furthermore, ta-siRNAs are 21 nt in length, associate with AGO1, and direct PTGS of their targets. *TAS1* ta-siRNAs target several uncharacterized genes and multiple mRNAs encoding pentatricopeptide repeat (PPR) proteins. *TAS2* ta-siRNAs similarly target *PPR* mRNAs. *PPR* genes are commonly found in eukaryotes and appear to have undergone a rapid expansion in plants; the *Arabidopsis* genome, for example, encodes over four hundred *PPR* genes. Some PPR proteins bind RNA and are predicted to regulate gene expression through RNA processing, editing, stability, and translation in mitochondria and chloroplasts. Although *PPR* genes are targeted by *TAS1* and *TAS2* ta-siRNAs, the biological relevance of these regulatory interactions remains unclear. Interestingly, *PPR* genes are targeted by both ta-siRNAs and miRNAs, and some transcripts contain multiple small RNA binding sites. Considering the large number of *PPR* genes in plants, ta-siRNAs and small RNAs may serve to dampen the detrimental effects caused by the rapid expansion of gene families.

TAS3 ta-siRNAs modulate auxin signaling networks by targeting *AUXIN RESPONSE FACTOR2* (*ARF2*), *ARF3*, and *ARF4* and are thus referred to as tasiR-ARF. Auxin is a major plant hormone and is involved in every phase of plant development. Although auxin may affect growth and development through numerous mechanisms, the mechanism that is best understood is auxin-mediated regulation of gene expression through the ARF and Aux/IAA proteins. In the basal condition, ARF proteins are bound and repressed by Aux/IAA proteins, and auxin-responsive genes are not expressed. When the level of auxin is increased, Aux/IAA is degraded by the ubiquitin-mediated proteasome pathway, and ARF proteins are released from Aux/IAA repression. ARFs recognize auxin-responsive elements in the promoters of downstream genes and activate their expression. Two of the diverse developmental processes affected by tasiR-ARF are phase transition and leaf pattern formation. When the tasiR-ARF binding site in *ARF3* is mutated to make *ARF3* resistant to tasiR-ARF, juvenile plants enter the adult phase prematurely, which is similarly observed in ta-siRNA biogenesis mutants such as *ago7*, *sgs3*, and *rdr6*. Thus, tasiR-ARF suppresses the juvenile-to-adult phase transition. In terms of leaf development, tasiR-ARF is expressed in the adaxial (upper) leaf region, and its movement to the abaxial (lower) region generates a concentration gradient of tasiR-ARF. *ARF3* is expressed throughout the leaf primordia, and *ARF4*RNA is detected in abaxial leaf tissue. Due to the higher concentration of tasiR-ARF in the adaxial region, ARF activity is higher in or restricted to the abaxial leaf region. Thus, the pattern of ARF activity across the adaxial and abaxial regions contrasts that of tasiR-ARF accumulation, and these distinct gradients are critical for polarized leaf pattern formation.

Lastly, *TAS4* ta-siRNAs are predicted to repress genes encoding MYB transcription factors. However, the *TAS4* ta-siRNAs were the last to be identified, likely owing to their low abundance, and their function is currently unknown.

Exogenous siRNAs

In addition to endogenously produced small RNAs, plants also contain small RNAs that derive from exogenous sources. In fact, small RNAs were first detected in plants that were infected with viruses and plants harboring transgenes. This pioneering discovery revealed the first clues that small RNAs play an important role in the repression of viruses and transgenes and revolutionized the field of RNA silencing.

Viral siRNAs (viRNAs)

Plants have adopted a small RNA-mediated repression mechanism to combat viral infection (reviewed in Ding and Voinnet 2007). After infection, plant DCL enzymes generate primary viRNAs from viral dsRNAs, which are produced by viral RDR during replication, by intramolecular hybridization, or by convergent transcription. Primary viRNAs elicit the biogenesis of secondary viRNAs in a manner similar to that of ta-siRNA biogenesis: viral target RNAs are cleaved, the cleavage products are made double stranded by plant RDRs, and DCL enzymes cleave the newly generated double-stranded precursors. Amplified viRNAs are incorporated into AGO proteins and repress the virus through PTGS. Multiple DCLs, RDRs, and AGOs in host plants have redundant functions, work in tandem, and/or perform a specialized function to defend plants against viral infection. The activities of these proteins also depend on the type of viral infection.

In response to the antiviral defense of the host plant, viruses have also developed mechanisms to counteract the host response. Numerous viruses encode viral suppressors of RNA silencing (VSRs) that oppose the repressive action of viRNA-mediated silencing in the host (reviewed in Burgyán and Havelda 2011). Specifically, VSRs intercept viral dsRNAs or silencing factors generated by the host plant. P19 from *Cymbidium ringspot virus* and P21 from *beet yellows virus* sequester short dsRNAs, effectively disrupting RISC assembly with viRNAs in the host. Other VSRs are capable of binding AGO proteins: 2b from *cucumber mosaic virus*, P0 from *beet polerovirus*, P1 from *sweet potato mild mottle virus*, and P38 from *turnip crinkle virus*. 2b inhibits the slicing activity of AGO1 in preassembled RISC. P0 contains a minimal F-box motif that may induce AGO1 degradation. P1 and P38 contain an AGO-hook GW/WG motif and may therefore compete with endogenous AGO-binding proteins in plants (Azevedo et al. 2011). Other components of the plant silencing machinery are also targeted by viruses (reviewed in Burgyán and Havelda 2011). For example, the binding of V2 from *tomato yellow leaf curl virus* to SGS3 compromises RDR6-mediated secondary viRNA biogenesis. Additionally, HC-Pro from *zucchini yellow mosaic virus* disrupts the methylation of small RNAs by HEN1.

siRNAs from Transgenes

Early studies in which plants were transformed with sense transgenes revealed the suppression of both transgenes and endogenous homologous genes in several transgenic lines, and the silencing phenomenon was termed co-suppression. Subsequent studies revealed that transgene-specific small RNAs accumulate in silenced plants and that proteins required for small RNA biogenesis and action are also involved in transgene silencing. In cases of transgene silencing, the ssRNA transcripts generated from the transgene are recognized by the plant machinery as aberrant and made double stranded by RDR6. The dsRNA subsequently triggers downstream events, including primary and secondary siRNA biogenesis. As a result, both the transgene and endogenous homologous genes are targeted for silencing.

Small RNA Turnover

Consistent with the critical roles of small RNAs in diverse biological processes, the abundance of small RNAs is also precisely regulated. Disrupting the homeostasis of small RNAs detrimentally affects developmental and metabolic processes. Because the abundance of small RNAs is affected by both internal and external signals, the balanced expression of small RNAs requires a precise regulatory mechanism. In plants, small RNA biogenesis and turnover are the critical phases for regulating the dynamics of small RNA populations (reviewed in Ji and Chen 2012).

The methylation of small RNAs during biogenesis is crucial for their stabilization. Small RNAs in *Arabidopsis* are methylated at the 2′-OH of the 3′ terminal ribose by HEN1. In *hen1* mutants, the abundance of small RNAs is dramatically reduced, and the residual small RNAs are tailed or trimmed. High-throughput sequencing of the small RNA population in *hen1* further revealed that the residual small RNAs are identical at their 5′ ends but heterogeneous at their 3′ ends. Specifically, the small RNAs were found to have oligonucleotide tails 1–7 nt in length, with a predominant enrichment of uridine among the four nucleotides. Furthermore, truncation from the 3′ terminus was observed for both intact and uridylated small RNAs in the *hen1* mutant. Thus, HEN1-mediated methylation at the 3′ end of small RNAs ultimately inhibits their degradation.

In *Arabidopsis*, the SMALL RNA-DEGRADING NUCLEASE (SDN) family of 3′–5′ exonucleases is responsible for small RNA degradation. When multiple *SDN* genes are simultaneously knocked down, increased miRNA levels and pleiotropic developmental defects are observed. SDN1 specifically degrades 17–27-nt single-stranded small RNAs, and its activity is partially inhibited by 2′-*O*-methylation at the 3′ end of small RNAs. Based on these observations, the removal of the 3′ most nucleotide by SDNs may be rate limiting and probably requires other assistant proteins or the combined activity of multiple SDNs.

From both forward and reverse genetic studies of the *hen1* mutant, HEN1 SUPPRESSOR1 (HESO1) was found to poly-uridylate small RNAs in *hen1* mutants. In contrast to the protective function of methylation, uridine tails at the 3′ end of small RNAs make miRNAs more susceptible to 3′–5′ exonuclease activity. Consistent with the hypothesis that a defect in uridylation activity should rescue the loss of methylation in *hen1* (i.e., unmethylated small RNAs that do not undergo uridylation should be less susceptible to 3′–5′ exonuclease activity), a mutation in *HESO1* suppresses the morphological defects of *hen1* mutants. Compared to *hen1*, miRNA levels are increased in *hen1 heso1*. However, tailed and trimmed miRNAs are still observed in the double mutant. Based on in vitro analysis, HESO1 has terminal nucleotidyl transferase activity with a preference for uridine substrates, and HESO1 function is completely inhibited by 2′-*O*-methylation at the 3′ end of small RNA substrates. High-throughput small RNA data for the *hen1 heso1* double mutant reveal shorter uridine tails in the double mutant compared to *hen1*, which further suggests that HESO1 is partially responsible for uridylation in the *hen1* mutant.

In addition to HEN1, SDN exonucleases, and HESO1, long RNA molecules may influence the rate of degradation of specific small RNAs. In a technical analysis of target mimicry, a short tandem target mimic (STTM), composed of two short sequences mimicking small RNA binding sites tandemly arrayed with an optimal spacing between them, was found to reduce the abundance of miRNAs whose binding sites were mimicked by the STTM. Interestingly, the reduction in miRNA abundance was dependent on SDN1 and SDN2 activity. Similarly, although less effectively, other artificial target mimicry transgenes led to reductions in the levels of cognate miRNAs. This suggests that target transcripts, especially those that cannot be cleaved by miRNAs, impact the stability of miRNAs. This raises the question of whether such targets exist naturally.

In *Arabidopsis*, miR399 recognizes two target RNAs: the mRNA transcript corresponding to the E2 ubiquitin conjugation enzyme PHOSPHATASE (PHO2) and the *INDUCED BY PHOSPHATASE STARVATION1* (*IPS1*) ncRNA. miR399 is induced under phosphate (Pi) starvation conditions and represses the activity of PHO2 through mRNA cleavage as an adaptive response that alters the metabolism of Pi in plants. In general, signaling cascades triggered by a certain event or treatment are eventually attenuated, and the recovery of steady expression levels facilitates the response to a prolonged stimulus. In a similar way, *PHO2* is temporarily silenced by miR399 under Pi starvation conditions but eventually achieves a steady level of activity, which is mediated by target mimicry. Long *IPS1* ncRNAs are also induced by Pi deficiency and sequester miR399 from *PHO2* mRNAs. Unlike *PHO2* mRNA, which is subject to miRNA-directed cleavage, *IPS1* ncRNAs are bound but not sliced by miR399 due to a mismatch at the cleavage site. Although *IPS1* ncRNAs do not alter the in vivo abundance of miR399, they suppress the effect of miR399 on *PHO2*.

A recent study identified many *IPS1*-like intergenic long ncRNAs that can pair with other miRNAs. Overexpression of some of the long ncRNAs led to a decrease in the abundance of the cognate miRNAs, raising the possibility that long ncRNAs regulate the stability of specific miRNAs in vivo.

Future Directions

While the overall framework of miRNA biogenesis is relatively well established, many aspects of the regulation of miRNA biogenesis remain unclear. The abundance of mature miRNAs is regulated by Pol II-mediated transcriptional regulation and during the processing of pri-miRNAs to mature miRNAs. However, it is also possible that the processing of miRNA precursors is also directly affected by the endogenous or exogenous signals integrated by Pol II. Furthermore, unique factors may differentially regulate certain miRNA species during the process of miRNA maturation. In terms of the activities directed by miRNAs, the molecular mechanisms of mRNA cleavage and translational inhibition require further study. It has been proposed that the extent of sequence complementarity between miRNAs and their targets dictates the mode of repression by miRNAs. However, this is unlikely because miRNAs with a high degree of sequence complementarity to their targets have also been shown to act via translational repression. In fact, the two modes of action may occur simultaneously for a given miRNA-target pair. The degree of miRNA-target complementarity that is required for translational repression has not been experimentally determined. If less extensive base pairing is sufficient to induce translational inhibition as observed in animals, the current views of the regulatory networks between miRNAs and their targets would need to be reevaluated. The translational repression activity of plant miRNAs also needs to be dissected at the mechanistic level. For instance, it is unknown how and at what step (e.g., ribosome loading, elongation, or termination) miRNAs inhibit protein synthesis carried out by ribosomes.

Through intensive genetic studies, the key players in hc-siRNA biogenesis and DNA methylation have been identified. Additionally, high-throughput methylome analysis has provided a wealth of information about targets methylated by RdDM at the nucleotide level. Nevertheless, major aspects of hc-siRNA biogenesis and cytosine methylation are not understood or require further experimental evidence. For example, although Pol IV is essential for the biogenesis of hc-siRNAs, Pol IV-dependent transcripts have not yet been detected. How Pol IV recognizes, and is recruited to, the promoters of these transcripts is also not known. The recruitment of DRM2 to target regions is known to be mediated by small RNAs and Pol V-dependent transcripts, but the underlying molecular mechanism remains to be elucidated. Along with cytosine methylation, there are other epigenetic modifications that undoubtedly contribute to TGS, including histone modification, histone variants, chromatin condensation, and higher-order chromatin structures. Future studies will need to establish the relationships between these different types of modification and address how cross talk among them governs the epigenetic landscape.

Although factors that favor ta-siRNA biogenesis have been uncovered, the biological function of ta-siRNAs and their targets remains enigmatic. Particularly, *PPR* genes are abundant in the *Arabidopsis* genome, but the underlying cause of the rapid expansion of this gene family is unclear, as is the functional relevance of the regulation of *PPR* genes by ta-siRNAs.

Mature small RNAs are loaded into AGO effector proteins to direct silencing activity, and the association with AGO proteins may protect small RNAs from harmful enzymatic activity, such as degradation by SDN1 or uridylation by HESO1. The molecular mechanism by which small RNAs are dislodged from AGO proteins and subsequently degraded is not well characterized. Another possibility is that small RNAs may be degraded while they are associated with AGO proteins. Both uridylation and 3′ truncation mechanisms that affect small RNAs warrant further study, not only in terms of the underlying molecular events but also with respect to whether and how these mechanisms are orchestrated in tandem. The fact that loss of HESO1 function reduces but does not eliminate the uridylation of miRNAs suggests the existence of other enzymes with overlapping functions. Moreover, there may be regulatory factors that determine the rate of degradation and sequester or degrade specific small RNAs in response to a signal or stress. Lastly, recent findings about small RNA turnover induced by small RNA target mimics challenge the current understanding of SDN exonuclease activity. Further study is required to address how SDN enzymes, which specifically degrade single-stranded small RNAs, may also be involved in the degradation of sequestered or bound small RNAs.

References

Allen E, Howell MD. miRNAs in the biogenesis of trans-acting siRNAs in higher plants. Semin Cell Dev Biol. 2010;21(8):798–804.

Azevedo J, Cooke R, Lagrange T. Taking RISCs with ago hookers. Curr Opin Plant Biol. 2011;14 (5):594–600.

Bazzini AA, Lee MT, Giraldez AJ. Ribosome profiling shows that miR-430 reduces translation before causing mRNA decay in zebrafish. Science. 2012;336(6078):233–7.

Burgyán J, Havelda Z. Viral suppressors of RNA silencing. Trends Plant Sci. 2011;16(5):265–72.

Castel SE, Martienssen RA. RNA interference in the nucleus: roles for small RNAs in transcription, epigenetics and beyond. Nat Rev Genet. 2013;14(2):100–12.

Chan SW, Henderson IR, Jacobsen SE. Gardening the genome: DNA methylation in *Arabidopsis thaliana*. Nat Rev Genet. 2005;6:351–60.

Chen X. Small RNAs and their roles in plant development. Annu Rev Cell Dev Biol. 2009;25 (1):21–44.

Ding S-W, Voinnet O. Antiviral immunity directed by small RNAs. Cell. 2007;130(3):413–26.

Djuranovic S, Nahvi A, Green R. miRNA-mediated gene silencing by translational repression followed by mRNA deadenylation and decay. Science. 2012;336(6078):237–40.

Haag JR, Pikaard CS. Multisubunit RNA polymerases IV and V: purveyors of non-coding RNA for plant gene silencing. Nat Rev Mol Cell Biol. 2011;12(8):483–92.

Ishizu H, Siomi H, Siomi MC. Biology of PIWI-interacting RNAs: new insights into biogenesis and function inside and outside of germlines. Genes Dev. 2012;26(21):2361–73.

Ji L, Chen X. Regulation of small RNA stability: methylation and beyond. Cell Res. 2012;22 (4):624–36.

Kim VN, Han J, Siomi MC. Biogenesis of small RNAs in animals. Nat Rev Mol Cell Biol. 2009;10 (2):126–39.

Kozomara A, Griffiths-Jones S. miRBase: integrating microRNA annotation and deep-sequencing data. Nucleic Acids Res. 2011;39 suppl 1:D152–7.

Krol J, Loedige I, Filipowicz W. The widespread regulation of microRNA biogenesis, function and decay. Nat Rev Genet. 2010;11(9):597–610.

Law JA, Jacobsen SE. Establishing, maintaining and modifying DNA methylation patterns in plants and animals. Nat Rev Genet. 2010;11(3):204–20.

Lisch D. Epigenetic regulation of transposable elements in plants. Annu Rev Plant Biol. 2009;60 (1):43–66.

Mosher RA, Schwach F, Studholme D, Baulcombe DC. PolIVb influences RNA-directed DNA methylation independently of its role in siRNA biogenesis. Proc Natl Acad Sci. 2008;105 (8):3145–50.

Slotkin RK, Martienssen R. Transposable elements and the epigenetic regulation of the genome. Nat Rev Genet. 2007;8(4):272–85.

Zhang X, Henderson IR, Lu C, Green PJ, Jacobsen SE. Role of RNA polymerase IV in plant small RNA metabolism. Proc Natl Acad Sci. 2007;104(11):4536–41.

Further Readings

Allen E, Xie Z, Gustafson AM, Sung G-H, Spatafora JW, Carrington JC. Evolution of microRNA genes by inverted duplication of target gene sequences in *Arabidopsis thaliana*. Nat Genet. 2004;36(12):1282–90.

Allen E, Xie Z, Gustafson AM, Carrington JC. microRNA-directed phasing during trans-acting siRNA biogenesis in plants. Cell. 2005;121(2):207–21.

Aravin AA, Hannon GJ, Brennecke J. The Piwi-piRNA pathway provides an adaptive defense in the transposon arms race. Science. 2007;318:761–4.

Aukerman MJ, Sakai H. Regulation of flowering time and floral organ identity by a microRNA and its APETALA2-like target genes. Plant Cell. 2003;15(11):2730–41.
Ausin I, Mockler TC, Chory J, Jacobsen SE. IDN1 and IDN2 are required for de novo DNA methylation in *Arabidopsis thaliana*. Nat Struct Mol Biol. 2009;16(12):1325–7.
Ausin I, Greenberg MVC, Simanshu DK, Hale CJ, Vashisht AA, Simon SA, Lee T-F, Feng S, Española SD, Meyers BC, Wohlschlegel JA, Patel DJ, Jacobsen SE. INVOLVED IN DE NOVO 2-containing complex involved in RNA-directed DNA methylation in *Arabidopsis*. Proc Natl Acad Sci. 2012;109(22):8374–81.
Axtell MJ. Evolution of microRNAs and their targets: are all microRNAs biologically relevant? Biochim Biophy Acta (BBA) Gene Regul Mech. 2008;1779(11):725–34.
Azevedo J, Garcia D, Pontier D, Ohnesorge S, Yu A, Garcia S, Braun L, Bergdoll M, Hakimi MA, Lagrange T, Voinnet O. Argonaute quenching and global changes in Dicer homeostasis caused by a pathogen-encoded GW repeat protein. Genes Dev. 2010;24(9):904–15.
Bartel DP. MicroRNAs: target recognition and regulatory functions. Cell. 2009;136(2):215–33.
Baumberger N, Baulcombe DC. *Arabidopsis* ARGONAUTE1 is an RNA Slicer that selectively recruits microRNAs and short interfering RNAs. Proc Natl Acad Sci USA. 2005;102 (33):11928–33.
Ben Amor B, Wirth S, Merchan F, Laporte P, d'Aubenton-Carafa Y, Hirsch J, Maizel A, Mallory A, Lucas A, Deragon JM, Vaucheret H, Thermes C, Crespi M. Novel long non-protein coding RNAs involved in *Arabidopsis* differentiation and stress responses. Genome Res. 2009;19(1):57–69.
Brodersen P, Voinnet O. Revisiting the principles of microRNA target recognition and mode of action. Nat Rev Mol Cell Biol. 2009;10(2):141–8.
Brodersen P, Sakvarelidze-Achard L, Bruun-Rasmussen M, Dunoyer P, Yamamoto YY, Sieburth L, Voinnet O. Widespread translational inhibition by plant miRNAs and siRNAs. Science. 2008;320(5880):1185–90.
Calarco JP, Borges F, Donoghue MT, Van Ex F, Jullien PE, Lopes T, Gardner R, Berger F, Feijó JA, Becker JD, Martienssen RA. Reprogramming of dna methylation in pollen guides epigenetic inheritance via small RNA. Cell. 2012;151(1):194–205.
Cao X, Aufsatz W, Zilberman D, Mette MF, Huang MS, Matzke M, Jacobsen SE. Role of the DRM and CMT3 methyltransferases in RNA-directed DNA methylation. Curr Biol. 2003;13 (24):2212–7.
Chapman EJ, Estelle M. Mechanism of auxin-regulated gene expression in plants. Annu Rev Genet. 2009;43(1):265–85.
Chen X. A microRNA as a translational repressor of APETALA2 in *Arabidopsis* flower development. Science. 2004;303(5666):2022–5.
Chen H-M, Li Y-H, Wu S-H. Bioinformatic prediction and experimental validation of a microRNA-directed tandem trans-acting siRNA cascade in *Arabidopsis*. Proc Natl Acad Sci. 2007;104(9):3318–23.
Chen H-M, Chen L-T, Patel K, Li Y-H, Baulcombe DC, Wu S-H. 22-nucleotide RNAs trigger secondary siRNA biogenesis in plants. Proc Natl Acad Sci. 2010;107(34):15269–74.
Chitwood DH, Timmermans MCP. Target mimics modulate miRNAs. Nat Genet. 2007;39 (8):935–6.
Chitwood DH, Nogueira FTS, Howell MD, Montgomery TA, Carrington JC, Timmermans MCP. Pattern formation via small RNA mobility. Genes Dev. 2009;23(5):549–54.
Choi Y, Gehring M, Johnson L, Hannon M, Harada JJ, Goldberg RB, Jacobsen SE, Fischer RL. DEMETER, a DNA glycosylase domain protein, is required for endosperm gene imprinting and seed viability in *Arabidopsis*. Cell. 2002;110(1):33–42.
Cokus SJ, Feng S, Zhang X, Chen Z, Merriman B, Haudenschild CD, Pradhan S, Nelson SF, Pellegrini M, Jacobsen SE. Shotgun bisulphite sequencing of the *Arabidopsis* genome reveals DNA methylation patterning. Nature. 2008;452(7184):215–9.
Cuperus J, Carbonell A, Fahlgren N, Garcia-Ruiz H, Burke R, Takeda A, Sullivan C, Gilbert S, Montgomery T, Carrington J. Unique functionality of 22-nt miRNAs in triggering RDR6

dependent siRNA biogenesis from target transcripts in *Arabidopsis*. Nat Struct Mol Biol. 2010;17:997–1003.

Cuperus JT, Fahlgren N, Carrington JC. Evolution and functional diversification of MIRNA genes. Plant Cell Online. 2011;23(2):431–42.

Dong Z, Han M-H, Fedoroff N. The RNA-binding proteins HYL1 and SE promote accurate in vitro processing of pri-miRNA by DCL1. Proc Natl Acad Sci. 2008;105(29):9970–5.

Du J, Zhong X, Bernatavichute YV, Stroud H, Feng S, Caro E, Vashisht AA, Terragni J, Chin HG, Tu A, Hetzel J, Wohlschlegel JA, Pradhan S, Patel DJ, Jacobsen SE. Dual binding of chromomethylase domains to H3K9me2-containing nucleosomes directs DNA methylation in plants. Cell. 2012;151(1):167–80.

El-Shami M, Pontier D, Lahmy S, Braun L, Picart C, Vega D, Hakimi M-A, Jacobsen SE, Cooke R, Lagrange T. Reiterated WG/GW motifs form functionally and evolutionarily conserved ARGONAUTE-binding platforms in RNAi-related components. Genes Dev. 2007;21(20):2539–44.

Eun C, Lorkovic ZJ, Naumann U, Long Q, Havecker ER, Simon SA, Meyers BC, Matzke AJM, Matzke M. AGO6 functions in RNA-mediated transcriptional gene silencing in shoot and root meristems in *Arabidopsis thaliana*. PLoS ONE. 2011;6(10):e25730.

Fahlgren N, Montgomery TA, Howell MD, Allen E, Dvorak SK, Alexander AL, Carrington JC. Regulation of AUXIN RESPONSE FACTOR3 by TAS3 ta-siRNA affects developmental timing and patterning in *Arabidopsis*. Curr Biol: CB. 2006;16(9):939–44.

Fahlgren N, Howell MD, Kasschau KD, Chapman EJ, Sullivan CM, Cumbie JS, Givan SA, Law TF, Grant SR, Dangl JL, Carrington JC. High-throughput sequencing of *Arabidopsis* microRNAs: evidence for frequent birth and death of *MIRNA* genes. PLoS ONE. 2007;2(2):e219.

Fenselau de Felippes F, Schneeberger K, Dezulian T, Huson DH, Weigel D. Evolution of *Arabidopsis thaliana* microRNAs from random sequences. RNA. 2008;14(12):2455–9.

Filipowicz W, Bhattacharyya SN, Sonenberg N. Mechanisms of post-transcriptional regulation by microRNAs: are the answers in sight? Nat Rev Genet. 2008;9(2):102–14.

Franco-Zorrilla JM, Valli A, Todesco M, Mateos I, Puga MI, Rubio-Somoza I, Leyva A, Weigel D, Garcia JA, Paz-Ares J. Target mimicry provides a new mechanism for regulation of microRNA activity. Nat Genet. 2007;39(8):1033–7.

Fusaro AF, Correa RL, Nakasugi K, Jackson C, Kawchuk L, Vaslin MFS, Waterhouse PM. The Enamovirus P0 protein is a silencing suppressor which inhibits local and systemic RNA silencing through AGO1 degradation. Virology. 2012;426(2):178–87.

Gandikota M, Birkenbihl RP, Höhmann S, Cardon GH, Saedler H, Huijser P. The miRNA156/157 recognition element in the 3′ UTR of the Arabidopsis SBP box gene SPL3 prevents early flowering by translational inhibition in seedlings. Plant J. 2007;49(4):683–93.

Gao Z, Liu H-L, Daxinger L, Pontes O, He X, Qian W, Lin H, Xie M, Lorkovic ZJ, Zhang S, Miki D, Zhan X, Pontier D, Lagrange T, Jin H, Matzke AJM, Matzke M, Pikaard CS, Zhu J-K. An RNA polymerase II- and AGO4-associated protein acts in RNA-directed DNA methylation. Nature. 2010;465(7294):106–9.

Gasciolli V, Mallory AC, Bartel DP, Vaucheret H. Partially redundant functions of *Arabidopsis* DICER-like enzymes and a role for DCL4 in producing trans-acting siRNAs. Curr Biol: CB. 2005;15(16):1494–500.

German MA, Pillay M, Jeong D-H, Hetawal A, Luo S, Janardhanan P, Kannan V, Rymarquis LA, Nobuta K, German R, De Paoli E, Lu C, Schroth G, Meyers BC, Green PJ. Global identification of microRNA-target RNA pairs by parallel analysis of RNA ends. Nat Biotech. 2008;26(8):941–6.

Ghildiyal M, Zamore PD. Small silencing RNAs: an expanding universe. Nat Rev Genet. 2009;10(2):94–108.

Giner A, Lakatos L, García-Chapa M, López-Moya JJ, Burgyán J. Viral protein inhibits RISC activity by Argonaute binding through conserved WG/GW motifs. PLoS Pathog. 2010;6(7):e1000996.

Glick E, Zrachya A, Levy Y, Mett A, Gidoni D, Belausov E, Citovsky V, Gafni Y. Interaction with host SGS3 is required for suppression of RNA silencing by tomato yellow leaf curl virus V2 protein. Proc Natl Acad Sci. 2008;105(1):157–61.

Gutierrez-Marcos JF, Dickinson HG. Epigenetic reprogramming in plant reproductive lineages. Plant Cell Physiol. 2012;53(5):817–23.

Hamilton AJ, Baulcombe DC. A species of small antisense RNA in posttranscriptional gene silencing in plants. Science. 1999;286(5441):950–2.

Han M-H, Goud S, Song L, Fedoroff N. The *Arabidopsis* double-stranded RNA-binding protein HYL1 plays a role in microRNA-mediated gene regulation. Proc Natl Acad Sci USA. 2004;101 (4):1093–8.

He X-J, Hsu Y-F, Zhu S, Wierzbicki AT, Pontes O, Pikaard CS, Liu H-L, Wang C-S, Jin H, Zhu J-K. An effector of RNA-directed DNA methylation in *Arabidopsis* is an ARGONAUTE 4- and RNA-binding protein. Cell. 2009;137(3):498–508.

Howell MD, Fahlgren N, Chapman EJ, Cumbie JS, Sullivan CM, Givan SA, Kasschau KD, Carrington JC. Genome-wide analysis of the RNA-DEPENDENT RNA POLYMERASE6/ DICER-LIKE4 pathway in *Arabidopsis* reveals dependency on miRNA- and tasiRNA-directed targeting. Plant Cell Online. 2007;19(3):926–42.

Hsieh T-F, Ibarra CA, Silva P, Zemach A, Eshed-Williams L, Fischer RL, Zilberman D. Genome-wide demethylation of *Arabidopsis* endosperm. Science. 2009;324(5933):1451–4.

Huang L, Jones AME, Searle I, Patel K, Vogler H, Hubner NC, Baulcombe DC. An atypical RNA polymerase involved in RNA silencing shares small subunits with RNA polymerase II. Nat Struct Mol Biol. 2009;16(1):91–3.

Hutvagner G, Simard MJ. Argonaute proteins: key players in RNA silencing. Nat Rev Mol Cell Biol. 2008;9:22–32.

Ibarra CA, Feng X, Schoft VK, Hsieh T-F, Uzawa R, Rodrigues JA, Zemach A, Chumak N, Machlicova A, Nishimura T, Rojas D, Fischer RL, Tamaru H, Zilberman D. Active DNA demethylation in plant companion cells reinforces transposon methylation in gametes. Science. 2012;337(6100):1360–4.

Isshiki M, Kodama H. Plant RNAi and Crop Improvement. Springer Netherlands: Molecular Techniques in Crop Improvement. S. M. Jain and D. S. Brar; 2009. p. 653–73.

Jackson JP, Lindroth AM, Cao X, Jacobsen SE. Control of CpNpG DNA methylation by the KRYPTONITE histone H3 methyltransferase. Nature. 2002;416:556–60.

Jamous RM, Boonrod K, Fuellgrabe MW, Ali-Shtayeh MS, Krczal G, Wassenegger M. The helper component-proteinase of the Zucchini yellow mosaic virus inhibits the Hua Enhancer 1 methyltransferase activity in vitro. J Gen Virol. 2011;92(9):2222–6.

Jeffrey AJ, Trevor LS, Eric JR. Maintenance of genomic methylation requires a SWI2/SNF2-like protein. Nat Genet. 1999;22:94–7.

Ji L, Liu X, Yan J, Wang W, Yumul RE, Kim YJ, Dinh TT, Liu J, Cui X, Zheng B, Agarwal M, Liu C, Cao X, Tang G, Chen X. *ARGONAUTE10* and *ARGONAUTE1* regulate the termination of floral stem cells through two microRNAs in *Arabidopsis*. PLoS Genet. 2011;7(3):e1001358.

Jin H, Zhu J-K. A viral suppressor protein inhibits host RNA silencing by hooking up with Argonautes. Genes Dev. 2010;24(9):853–6.

Jones-Rhoades MW, Bartel DP, Bartel B. MicroRNAs and their regulatory roles in plants. Annu Rev Plant Biol. 2006;57(1):19–53.

Jullien PE, Mosquna A, Ingouff M, Sakata T, Ohad N, Berger F. Retinoblastoma and its binding partner MSI1 control imprinting in *Arabidopsis*. PLoS Biol. 2008;6(8):e194.

Kapranov P, Cheng J, Dike S, Nix DA, Duttagupta R, Willingham AT, Stadler PF, Hertel J, Hackermüller J, Hofacker IL, Bell I, Cheung E, Drenkow J, Dumais E, Patel S, Helt G, Ganesh M, Ghosh S, Piccolboni A, Sementchenko V, Tammana H, Gingeras TR. RNA maps reveal new RNA classes and a possible function for pervasive transcription. Science. 2007;316 (5830):1484–8.

Kasschau K, Fahlgren N, Chapman E, Sullivan C, Cumbie J, Givan S, Carrington J. Genome-wide profiling and analysis of *Arabidopsis* siRNAs. PLoS Biol. 2007;5:e57.

Kato M, Miura A, Bender J, Jacobsen SE, Kakutani T. Role of CG and Non-CG methylation in immobilization of transposons in *Arabidopsis*. Curr Biol. 2003;13(5):421–6.

Khraiwesh B, Zhu J-K, Zhu J. Role of miRNAs and siRNAs in biotic and abiotic stress responses of plants. Biochim Biophys Acta (BBA) Gene Regul Mech. 2012;1819(2):137–48.

Kim YJ, Zheng B, Yu Y, Won SY, Mo B, Chen X. The role of Mediator in small and long noncoding RNA production in *Arabidopsis thaliana*. EMBO J. 2011;30(5):814–22.

Kinoshita T, Miura A, Choi Y, Kinoshita Y, Cao X, Jacobsen SE, Fischer RL, Kakutani T. - One-way control of FWA imprinting in *Arabidopsis* endosperm by DNA methylation. Science. 2004;303(5657):521–3.

Kurihara Y, Watanabe Y. *Arabidopsis* micro-RNA biogenesis through Dicer-like 1 protein functions. Proc Natl Acad Sci U S A. 2004;101(34):12753–8.

Kurihara Y, Takashi Y, Watanabe Y. The interaction between DCL1 and HYL1 is important for efficient and precise processing of pri-miRNA in plant microRNA biogenesis. RNA. 2006;12 (2):206–12.

Lakatos L, Szittya G, Silhavy D, Burgyan J. Molecular mechanism of RNA silencing suppression mediated by p19 protein of tombusviruses. EMBO J. 2004;23(4):876–84.

Lares MR, Rossi JJ, Ouellet DL. RNAi and small interfering RNAs in human disease therapeutic applications. Trends Biotechnol. 2010;28(11):570–9.

Law JA, Ausin I, Johnson LM, Vashisht AA, Zhu J-K, Wohlschlegel JA, Jacobsen SE. A protein complex required for polymerase V transcripts and RNA- directed DNA methylation in *Arabidopsis*. Curr Biol: CB. 2010;20(10):951–6.

Law JA, Vashisht AA, Wohlschlegel JA, Jacobsen SE. SHH1, a homeodomain protein required for DNA methylation, as well as RDR2, RDM4, and chromatin remodeling factors, associate with RNA polymerase IV. PLoS Genet. 2011;7(7):e1002195.

Lee JT. Epigenetic regulation by long noncoding RNAs. Science. 2012;338(6113):1435–9.

Lee RC, Feinbaum RL, Ambros V. The *C. elegans* heterochronic gene lin-4 encodes small RNAs with antisense complementarity to lin-14. Cell. 1993;75(5):843–54.

Lee Y, Kim M, Han J, Yeom K-H, Lee S, Baek SH, Kim VN. MicroRNA genes are transcribed by RNA polymerase II. EMBO J. 2004;23(20):4051–60.

Li J, Yang Z, Yu B, Liu J, Chen X. Methylation protects miRNAs and siRNAs from a 3'-end uridylation activity in *Arabidopsis*. Curr Biol. 2005;15(16):1501–7.

Li S, Liu L, Zhuang X, Yu Y, Liu X, Cui X, Ji L, Pan Z, Cao X, Mo B, Zhang F, Raikhel N, Jiang L, Chen X. MicroRNAs inhibit the translation of target mRNAs on the endoplasmic reticulum in *Arabidopsis*. Cell. 2013;153(3):562–74.

Lindroth AM, Cao X, Jackson JP, Zilberman D, McCallum CM, Henikoff S, Jacobsen SE. Requirement of CHROMOMETHYLASE3 for maintenance of CpXpG methylation. Science. 2001;292(5524):2077–80.

Lister R, O'Malley RC, Tonti-Filippini J, Gregory BD, Berry CC, Millar AH, Ecker JR. Highly integrated single-base resolution maps of the epigenome in *Arabidopsis*. Cell. 2008;133 (3):523–36.

Llave C, Xie Z, Kasschau KD, Carrington JC. Cleavage of scarecrow-like mRNA targets directed by a class of *Arabidopsis* miRNA. Science. 2002;297(5589):2053–6.

Lobbes D, Rallapalli G, Schmidt DD, Martin C, Clarke J. SERRATE: a new player on the plant microRNA scene. EMBO Rep. 2006;7(10):1052–8.

Luo Z, Chen Z. Improperly terminated, unpolyadenylated mRNA of sense transgenes is targeted by RDR6-Mediated RNA silencing in *Arabidopsis*. Plant Cell Online. 2007;19(3):943–58.

Lurin C, Andrés C, Aubourg S, Bellaoui M, Bitton F, Bruyère C, Caboche M, Debast C, Gualberto J, Hoffmann B, Lecharny A, Le Ret M, Martin-Magniette M-L, Mireau H, Peeters N, Renou J-P, Szurek B, Taconnat L, Small I. Genome-wide analysis of *Arabidopsis* pentatricopeptide repeat proteins reveals their essential role in organelle biogenesis. Plant Cell Online. 2004;16(8):2089–103.

Maher C, Stein L, Ware D. Evolution of *Arabidopsis* microRNA families through duplication events. Genome Res. 2006;16(4):510–9.

Malik S, Roeder RG. The metazoan Mediator co-activator complex as an integrative hub for transcriptional regulation. Nat Rev Genet. 2010;11(11):761–72.

Mallory AC, Vaucheret H. Functions of microRNAs and related small RNAs in plants. Nat Genet. Suppl:S31-6, 2006.

Manavella PA, Koenig D, Weigel D Plant secondary siRNA production determined by microRNA-duplex structure. Proc Nat Acad Sci. 2012;109(7):2461–6.

Mercer TR, Mattick JS. Structure and function of long noncoding RNAs in epigenetic regulation. Nat Struct Mol Biol. 2013;20(3):300–7.

Meyers BC, Axtell MJ, Bartel B, Bartel DP, Baulcombe D, Bowman JL, Cao X, Carrington JC, Chen X, Green PJ, Griffiths-Jones S, Jacobsen SE, Mallory AC, Martienssen RA, Poethig RS, Qi Y, Vaucheret H, Voinnet O, Watanabe Y, Weigel D, Zhu J-K. Criteria for annotation of plant microRNAs. Plant Cell Online. 2008;20(12):3186–90.

Miura A, Yonebayashi S, Watanabe K, Toyama T, Shimada H, Kakutani T. Mobilization of transposons by a mutation abolishing full DNA methylation in *Arabidopsis*. Nature. 2001;411(6834):212–4.

Montgomery T, Howell M, Cuperus J, Li D, Hansen J, Alexander A, Chapman E, Fahlgren N, Allen E, Carrington J. Specificity of ARGONAUTE7-miR390 interaction and dual functionality in TAS3 trans-acting siRNA formation. Cell. 2008;133:128–41.

Napoli C, Lemieux C, Jorgensen R. Introduction of a chimeric chalcone synthase gene into petunia results in reversible co-suppression of homologous genes in trans. Plant Cell Online. 1990;2 (4):279–89.

Olmedo-Monfil V, Duran-Figueroa N, Arteaga-Vazquez M, Demesa-Arevalo E, Autran D, Grimanelli D, Slotkin RK, Martienssen RA, Vielle-Calzada J-P. Control of female gamete formation by a small RNA pathway in *Arabidopsis*. Nature. 2010;464(7288):628–32.

Park W, Li J, Song R, Messing J, Chen X. CARPEL FACTORY, a Dicer homolog, and HEN1, a novel protein, act in microRNA metabolism in *Arabidopsis thaliana*. Curr Biol. 2002;12(17):1484–95.

Park MY, Wu G, Gonzalez-Sulser A, Vaucheret H, Poethig RS. Nuclear processing and export of microRNAs in *Arabidopsis*. Proc Natl Acad Sci U S A. 2005;102(10):3691–6.

Pekker I, Alvarez JP, Eshed Y. Auxin response factors mediate *Arabidopsis* organ asymmetry via modulation of KANADI activity. Plant Cell Online. 2005;17(11):2899–910.

Peragine A, Yoshikawa M, Wu G, Albrecht HL, Poethig RS. SGS3 and SGS2/SDE1/RDR6 are required for juvenile development and the production of trans-acting siRNAs in *Arabidopsis*. Genes Dev. 2004;18(19):2368–79.

Pulido A, Laufs P. Co-ordination of developmental processes by small RNAs during leaf development. J Exp Bot. 2010;61(5):1277–91.

Qi Y, Denli AM, Hannon GJ. Biochemical specialization within *Arabidopsis* RNA silencing pathways. Mol Cell. 2005;19(3):421–8.

Qi Y, He X, Wang X-J, Kohany O, Jurka J, Hannon GJ. Distinct catalytic and non-catalytic roles of ARGONAUTE4 in RNA-directed DNA methylation. Nature. 2006;443(7114):1008–12.

Rajagopalan R, Vaucheret H, Trejo J, Bartel DP. A diverse and evolutionarily fluid set of microRNAs in *Arabidopsis thaliana*. Genes Dev. 2006;20(24):3407–25.

Ramachandran V, Chen X. Degradation of microRNAs by a family of exoribonucleases in *Arabidopsis*. Science. 2008;321(5895):1490–2.

Ream TS, Haag JR, Wierzbicki AT, Nicora CD, Norbeck AD, Zhu J-K, Hagen G, Guilfoyle TJ, Paša-Tolić L, Pikaard CS. Subunit COmpositions of the RNA-silencing enzymes pol IV and pol V reveal their origins as specialized forms of RNA polymerase II. Mol Cell. 2009;33 (2):192–203.

Reinhart BJ, Weinstein EG, Rhoades MW, Bartel B, Bartel DP. MicroRNAs in plants. Genes Dev. 2002;16(13):1616–26.

Ren G, Chen X, Yu B. Uridylation of miRNAs by HEN1 SUPPRESSOR1 in *Arabidopsis*. Curr Biol. 2012;22(8):695–700.

Rymarquis LA, Souret FF, Green PJ. Evidence that XRN4, an Arabidopsis homolog of exoribonuclease XRN1, preferentially impacts transcripts with certain sequences or in particular functional categories. RNA. 2011;17(3):501–11.

Shen B, Goodman HM. Uridine addition after microRNA-directed cleavage. Science. 2004;306 (5698):997.

Siomi MC, Sato K, Pezic D, Aravin AA. PIWI-interacting small RNAs: the vanguard of genome defence. Nat Rev Mol Cell Biol. 2011;12(4):246–58.

Slotkin RK, Vaughn M, Borges F, Tanurdzić M, Becker JD, FeijÛ JA, Martienssen RA. Epigenetic reprogramming and small RNA silencing of transposable elements in pollen. Cell. 2009;136 (3):461–72.

Smith LM, Pontes O, Searle I, Yelina N, Yousafzai FK, Herr AJ, Pikaard CS, Baulcombe DC. An SNF2 protein associated with nuclear RNA silencing and the spread of a silencing signal between cells in *Arabidopsis*. Plant Cell Online. 2007;19(5):1507–21.

Song M-G, Kiledjian M. 3′ terminal oligo U-tract-mediated stimulation of decapping. RNA. 2007;13(12):2356–65.

Soppe WJJ, Jacobsen SE, Alonso-Blanco C, Jackson JP, Kakutani T, Koornneef M, Peeters AJM. The late flowering phenotype of fwa mutants is caused by gain-of-function epigenetic alleles of a homeodomain gene. Mol Cell. 2000;6(4):791–802.

Souret FF, Kastenmayer JP, Green PJ. AtXRN4 degrades mRNA in *Arabidopsis* and its substrates include selected miRNA targets. Mol Cell. 2004;15(2):173–83.

Spitz F, Furlong EEM. Transcription factors: from enhancer binding to developmental control. Nat Rev Genet. 2012;13(9):613–26.

Sunkar R, Zhu J-K. Novel and stress-regulated microRNAs and other small RNAs from *Arabidopsis*. Plant Cell Online. 2004;16(8):2001–19.

Tsukahara S, Kobayashi A, Kawabe A, Mathieu O, Miura A, Kakutani T. Bursts of retrotransposition reproduced in *Arabidopsis*. Nature. 2009;461(7262):423–6.

van der Krol AR, Mur LA, Beld M, Mol JN, Stuitje AR. Flavonoid genes in petunia: addition of a limited number of gene copies may lead to a suppression of gene expression. Plant Cell Online. 1990;2(4):291–9.

Vaucheret H. Plant ARGONAUTES. Trends Plant Sci. 2008;13:350–8.

Vaucheret H, Vazquez F, Crété P, Bartel DP. The action of ARGONAUTE1 in the miRNA pathway and its regulation by the miRNA pathway are crucial for plant development. Genes Dev. 2004;18(10):1187–97.

Vazquez F, Gasciolli V, CrÈtÈ P, Vaucheret H. The nuclear dsRNA binding protein HYL1 is required for microRNA accumulation and plant development, but not posttranscriptional transgene silencing. Curr Biol: CB. 2004a;14(4):346–51.

Vazquez F, Vaucheret H, Rajagopalan R, Lepers C, Gasciolli V, Mallory AC, Hilbert J-L, Bartel DP, Crété P. Endogenous trans-acting siRNAs regulate the accumulation of *Arabidopsis* mRNAs. Mol Cell. 2004b;16(1):69–79.

Vongs A, Kakutani T, Martienssen RA, Richards EJ. *Arabidopsis thaliana* DNA methylation mutants. Science. 1993;260:1926–8.

Wierzbicki A, Haag J, Pikaard C. Noncoding transcription by RNA polymerase Pol IVb/Pol V mediates transcriptional silencing of overlapping and adjacent genes. Cell. 2008;135:635–48.

Wierzbicki AT, Ream TS, Haag JR, Pikaard CS. RNA polymerase V transcription guides ARGONAUTE4 to chromatin. Nat Genet. 2009;41:630–4.

Wu H-J, Wang Z-M, Wang M, Wang X-J. Widespread long noncoding RNAs as endogenous target mimics for microRNAs in plants. Plant Physiol. 2013;161(4):1875–84.

Xie Z, Kasschau KD, Carrington JC. Negative feedback regulation of Dicer-like1 in *Arabidopsis* by microRNA-guided mRNA degradation. Curr Biol. 2003;13(9):784–9.

Xie Z, Johansen LK, Gustafson AM, Kasschau KD, Lellis AD, Zilberman D, Jacobsen SE, Carrington JC. Genetic and functional diversification of small RNA pathways in plants. PLoS Biol. 2004;2(5):e104.

Xie Z, Allen E, Fahlgren N, Calamar A, Givan SA, Carrington JC. Expression of *Arabidopsis* MIRNA genes. Plant Physiol. 2005a;138(4):2145–54.

Xie Z, Allen E, Wilken A, Carrington JC. DICER-LIKE 4 functions in trans-acting small interfering RNA biogenesis and vegetative phase change in *Arabidopsis thaliana*. Proc Natl Acad Sci USA. 2005b;102(36):12984–9.

Yan J, Gu Y, Jia X, Kang W, Pan S, Tang X, Chen X, Tang G. Effective small RNA destruction by the expression of a short tandem target mimic in *Arabidopsis*. Plant Cell. 2012;24:415.

Yang L, Liu Z, Lu F, Dong A, Huang H. SERRATE is a novel nuclear regulator in primary microRNA processing in *Arabidopsis*. Plant J. 2006;47(6):841–50.

Yang L, Wu G, Poethig RS. Mutations in the GW-repeat protein SUO reveal a developmental function for microRNA-mediated translational repression in *Arabidopsis*. Proc Natl Acad Sci. 2012;109(1):315–20.

Ye K, Patel DJ. RNA silencing suppressor p21 of beet yellows virus forms an RNA binding octameric ring structure. Structure. 2005;13(9):1375–84.

Ye R, Wang W, Iki T, Liu C, Wu Y, Ishikawa M, Zhou X, Qi Y. Cytoplasmic assembly and selective nuclear import of *Arabidopsis* ARGONAUTE4/siRNA complexes. Mol Cell. 2012;46(6):859–70.

Yoshikawa M, Peragine A, Park MY, Poethig RS. A pathway for the biogenesis of trans-acting siRNAs in *Arabidopsis*. Genes Dev. 2005;19(18):2164–75.

Yu B, Yang Z, Li J, Minakhina S, Yang M, Padgett RW, Steward R, Chen X. Methylation as a crucial step in plant microRNA biogenesis. Science. 2005;307(5711):932–5.

Yu B, Bi L, Zheng B, Ji L, Chevalier D, Agarwal M, Ramachandran V, Li W, Lagrange T, Walker JC, Chen X. The FHA domain proteins DAWDLE in *Arabidopsis* and SNIP1 in humans act in small RNA biogenesis. Proc Natl Acad Sci. 2008;105(29):10073–8.

Zhang X, Yuan Y-R, Pei Y, Lin S-S, Tuschl T, Patel DJ, Chua N-H. Cucumber mosaic virus-encoded 2b suppressor inhibits *Arabidopsis* Argonaute1 cleavage activity to counter plant defense. Genes Dev. 2006;20(23):3255–68.

Zhang C, Ng DWK, Lu J, Chen ZJ. Roles of target site location and sequence complementarity in trans-acting siRNA formation in *Arabidopsis*. Plant J. 2012a;69(2):217–26.

Zhang C-J, Ning Y-Q, Zhang S-W, Chen Q, Shao C-R, Guo Y-W, Zhou J-X, Li L, Chen S, He X-J. IDN2 and its paralogs form a complex required for RNA–directed DNA methylation. PLoS Genet. 2012b;8(5):e1002693.

Zhao Y, Yu Y, Zhai J, Ramachandran V, Dinh TT, Meyers BC, Mo B, Chen X. The *Arabidopsis* nucleotidyl transferase HESO1 uridylates unmethylated small RNAs to trigger their degradation. Curr Biol. 2012;22(8):689–94.

Zheng X, Zhu J, Kapoor A, Zhu J-K. Role of *Arabidopsis* AGO6 in siRNA accumulation, DNA methylation and transcriptional gene silencing. EMBO. 2007;J26(6):1691–701.

Zheng B, Wang Z, Li S, Yu B, Liu J-Y, Chen X. Intergenic transcription by RNA Polymerase II coordinates Pol IV and Pol V in siRNA-directed transcriptional gene silencing in *Arabidopsis*. Genes Dev. 2009;23:2850–60.

Zhu H, Hu F, Wang R, Zhou X, Sze S-H, Liou LW, Barefoot A, Dickman M, Zhang X. *Arabidopsis* Argonaute10 specifically sequesters miR166/165 to regulate shoot apical meristem development. Cell. 2011;145(2):242–56.

Zilberman D, Cao X, Jacobsen SE. ARGONAUTE4 control of locus-specific siRNA accumulation and DNA and histone methylation. Science. 2003;299(5607):716–9.

Plant Translational Machinery

5

Karen S. Browning

Contents

K.S. Browning (✉)
Department of Molecular Biosciences and the Insitute for Cellular and Molecular Biology, The University of Texas at Austin, Austin, TX, USA
e-mail: kbrowning@cm.utexas.edu; kbrowning@mail.utexas.edu

S.H. Howell (ed.), *Molecular Biology, The Plant Sciences 2*,
DOI 10.1007/978-1-4614-7570-5_8

Abstract

- Components of the translational machinery
- Key steps in initiation, elongation, and termination
- Regulation of translation

Introduction

"It's complicated." This statement applies to all aspects of life, whether it refers to the complexity and diversity of life forms that have evolved on this planet or to the inner workings of cells. For both single-cell and multicellular organisms, "it's complicated" can also refer to any of the many processes of "doing business" as a cell, such as energy metabolism, synthesis of amino acids or nucleotides, DNA replication, RNA transcription, and protein translation. All of these processes are highly integrated and, like many complex systems, are connected in numerous ways, some that we understand, some that we are only beginning to understand, and many that we have yet to know even exist.

Photosynthetic organisms evolved to harness the abundant energy provided by the sun and turn it into storable energy that could be used to survive when light energy was not present. As photosynthetic organisms turned the sun's energy into storable energy, the process also produced oxygen, which further shaped the evolution of aerobic life forms. In turn, these new life forms began consuming stored energy provided by photosynthetic organisms. Thus, all animal life on the planet is dependent upon plants not only for stored energy in the form of starch but also for the oxygen that is produced in the process.

This review will focus on one of a plant cell's many "businesses," the process of synthesizing proteins. The central dogma defines a linear progression of the genetic material (DNA) to the "messenger" (RNA) that is then fashioned into the "workers" (proteins). The protein "workers" take part in all aspects of the cell's business, making and breaking of small molecules for energy or other metabolic processes, as well as the synthesis and degradation of macromolecules (DNA, RNA, protein). It is known that this process is not always linear and that there are many aspects of the regulation of translation, particularly in plants, that are only beginning to be fully appreciated and understood. The chemistry of life is exquisitely complicated.

Translation of Proteins

The basics of how messenger RNA (mRNA) is read and the genetic code is used to signal the type and order of amino acids to make a protein is similar in both prokaryotic and eukaryotic organisms. This similarity suggests that translation arose in the last common ancestor as a fundamental process in moving from nucleotide-based catalysis (RNA world) to proteins providing the catalytic power for utilization of energy and metabolic processes. The machine for this process of reading the code and assembling amino acids into protein is the ribosome. The modern ribosome is made of ribosomal RNA (rRNA) and proteins; however, the catalytic site for synthesis of the peptide bond is the 28S rRNA, suggesting that the earliest "ribosome-like" machine was probably a catalytic RNA.

Translation is divided into three phases: initiation, elongation, and termination. Each phase requires, in addition to ribosomes, many components of the "translational machinery." The number of ancillary components of the translational machinery varies between prokaryotes and eukaryotes, and even among eukaryotes, there are differences in the type and number of these components. Since translation is very expensive in terms of energy needs (ATP and GTP), organisms have evolved various methods of regulation to prevent unnecessary expenditures of energy by not making proteins until they are needed.

The process of selecting an mRNA for translation is much like a dance: the correct partners must find each other (mRNA and ribosome), know how to do the correct first steps (locate the correct initiation codon), continue the dance (assemble functional ribosome and elongate the protein), and finally, end the dance (terminate, release the protein, and dissociate the ribosomal subunits). In prokaryotes, that process requires only three initiation factors; but in eukaryotes, many initiation factors (see Table 1) are needed to just get up to the first "dance steps." Conversely, elongation and termination are more similar in both prokaryotes and eukaryotes requiring similar numbers of proteins to elongate the polypeptide and then complete the "protein dance."

There are a number of excellent detailed reviews on eukaryotic translation initiation (Aitken and Lorsch 2012; Hinnebusch and Lorsch 2012; Jackson et al. 2010; Sonenberg and Hinnebusch 2009; Valasek 2012), as well as several recent reviews specific for plant translational regulation (Echevarria-Zomeno et al. 2013; Gallie 2007; Muench et al. 2012; Munoz and Castellano 2012). The basics of translation in plants and other eukaryotes are remarkably similar.

Table 1 Eukaryotic initiation factors (eIFs) in plants

Factor	M_r[a]	Function[b]	Plant protein studied	Other names
eIF1	~12,600	Participates in pre-initiation complex (PIC) formation, scanning, and start-site selection; controls GAP activity of eIF5	Yes	
eIF1A	~17,000	Participates in PIC complex formation, scanning, and start-site selection	Yes	eIF4C
eIF2 complex		Ternary complex formation with GTP, Met-tRNA$_i^{Met}$; PIC formation	Yes	
eIF2α	~38,700		Yes	
eIF2β	~30,500		Yes	
eIF2γ	~50,000		Yes	
eIF2A	~56,200	Alternative to eIF2 for delivery of Met-tRNA; functions in IRES mediated initiation	No	
eIF2B complex		Promotes exchange of GTP for eIF2•GDP	No	
eIF2Bα	~42,100		No	
eIF2Bβ	~43,800		No	
eIF2Bγ	~49,000		No	
eIF2Bδ	~69,500		No	
eIF2Bε	~76,300		No	
eIF3 complex		PIC formation; binding of mRNA	Yes	
eIF3a	~114,300		Yes	TIF32; RPG1
eIF3b	~84,500		Yes	PRT1
eIF3c	~102,900		Yes	NIP1
eIF3d	~66,700		Yes	
eIF3e	~51,700		Yes	Int6
eIF3f	~31,800		Yes	
eIF3g	~35,500		Yes	TIF35
eIF3h	~38,300		Yes	
eIF3i	~36,300		Yes	TRIP1; TIF34
eIF3j[c]	~25,100		No	HCR1
eIF3k	~25,700		Yes	
eIF3l	~60,100		Yes	
eIF3m	~49,700		Yes	
eIF4A	~46,000	DEAD-box RNA/ATP-dependent helicase; interacts with eIF4G	Yes	
eIF4B	~57,700	RNA-binding protein; interacts with eIF4G; enhances helicase activity of eIF4A	Yes	

(*continued*)

Table 1 (continued)

Factor	M_r[a]	Function[b]	Plant protein studied	Other names
eIF4F complex		mRNA cap-binding complex; participates in helicase activity; participates in binding mRNA to PIC	Yes	CBP complex
eIF4E	~26,000	Cap-binding protein (CBP)	Yes	CUM1
eIF4G	~187,900	Scaffold for assembly of other factors	Yes	CUM2
eIFiso4F complex		Similar to eIF4F, plant-specific form	Yes	
eIFiso4E	~22,000	Cap-binding protein (CBP)	Yes	LSP1
eIFiso4G	~85,500	Scaffold for assembly of other factors	Yes	
eIF5	~48,600	GAP activity for the hydrolysis of GTP bound to eIF2; participates in scanning, start-site selection	Yes	
eIF5A[d]	~17,100		Yes	eIF4D
eIF5B	~142,100	GTPase that participates in subunit joining	No	
eIF6	~26,400	Prevents premature joining of subunits	Yes	
PABP	~72,000	Poly(A) binding protein; Interaction with mRNA 3′ poly(A); interaction with eIF4G	Yes	

[a]Based on TAIR database molecular weight predictions from genes identified in *A. thaliana* by NCBI *HomoloGene* and, unless otherwise indicated (plant protein studied), the protein(s) is only predicted to be present. In most cases, more than one gene in plants may encode the protein
[b]Functions were primarily determined from mammalian and yeast systems (Jackson et al. 2010). Mechanistic details from plants are only known for a few of these factors
[c]Not yet shown to be part of plant eIF3 complex, although gene is present
[d]This factor was originally named and identified as an initiation factor although the role is now known to be in elongation

Differences do occur in some components of the plant translational machinery that likely evolved due to selective pressures by the environment. Plants, being sessile, have many strategies to deal with their biotic and abiotic stresses and appear to have evolved strategies to regulate protein synthesis that differ from other eukaryotic organisms.

Initiation of Translation: Shall We Dance?

The nomenclature of the eukaryotic initiation factors (eIFs) involved in translation is often very confusing and can be "alphabet soup" for those new to the field (Browning et al. 2001; Safer 1989). Table 1 and Fig. 1 should be referred to often to reduce confusion and help in becoming familiar with the various factors and their roles in translation. Some names are very similar but refer to very different components. In addition, some factors have multiple subunits adding further complexity to the "soup."

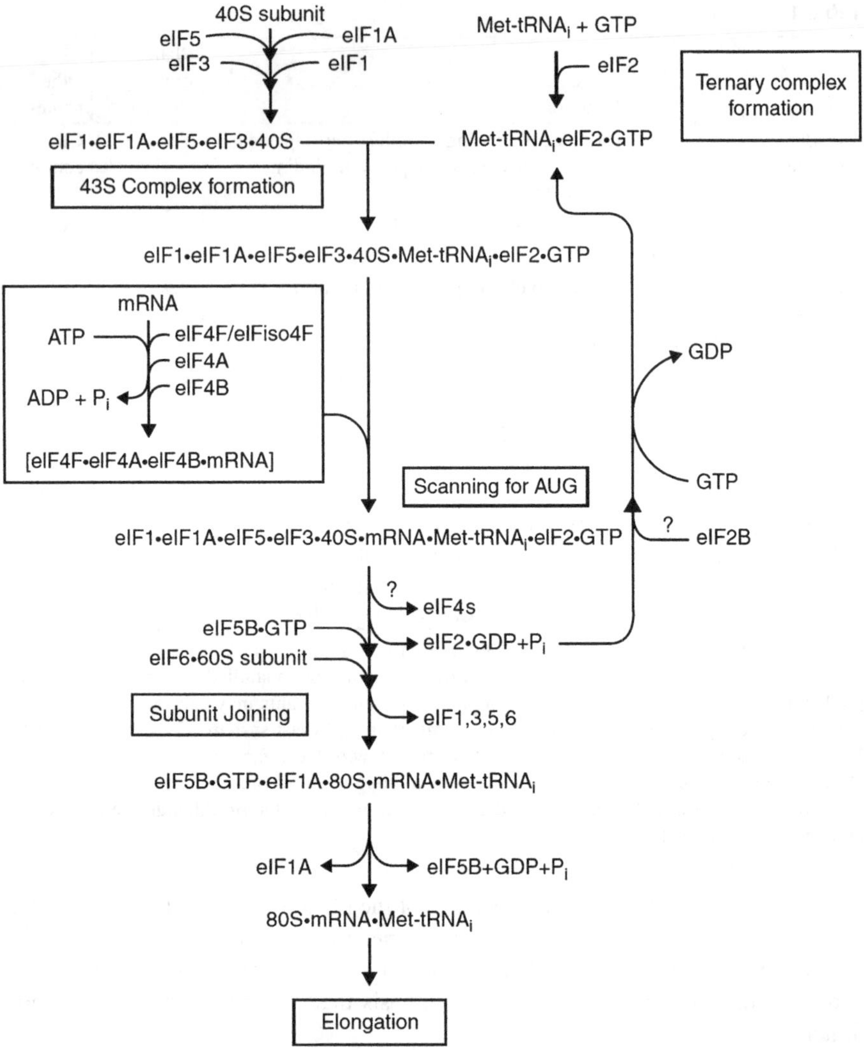

Fig. 1 Initiation of translation in plants. Initiation factors eIF4F, eIF4A, eIF4B, and PABP bind mRNA via the interaction of the m^{7}G cap and the eIF4E (or eIFiso4E) subunit of eIF4F (or eIFiso4F). The eIF4G (or eIFiso4G) subunit of eIF4F (or eIFiso4F) provides the scaffold for assembly of the factors on the mRNA. eIFs 1, 1A, and 3 bind the 40S ribosome and stimulate dissociation of 80S ribosomes. eIF6 binding to the 60S ribosome prevents reassociation of the subunits. eIF5 and ternary complex (eIF2•GTP•Met-tRNA$_i^{Met}$) assemble with the 40S ribosome and its associated factors (eIFs 1, 1A, and 3) forming the 43S pre-initiation complex (PIC). The PIC associates with the mRNA near the cap using interactions with eIF4F/eIF3/eIF5 and begins scanning 5′ to 3′ using eIF4F/eIF4A/eIF4B in an ATP-dependent manner to find the correct AUG codon. The ternary-complex-bound GTP is partially hydrolyzed to eIF2•GDP•P$_i$ during the scanning process. Upon arrival at the correct AUG, eIF1 dissociates triggering the releases of P$_i$ and eIF2•GDP. It is not clear when the eIF4 factors exit the ribosome. The 60S ribosome joining is stimulated by GTP hydrolysis by eIF5B, and eIFs 1, 3, 5, and 6 are released followed by the release

The Dancers: mRNA and Associated Factors

The mRNA

Eukaryotic messenger RNAs (mRNAs) are transcribed from genes in the nucleus. Prior to export from the nucleus, introns are removed by splicing and the mRNA is further processed to add a m^{7}GpppX group or "cap" to the 5′ end (Topisirovic et al. 2011) and variable lengths and sites of adenylate residues ("poly(A) tail") are added to the 3′ end (Hunt et al. 2012). Alternative splicing of the mRNA occurs in up to 60 % of all transcripts in plants and likely plays a significant role in gene expression in response to stress (Syed et al. 2012). The cap and poly(A) tail are features of all cellular eukaryotic mRNAs and are the primary molecular signals that the mRNA uses to indicate that it is ready to engage the initiation factor/ribosome machinery and start the "dance."

The "scanning model" proposed by Kozak in 1989 is still the preferred model for initiation of cellular mRNAs (Alekhina and Vassilenko 2012), whereas viral RNAs have evolved many alternative methods for attracting the initiation factors/machinery (Jackson 2012; Simon and Miller 2013). The scanning model has the 48S ribosomal PIC (see below) binding to the 5′ end of the mRNA via the cap and its associated initiation factors. This complex moves along the mRNA "scanning" for the correct start site (initiation codon or initiator AUG) in an ATP and initiation factor-dependent manner (Alekhina and Vassilenko 2012; Hinnebusch 2011). Upon recognition of the correct AUG, initiation factors are released and the 60S ribosomal subunit joins to form the elongating 80S ribosome. A molecular description of many aspects of this process is still unknown, and although some of the details and factors are known, others are still to be discovered as a more complete molecular model emerges (Aitken and Lorsch 2012).

The mRNA-Binding Factors

A set of initiation factors (eIF4F, eIF4A, eIF4B, PAPB) work together to bind and prepare the mRNA to interact with the PIC consisting of the 40S ribosomal subunit and another set of translation initiation factors (eIF1, eIF1A, eIF2, eIF3, eIF5, eIF5B) that assist the 40S ribosomal subunit with the selection of the correct start site and subsequent joining of the 60S ribosomal subunit for protein synthesis (see Table 1 and Fig. 1).

Fig. 1 (continued) of eIF5B. eIF1A is the last initiation factor to leave. The functional 80S ribosome is now ready for elongation. eIFiso4F is a plant-specific complex. The roles of eIF2α kinases (other than GCN2) and recycling of ternary complex by eIF2B have not been shown to function in plants to date. It is also doubtful that 4EBP-type regulation functions in plants. The mechanisms of regulation of plant protein synthesis are still not fully elucidated and are an area of intense research

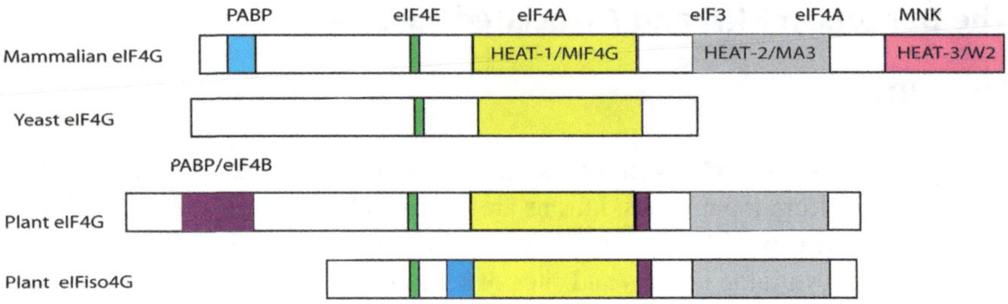

Fig. 2 **The domain organization of eIF4G and eIFiso4G**. Plant eIF4G and eIFiso4G have HEAT1 and HEAT2 domains in common with vertebrate eIF4G. The third vertebrate HEAT3 domain, containing the interaction site for MNK kinase, has been lost in plant lineages. Binding sites for eIF4E (*green*), eIF4A (*yellow and gray*), eIF3 (*gray*), PABP (*blue*), and PABP/eIF4B (*purple*) are indicated. Note the absence of the N-terminal extension in plant eIFiso4G compared to eIF4G

eIF4F (eIF4E + eIF4G)

The first initiation factor that recognizes and binds to the cap of the mRNA is the cap-binding complex or eukaryotic initiation factor 4F (eIF4F). The cap-binding complex is comprised of two proteins, eIF4G and eIF4E. Depending upon the organism and method of purification, another factor, eIF4A, may be present in this complex.

eIF4E is the cap-binding protein (CBP) that binds directly to the m^{7}GpppX cap of the mRNA (Topisirovic et al. 2011). The eIF4E cap-binding site has two highly conserved tryptophans, in addition to other conserved residues, that interact with the m^{7}GpppX cap of the mRNA. eIF4E is conserved across all eukaryotes from single-cell organisms (protists, yeasts) to higher eukaryotes, including plants.

eIF4G serves as a platform to assemble other components of the translational machinery on the mRNA (Aitken and Lorsch 2012; Hinnebusch and Lorsch 2012; Valasek 2012). The binding site for eIF4E is located near the N-terminal side of the first HEAT domain (**H**untington, **e**longation factor 3, **p**rotein phosphatase 2A, **T**OR1 kinase) in all eIF4G molecules (see Fig. 2). HEAT domains are helical regions that may interact with other proteins. eIF4G, depending upon the organism, may contain up to three HEAT domains. Yeast eIF4G has only one HEAT domain, plants have two, and mammals have three. The HEAT domains serve as docking sites for other initiation factors, such as eIF4A, eIF4B, eIF3, and poly(A)-binding protein (PABP). The N-terminal half of eIF4G is highly diverged in sequence between organisms, and its role beyond facilitating mRNA binding is unknown (Valasek 2012).

A second form of eIF4F is found only in plants (Browning 2004). This isozyme form, eIFiso4F, also has two subunits, eIFiso4G and eIFiso4E. These subunits are genetically distinct from the plant eIF4F subunits. eIFiso4E is similar in size to eIF4E; however, eIFiso4G is roughly half the size of eIF4G. The size difference is due to the presence of the N-terminal extension in eIF4G that is absent in eIFiso4G (see Fig. 2). eIFiso4F and eIF4F have similar functions in vitro, suggesting that the N-terminal domain of eIF4G is not necessary for function in initiation of protein synthesis per se, but may have a role in regulation or another as yet unknown role (Browning 2004). *Arabidopsis* mutants lacking a functional eIFiso4F complex show developmental and growth defects suggesting an important role for this specific translation initiation factor in plants (Muench et al. 2012; Munoz and Castellano 2012).

eIF4A

eIF4A is the prototype for the DEAD-box helicase family (Andreou and Klostermeier 2012; Marintchev 2013) and is a highly conserved protein across all eukaryotes. This protein binds to eIF4G and promotes RNA-dependent RNA helicase activity that presumably unwinds any structure in the mRNA to allow the ribosome to scan along the mRNA and select the correct AUG start site (see below for scanning and start-site selection initiation factors). Other RNA helicases appear to have a role in the scanning process. Ded1 (yeast) is a DEAD-box helicase that is proposed to be a member of the canonical translational machinery based on its role in stimulating translation of mRNAs with long 5′ UTRs or secondary structure in yeast extracts and in vivo (Marintchev 2013). Mammalian DHX29 is a DExH-box helicase that is thought to participate in initiation and promote scanning of long 5′ UTRs (Marintchev 2013). Plants have several proteins that are similar to eIF4A-like helicases such as Ded1 or DHX29; however, their roles in plant translation have not been systematically studied. Many of these uncharacterized helicases will likely be found to have both general and specialized roles in translation of mRNAs.

eIF4B and eIF4H

These two factors are the least conserved of all the canonical initiation factors. eIF4B in plants, yeast and mammals is not well conserved at the sequence level, although there appears to be functional conservation (Browning 2004). In general, eIF4B binds to eIF4G and appears to stimulate the RNA-unwinding activities of the eIF4A and eIF4F (or plant eIFiso4F) complex on the 48S ribosome prior to start-site selection. Deletion of eIF4B in yeast is not lethal, which suggests an ancillary role in the initiation process (Valasek 2012). eIF4H in mammals (there is no known yeast eIF4H) shares RNA-binding activity and sequence similarity to eIF4B (Valasek 2012). eIF4H does not appear to be present in plants, as no genes have been identified; however, given the divergence of eIF4B, it is possible that there are unidentified RNA-binding proteins that may serve the same role as mammalian eIF4H in plants.

Poly(A)-Binding Protein (PABP)

All eukaryotic mRNAs possess at their 3′ ends a region of variable length of adenine residues, or the poly(A) "tail." This region of adenine residues is added posttranscriptionally. Multiple PABP molecules bind to the poly(A) tail of a mRNA, and it is believed that one of the roles of PABP is to protect the mRNA from degradation at the 3′ end, much as the 5′ cap is thought to protect the mRNA from degradation. The finding that PABP binds to eIF4G suggested a role for PABP in initiation and led to the idea that the mRNA is circularized through the eIF4G/PABP interaction. However, recent evidence in yeast suggests that circularization is not required in vivo (Goss and Kleiman 2013). PABPs have been implicated in many aspects of RNA metabolism and clearly have complex roles in these processes through interactions with many proteins, as well as RNA. Although yeast and other single-cell eukaryotes possess only a single copy of PABP, plants and higher eukaryotes have as many as eight copies that have specific tissue or organelle expression. In addition, PAPB proteins are modified by phosphorylation, methylation, and acetylation, which may affect the types of interactions with other proteins, such as translation factors and RNA (Goss and Kleiman 2013).

The Dancers: Ribosomes and Associated Factors

The Ribosome

The ribosome is a "machine" that does the work of decoding the mRNA, assembling the correct amino acids brought in by the appropriate aminoacylated transfer RNA (tRNA), and then catalyzing the formation of the peptide bond (Melnikov et al. 2012). The basic structure, components, and mechanisms of ribosomes are highly conserved across eukaryotic organisms and even share similarity to prokaryotic ribosomes (Melnikov et al. 2012).

Ribosomes play a critical role in many aspects of a plant life cycle including development and response to the environment (Byrne 2009; Carroll 2013; Horiguchi et al. 2012). The *Arabidopsis* 40S ribosomal subunit contains the 18S rRNA and 30 ribosomal proteins, and it decodes the mRNA as its main role. The 60S subunit contains the 28S, 5.8S, and 5S rRNAs and 44 ribosomal proteins, and it forms the peptide bond using the peptidyltransferase center which is comprised of the 28S rRNA (Melnikov et al. 2012; Valasek 2012). Plants also have a unique member of the acidic phosphoprotein family (P3) present in the 60S ribosomal subunit.

Eukaryotic ribosomes vary in size due to expansion regions in the rRNAs. In addition, many of the eukaryotic ribosomal proteins have unusually long tails or loops extending from the folded core. The role of these expansions in the rRNA and the protein tails/loops is not clear, but it is thought that they may facilitate interactions with eukaryotic-specific protein factors during initiation, elongation, termination, and/or ribosome dissociation (Melnikov et al. 2012). Plants may have up to four copies of their ribosomal genes compared to mammalian (one copy) or yeast (one to two copies) ribosomal protein genes and may vary the ribosomal protein composition in response to various stress conditions (Carroll 2013).

The Ribosome-Associated Initiation Factors

This set of initiation factors binds (except for eIF6 which binds to the 60S ribosomal subunit) to the 40S ribosomal subunit prior to the binding of the mRNA to form the pre-initiation complex (PIC). These conserved factors work to either interact with the mRNA and its associated initiation factors (eIF3) or participate in scanning and start-site selection (eIF1, eIF1A, eIF5) or deliver the Met-tRNA (eIF2) to the ribosome (Aitken and Lorsch 2012; Hinnebusch 2011; Valasek 2012).

eIF1

eIF1 has functional similarity to IF3 in prokaryotic initiation. eIF1 interacts near the peptidyl (P)-site of the small ribosomal subunit where the Met-tRNA$_i^{Met}$ will ultimately bind. eIF1 participates in pre-initiation complex formation, scanning, and start-site selection (Aitken and Lorsch 2012; Hinnebusch 2011; Valasek 2012).

eIF1A

eIF1A (eIF4C in older literature) shows both structural and sequence conservation to prokaryotic IF1. It binds to the aminoacyl (A)-site in the small ribosomal subunit and facilitates the assembly of the pre-initiation complex, scanning, and correct start-site selection. eIF1A has both N-terminal and C-terminal extensions that have roles in start-site selection (Aitken and Lorsch 2012; Hinnebusch 2011; Valasek 2012).

eIF2A

This factor is believed to be involved in an alternative path for delivery of the Met-tRNA$_i^{Met}$ to the ribosome in the case of noncanonical cap-dependent initiation in yeast and mammals; however, the role of this protein in plant translation is not known.

eIF2

This complex is comprised of 3 subunits (termed eIF2α, eIF2β, eIF2γ) and binds to the initiator Met-tRNA$_i^{Met}$ and GTP to form the "ternary complex." The ternary complex delivers the Met-tRNA$_i^{Met}$ to the 40S ribosomal subunit during canonical cap-dependent initiation using contacts with eIF3, eIF5, eIF1A, and eIF1. The GTP hydrolysis activity of eIF2 is stimulated by the GTPase-activating protein (GAP) activity of eIF5 (Aitken and Lorsch 2012; Hinnebusch 2011; Immanuel et al. 2012; Valasek 2012).

eIF3

This large complex is comprised of 13 subunits (a–m) in most eukaryotes (see Table 1), including plants, with a considerable degree of conservation at the sequence level (Aitken and Lorsch 2012; Browning et al. 2001; Valasek 2012). The 26S proteasome lid (**P**) and the COP9 signalosome (**C**) and eIF3 (**I**) are collectively known as **PCI** complexes (Valasek 2012). The subunits of these complexes share a PCI signature domain that is thought to promote complex formation and protein interactions, as well as subunits with MPN domains associated with metal binding. In the case of eIF3, six subunits (eIF3a, eIF3c, eIF3e, eIF3k, eIF3l, eIF3m) contain the PCI domain and two subunits have MPN domains (eIF3f, eIF3h). Interestingly, *S. cerevisiae* has taken a different path for eIF3 requiring only 5 subunits for eIF3 core function (eIF3a, eIF3b, eIF3c, eIF3i, and eIF3g) and one noncore subunit (eIF3j). Even other fungi, such as *S. pombe*, have 13 subunits in their eIF3. eIF3 binds to the 40S ribosomal subunit and interacts with multiple initiation factors (eIF5, eIF1, and eIF4G) as well as directly contacting the mRNA. eIF3 appears to play roles in every aspect of the initiation process including mRNA binding to the 40S ribosomal subunit, scanning, start-site selection, reinitiation, and recruitment of regulatory proteins such as S6 kinase (Valasek 2012). There is evidence in yeast that eIF3 remains bound to the elongating ribosome for several rounds of elongation, suggesting there may be post-initiation/reinitiation roles for this complex (Valasek 2012).

eIF5

eIF5 functions as a GTPase-accelerating protein (GAP) for the hydrolysis of eIF2•GTP during start-site recognition. eIF5 bound to eIF2•GDP is released during the joining of the 60S subunit (Jennings and Pavitt 2010). Yeast eIF5 is also thought to be a GDP dissociation inhibitor (GDI) for eIF2•GDP after release from the pre-initiation complex (Jennings and Pavitt 2010). These two activities are independent of each other and occur in separate domains, GAP in the N-terminal domain and GDI in the C-terminal domain (Jennings and Pavitt 2010). eIF5 interacts with eIF2 through both its N-terminal (with eIF2γ-subunit) and C-terminal (with eIF2β-subunit) domains. It may also interact with other initiation factors, such as eIF4G or eIF3, but these interactions may be organism specific.

eIF5B

The final step of the initiation process is the joining of the 60S ribosomal subunit to the 48S pre-initiation complex to form the functional 80S ribosome. This step requires the hydrolysis of GTP and is catalyzed by eIF5B, a structural homolog of bacterial IF2. eIF5B shows structural similarity to two domains of the eEF1A family, also a GTPase (see below). eIF5B likely interacts with and stabilizes the Met-tRNA$_i^{Met}$ in preparation for subunit joining and elongation. Interaction of the C-terminal domain of eIF1A with eIF5B facilitates joining after release of the eIF2•GDP/eIF5 complex (Aitken and Lorsch 2012; Valasek 2012).

RACK1

RACK1 (receptor for activated C kinase 1) is a multifunctional scaffolding WD-repeat protein that, in addition to roles in various other pathways, is associated with the 40S ribosomal subunit. Although not considered either a ribosomal protein or translation factor, it plays an important role as a scaffolding protein to recruit other protein factors that function in translation regulation (Gibson 2012). Mammalian RACK1 recruits protein kinase C (PKC) to the ribosome where during the joining of the 40S and 60S ribosome, RACK1-associated PKC presumably phosphorylates eIF6 on the 60S ribosome stimulating release of eIF6 and ribosome joining (Gibson 2012). Interestingly, although RACK1 and eIF6 are conserved in plants, the PKC phosphorylation site on plant eIF6 is not. RACK1 is linked to abscisic acid (ABA) signaling, and a direct interaction between RACK1 and eIF6 has been demonstrated in plants (Guo et al. 2011).

eIF6

eIF6 is an anti-association factor on 60S ribosomal subunits that prevents premature association with the 48S PIC. eIF6 functions also during the assembly and export of 60S ribosomal subunits in the nucleus. It is conserved in eukaryotes and in *Archaea*, but not in eubacteria (Guo et al. 2011).

Formation of the Pre-initiation complex (PIC), Scanning, and Subunit Joining

The PIC consists of the 40S ribosome, ternary complex (eIF2•GTP•Met-tRNA$_i^{Met}$), eIF1, eIF1A, eIF3, and eIF5. eIF1, eIF1A, eIF3, and eIF5 all function to recruit the ternary complex to the 40S ribosome (Aitken and Lorsch 2012). Although it has been shown that many of these factors can associate in vitro in the absence of the 40S ribosome to form a "multifactor complex" or MFC, it is not clear that the MFC forms in vivo prior to interactions with the 40S ribosome (Aitken and Lorsch 2012; Valasek 2012). The locations of these factors on the 40S ribosome have been determined using co-crystal structures and cryo-EM (Valasek 2012). eIF1 is located at or near the P-site (peptidyl) in the ribosome where the initiator Met-tRNA will be bound, whereas eIF1A is located near the A-site (aminoacyl) where the mRNA will enter (Aitken and Lorsch 2012; Valasek 2012). eIF1A also has N- and C-terminal extensions that interact with the P-site prior to full engagement of the initiator Met-tRNA$_i^{Met}$ upon the PIC arriving at the proper initiator AUG codon (Aitken and Lorsch 2012; Valasek 2012). Both eIF1 and eIF1A are crucial for binding of ternary complex, and eIF1 appears to promote an "open" orientation of the PIC that allows binding of the mRNA and subsequent scanning for the correct AUG start codon (Aitken and Lorsch 2012; Valasek 2012). Protein-protein interactions between eIF3, the N-terminal extension of eIF1A, and eIF2 are thought to stabilize the ternary complex on the 40S ribosomal subunit. The PIC in the "open" conformation is ready to receive the mRNA and the associated factors (eIF4F, eIF4A, eIF4B, and PABP) that have presumably cleared any structures from the 5′ end and made an "open area" for the PIC to bind and engage the mRNA for its "dance."

Although the exact mechanistic details of interaction with the 43S PIC and mRNA associated factors are not known, eIF3 appears to be the most likely facilitator in this step with its many subunits and known interactions with other initiation factors such as eIF5 and eIF4G; however, additional, as yet unknown factors may also participate in this process (Aitken and Lorsch 2012). Several models are proposed to describe the journey from the attachment of the 43S PIC at the 5′ end of the mRNA to the correct initiation codon, but the molecular details of this process have not been elucidated (Aitken and Lorsch 2012).

Yeast genetics has been a powerful tool in understanding many of the molecular details of start-site selection and events leading up to joining of the 60S ribosome (Hinnebusch 2011). The recognition of the AUG start site requires eIF1, eIF1A, ternary complex (eIF2•GTP•Met-tRNA$_i^{Met}$), and eIF5. In a series of poorly understood molecular signals, these factors and Met-tRNA$_i^{Met}$ participate in the selection of the start site by the scanning 48S PIC held in the "open" conformation by eIF1 and eIF1A (Aitken and Lorsch 2012; Hinnebusch 2011; Valasek 2012). Met-tRNA$_i^{Met}$ is positioned by eIF2•GTP but not completely lodged into the P-site until the correct start site is reached. During this process, the eIF2•GTP is hydrolyzed to eIF2•GDP•P$_i$ by the GAP activity of eIF5; however, the P$_i$ is held in place by eIF1 until the AUG start codon is reached. Upon reaching the correct AUG codon, a series of events take place in which several molecular signals result in conformational changes of various components. These signals include the formation of a perfect fit for the Met-tRNA$_i^{Met}$ in the P-site and changes in conformation of the Met-tRNA$_i^{Met}$ that lead to eIF1 being released along with P$_i$ from eIF2•GDP. All of these changes result in the 48S PIC switching from an "open" to a "closed" conformation. Structural changes from the "open" to "closed" 48S PIC also include structural alterations in the domains of eIF1A and eIF5 that may affect the release of P$_i$ from eIF2•GDP. This release of P$_i$ appears to be the most important step in the transition from the "open" scanning PIC to the "closed" 48S PIC that is ready to join with the 60S ribosomal subunit.

Upon formation of the "closed" 48S PIC, eIF2•GDP and eIF5 must dissociate from the interface surface where the 60S ribosomal subunit will bind. eIF5B catalyzes the hydrolysis of GTP in this process that results in conformational changes to form the functional 80S ribosome. eIF1A also participates in this process and facilitates subunit joining and thus is an important participant at several stages of the initiation process. eIF1A is the last remaining initiation factor on the 48S PIC interface to leave. eIF3 binds on the non-ribosomal interface side and may stay bound to the functional 80S ribosome for a while and have a role in reinitiation events (Aitken and Lorsch 2012; Hinnebusch 2011; Valasek 2012). The role of eIF6 in this process is to prevent premature association of 40S and 60S ribosomal subunits and to interact with RACK1 on the 40S ribosomal subunit to facilitate the joining of the 60S ribosomal subunit (Guo et al. 2011).

The dancers, mRNA and ribosome, are now fully engaged and ready to begin the process of reading the mRNA and joining amino acids in the correct order to form a protein in another series of well-orchestrated dance steps.

Elongation

Eukaryotic elongation, in comparison to initiation, is much simpler and bears strong similarity to the process in bacteria (see Table 2 and Fig. 3). Like bacterial elongation, eukaryotic elongation has only three factors: eEF1A, eEF1B, and eEF2. Detailed studies of plant elongation and termination have not been done.

eEF1A

eEF1A (eEF1α in older literature) is comparable to bacterial EF-Tu and its role is to bring the charged tRNA, in a GTP-dependent manner, to the A-site in the 80S ribosome for the formation of the peptidyl bond. Hydrolysis of GTP occurs when a match occurs between the mRNA and anticodon of the tRNA, releasing the eEF1A•GDP and allowing the tRNA to lodge completely in the A-site (Dever and Green 2012). Once the tRNA is fully lodged in the A-site, the peptidyltransferase center (PTC) of the 28S rRNA catalyzes peptide bond formation. eEF1A is among the most abundant proteins in all organisms and appears to have roles in addition to translation elongation that include association with cytoskeleton, nuclear export, proteolysis, apoptosis, and viral propagation (Sasikumar et al. 2012). eEF1A is also a target for several kinases and other types of posttranslational control suggesting multiple mechanisms for its regulation (Sasikumar et al. 2012).

eEF1B

The role of the eEF1B complex is to remove GDP bound to eEF1A so that it may bind GTP and participate in another round of elongation. This function is analogous to EF-Ts in bacteria. eEF1B is a complex of 2–3 subunits depending upon the organism. eEF1B subunits are targets for various kinases suggesting many levels of

Table 2 Eukaryotic elongation (eEF) and termination (eRF) factors in plants

Factor	Mr[a]	Function[b]	Plant protein studied	Other names
Elongation				
eEF1A	~49,000	Binding/delivery of aminoacylated tRNA to ribosome; GTPase activity	Yes	eEF1α
eEF1B complex		Recycling of GDP from eEF1A	Yes	
eEF1Bα	~24,700			eEF1β′
eEF1Bβ	~28,700			eEF1β
eEF1Bγ	~46,600			eEF1γ
eEF2	~93,200	GTP-dependent translocation of mRNA	Yes	LOS1
eIF5A	~17,100	Stimulates peptide bond formation; hypusine modification required for activity	Yes	eIF4D
Termination				
eRF1	~48,700	High-fidelity stop codon recognition; peptidyl-tRNA hydrolysis	Yes	
eRF3	~60,500	Interacts with eRF1 to promote termination/peptide release; GTPase activity	No	

[a]Based on TAIR database molecular weight predictions from genes identified in *A. thaliana* by NCBI *HomoloGene* and, unless otherwise indicated (plant protein studied), the protein(s) is only predicted to be present. In most cases, more than one gene in plants may encode the protein
[b]Functions primarily determined from mammalian and yeast systems (Dever and Green 2012; Sasikumar et al. 2012)

control and have putative functions outside of the canonical role of recycling of eEF1A•GDP (Dever and Green 2012; Sasikumar et al. 2012).

eEF2

eEF2 is comparable to EF-G in bacteria. eEF2 is a GTPase that stimulates translocation of the peptidyl-tRNA to the P-site and the empty tRNA to the E-site. Unlike eEF1A, eEF2 does not require a specific factor for recycling of the GDP (Dever and Green 2012). eEF2 in mammalian cells is modified by a posttranslational modification with diphthamide, as well as regulation by kinases; however, it is not known if similar modifications/regulation occur in vivo in plants.

eIF5A

eIF5A (eIF4D in older literature) is the only protein known to contain hypusine, a posttranslational modification of a specific lysine in this factor that is highly conserved. This highly specific modification to hypusine is required for activity. This

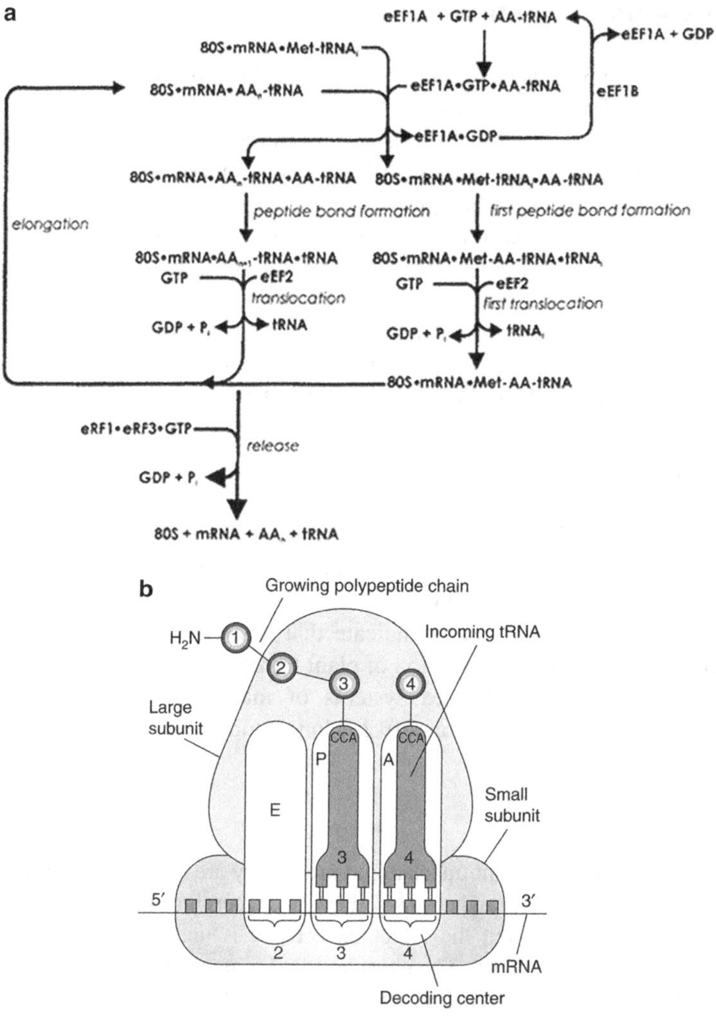

Fig. 3 **Steps in elongation.** (**a**) Summary of steps in elongation. (**b**) **Details of the tRNA binding sites (A, P, E) of the ribosome.** eEF1A•GTP•aminoacyl-tRNA enters the A-site of the ribosome. Peptide bond formation is catalyzed by the 28S rRNA and transfers the nascent polypeptide to the aminoacyl-tRNA in the A-site. Translocation of the mRNA to the next codon is catalyzed by eEF2 hydrolysis of GTP, moving the peptidyl-tRNA over to the P-site and pushing the deacylated tRNA into the E-site. The A-site is now open for another round of elongation. eEF1A•GDP requires eEF1B to remove the GDP to allow GTP to bind and eEF1A•GTP to acquire another tRNA to deliver to the ribosome. The ribosome shown in part B is prior to the formation of the peptide bond for the incoming amino acid 4

protein is functionally similar to the bacterial EF-P in promoting the formation of the peptide bond during elongation (Dever and Green 2012). Both EF-P and eIF5A have recently been shown to be necessary for efficient elongation of multiple prolines and glycines and may have other amino acid specific functions in elongation.

Termination/Dissociation

Multiple rounds of peptide bond formation and translocation occur as the polypeptide encoded by the mRNA is formed by the ribosome. Upon arrival of the ribosome at the termination codon of the mRNA, the dancers must now end their dance and separate (Dever and Green 2012; Jackson et al. 2012). Detailed studies of this process in plants have not been carried out, but presumably, plant termination is similar to the mammalian and yeast termination process (Dever and Green 2012; Jackson et al. 2012). Two factors, eRF1 and eRF3, participate in this process in a complex with GTP. eRF1 has a structure that resembles a tRNA, which docks with the ribosome in the A-site, and eRF3 stimulates hydrolysis of the GTP and dissociates leaving eRF1 on the ribosome. eRF1 catalyzes hydrolysis of the peptidyl-tRNA to release the polypeptide. Thus, the dance ends, releasing a newly synthesized protein.

Regulation

Plant translation initiation, elongation, and termination and ribosome dissociation are assumed to bear functional similarity to yeast and mammalian systems in which most of the mechanistic studies have been carried out. Several recent reviews on the regulation of translation in plants indicate that despite similarities in the factors involved in translation, the regulation of plant translation does not appear to utilize several of the well-characterized systems of mammals or yeast (Echevarria-Zomeno et al. 2013; Muench et al. 2012; Munoz and Castellano 2012).

Phosphorylation

Many components of the plant translational machinery are targets of phosphorylation. eIF2, eIF4A, eIF4B, eIF4E, and eIFiso4E become hyperphosphorylated during development or abiotic stress, but the effect on activity is unclear (Echevarria-Zomeno et al. 2013; Muench et al. 2012). The phosphorylation states of plant PABP, eIF4G, and eIF4B appear to play a role in the regulation of assembly of the mRNA complex prior to association with the 48S PIC (Muench et al. 2012). The kinases responsible for many of these modifications in plants have not been identified; however, some of the components (eIF2α, eIF2β, eIF3c, eIF4B, and eIF5) of the PIC and eIF5A are targets of casein kinase 2 (CK2) suggesting that the phosphorylation state may also play a role in the assembly of the plant PIC and subsequent steps (Muench et al. 2012). Although likely to occur, phosphorylation and other posttranslational modifications of the plant elongation factors have not been extensively studied.

eIF2α Phosphorylation

Among the most studied regulatory mechanisms is the phosphorylation of eIF2α by a variety of stress kinases in mammals (GCN2, PERK, HRI, and PKR).

Phosphorylation of a conserved serine of eIF2α by these kinases inhibits the recycling of eIF2•GDP by the eIF2B complex (5 subunits) and effectively shuts down translation in mammalian cells. Similarly, GCN2 kinase in yeast has been shown to regulate translation of GCN4 upstream open reading frames during nutrient deprivation. Although the presence of a GCN2 kinase has been shown to exist in plants and appears to alter polysome profiles, the effect of eIF2α phosphorylation during stress on translation does not appear to be comparable to that of mammalian or yeast systems (Echevarria-Zomeno et al. 2013; Immanuel et al. 2012; Muench et al. 2012; Munoz and Castellano 2012). Moreover, functional eIF2B has yet to be purified from a plant system, and additional eIF2α kinases have not been identified yet (Immanuel et al. 2012). Thus, plants appear to lack this mechanism of regulation or utilize it in a very different manner.

4E-Binding Proteins (4EBP)

Another well-characterized regulatory pathway in mammals and yeast is the presence of proteins that compete with eIF4G for the binding of eIF4E. In mammals, the 4EBP are regulated through phosphorylation via signaling from growth factors and the mTORC1 signaling cascade (Hinnebusch and Lorsch 2012). Plants do not appear to have a 4EBP-like regulatory system, although there is a mTOR-like kinase present in plants (Echevarria-Zomeno et al. 2013; Muench et al. 2012; Munoz and Castellano 2012; Robaglia et al. 2012). Also lacking is apparent regulation of plant eIF4E via phosphorylation by Mnk signaling pathways. Plant eIF4G and eIFiso4G lack the third HEAT domain of mammalian eIF4G that specifically binds Mnk1 kinases and promotes phosphorylation of mammalian eIF4E. Again, plants either lack or have significantly altered these mechanisms for regulation in ways that have yet to be discovered.

eIF6 Phosphorylation

The availability of 60S ribosomes for formation of the functional 80S ribosome relies upon fine tuning of the amount of eIF6 as well as its phosphorylation by PKC in mammals. Plant eIF6 apparently lacks PKC phosphorylation sites and thus lacks regulation via the Ras/PKC signaling cascades used by mammals. Given the apparent connection with abscisic acid signaling between eIF6 and RACK1 (Guo et al. 2011), there are likely alternative mechanisms for regulation of eIF6 and subunit joining in plants that have yet to be elucidated.

S6 Kinase and mTOR Signaling

Although plants appear to lack 4EBPs that are regulated through the mTORC1 signaling network, plants do have a S6 kinase (S6K) that functions during stress and may signal the energy state of the chloroplast and to regulate translation via phosphorylation of the small ribosomal protein S6 (Robaglia et al. 2012). Mammalian and yeast S6 targets many other translational components and alters their activity, but it is not known if plant S6 has similar targets and effects.

Plant Viruses

A number of recent reviews on plant viruses and the translational machinery highlight some interesting features of the viral RNAs and interactions with the cap-binding complex subunits (Echevarria-Zomeno et al. 2013; Jiang and Laliberte 2011; Muench et al. 2012; Munoz and Castellano 2012; Nicholson and White 2011; Simon and Miller 2013). The positive-strand plant viruses lacking a cap group at their 5′ end have evolved a series of novel structures/shapes in their 3′ UTRs to recruit various components of the eIF4F complex and to form RNA to RNA interactions between the 3′ and 5′ ends of their mRNAs. The structures at the 3′ ends are termed cap-independent translational enhancers or 3′ CITEs (Nicholson and White 2011; Simon and Miller 2013). Many classes of plant viruses also have, in lieu of a m^7GpppX cap group, a viral-encoded protein (VPg) covalently linked to the 5′ end of their RNA genome (Jiang and Laliberte 2011). It has been shown that these VPg proteins interact with plant cap-binding proteins (eIF4E or eIFiso4E) among other host proteins; however, the mechanistic details of the VPg-cap-binding protein interaction and its relationship to viral replication are still unknown (Jiang and Laliberte 2011). Interestingly, plants have naturally developed mutations in their subunits for eIF4E, eIFiso4E, eIF4G, or eIFiso4G that provide resistance to a wide variety of these viruses, but these mutations do not appear to affect their ability to function in canonical host translation (Wang and Krishnaswamy 2012). There is still much to be learned about this unusual relationship between host translation factors and plant viruses.

Future Directions

Mechanistic studies of the plant translational machinery are still in their infancy. Much has been learned from yeast and mammalian systems, but there are likely to be subtle (or not so subtle) differences found, both in the machinery (e.g., eIFiso4F) and the regulation (e.g., apparent lack of 4EBP) as further studies on plant translation are carried out. Although *Arabidopsis* is a great model system, there are likely differences in protein translation even among plants such as monocots versus dicots or long-lived plants such as trees versus annuals. There is no doubt that there is much to be discovered as more plant genomes are sequenced, and we may be able to find these differences.

Among the insights for plant translation that we look forward to are the structures of plant ribosomes, additional molecular details of the formation of the pre-initiation complex, scanning and joining processes, posttranslational modifications of initiation factors, as well as molecular details about plant elongation and termination processes.

We have very little understanding of how plants regulate translation at this point in time. We only know that it appears to differ significantly from mammalian and yeast regulation in many respects. Future work should begin to reveal the connections that we know must exist in various plant signaling pathways and how those must regulate translation at many levels. We particularly need to know if the eIF2 phosphorylation/eIF2B recycling pathway is functional in plants or if there is an alternative method(s) of regulation and if there is a process by which cap-binding protein availability (eIF4E and/or eIFiso4E) is regulated comparable to the 4EBP system in other eukaryotes. Plants may have as yet undiscovered novel pathways for regulation of translation.

The methods that plant viruses use to appropriate the translational machinery suggest that cellular plant mRNAs may also have as yet unrecognized similar features such as 3′ CITEs that may regulate or facilitate cellular mRNA initiation. These are intriguing questions, for which answers do not yet exist.

One last area not included in this review, but that will certainly become important in the near future, is the role of small RNAs in regulation of translation. There is increasing evidence that small RNAs play many roles in all aspects of "doing business" as a cell, so it will not come as a surprise when we discover the many ways that small RNAs are involved in the regulation of plant translation.

Acknowledgment The research in Dr. Browning's laboratory is supported by grants from the National Science Foundation (MCB1052530 and *Arabidopsis* 2010S-0000335).

References

Aitken CE, Lorsch JR. A mechanistic overview of translation initiation in eukaryotes. Nat Struct Mol Biol. 2012;19:568–76.

Alekhina OM, Vassilenko KS. Translation initiation in eukaryotes: versatility of the scanning model. Biochemistry (Mosc). 2012;77:1465–77.

Andreou AZ, Klostermeier D. The DEAD-box helicase eIF4A: paradigm or the odd one out? RNA Biol. 2012;10:19–32.

Browning KS. Plant translation initiation factors: it is not easy to be green. Biochem Soc Trans. 2004;32:589–91.

Browning KS, Gallie DR, Hershey JWB, Hinnebusch AG, Maitra U, Merrick WC, Norbury C. Unified nomenclature for the subunits of eukaryotic initiation factor 3. Trends Biochem Sci. 2001;26:284.

Byrne ME. A role for the ribosome in development. Trends Plant Sci. 2009;14:512–9.

Carroll AJ. The *Arabidopsis* cytosolic ribosomal proteome: from form to function. Front Plant Sci. 2013;4:32–56.

Dever TE, Green R. The elongation, termination, and recycling phases of translation in eukaryotes. Cold Spring Harb Perspect Biol. 2012;4:55–70.

Echevarria-Zomeno S, Yanguez E, Fernandez-Bautista N, Castro-Sanz AB, Ferrando A, Castellano MM. Regulation of translation initiation under biotic and abiotic stresses. Int J Mol Sci. 2013;14:4670–83.

Gallie DR. Translational control in plants and chloroplasts. In: Translational control in biology and medicine, vol 26. 2007. p. 747–74

Gibson TJ. RACK1 research – ships passing in the night? FEBS Lett. 2012;586:2787–9.

Goss DJ, Kleiman FE. Poly(A) binding proteins: are they all created equal? Wiley Interdiscip Rev RNA. 2013;4:167–79.

Guo J, Jin Z, Yang X, Li JF, Chen JG. Eukaryotic initiation factor 6, an evolutionarily conserved regulator of ribosome biogenesis and protein translation. Plant Signal Behav. 2011;6:766–71.

Hinnebusch AG. Molecular mechanism of scanning and start codon selection in eukaryotes. Microbiol Mol Biol Rev. 2011;75:434–67.

Hinnebusch AG, Lorsch JR. The mechanism of eukaryotic translation initiation: new insights and challenges. Cold Spring Harb Perspect Biol. 2012;4:29–54.

Horiguchi G, Van LM, Candela H, Micol JL, Tsukaya H. Ribosomes and translation in plant developmental control. Plant Sci. 2012;191–192:24–34.

Hunt AG, Xing D, Li QQ. Plant polyadenylation factors: conservation and variety in the polyadenylation complex in plants. BMC Genomics. 2012;13:641–53.

Immanuel TM, Greenwood DR, MacDiarmid RM. A critical review of translation initiation factor eIF2α kinases in plants–regulating protein synthesis during stress. Funct Plant Biol. 2012;39:717–35.

Jackson RJ. The current status of vertebrate cellular mRNA IRESs. Cold Spring Harb Perspect Biol. 2012;4:89–108.

Jackson RJ, Hellen CU, Pestova TV. The mechanism of eukaryotic translation initiation and principles of its regulation. Nat Rev Mol Cell Biol. 2010;11:113–27.

Jackson RJ, Hellen CU, Pestova TV. Termination and post-termination events in eukaryotic translation. Adv Protein Chem Struct Biol. 2012;86:45–93.

Jennings MD, Pavitt GD. eIF5 is a dual function GAP and GDI for eukaryotic translational control. Small GTPases. 2010;1:118–23.

Jiang J, Laliberte JF. The genome-linked protein VPg of plant viruses-a protein with many partners. Curr Opin Virol. 2011;1:347–54.

Marintchev A. Roles of helicases in translation initiation: a mechanistic view. Biochim Biophys Acta. 2013;1829:799–809.

Mathews MB, Sonenberg N, Hershey JWB, eds. Cold Spring Harbor Laboratory Press, Cold Spring Harbor, NY.

Melnikov S, Ben-Shem A, Garreaude LN, Jenner L, Yusupova G, Yusupov M. One core, two shells: bacterial and eukaryotic ribosomes. Nat Struct Mol Biol. 2012;19:560–7.

Muench DG, Zhang C, Dahodwala M. Control of cytoplasmic translation in plants. Wiley Interdiscip Rev RNA. 2012;3:178–94.

Munoz A, Castellano MM. Regulation of translation initiation under abiotic stress conditions in plants: is it a conserved or not so conserved process among eukaryotes? Comp Funct Genomics. 2012;2012:406357–65.

Nicholson BL, White KA. 3′ Cap-independent translation enhancers of positive-strand RNA plant viruses. Curr Opin Virol. 2011;1:373–80.

Robaglia C, Thomas M, Meyer C. Sensing nutrient and energy status by SnRK1 and TOR kinases. Curr Opin Plant Biol. 2012;15:301–7.

Safer B. Nomenclature of initiation, elongation and termination factors for translation in eukaryotes. Eur J Biochem. 1989;186:1–3.

Sasikumar AN, Perez WB, Kinzy TG. The many roles of the eukaryotic elongation factor 1 complex. Wiley Interdiscip Rev RNA. 2012;3:543–55.

Simon AE, Miller WA. 3′ Cap-independent translation enhancers of plant viruses. Annu Rev Microbiol. 2013;67:21–42.

Sonenberg N, Hinnebusch AG. Regulation of translation initiation in eukaryotes: mechanisms and biological targets. Cell. 2009;136:731–45.

Syed NH, Kalyna M, Marquez Y, Barta A, Brown JW. Alternative splicing in plants–coming of age. Trends Plant Sci. 2012;17:616–23.

Topisirovic I, Svitkin YV, Sonenberg N, Shatkin AJ. Cap and cap-binding proteins in the control of gene expression. Wiley Interdiscip Rev RNA. 2011;2:277–98.

Valasek LS. 'Ribozoomin'–translation initiation from the perspective of the ribosome-bound eukaryotic initiation factors (eIFs). Curr Protein Pept Sci. 2012;13:305–30.

Wang A, Krishnaswamy S. Eukaryotic translation initiation factor 4E-mediated recessive resistance to plant viruses and its utility in crop improvement. Mol Plant Pathol. 2012;13:795–803.

Further Reading

Hwang J, Oh CS, Kang BC. Translation elongation factor 1B (eEF1B) is an essential host factor for Tobacco mosaic virus infection in plants. Virology. 2013;439:105–14.

Juntawong P, Bailey-Serres J. Dynamic light regulation of translation status in *Arabidopsis thaliana*. Front Plant Sci. 2012;3:66.

Patrick RM, Browning KS. The eIF4F and eIFiso4F complexes of plants: an evolutionary perspective. Comp Funct Genomics. 2012;2012:287814.

Rosado A, Li R, van de Van W, Hsu E, Raikhel NV. *Arabidopsis* ribosomal proteins control developmental programs through translational regulation of auxin response factors. Proc Natl Acad Sci U S A. 2012;109:19537–44.

Schepetilnikov M, Dimitrova M, Mancera-Martinez E, Geldreich A, Keller M, Ryabova LA. TOR and S6K1 promote translation reinitiation of uORF-containing mRNAs via phosphorylation of eIF3h. EMBO J. 2013;32:1087–102.

Chloroplast Signaling in Plants

6

Robert M. Larkin

Contents

R.M. Larkin (✉)
Michigan State University-Department of Energy Plant Research Laboratory, Michigan State University, East Lansing, MI, USA
e-mail: larkinr@msu.edu

S.H. Howell (ed.), *Molecular Biology, The Plant Sciences 2*,
DOI 10.1007/978-1-4614-7570-5_10

Abstract

- Plastids are not only major centers of plant metabolism, plastids also emit signals in response to endogenous and environmental cues.
- Plastid signals are major regulators of nuclear gene expression.
- Several plastid signals are known.
- Plastid signals appear to regulate a complex network of signaling mechanisms.
- Plastid signals contribute to chloroplast biogenesis, chloroplast function, and chloroplast stress tolerance.
- Plastid signals contribute to extraplastidic processes such as the circadian rhythm, development, biotic and abiotic stress tolerance, and programmed cell death.

Introduction

Photosynthesis is the conversion of solar energy into biological useful forms of energy. This process underpins plant growth, development, reproduction, crop yields, and life on earth. Photosynthesis and a considerable amount of nonphotosynthetic metabolism are performed by chloroplasts. Chloroplasts evolved from an endosymbiosis between the cyanobacterial progenitor of modern chloroplasts and a heterotrophic eukaryote that had already acquired mitochondria from a previous endosymbiosis with a proteobacterium. As endosymbiosis was established, many of the cyanobacterial genes were either lost or transferred to the nucleus. Indeed, most of the approximately 3,000 chloroplast proteins of plants are encoded by nuclear genes. Nonetheless, the chloroplasts of plants do retain relatively small genomes that encode less than 100 proteins. Many of these chloroplast genes contribute to either the expression of the chloroplast genome or photosynthesis. To cope with this separation of genetic material, chloroplasts acquired capabilities such as the ability to import chloroplast proteins that are encoded by nuclear genes and the ability to communicate with other cellular compartments.

Signaling mechanisms that are localized entirely outside of chloroplasts are major regulators of nuclear genes that encode chloroplast proteins. These signaling mechanisms are regulated by environmental and endogenous cues such as qualities and quantities of light, cell type, hormones, carbohydrates, and the circadian rhythm. The chloroplast is not completely submissive to this anterograde flow of information from the nucleus to the chloroplast. The chloroplast emits signals that can regulate the expression of nuclear genes. Thus, chloroplasts are not only major centers of plant metabolism, they are also important sensors. Retrograde plastid-to-nucleus signaling appears complex. Plastid-to-nucleus signaling mechanisms are integrated with other plastid-to-nucleus signaling mechanisms and with extraplastidic signaling mechanisms. Plastid-to-nucleus signaling regulates the expression of nuclear genes that encode chloroplast proteins and nuclear genes that encode extraplastidic proteins. Thus, plastid-to-nucleus signaling is not only a feedback mechanism that optimizes chloroplast function but also a signaling system that broadly affects plants. At present, our knowledge of plastid-to-nucleus signaling is incomplete. The central focus of this field of research is to identify plastid signals, determine plastid-to-nucleus signaling mechanisms, and assign biological functions to particular plastid-to-nucleus signaling mechanisms (Susek and Chory 1992; Gray et al. 2003; Larkin and Ruckle 2008; Mullineaux and Baker 2010; Inaba et al. 2011; Karpiński et al. 2012; Krause et al. 2012; Leister 2012; Pfalz et al. 2012; Rüdiger and Oster 2012; Tikkanen et al. 2012; Chi et al. 2013; Larkin 2014).

Early Findings

The first evidence that chloroplasts can emit signals that affect nuclear gene expression was reported by Bradbeer et al. 1979, who performed experiments with two barley mutants that develop green and white striped tissues and also completely white leaves and shoots. The activities of plastid enzymes that are encoded by nuclear genes were reduced in the white tissues of both mutants relative to wild type. These findings were consistent with dysfunctional chloroplasts emitting signals that downregulate the expression of nuclear genes encoding chloroplast proteins or functional chloroplasts emitting signals that induce the expression of these genes.

Another set of influential early experiments were performed by Mayfield and Taylor 1984. These experiments utilized maize mutants that accumulate phytoene (a carotenoid precursor) but do not accumulate colored carotenoids. Colored carotenoids perform a number of functions that support photosynthesis such as stabilizing antenna proteins, harvesting light, quenching both triplet chlorophyll and singlet oxygen ($^1O_2$), contributing to the thermal dissipation of excess-light energy from singlet chlorophyll (i.e., nonphotochemical quenching), and serving as an antioxidant for the lipids of the thylakoid membranes. When carotenoids do not accumulate, chloroplast biogenesis is blocked. The mechanism that blocks chloroplast biogenesis when plants are deficient in carotenoids is not completely understood. Some data is consistent with carotenoid deficiencies blocking chloroplast biogenesis by promoting collisions between O_2 and triplet chlorophyll, which induces the levels of $^1O_2$ – a toxic reactive oxygen species (Susek and Chory 1992; Gray et al. 2003). Other data is consistent with some other mechanism such as the misfolding of carotenoid-binding proteins that localize to the thylakoid membranes blocking chloroplast biogenesis without the accumulation of $^1O_2$ (Kim and Apel 2013a). Regardless of the mechanism, carotenoid-deficient mutants contain nonphotosynthetic plastids that resemble proplastids rather than chloroplasts and exhibit albino phenotypes. These carotenoid-deficient mutants do not accumulate detectable levels of mRNAs that encode the type I proteins of the major light-harvesting complex of photosystem II (Lhcb1) relative to their wild-type siblings. The Lhcb1 proteins are encoded by nuclear genes and reside in the thylakoid membranes where they bind chlorophylls and carotenoids and serve as antennae for photosystem II (PSII). These findings provide evidence that chloroplast biogenesis is required for the accumulation of mRNAs that are transcribed from *Lhcb1* (Mayfield and Taylor 1984).

Although carotenoid-deficient mutants were useful for early experiments, they are difficult to maintain and none are conditional. Thus, researchers began using inhibitors that specifically block chloroplast biogenesis to study plastid-to-nucleus signaling. Blocking chloroplast biogenesis with inhibitors downregulates the expression of *Lhcb1* genes and other photosynthesis-associated nuclear genes (PhANGs) (Fig. 1). In general, these inhibitors cannot convert mature chloroplasts to proplastids, but they can block the conversion of proplastids to chloroplasts, which occurs during germination and leaf development.

In the early days of plastid-to-nucleus signaling research, norflurazon was the most widely used inhibitor of chloroplast biogenesis. Norflurazon specifically blocks chloroplast biogenesis by specifically inhibiting phytoene desaturase, an enzyme that is essential for carotenoid biosynthesis. Phytoene desaturase contributes to carotenoid biosynthesis by catalyzing two sequential dehydrogenation reactions on phytoene yielding ζ-carotene. Without carotenoids, proplastids are not converted to chloroplasts as described above. When plant tissues are treated

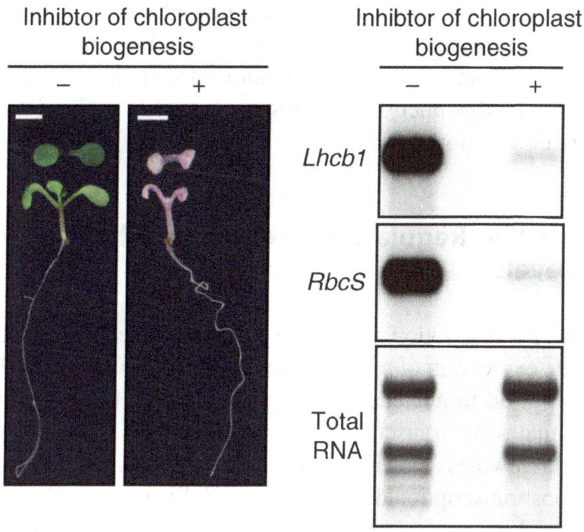

Fig. 1 Photosynthesis-associated nuclear gene (PhANG) expression in Arabidopsis seedlings treated with an inhibitor of chloroplast biogenesis. Images of Arabidopsis seedlings (*left*). Arabidopsis seedlings were germinated and grown for several days on media that contains no inhibitor (−) or media that contains lincomycin (+), an inhibitor of chloroplast biogenesis. Lincomycin treated seedlings do not accumulate chlorophyll but they do accumulate anthocyanins. Thus, lincomycin-treated seedlings are purple. Bar = 2 mm. Expression of *Lhcb1* and *RbcS* in untreated and lincomycin-treated seedlings (*right*). Total RNA was extracted from Arabidopsis seedlings that were treated (+) or not treated (−) with lincomycin. The levels of mRNAs that are transcribed from *Lhcb1* (*top*) and *RbcS* (*middle*) were quantified by RNA blot hybridization. Methylene blue staining of total RNA demonstrates equivalent loading (*bottom*) (Reproduced with permission from Larkin and Ruckle (2008))

with norflurazon, their plastids are arrested at an early stage of chloroplast biogenesis and do not contain chlorophyll or thylakoid membranes.

Although norflurazon is widely used for blocking chloroplast biogenesis, a variety of inhibitors that block distinct molecular processes are also useful for blocking chloroplast biogenesis. These include inhibitors of chloroplast translation such as chloramphenicol, lincomycin, erythromycin, streptomycin, and heat treatments; inhibitors of chloroplast transcription such as tagetitoxin and rifampicin; nalidixic acid, a prokaryotic DNA gyrase inhibitor that affects plastid DNA replication and transcription; and various mutant alleles that cause albinism. In general, mutants with variegated leaves (i.e., leaves that contain green and yellow/white sectors) express normal levels of PhANGs in green sectors and reduced levels of PhANGs in yellow and white sectors. The finding that diverse inhibitors of chloroplast biogenesis downregulate the expression of PhANGs when applied before chloroplast biogenesis occurs but not after chloroplast is completed provides evidence that the expression of PhANGs depends on chloroplast biogenesis (Susek and Chory 1992; Gray et al. 2003).

Recently, lincomycin treatments were found to inhibit the expression of genes that contribute to starch metabolism during amyloplast biogenesis. Amyloplasts are nonpigmented plastids that synthesize and store starch (Chi et al. 2013). Thus, the phenomenon of plastid dysfunction affecting nuclear gene expression is observed in both photosynthetic and nonphotosynthetic cells.

Plastid Signals Can Regulate Expression of PhANGs at Multiple Levels

Results from run-on transcription assays that were performed with purified nuclei and quantitation of the expression of reporter genes that are driven by PhANG promoters indicate that the plastid-to-nucleus signaling activated by blocking chloroplast biogenesis downregulates the transcription of most PhANGs (Susek and Chory 1992; Gray et al. 2003). However, plastid signals appear to regulate the expression of one PhANG by a posttranscriptional mechanism (Sullivan and Gray 2002; Brown et al. 2005). Plastid-responsive promoter elements were identified in norflurazon-treated transgenic plants using reporter genes in which expression is driven by PhANG promoters with mutations in particular promoter elements. In all plastid-responsive promoters studied to date, light and plastid signaling utilize common or overlapping promoter elements. A combination of two fairly common promoter elements is required for both light- and plastid-responsive transcription. Thus, there is no single light- and plastid-responsive promoter element (Gray et al. 2003).

Light-responsive promoter elements are more thoroughly studied than plastid-response promoter elements. Sequences that can contribute to light responsiveness include the GT element (GR(T/A)AA(T/A)), G box (CACGTG), I box (GATA), AT-rich motifs, CCAAT motif, and the Gap box (ATGAA(A/G)A). The diversity of light- and plastid-responsive promoter elements may contribute to the diverse response of these genes to light quality, light intensity, and plastid dysfunction (Terzaghi and Cashmore 1995). Although light and plastid signals utilize common or overlapping promoter elements, light and plastid signals appear distinct because plastid dysfunction downregulates PhANG expression in the dark (Gray et al. 2003).

The First Proposals for Plastid Signals

Bradbeer et al. (1979) proposed that a chloroplastic RNA might serve as a plastid signal. Currently, no evidence exists for a chloroplastic RNA affecting nuclear gene expression. Other early proposals for plastid signals included proteins that move between the plastids and the nucleus such as proteins that might dissociate from dysfunctional plastids or that dysfunctional plastids might inefficiently import (Susek and Chory 1992). There is now evidence for such mechanisms (sections "Loss-of-Function Alleles of *GUN1* Broadly Disrupt Plastid Signaling" and "Chloroplast-Localized Transcription Factors Contribute to Plastid-to-Nucleus Signaling").

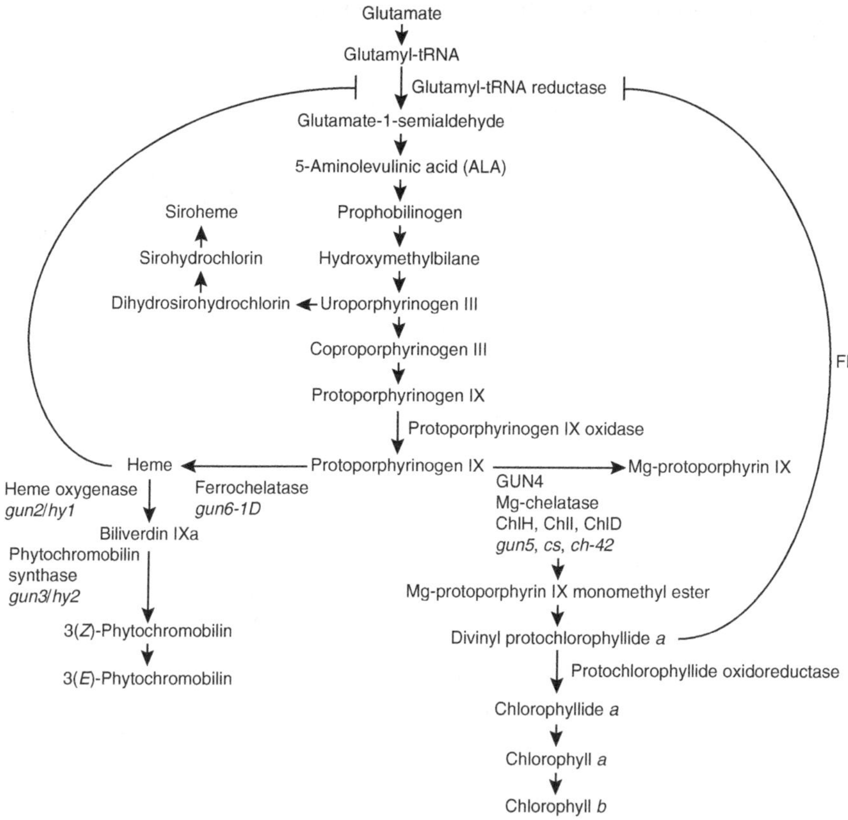

Fig. 2 The plastid-localized tetrapyrrole biosynthetic pathway. The names of each metabolite are indicated. Enzymatic reactions are indicated with *arrows*. Feedback inhibition is indicated with *T bars*. The names of the enzymes, enzyme subunits, and mutant alleles that are discussed in the text are indicated

Porphyrins such as chlorophyll precursors and heme were also considered attractive candidates for plastid signals because chlorophyll precursors were reported to regulate the expression of *Lhcb* genes in *Chlamydomonas reinhardtii* and heme was known to regulate the expression of mitochondria-related genes in *Saccharomyces cerevisiae* (Susek and Chory 1992). Porphyrins are heterocyclic macrocycles that contain four pyrrole groups. These and other tetrapyrroles such as chlorophylls, siroheme, and phytochromobilin are synthesized in the plastid (Fig. 2). The chloroplast can export tetrapyrroles into the cytosol. Indeed, heme is transported from the chloroplast to several distinct cellular compartments where it serves as a cofactor for diverse proteins. Based on this information, dysfunctional chloroplasts were proposed to export porphyrins into the cytosol. In the cytosol, porphyrins were proposed to activate a signaling mechanism that regulates the transcription of PhANGs (Susek and Chory 1992). Indeed, heme is known to activate signaling in bacteria, fungi, and animals (Terry and Smith 2013).

Porphyrins Can Regulate Nuclear Gene Expression in *C. reinhardtii* and Plants

The first evidence that a chlorophyll precursor might regulate PhANG expression in *C. reinhardtii* was reported by Johanningmeier and Howell (1984). An inverse correlation was reported between the levels of Mg-protoporphyrin IX monomethyl ester (Mg-ProtoME) and the levels of the mRNAs that encode Lhcb and the small subunit of ribulose-1,5-bisphosphate carboxylase/oxygenase (RbcS). In these experiments, mutant alleles and inhibitors were used to alter the levels of particular chlorophyll precursors.

Subsequent work in *C. reinhardtii* took advantage of this alga's ability to import porphyrins from growth media. Feeding porphyrins such as hemin and Mg-protoporphyrin IX (Mg-Proto) to *C. reinhardtii* was found to regulate the expression of nuclear genes. Hemin is protoporphyrin IX chelating a ferric ion (i.e., Fe^{3+}) with a chloride ligand. In contrast, heme (aka, heme B or protoheme) is protoporphyrin IX chelating a ferrous ion (Fe^{2+}). The ferrous ion readily oxidizes to the ferric ion. Thus, feeding heme to *C. reinhardtii* is not practical. Although porphyrins are not toxic when they are fed to *C. reinhardtii* in the dark, feeding porphyrins to *C. reinhardtii* in the light causes lethal photooxidative damage because when these porphyrins are excited by light, they can transfer energy to O_2 yielding $^1O_2$.

Feeding Mg-Proto or hemin to *C. reinhardtii* in the dark was found to transiently induce the expression of nuclear genes such as genes that encode glutamyl-tRNA reductase (HEMA) and the heat shock proteins HSP70A and HSP70B. These effects of Mg-Proto and hemin feeding on the expression of *HSP70A* were observed with low concentrations of either Mg-Proto or hemin. Both of these porphyrins were found to rapidly induce the expression of *HSP70A*. Light was found to induce a transient increase in Mg-Proto levels and to promote the export Mg-Proto from the chloroplast to the cytosol. Mg-Proto-, hemin-, and light-mediated induction of *HSP70A* expression depend on a single promoter element named the plastid response element ($GCGACNAN_{15}TA$). Thus, the transient light-induced expression of *HSP70A* was suggested to depend on the light-induced biosynthesis of Mg-Proto (Rüdiger and Oster 2012; Chi et al. 2013).

Based on transcriptome analysis, feeding either Mg-Proto or hemin to *C. reinhardtii* in the dark regulates the expression of nearly 1,000 genes that contribute to diverse processes. These experiments demonstrate that few genes contributing to photosynthesis are regulated by feeding either Mg-Proto or hemin to *C. reinhardtii*. These data conflict with the earlier finding that the accumulation of Mg-ProtoMe correlates with the downregulation of *Lhcb* and *RbcS* expression. However, the early work was carried out with light-grown cultures. Light regulates gene expression in *C. reinhardtii* and porphyrins can act as photosensitizers. The rationale for this discrepancy is not yet determined. The largest group of genes that were regulated when Mg-Proto and hemin were fed to *C. reinhardtii* in the dark contributes to proteolysis and protein folding. The expression of more than half of these genes is also regulated by heat shock. Based on these data, porphyrin signaling is proposed to help *C. reinhardtii* adapt to changing environmental conditions (Chi et al. 2013).

The finding that Mg-Proto can affect gene expression in *C. reinhardtii* was corroborated by experiments with land plants. To modulate porphyrin levels *in planta*, researchers feed the porphyrin precursor 5-aminolevulinic acid (ALA) to plants (Fig. 2). ALA is taken up by the roots. ALA accumulates in plastids where it serves as a precursor for porphyrin biosynthesis. Feeding ALA to plants can flood the tetrapyrrole biosynthetic pathway with intermediates and cause a massive accumulation of numerous chlorophyll precursors.

To circumvent the potential photosensitizing effects of porphyrins in these experiments, the levels of PhANG mRNA were monitored after a brief, non-photosensitizing pulse of red light, during continuous illumination with non-photosensitizing far-red light, or in complete darkness. In garden cress (*Lepidium sativum L.*) and Arabidopsis, ALA feeding was found to either upregulate or downregulate PhANG expression depending on the concentration of ALA and the age of the seedlings (Rüdiger and Oster 2012; Woodson et al. 2012).

Particular inhibitor treatments induce higher levels of Mg-Proto and Mg-ProtoME. β-thujaplicin and amitrole treatments that cause Mg-ProtoMe and Mg-Proto accumulation, respectively, were found to specifically inhibit *Lhcb1* and/or *RbcS* expression. Similar to ALA-feeding experiments, the levels of mRNAs transcribed from *Lhcb1* and *RbcS* were reduced after Mg-Proto or Mg-ProtoMe levels were induced by these inhibitor treatments in light conditions that do not cause photooxidative stress such as dim white light, far-red light, and darkness. Based on these data, the accumulation of tetrapyrroles was suggested to affect the expression of PhANGs in plants (Rüdiger and Oster 2012).

Recently, directly watering Arabidopsis roots with solutions that contain Mg-Proto was reported to induce Mg-Proto accumulation in green aerial tissues and to downregulate the expression of *Lhcb* and *RbcS*. The Mg-Proto feeding that downregulates the expression of *Lhcb* depends on a chloroplastic protein named GUN1 (section "*cry1* Mutants Are *gun* Mutants"). Thus, Mg-Proto feeding appears to downregulate the expression of *Lhcb* by affecting a plastid-to-nucleus signaling mechanism in the chloroplast (Larkin 2014).

Mg-Proto Helps to Coordinate DNA Replication in Nuclei, Chloroplasts, and Mitochondria

In the red alga *Cyanidioschyzon merolae* and potentially in cultured tobacco cells, Mg-Proto binds and inhibits an F-box protein named Fbx3 that contributes to an S-phase kinase-associated protein 1 (Skp1)–Cullin–F-box protein (SCF) complex. SCF-type ubiquitin ligases regulate diverse cellular processes. In this type of regulation, the F-box protein of the SCF complex binds a target protein and conjugates ubiquitin to the target protein. Ubiquitinylation of proteins targets them for rapid degradation by the proteasome. Fbx3 is essential for the

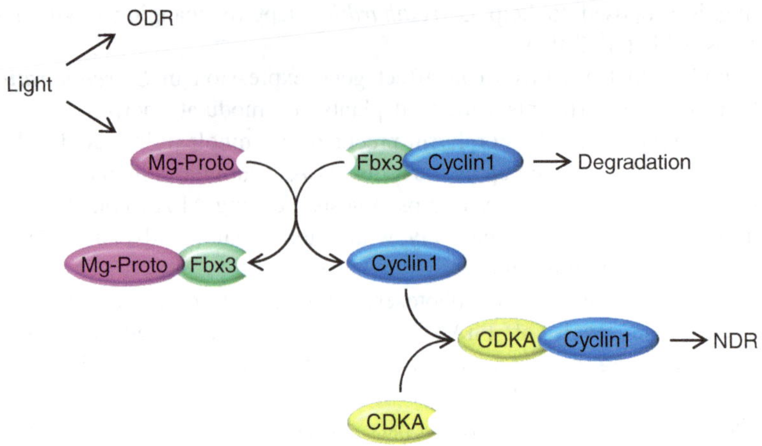

Fig. 3 Coordination of DNA replication by Mg-Proto. Light induces a rise in the levels of Mg-Proto and induces DNA replication in chloroplasts and mitochondria (ODR). Mg-Proto activates cyclin-dependent kinase A (CDKA) by binding and inhibiting Fbx3. Fbx3 targets cyclin1 for ubiquitinylation and degradation by the proteasome. Mg-Proto binds Fbx3 and releases both Fbx3 and cyclin1 from the SCF complex. Cyclin1 binds and activates CDK2. CDK2 activates nuclear DNA replication (NDR)

ubiquitinylation of cyclin1 and thus targets cyclin1 for degradation. Cyclin1 stimulates nuclear DNA replication by binding and activating cyclin-dependent kinase A (CDKA). Light induces the replication of DNA in chloroplasts and mitochondria and the levels of Mg-Proto, which binds and inhibits Fbx3. Thus, Mg-Proto protects cyclin 1 from degradation by the proteasome and, therefore, stimulates nuclear DNA replication. This regulation of targeted proteolysis by Mg-Proto helps to coordinate DNA replication in the nucleus with DNA replication in other organelles (Fig. 3) (Tanaka and Hanaoka 2013).

The *genomes uncoupled* (*gun*) Mutant Screen

Plastid signals and the components of plastid-to-nucleus signaling mechanisms are difficult to predict. Unbiased mutant screens are attractive approaches for studying such processes. The first screen for mutant alleles that disrupt plastid-to-nucleus signaling was one that yielded *genomes uncoupled* (*gun*) mutants (Susek and Chory 1992; Susek et al. 1993). When wild-type Arabidopsis seedlings were treated with inhibitors that block chloroplast biogenesis, the expression of PhANGs was downregulated, but when *gun* mutants were treated with these same inhibitors, the expression of PhANGs was upregulated relative to wild type. Thus, *gun* alleles uncouple the expression of PhANGs from chloroplast function. *gun* alleles may attenuate the activity of negative regulators of PhANG expression that are activated when chloroplast biogenesis is blocked, or *gun* alleles may induce the activity of positive regulators of PhANG expression.

gun mutant screens used reporter genes to identify mutants that express higher levels of PhANGs when chloroplast biogenesis was blocked. The expression of these reporter genes was driven by promoters from *Lhcb1* genes or other PhANGs. In *gun* mutant screens, researchers blocked chloroplast biogenesis by germinating seeds on media that contained norflurazon and screened for mutant seedlings that express elevated levels of PhANG-driven reporter genes relative to wild type after several days of growth.

In general, researchers have performed mutant screens in various organisms using a variety of experimental conditions. These conditions include optimal laboratory conditions, stressful conditions in the laboratory, and specialized laboratory conditions that are neither optimal nor stressful. Whether growth conditions for screening are natural, stressful, or unnatural and whether the mutant screen is performed in a wild-type or mutant organism does not matter. When researchers perform a mutant screen, they do so assuming that they can specifically disrupt a particular process or some small number of processes with their screen. If this is the case, then the researcher will repeatedly isolate alleles of genes that contribute to a particular process or a small number of processes. If, however, a researcher isolates only single alleles of a large number of different genes that contribute to diverse processes, that would indicate that the screen does not specifically disrupt a particular process. *gun* mutant screens yielded large numbers of alleles of a particular set of genes that contribute to a few distinct processes. This indicates that the *gun* mutant screens specifically disrupt a few distinct mechanisms that downregulate the expression of PhANGs when chloroplast biogenesis is blocked.

Particular *gun* Alleles Disrupt Tetrapyrrole Metabolism

The *gun* screen yielded several alleles of genes that contribute to tetrapyrrole metabolism in chloroplasts. This finding indicated that the *gun* mutant screen specifically disrupted a plastid-to-nucleus signaling mechanism that depends on tetrapyrrole metabolism. *gun2* and *gun3* are loss-of-function alleles of previously described genes, *HY1* and *HY2*. *HY1/GUN2* encodes heme oxygenase (Fig. 2). *HY2/GUN3* encodes phytochromobilin synthase (Fig. 2). Both *HY1/GUN2* and *HY2/GUN3* contribute to the biosynthesis of phytochromobilin (Fig. 2). Loss-of-function alleles of these genes lower the levels of chlorophyll and chlorophyll precursors relative to wild type in Arabidopsis, tomato, and pea by inhibiting heme catabolism and thus inducing a rise in heme levels. Heme inhibits glutamyl-tRNA reductase (GluTR), the second enzyme in the plastid-localized tetrapyrrole biosynthetic pathway (Fig. 2). Inhibiting GluTR lowers the levels of chlorophyll.

GUN5 encodes the 140-kDa subunit of Mg-chelatase. Mg-chelatase catalyzes the insertion of Mg^{2+} into Proto, yielding Mg-Proto (Fig. 2). Mg-chelatase requires three subunits in vitro and in vivo. In plants, these subunits are commonly referred to as ChlH/GUN5, ChlD, and ChlI. ChlH/GUN5 is the Proto- and Mg-Proto-binding subunit and is likely the Mg^{2+}-binding subunit of Mg-chelatase. ChlI and ChlD are related to AAA-type ATPases that form interacting oligomeric rings. The rings associate with ChlH and drive the ATP-dependent metalation of Proto, yielding Mg-Proto. GUN4 is a major activator of Mg-chelatase (Fig. 2) binding to Proto, Mg-Proto, and the 140-kDa subunit of Mg-chelatase (i.e., GUN5). GUN4 activates Mg-chelatase and helps to channel porphyrins into chlorophyll biosynthesis on chloroplast membranes.

As described in section "Early Findings", norflurazon inhibits carotenoid biosynthesis and blocks chloroplast biogenesis. The mechanism of norflurazon action is debated. If norflurazon blocks chloroplast biogenesis by promoting collisions between O_2 and chlorophyll that yield $^1O_2$, the attenuated chlorophyll biosynthesis in the leaky mutants *gun2*, *gun3*, *gun4-1*, and *gun5* provides sufficient photooxidative stress to block chloroplast biogenesis and downregulate the expression of *Lhcb1* to the same degree as in wild type. This conclusion is based on the analysis of a leaky loss-of-function allele for the gene that encodes the ChlI subunit of Mg-chelatase named *cs*. The *cs* and *gun4-1* mutants accumulate approximately 40 % of the chlorophyll found in wild type, while *gun5* accumulates approximately 70 % of the chlorophyll found in wild type. In contrast to *gun4-1* and *gun5*, *cs* does not accumulate more *Lhcb1* mRNA than wild type when chloroplast biogenesis is blocked with norflurazon. Thus, *cs* is not a *gun* mutant. This finding indicates that norflurazon does not require robust chlorophyll biosynthesis to block chloroplast biogenesis and downregulate PhANG expression in Arabidopsis and that the attenuated chlorophyll biosynthesis of *cs* is sufficient to block chloroplast biogenesis and to downregulate the expression of PhANGs. Therefore, the attenuated chlorophyll metabolism in *gun5* is also sufficient to block chloroplast biogenesis and downregulate the expression PhANGs to the levels that are observed in wild type. Thus, *gun2*, *gun3*, *gun4-1*, and *gun5* are proposed not to cause resistance to norflurazon. Norflurazon is proposed to block chloroplast biogenesis and to activate a plastid-to-nucleus signaling mechanism that depends on porphyrin metabolism (Inaba et al. 2011; Chi et al. 2013). Consistent with this interpretation, none of the abovementioned *gun* mutants were obtained from screens for norflurazon-resistant mutants (i.e., the *happy on norflurazon* (*hon*) mutants). In particular, *gun4-1* and *gun5* were found to not cause resistance to low concentrations of norflurazon, in contrast to the *hon* mutants (Saini et al. 2011).

More support for the interpretation that perturbations in tetrapyrrole biosynthesis caused by *gun* alleles affect plastid-to-nucleus signaling came from an analysis of Arabidopsis sigma factor (*SIG*) gene mutants. The nuclear genome of Arabidopsis encodes six sigma factors that are required by the plastid-encoded DNA-dependent RNA polymerase (PEP) for accurate transcription of plastid genes. PEP and the nuclear encoded DNA-dependent RNA polymerase (NEP) transcribe the plastid genome. The *sig2* and *sig6* null mutants were found to accumulate less than 20 % of the chlorophyll that was found in wild type. Like norflurazon treatments, null alleles of *SIG2* and *SIG6* reduced the expression of PhANGs relative to wild type. The *gun5* allele upregulated the expression of *Lhcb2.2* and other PhANGs in the *gun5 sig2* double mutant relative to the *sig2* single mutant. These data provide evidence that the *gun5* allele attenuates a plastid-to-nucleus signaling mechanism that is activated by a *sig2* allele (Woodson et al. 2012). In summary, the current data indicate that inhibiting chloroplast function with either a norflurazon treatment or a *sig2* allele activates a plastid-to-nucleus signaling mechanism that depends on porphyrin metabolism.

GUN4 also promotes chlorophyll biosynthesis and plastid-to-nucleus signaling in *C. reinhardtii*. Although the *gun4* mutant of *C. reinhardtii* accumulates 50 % of the chlorophyll of wild type in the dark and in low-fluence-rate light, this mutant accumulates higher levels of mRNAs that are transcribed from *LhcbM1* and *LhcbM5* than wild type under optimal growth conditions (i.e., without an inhibitor of chloroplast biogenesis). *LhcbM1* and *LhcbM5* encode antennae proteins of PSII. Excess light that induces ROS levels downregulates the expression of these genes in this mutant, presumably by activating ROS-responsive plastid-to-nucleus signaling mechanisms (sections "Excess Light Activates Plastid-to-Nucleus Signaling," "Plastid-to-Nucleus Signaling Contributes to Systemic-Acquired Acclimation," "Excess Light Can Induce Acclimation or Cell Death," "Moderate Increases in Light Intensity Can Induce the Expression of *APX2*," "PSII Can Produce $^1O_2$," "$^1O_2$ Activates Plastid-to-Nucleus Signaling in the Arabidopsis *flu* Mutant," "In the *flu* Mutant, Plastid-to-Nucleus Signaling Requires *EXECUTER1* and *EXECUTER2*," "Other *flu*-Based Screens for Mutant Alleles that Disrupt Plastid-to-Nucleus Signaling," "$^1O_2$ and cry1 Signaling Are Required for Light-Dependent PCD," "*flu*-Derived $^1O_2$ Can Affect Chloroplast Biogenesis," "$^1O_2$-Dependent Plastid-to-Nucleus Signaling Induces Either Acclimation or Cell Death," and "β-Cyclocitral Activates Plastid-to-Nucleus Signaling") (Formighieri et al. 2012). Thus, porphyrins and ROS appear to activate distinct plastid-to-nucleus signaling mechanisms in *C. reinhardtii*.

One gain-of-function allele was isolated from a *gun* mutant screen, *gun6-1D*. *gun6-1D* is a gain-of-function allele for the gene that encodes ferrochelatase 1 (FC1). Ferrochelatase inserts a ferrous ion (Fe^{2+}) into Proto to yield heme (aka protoheme and heme B) (Fig. 2). Ferrochelatase and Mg-chelatase have a common substrate, Proto (Fig. 2). Thus, *gun6-1D* is thought to induce heme biosynthesis and decrease Mg-Proto biosynthesis by promoting *FC1* expression. Arabidopsis contains two ferrochelatase genes, *FC1* and *FC2*. *FC1* is expressed in all tissues and is induced by stress. *FC2* is coexpressed with PhANGs. Overexpression of *FC2* does not appear to disrupt plastid-to-nucleus signaling. Based on these data, *FC1* and *FC2* would appear to synthesize physiologically distinct pools of heme. The pool of heme synthesized by FC1 affects plastid-to-nucleus signaling.

The identity of the porphyrin signal that regulates nuclear gene expression in plants is not yet known. Popular candidates include Mg-Proto and heme, which are proposed to downregulate or upregulate nuclear gene expression, respectively. Some researchers propose that plastid-to-nucleus signaling is activated by local or transient concentrations of particular porphyrins.

Porphyrin signaling appears to broadly affect plant cells. In addition to contributing to the plastid-to-nucleus signaling that regulates the expression of PhANGs, heme signaling contributed to the expression of genes that contribute to starch biosynthesis during amyloplast biogenesis in nonphotosynthetic cultured tobacco cells that do not express PhANGs (Chi et al. 2013). Additionally, porphyrin signaling may also contribute to stress tolerance. Overexpressing protoporphyrinogen IX oxidase

(Fig. 2) in *Oryza sativa* enhanced drought tolerance relative to wild type. Additionally, *gun5* had less basal thermotolerance than wild type (Larkin 2014).

Loss-of-Function Alleles of *GUN1* Broadly Disrupt Plastid Signaling

gun mutant screens yielded 13 loss-of-function alleles of *GUN1*. Thus, the *gun* mutant screen specifically disrupts a plastid-to-nucleus signaling mechanism that depends on *GUN1*. *GUN1* encodes a pentatricopeptide repeat (PPR) protein that colocalizes with nucleoids in chloroplasts (Woodson and Chory 2008; Chi et al. 2013). PPR proteins are defined by a degenerate 35-amino acid residue sequence that is repeated in tandem from 2 to 30 times. These PPR tracts bind RNA. In plants, PPR proteins make essential contributions to posttranscriptional processes in chloroplasts and mitochondria such as reactions that splice, edit, process, stabilize, and translate RNA. Thus, PPR proteins make numerous contributions to the expression of the genomes of chloroplasts and mitochondria. Genes that encode PPR proteins experienced a major expansion in plants. The genomes of protists, fungi, and metazoans encode approximately 5–50 PPR proteins. Plant genomes encode more than 400 PPR proteins (Fujii and Small 2011). The GUN1 protein belongs to a subfamily of 5 PPR proteins that contain an SMR domain that is found in proteins that contribute to DNA repair and recombination (Woodson and Chory 2008).

Loss-of-function alleles of *GUN1* were shown to upregulate the expression of PhANGs regardless of whether chloroplast biogenesis is blocked with norflurazon, chloramphenicol, lincomycin, or high levels of glucose (Susek et al. 1993; Woodson and Chory 2008). These alleles of *GUN1* also upregulated the expression of PhANGs in the albino sectors of variegated leaves (Susek et al. 1993). *gun1* mutants are more sensitive to inhibitors of chloroplast biogenesis than wild type. For example, lower concentrations of lincomycin are required to block chloroplast biogenesis in *gun1* mutants than in wild type (Ruckle et al. 2012). Thus, *gun1* mutants are neither resistant to particular inhibitors of chloroplast biogenesis nor do they contain chloroplasts that are more resistant to diverse inhibitors of chloroplast biogenesis than wild type because of enhanced acclimation to chloroplast stress. Most likely, these *GUN1* alleles upregulate the expression of PhANGs by disrupting a plastid-to-nucleus signaling mechanism. Transcriptome analyses were consistent with loss-of-function alleles of *GUN1* causing the misregulation of more than 1,000 genes (Woodson and Chory 2008; Chi et al. 2013).

Loss-of-function alleles of *GUN1* not only disrupt the plastid-to-nucleus signaling that is activated when chloroplast biogenesis is blocked, they also disrupt the plastid-to-nucleus signaling that is activated when chloroplast functions are attenuated. For instance, the expression of PhANGs was upregulated in the *gun1 sig2* and *gun1 sig6* double mutants relative to the *sig2* and *sig6* single mutants, which accumulate less chlorophyll than wild type (Woodson et al. 2012). Additionally, loss-of-function alleles of *GUN1* attenuated the downregulated expression of *Lhcb1* caused by excess light in green seedlings that contain normal levels of chlorophyll.

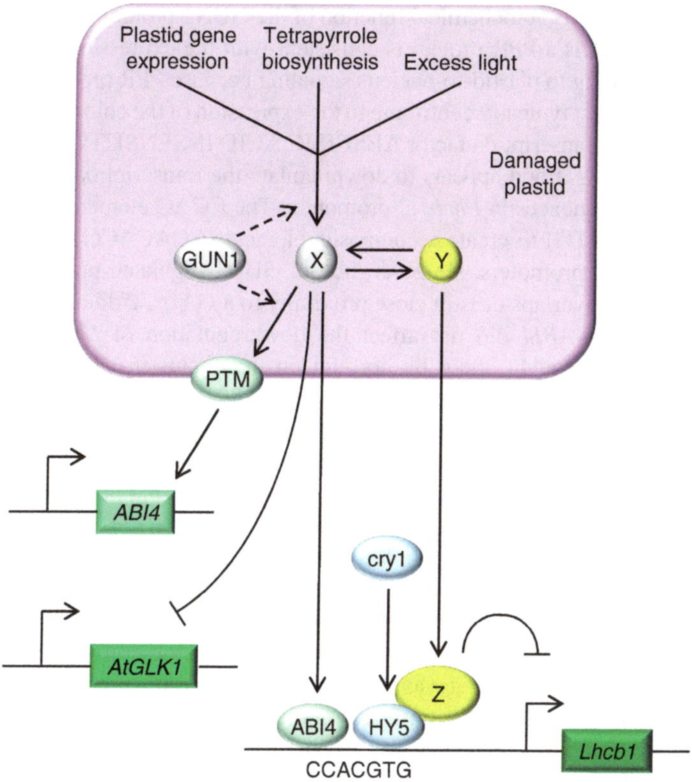

Fig. 4 *GUN* gene-dependent plastid-to-nucleus signaling mechanisms. When chloroplast biogenesis is blocked, plastid-to-nucleus signaling regulates the expression of more than 1,000 nuclear genes. In particular, the expression of *ABI4* is up-regulated and the expression of both *AtGLK1* and *Lhcb* is down-regulated. This plastid-to-nucleus signaling is activated by changes in the expression of the plastid genome, porphyrin metabolism, and excess light. GUN1 acts downstream of these signals by contributing to the biosynthesis or transduction of an unknown plastid signal (X) that triggers the partial proteolysis and release of PTM from the cytoplasmic surface of the chloroplast envelope. PTM induces the expression of *ABI4*. ABI4 helps to down-regulate the expression of *Lhcb1* by binding a CCAC element that overlaps with the G-box (*CACGTG*). A distinct signaling mechanism that depends on GUN1 down-regulates the expression of *AtGLK1* and *Lhcb1*. A plastid signal (Y) can convert cry1 from a positive to a negative regulator of *Lhcb1* expression by converting HY5 that is bound to the G box from a positive to a negative regulator of *Lhcb1*. This conversion may depend on a plastid signal (Y) promoting the association of a corepressor (Z) with HY5. The plastid-to-nucleus signaling mechanism that depends on X affects that plastid-to-nucleus signaling mechanism that depends on Y and vice versa. This signal integration may occur in the plastid or in a distinct cellular compartment

Thus, the current data indicate that the plastid-to-nucleus signaling that is activated by diverse mechanisms depends on *GUN1*. Additionally, double mutant studies and Mg-Proto feeding experiments provided evidence that porphyrins affect plastid-to-nucleus signaling by acting upstream of GUN1. Based on these data, *GUN1* is proposed to help integrate a number of distinct plastid-to-nucleus signaling mechanisms (Fig. 4) (Woodson and Chory 2008; Chi et al. 2013; Larkin 2014).

Although the precise biochemical function of the GUN1 protein is not known, the finding that GUN1 is a PPR protein is consistent with the expression of the plastid genome contributing to plastid-to-nucleus signaling because PPR proteins that localize to the chloroplast typically contribute to the expression of the chloroplast genome. The Apetala-type transcription factor ABSCISIC ACID INSENSITIVE4 (ABI4) acts downstream of GUN1 and appears to downregulate the transcription of *Lhcb1.2* by binding CCAC sequences in *Lhcb1.2* promoters. The CCAC element overlaps with the G box (CACGTG) to create a composite element (CCACACGTG) in *Lhcb1.2* and other PhANG promoters. However, not all GUN1-regulated promoters have a CCAC motif that overlaps or is in close proximity to a G box. Additionally, loss-of-function alleles of *ABI4* did not affect the downregulation of *Lhcb1.1* or *RbcS* expression that is brought about by norflurazon treatments that block chloroplast biogenesis (Woodson and Chory 2008; Chi et al. 2013; Larkin 2014). Thus, the degree that ABI4 contributes to plastid-to-nucleus signaling varies among PhANGs.

GUN1-dependent plastid-to-nucleus signaling that regulates *Lhcb1* expression also requires a transcription factor tethered to chloroplasts that is named PHD-type transcription factor with transmembrane domains (PTM). PTM was found on the cytoplasmic surface of the chloroplast outer envelope. Double mutant studies indicated that loss-of-function alleles of *GUN1* and *ABI4* disrupt a process that involves PTM. Proteolysis appears to release the transcription factor domain of PTM from its chloroplast tether. This transcription factor domain is thought to mobilize to the nucleus where it was shown to bind the promoter of *ABI4*. However, overexpression of the transcription factor domain of PTM in transgenic seedlings did not downregulate the expression of *Lhcb1* in seedlings that contain functional chloroplasts. These findings provide evidence that this signaling requires other factors besides PTM (Fig. 4) (Woodson and Chory 2008; Chi et al. 2013).

GUN1 alleles produced a variety of chloroplast phenotypes in the absence of any inhibitor and resulted in partial or complete chlorophyll deficiencies in cotyledons and leaves during chloroplast biogenesis. These are low penetrance phenotypes that were enhanced by loss-of-function alleles of genes that encoded the blue-light photoreceptor cryptochrome 1 (cry1) (section "*cry1* Mutants Are *gun* Mutants"). De-etiolating assays provide another approach for testing the efficiency of chloroplast biogenesis. Dark-grown seedlings contain etioplasts rather than chloroplasts and do not contain chlorophyll. When dark-grown seedlings are transferred to the light, etioplasts are converted into chloroplasts. De-etiolation was inefficient in *gun1*. Thus, *GUN1* promotes chloroplast biogenesis (Woodson and Chory 2008; Larkin 2014).

gun1 mutants exhibited other phenotypes. For instance, *gun1* mutants had abnormal circadian rhythms. A loss-of-function allele of *GUN1* increased the amplitude of the circadian oscillations of a model circadian-regulated gene (Larkin and Ruckle 2008). Also, blocking chloroplast biogenesis with either norflurazon or lincomycin increased the period length for model circadian-regulated genes. *GUN1*-dependent plastid signals also play a role in development (section "*gun* Mutants Can Exhibit Abnormal Development"). *gun1* and *abi4* mutants had less basal thermotolerance than wild type (Larkin 2014).

cry1 Mutants Are *gun* Mutants

A *gun* mutant screen yielded four alleles of *CRYPTOCHROME1* (*CRY1*), which encodes a well-characterized blue-light receptor (Larkin 2014). Thus, the *gun* mutant screen specifically disrupts a plastid-to-nucleus signaling mechanism that depends on *CRY1*. The genome of Arabidopsis contains three genes that encode cryptochromes: *CRY1*, *CRY2*, and *CRY3*. cry1 and cry2 function in the nucleus. cry3 probably functions in the chloroplasts and mitochondria. cry1 and cry2 broadly contribute to growth, development, and the interactions between plants and their environments (Liu et al. 2011). Null alleles of *CRY1* upregulated the expression of *Lhcb1* regardless of whether chloroplast biogenesis was blocked by treating seedlings with norflurazon, lincomycin, or erythromycin (section "Early Findings"). Thus, *CRY1* alleles do not cause resistance to inhibitors of chloroplast biogenesis. Additionally, *cry1* exhibits chlorophyll deficiencies that are enhanced by intense light. Thus, null alleles of *CRY1* cause chloroplast dysfunction, not acclimation responses that promote chloroplast stress tolerance.

The finding that *cry1* mutants are *gun* mutants was unexpected because in general, cry1 and other photoreceptors are positive regulators of *Lhcb1* and other PhANGs in plants that are not treated with inhibitors of chloroplast biogenesis. These findings indicate that plastid dysfunction converts cry1 signaling from a positive to a negative regulator of *Lhcb1* expression (Fig. 4). Transcriptome analyses indicated that blocking chloroplast biogenesis severely attenuates light-regulated expression of PhANGs but does not convert light signals from positive to negative regulators of most PhANGs (Larkin 2014).

Based on double mutant studies, the interactions between light and the plastid signaling that convert light signals from positive to negative regulators of *Lhcb1* expression appear to largely result from plastid signals converting LONG HYPOCOTYL5 (HY5) from a positive to a negative regulator of *Lhcb1* expression (Fig. 4) (Larkin 2014). HY5 is a bZIP-type transcription factor that acts downstream of cry1 and other photoreceptors. HY5 contributes to photomorphogenesis, numerous light-regulated processes, and the expression of approximately 20 % of light-regulated genes. HY5 binds the G box and its variants. Indeed, HY5 was shown to bind more than 9,000 genes and contributed to the expression of more than 1,100 different genes. These genes include a high proportion of early light-inducible genes. Thus, HY5 occupies a high position in the transcriptional cascade that drives photomorphogenesis (Jiao et al. 2007; Zhang et al. 2011). The mechanism by which HY5 is converted from a positive regulator to a negative regulator of *Lhcb1* expression is an open question and may depend on posttranslational modifications, associations with distinct proteins, or some other mechanism. There is precedence for one of these mechanisms. Transcription factors from animals such as nuclear receptors and Myc – a basic helix-loop-helix leucine zipper transcription factor – are converted from positive to negative regulators or vice versa by interacting with distinct coactivators. The finding that plastid signals depend on HY5 – a G box-binding factor – to downregulate the expression of *Lhcb1* is consistent with both HY5 and AB4 helping to downregulate the expression of *Lhcb1* when chloroplasts are dysfunctional (Fig. 4).

When chloroplast biogenesis was blocked by lincomycin treatment in either *gun1* or *cry1*, plants accumulated from 5 % to 50 % of the Lhcb1 mRNA in untreated wild type. Under these same conditions, *gun1 cry1* double mutants accumulated 100 % of the Lhcb1 mRNA in untreated wild-type seedlings. Similar results were obtained with *gun1 hy5* double mutants. Based on these data, we know that (1) the plastid regulation of *Lhcb1* depends on GUN1 and HY5 acting downstream of CRY1. (2) *GUN1*-dependent plastid-to-nucleus signaling and *HY5*-dependent plastid-to-nucleus signaling are distinct signaling mechanisms. Consistent with this interpretation, *GUN1*-dependent plastid-to-nucleus signaling occurs in the dark. The synergistic increase in the levels of *Lhcb1* mRNA in *gun1 cry1* and *gun1 hy5* is consistent with some type of interaction between these two signaling mechanisms (Fig. 4) (Larkin and Ruckle 2008; Larkin 2014).

The plastid-to-nucleus signaling that "rewires" light signaling probably does not depend on porphyrin biosynthesis because *hy5* was a *gun* mutant when chloroplast biogenesis was blocked by lincomycin. Blocking chloroplast biogenesis with lincomycin does not activate porphyrin-dependent plastid-to-nucleus signaling (Gray et al. 2003). Based on these data, a novel plastid signal appears to "rewire" light signaling (Fig. 4). Feeding Mg-Proto to intact plants was shown to downregulate the expression of *RbcS* and/or *Lhcb* in wild type but not in *gun1*, *abi4*, and *hy5*. These data and double mutant studies are consistent with porphyrins acting upstream of GUN1-dependent plastid signals and affecting the integration of light and plastid signaling (Fig. 4).

Analysis of transcriptomes provides evidence that light signaling promotes chloroplast function by inducing the expression of genes that promote chloroplast function. When chloroplast biogenesis was blocked, plastid-to-nucleus signaling appeared to "rewire" light signaling to promote chloroplast function (1) by attenuating the expression of genes that promote chloroplast function and (2) by inducing the expression of genes that contribute to the protection and repair of dysfunctional chloroplasts (Larkin 2014).

Plastid Signals that Depend on cry1, HY5, and GUN1 Promote the Accumulation of Anthocyanins

Anthocyanins accumulate when plants experience stress. Anthocyanins absorb the yellow and green wavebands and help to protect the chloroplasts from excess-light-induced photooxidative stress and photoinhibition by serving as a "sunscreen" (Gould 2004). *CRY1*- and *HY5*-dependent accumulation of anthocyanins was observed when Arabidopsis seedlings were grown on media that contained sucrose or when plants were exposed to excess light (Fankhauser and Casal 2004; Larkin 2014). Like excess light, sucrose and other carbohydrates downregulate PhANG expression and attenuate chloroplast function (To et al. 2003; Rook et al. 2006). These findings are consistent with plastid signals that are activated by chloroplast dysfunction "rewiring" the cry1 signaling to induce anthocyanin biosynthesis. *GUN1* also contributes to the accumulation of anthocyanins. When Arabidopsis seedlings are treated with inhibitors of chloroplast biogenesis, GUN1 made major contributions and both *CRY1* and *HY5* made minor contributions to the accumulation of anthocyanins. *CRY1*, *HY5*, and *GUN1* appeared to enhance anthocyanin biosynthesis by promoting the expression of genes that encode anthocyanin biosynthetic enzymes (Larkin 2014).

Blocking the Import of Proteins into the Plastid Activates Plastid-to-Nucleus Signaling

The *plastid protein import2* (*ppi2*) mutant of Arabidopsis lacks the Toc159 protein import receptor, which is required for the proper import of photosynthesis-related proteins into plastids. *ppi2* did not efficiently import proteins that contribute to photosynthesis but did efficiently import proteins that contribute to nonphotosynthetic functions. Thus, null alleles of *ppi2* block chloroplast biogenesis. Although *ppi2* alleles block chloroplast biogenesis, *ppi2* alleles appeared to activate a plastid-to-nucleus signaling mechanism that differs from the plastid-to-nucleus signaling mechanism associated with norflurazon and lincomycin treatments. Although GUN1 contributed to the downregulation of PhANG expression in *ppi2*, ABI4 did not contribute to the downregulation of PhANG expression in *ppi2*. Also, the *ppi2-2 gun1* double mutant exhibited an embryo lethal phenotype. In contrast, neither norflurazon nor lincomycin treatments caused embryo lethal phenotypes in *gun1* mutants.

ppi2 alleles and treatments with either norflurazon or lincomycin downregulated the expression of *AtGLK1* and *AtGLK2* (Inaba et al. 2011). *AtGLK1* and *AtGLK2* encode the GOLDEN2-LIKE (GLK) transcription factors of Arabidopsis. GLKs promote chloroplast function in all plants tested. As such, GLKs induce the expression of genes that encode the antennae proteins of the photosystems (e.g., *Lhcb1*) and the enzymes that contribute to chlorophyll biosynthesis (Waters and Langdale 2009). *GUN1*-dependent plastid-to-nucleus signaling appears to downregulate the expression of *Lhcb1* and chloroplast function in part by downregulating the expression of *AtGLK1* when chloroplast biogenesis is blocked (Inaba et al. 2011) (Fig. 4).

Photosynthetic Electron Transport Can Activate Plastid-to-Nucleus Signaling

Perturbations of photosynthesis that routinely occur in nature can activate plastid-to-nucleus signaling. For instance, environmental stresses such as suboptimal temperature, drought, variations in the quantity and quality of light, availability of CO_2, and insufficient availability of nutrients can attenuate photosynthesis. The plastid-to-nucleus signaling that is activated by perturbations of photosynthesis

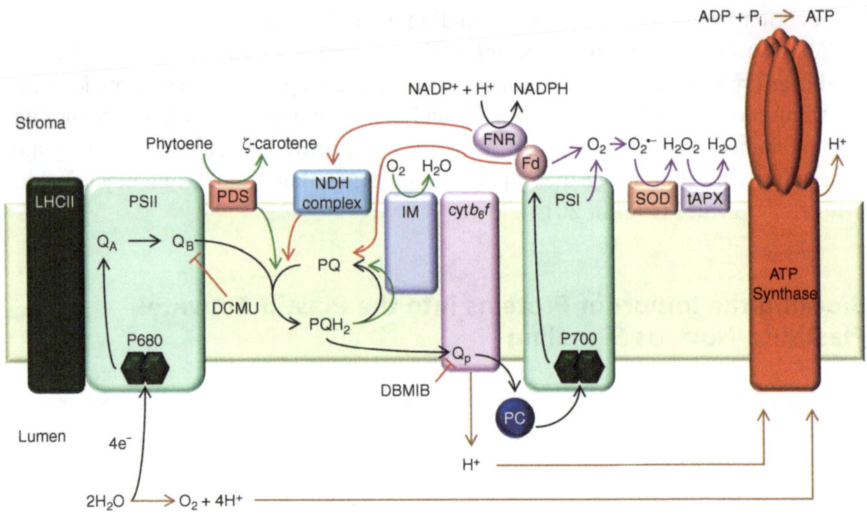

Fig. 5 The light reactions of photosynthesis. PSII associated with LHCII oxidizes H_2O, yielding O_2, H^+, and electrons (e−) that are ultimately transferred to $NADP^+$ (*black arrows*). Phytoene desaturase (*PDS*) uses PQ and IM to transfer electrons from phytoene to O_2, yielding H_2O (*green arrows*). PSI reduces O_2 yielding $O_2^{•-}$. $O_2^{•-}$ is disproportionated to H_2O_2, which is reduced to H_2O (*purple arrows*). Electrons from PSI can reduce PQ by cyclic electron flow (*red arrows*). The water oxidase of PSII and the cyt*b₆f* complex generate a proton motive force that is used to synthesize ATP (*brown arrows*). The inhibition of the reduction of PQ and the oxidation of PQH_2 by DCMU and DBMIB, respectively, are indicated with *red T bars*. The lipids of the thylakoid membrane are represented by a *yellow rectangle*. Abbreviations that are not defined in the text: *NDH complex* NAD(P)H dehydrogenase complex, *cytb6f* cytochrome *b6f* complex, *PC* plastocyanin, *FNR* ferredoxin-$NADP^+$ reductase, and *Fd* ferredoxin

facilitates plant growth and development by helping plants acclimate their photosynthetic machinery to the prevailing environmental conditions.

The light reactions of photosynthesis convert the free energy of absorbed light into biologically useful forms of energy. Briefly, these reactions use the free energy of absorbed light to drive the transfer of electrons from H_2O to $NADP^+$, yielding O_2, NADPH, reduced ferredoxin, and a *trans*-thylakoid pH gradient. The *trans*-thylakoid voltage and the pH gradient create a proton motive force that is used by the ATP synthase to generate ATP (Fig. 5). ATP, ferredoxin, and NADPH drive metabolic processes such as the reduction of CO_2 to carbohydrate and the assimilation of nitrate and sulfate. To meet the needs of the cell, the light reactions of photosynthesis can adjust the ratio of ATP to reductant (i.e., NADPH and ferredoxin) by inducing cyclic electron transport (Fig. 5). Cyclic electron transport generates a proton motive force but does not generate reductant (Foyer et al. 2012). Photosynthesis-derived plastid signals can help to optimize these processes by adjusting the ratio of PSI to PSII and by promoting chloroplast biogenesis (Foudree et al. 2010; Pfalz et al. 2012).

When light harvesting exceeds the capacity of photosynthetic electron transport (PET), the photosynthetic machinery dissipates the excess free energy of absorbed light as heat to avoid potential damage to PSII using processes referred to as nonphotochemical quenching. Nonetheless, PSII suffers photooxidative damage at optimal fluence rates and is continually repaired. In particular, the D1 reaction center protein of PSII is continuously damaged by the $^1O_2$ that is produced by the reaction center of PSII. Indeed, the reaction center of PSII is rebuilt every 20–30 min because of this damage. Excess light can cause PSII to suffer photooxidative damage at a rate that is greater than a plants ability to repair PSII. Such damage causes photoinhibition. A number of processes that do not depend on plastid-to-nucleus signaling help to prevent photoinhibition by regulating PET and by causing the dissipation of excess-light energy (Foyer et al. 2012). These mechanisms are beyond the scope of this chapter and are only mentioned briefly. When plants are exposed to excess light, plastid-to-nucleus signaling can promote acclimation responses that protect the photosynthetic machinery from photoinhibition and the cell from photooxidative stress. Alternatively, plastid-to-nucleus signaling can promote cell death (Mullineaux and Baker 2010; Karpiński et al. 2012; Kim and Apel 2013b).

Photosynthesis Activates Plastid-to-Nucleus Signaling in Algae

Early evidence that photosynthetic electron transport (PET) can activate plastid-to-nucleus signaling came from studying the photoacclimation responses in green algae. Transferring cultures of *Dunaliella tertiolecta* and *D. salina* from high-intensity light (e.g., 700 μmol m^{-2} s^{-1}) to low-intensity light (e.g., 70 μmol m^{-2} s^{-1}) increases the size of the antennae of PSII by inducing the levels of chlorophyll and Lhcbs. Indeed, reducing the fluence rate increased the chlorophyll content up to three fold within 24 h. This increase in chlorophyll content was accompanied by an increase of the transcription of the *Lhcb* genes, which resulted in a three- to fourfold increase in the levels of *Lhcb* mRNA and, subsequently, an increase in Lhcb protein levels. With substantially enhanced antennae, these algae are more capable of harvesting light energy from low fluence rates of light. Transferring these green algae from low-intensity light to high-intensity light reduced the levels of chlorophyll and Lhcb proteins (i.e., a reduction in antenna size). Reducing the size of the antennae in high-intensity light protects the photosynthetic reaction centers from photoinhibition.

Inhibitors of PET mimicked these photoacclimation responses in *D. tertiolecta* by affecting the levels of chlorophyll and the transcription rates of *Lhcb* genes. These inhibitors are 3-(3′,4′-dichlorophenyl)-1,1-dimethylurea (DCMU) and 2,5-dibromo-3-methyl-6-isopropylbenzoquinone (DBMIB). DCMU inhibits the reduction of plastoquinone (PQ) to plastoquinol (PQH_2) by binding the Q_B site of photosystem II (PSII) (Fig. 5). DBMIB inhibits the oxidation of PQH_2 to PQ by binding the plastoquinol oxidation site on the cytochrome b_6f complex (Fig. 5). Increasing the proportion of PQ relative to PQH_2 with DCMU treatments caused a twofold increase in the levels of chlorophyll and *Lhcb* mRNA. Thus, oxidizing PQH_2 mimicked a shift to low-intensity light. In contrast, increasing the proportion of PQH_2 relative to PQ with DBMIB treatments caused a 25 % decrease in chlorophyll levels and a 75 % decrease in *Lhcb* mRNA levels. Thus, reducing PQ mimicked a shift to high-intensity light. Based on these data, the ratio of PQ to PQH_2 was proposed to activate a plastid-to-nucleus signaling mechanism that optimizes the light-harvesting capacity of PSII to light intensity in *D. tertiolecta* (Escoubas et al. 1995).

Varying the excitation pressure of PSII caused essentially the same effects on the levels of chlorophyll, *Lhcb* mRNA, and Lhcb protein as treating these algal cells with DCMU and DBMIB. The excitation pressure is a relative measure of the reduction of the first stable electron acceptor of PSII, Q_A (Fig. 5). Researchers can vary the excitation pressure by varying light intensities or by varying other parameters that attenuate the biochemical reactions of photosynthesis such as temperature. Downregulating the biochemical reactions by lowering the temperature increases the excitation pressure and the ratio of PQH_2 to PQ (Maxwell et al. 1995).

Photosynthesis Activates Plastid-to-Nucleus Signaling in Plants

In plants, a number of strategies were employed to test whether particular components of the PET chain can affect the expression of PhANGs. These strategies included varying light intensity, varying some other parameter such as temperature, varying light quality, and using mutant alleles or chemicals to specifically inhibit one step in the PET chain. Results from these experiments provided evidence that distinct molecules reduced by PET activate plastid-to-nucleus signaling. This signaling regulates the expression of nuclear genes at both the transcriptional and posttranscriptional levels.

In tobacco plants, *FED1* and *Lhcb* mRNAs are loaded onto polyribosomes in the light but not in the dark or in DCMU-treated plants. *FED1* is a nuclear gene that encodes ferredoxin (Fig. 5). The dark and DCMU treatments decrease the stability of *FED1* mRNA but have no effect on the stability of *Lhcb* mRNA. These data indicate that the reduction of the PET chain can regulate the expression of *FED1* by a posttranscriptional mechanism.

In Arabidopsis suspension cultures capable of photoautotrophic growth, PhANG expression was upregulated as carbohydrates were depleted from the growth media. This upregulation was not simply caused by removing carbohydrates that downregulate the expression of PhANGs. Inhibiting PET with DCMU attenuated the upregulation of *Lhcb* and *RbcS* expression that occurred as carbohydrates were depleted from the growth medium. Based on these findings, the reduction of the entire PET chain and, thus, possibly the redox state of the stroma appears to upregulate the expression of *Lhcb* and *RbcS*. Other studies provided evidence that reducing the PET chain does not always induce the expression of these PhANGs. Increasing the excitation pressure of PSII by transferring winter rye to either higher intensity light or lower temperatures attenuated the expression of both *Lhcb* and *RbcS*.

Light qualities that favor the activities of either PSI or PSII appear to activate plastid-to-nucleus signaling by affecting PET. (Note: The absorbance maximum of PSI is 700 nm, while that for PSII is 680 nm). This signaling can change the ratio of PSI to PSII in hours to days. Evidence that distinct PET-related redox signals regulate the transcription of PhANGs was obtained by transferring transgenic tobacco plants that harbor reporter genes from light conditions that favor PSI activity to light conditions that favor PSII and vice versa. These reporter genes were driven by promoters from Arabidopsis genes that encode nitrate reductase (*NIA2*) and Arabidopsis genes that directly contribute to the activity of PSI: *PSAD*, *PSAF*, and *PETE*. *PSAD* and *PSAE* encode subunits of PSI. *PETE* encodes plastocyanin, a small copper-containing protein that transfers electrons from the cytochrome b_6f complex to PSI (Fig. 5). The reporter gene that was driven by the promoter from *NIA2* was expressed at higher levels in light conditions that favor the activity of PSI than in light conditions that favor the activity of PSII. Contrariwise, the reporter genes that were driven by promoters from *PSAD*, *PSAF*, and *PETE* were expressed at higher levels in light conditions that favor the activity of PSII than in light conditions that favor the activity of PSI. Experiments with DCMU and DBMIB indicated that the redox state of PQ regulates the expression of *PETE* and that the redox state of some component of the PET chain downstream of PQ or the redox state of the stroma can regulate the expression of *PSAD*, *PSAF*, and *NIA2* (Pfalz et al. 2012).

Development has a major impact on the plastid-to-nucleus signaling that is regulated by PET. Plastid-to-nucleus signaling activated by DCMU treatments regulated the expression of *PETE* in 4-week-old tobacco plants but not in 7-day-old tobacco seedlings. DCMU treatments downregulated the expression of *PETE* by means of a posttranscriptional mechanism (Sullivan and Gray 2002).

After these early studies with individual genes, advances in technology allowed researchers to test the effects of PET on the expression on large numbers of genes from Arabidopsis. Some of these early studies produced conflicting results. Some studies were consistent with PET activating plastid-to-nucleus signaling. Others reported that the redox state of the electron carriers that reside on the reducing side of PSI, the redox state of the stroma, and changes in CO_2 fixation rates correlated with changes in nuclear gene expression.

A subsequent study analyzed the expression of 3,292 Arabidopsis genes using a macroarray. Gene expression was analyzed at various times after plants were transferred from light conditions that favor the activity of PSI to light qualities that favor the activity of PSII and vice versa. In contrast, earlier studies analyzed transcriptomes at single time points after plants were exposed to a different light quality. Based on an analysis of these kinetic data, light quality shifts that favor either PSI or PSII were found to cause dynamic changes in gene expression. This discovery may explain some of the earlier conflicting results. Based on these kinetic data, the redox state of both the PQ pool and the stroma appears to regulate the expression of nuclear genes. These data provide evidence that genes that encode proteins active in metabolism – including PhANGs – are major targets of the plastid-to-nucleus signaling that is activated by the PET chain (Pfalz et al. 2012).

The Plastid-to-Nucleus Signaling that Changes the Ratio of PSI to PSII may Depend on STN7

Although the plastid-to-nucleus signaling mechanism that adjusts the ratio of PSI to PSII is poorly understood, there is evidence that this mechanism depends on chloroplastic serine-threonine kinases. Although some of the mechanisms that help to balance the activities of PSI and PSII are localized entirely within the chloroplast, plastid-to-nucleus signaling is thought to collaborate with these chloroplast-localized mechanisms.

Chloroplast-localized responses that are regulated by the redox state of the PQ pool include state transitions. State transitions are the movement of the light-harvesting complex II (LHCII) proteins from PSII to PSI. This movement is activated by an increase in the ratio of PQH_2 to PQ – the movement takes minutes and requires the phosphorylation of the LHCII proteins by a chloroplastic serine-threonine kinase. This protein kinase was originally discovered in *C. reinhardtii* and is named state transition-deficient mutant 7 (Stt7). Stt7 contains a single transmembrane helix. The amino-terminal domain of Stt7 localizes to the lumen of the thylakoid membranes, and the carboxy-terminal kinase domain localizes to the chloroplast stroma. Subsequently, a relative of Stt7 was discovered in Arabidopsis, STATE TRANSITION 7 (STN7). In Arabidopsis, state transitions require STN7. Like Stt7, STN7 phosphorylates LHCII proteins. The association of phosphorylated LHCII with PSI increases the light-harvesting activity of PSI and, therefore, leads to a decrease in the ratio of PQH_2 to PQ. The dephosphorylation of LHCII by a phosphatase named PPH1/TAP38 appears to cause the return of LHCII proteins to PSII (Rochaix et al. 2012). The redox state of the PQ pool can also regulate the expression of chloroplast genes. Redox-regulated protein kinases in the chloroplast affect chloroplastic gene expression by phosphorylating the plastid-encoded DNA-dependent RNA polymerase and sigma factors (Schönberg and Baginsky 2012).

In Arabidopsis, light that favors the activity of PSII caused an increase in the abundance of PSI due to an increase in the expression of the nuclear and chloroplast genes that encode subunits of PSI. Loss-of-function alleles of *STN7* prevented this adjustment in the ratio of PSI to PSII. Mutants that are deficient in their ability to perform state transitions because they accumulate only low levels of Lhcb proteins or because they harbor a loss-of-function allele for the *psaL* gene, which contributes to the docking of LHCII on PSI, were used to demonstrate that plastid-to-nucleus signaling that changes the ratio of PSI to PSII requires *STN7* but does not require state transitions (Pfalz et al. 2012). Nonetheless, the idea that STN7 is essential for plastid-to-nucleus signaling is debated (Tikkanen et al. 2012).

The Redox State of PQ Can Affect Chloroplast Biogenesis

Analyses of the *immutans* (*im*) mutant provided evidence that the redox state of the PQ pool activates a plastid-to-nucleus signaling mechanism that contributes to chloroplast biogenesis. *im* is a variegated mutant of Arabidopsis. The leaves of variegated mutants have green and yellow/white sectors. *IM* is a nuclear gene that encodes a protein that resides in the stromal lamellae of the thylakoid membranes. Several functions are attributed to the IM protein. One of these functions is to promote carotenoid biosynthesis by serving as the terminal oxidase in the O_2-dependent oxidation of phytoene (Fig. 5). Phytoene desaturase oxidizes phytoene by transferring electrons to PQ, yielding ζ-carotene and PQH_2. These reactions increase excitation pressure by reducing PQ to PQH_2. IM relieves this excitation pressure by transferring electrons from PQH_2 to O_2, yielding PQ and H_2O (Foudree et al. 2012; Fig. 5).

The IM protein is especially important during chloroplast biogenesis because during chloroplast biogenesis, the thylakoid membranes are not completely assembled, and therefore, the PET chain cannot relieve excitation pressure by transferring electrons from PQH_2 to downstream components of the PET chain. In *im*, the PQ pool becomes overly reduced. This overreduction of the PQ pool attenuates the activity of phytoene desaturase, which reduces the production of colored carotenoids. Insufficient quantities of colored carotenoids can block chloroplast biogenesis (section "Early Findings"). The green sectors of *im* leaves appear to escape photooxidation because undefined factors compensate for IM deficiencies in a fraction of plastids during chloroplast biogenesis. The redox state of the PQ pool is proposed to activate a plastid-to-nucleus signaling mechanism that promotes chloroplast biogenesis by regulating the expression of nuclear genes that encode these compensating factors (Foudree et al. 2010).

Excess Light Activates Plastid-to-Nucleus Signaling

Although the midday sun can deliver fluence rates of at least 2,000 μmol m^{-2} s^{-1}, such fluence rates exceed the photochemical capacity of the light reactions of photosynthesis. Such fluence rates are sometimes referred to as excess light. Transferring Arabidopsis plants from low-intensity light (e.g., 200 μmol m^{-2} s^{-1}) to high-intensity light (e.g., 2,000 μmol m^{-2} s^{-1}) was found to causes significant photoinhibition and to induce the expression of nuclear genes that defend the cell against oxidative stress such as glutathione reductase (GOR2), ascorbate peroxidases (APX1 and APX2), and catalase (CAT1).

Based on the data that was obtained from experiments with DCMU and DBMIB, the photoinhibition that is caused by exposing Arabidopsis plants to 2,000 μmol m^{-2} s^{-1} white light was proposed to activate a plastid-to-nucleus signaling mechanism in part by reducing the pool of PQ. The size or the redox state of the glutathione pool was proposed to affect this plastid-to-nucleus signaling mechanism because feeding oxidized or reduced glutathione to leaf disks prevented this fluence-rate shift from upregulating the expression of *APX1* and *APX2* and because a loss-of-function allele of the Arabidopsis gene *REGULATOR OF APX2 1* (*RAX1*) was found to induce high levels of *APX2* expression when plants are not exposed to excess light. *RAX1* encodes γ-glutamylcysteine synthetase1, a plastid-localized enzyme that catalyzes the first step in glutathione biosynthesis. This signaling mechanism was proposed to help protect the cytosol from oxidative stress by inducing the levels of cytosolic enzymes that scavenge hydrogen peroxide (H_2O_2) that leaks from chloroplasts into the cytosol when plants are exposed to excess light.

More support for glutathione affecting this signaling comes from the finding that the jasmonic acid precursor (+)-12-oxo-phytodienoic acid (OPDA) accumulates in the chloroplast when plants are exposed to excess light and induces the expression of stress-related genes. The OPDA receptor is a chloroplastic cyclophilin, cyclophilin 20-3 (CYP20-3). Cyclophilins are peptidyl-prolyl *cis-trans* isomerases that affect enzyme activity by altering enzyme conformation or by protein-protein interactions. The OPDA-CYP20-3 complex induced the biosynthesis of glutathione in the chloroplast by binding the cysteine synthase complex, which promoted the biosynthesis of the glutathione precursor cysteine. Thiols such as glutathione are thought to induce the expression of OPDA-inducible genes by affecting the redox state of the cell (Galvez-Valdivieso and Mullineaux 2010; Kopriva 2013).

H_2O_2 is a major ROS that accumulates in leaves that experience excess-light stress. Two major mechanisms produce H_2O_2 when leaves are exposed to excess light. One mechanism involves PSI transferring electrons to O_2 yielding superoxide ($O_2^{\bullet-}$). Superoxide dismutase catalyzes the disproportionation of $O_2^{\bullet-}$ yielding H_2O_2 and O_2 ($2O_2^{\bullet-} + 2H^+ \rightarrow H_2O_2 + O_2$). The photoreduction of O_2 to H_2O_2 is often referred to as the Mehler reaction. H_2O_2 is subsequently reduced to H_2O by chloroplastic APX, which uses ascorbate as an electron donor. Superoxide dismutase and APX are closely associated with PSI, which presumably enhances the efficiency of this process (Fig. 5). The other mechanism is the consumption of photosynthetic reducing equivalents by photorespiration, a process that yields H_2O_2 in the peroxisome. Excess H_2O_2 that leaks into the cytosol is reduced to H_2O by cytosolic APX. In Arabidopsis, the cytosol contains two isozymes of ascorbate peroxidase encoded by *APX1* and *APX2*. The expression of the *APX2* gene is induced in response to excess-light stress (Karpiński et al. 2012).

Excess light increases the levels of various signals that have distinct effects on the expression a number of nuclear genes. These signals include signals derived from PET, reduced glutathione, chloroplastic ROS, extraplastidic H_2O_2, and ABA. Additionally, the effect of excess light on the expression of particular genes can vary among different cell types. For example, particular excess-light-inducible genes are more prominently expressed in the upper third of the leaf blade than in petioles. Thus, the signaling that is activated by excess light appears complex (Bechtold et al. 2008).

Plastid-to-Nucleus Signaling Contributes to Systemic-Acquired Acclimation

Exposing approximately one third of Arabidopsis rosette leaves to excess light was found to induce the expression of *APX2*, accumulation of H_2O_2, and acclimation of photosynthesis to excess light in distal leaves that are not exposed to excess light. Indeed, when these distal leaves are subsequently exposed to excess light, the photosynthetic efficiency (F_v/F_m) and photochemical quenching (q_P) suffer only minor reductions. Based on these and other data, excess light was proposed to induce the production of a systemic signal that helps leaves acclimate to excess light. The process known as systemic-acquired acclimation (SAA) was proposed to depend on the redox state of either the Q_B site of PSII or the PQ pool and H_2O_2 from leaves exposed to excess light. Such a response could protect leaves from the stress that is induced by large variations in light intensity that routinely occur from changes in the position of the sun throughout the day and from transient shading effects of clouds and leaves.

Exposure of leaves to excess light induces the expression of hundreds of genes in distal leaves that are not exposed to excess light within 30 min. Many of the genes in which expression is upregulated by excess light were also upregulated in distal leaves that were not exposed to excess light. Likewise, many of the genes in which expression was downregulated by excess light were also downregulated in distal leaves not exposed to excess light. In both types of leaves, excess light regulated the expression of many genes induced by excess light, ROS, and drought. Thus, SAA appears to protect distal leaves that are not exposed to excess light from subsequent exposures to excess light at least in part by driving an excess-light response in distal leaves that are not exposed to excess light. The nuclear zinc-finger transcription factor ZAT10 contributes to approximately 20 % of this response (Karpiński et al. 2012).

Subsequently, increasing the fluence rate from 150 μmol m^{-2} s^{-1} to only 250 μmol m^{-2} s^{-1} light was found to induce components of SAA. Thus, SAA does not depend on high-level light stress. These findings are consistent with the idea that plastid-to-nucleus signaling underpins SSA rather than the cellular damage that is induced by severe oxidative stress. Nonetheless, the intensity of SAA can vary and depends on the severity of the excess-light stress (Gordon et al. 2012).

Systemic-acquired resistance (SAR) is similar to SAA in the sense that SAR is a systemic response that occurs when a small part of the plant experiences stress. SAR nonspecifically restricts the growth of pathogens and is thus analogous to the innate immunity of animals. There is significant overlap between SAA and SAR because when leaves were treated with excess light, both leaves that were exposed to excess light and distal leaves that were not exposed to excess light limit the growth of the virulent biotrophic pathogen *Pseudomonas syringae* DC3000 relative to control plants that were not exposed to excess light. Thus, excess light induces both SAA and SAR. Plastid-to-nucleus signaling activated by the redox state of the PQ pool appears to promote SAR because treating plants with DCMU prior to excess light or light that favors PSI activity (i.e., two treatments that oxidize the PQ pool) did not restrict the growth of *P. syringae* DC3000 and treating leaves with either DBMIB or light that favors PSII activity (i.e., two treatments that reduce the PQ pool) restricted the growth of *P. syringae* DC3000 (Mullineaux and Baker 2010; Karpiński et al. 2012).

Excess Light Can Induce Acclimation or Cell Death

In addition to inducing SAA, excess light can induce programmed cell death (PCD) in leaves and in adjacent leaves that are not exposed to excess light, yielding foliar lesions. Plastid-to-nucleus signaling that is activated by the redox state of the PQ pool appears to induce PCD because treating plants with either DCMU or DBMIB and treating plants with light qualities that favor the activities of either PSI or PSII affected the PCD response. Thus, the redox state of the PQ pool appears to activate a plastid-to-nucleus signaling mechanism that contributes to SAA, SAR, and PCD. Although excess light induced SAA and SAR, excess light that exceeds a particular threshold activated PCD yielding foliar lesions. At the level of an organ or an entire plant, both SAA and PCD are essential for a successful response to excess light. Thus, the cell death caused by oxidative stress is often under genetic control and is not simply a consequence of excessive necrosis that is induced by ROS. The mechanisms that determine whether plants cells acclimate or die is not completely understood.

Some of the genes that excess light requires to drive cell acclimation and cell death responses were originally discovered during the study of the plant response to pathogens. These genes include *LESION SIMULATING DISEASE 1* (*LSD1*), *ENHANCED DISEASE SUSCEPTIBILITY 1*(*EDS1*), and *PHYTOALEXIN DEFICIENT 4* (*PAD4*). The LSD1 protein was proposed to function as a transcription factor. The EDS1 and PAD4 proteins have sequence similarities to lipases. Thus, hydrolase activities were proposed to underpin the biological functions of EDS1 and PAD4. LSD1, EDS1, and PAD4 appear to promote redox signaling in response to a variety of plant stresses. EDS1 seems particularly important for plant responses to chloroplastic ROS and for the switch between acclimation and death.

Excess light induces the production of H_2O_2 from the Mehler reaction. Excess light also induces increases in transpiration rates. Increases in the transpiration rates can close stomata, which induces photorespiration by reducing the availability of CO_2 for photosynthesis. Thus, excess light induces the levels of H_2O_2 derived from both the Mehler reaction and photorespiration. Characterization of loss-of-function alleles of *EDS1* and *PAD4* indicates that these genes upregulate ethylene biosynthesis when plants are exposed to excess light. Thus, excess light can ultimately lead to the production of ethylene and ROS that can induce SAA, SAR, or PCD depending on the severity of the stress. The NUDIX HYDROXYLASE7 (NUDT7) counteracts these effects. NUDT7 belongs to a family of pyrophosphohydrolases that helps to downregulate oxidative stress. In leaves that are exposed to excess light, EDS1 attenuated the protective effect of NUDT7 and promoted PCD. Mitogen-activated kinases appear to help initiate PCD in response to chloroplastic H_2O_2.

ESD1-dependent signaling also promotes an acclimation response that prevents the spread of foliar lesions. The respiratory burst oxidase homolog D (rbohD) contributes to this response. The Arabidopsis genome contains 10 *Rboh* genes, *RbohA* to *RpohJ*. These genes encode plasma membrane-localized NADPH oxidases. They were discovered based on their sequence similarity to the mammalian respiratory burst NADPH oxidase subunit gp91[phox]. RbohD catalyzes a burst of $O_2^{\bullet -}$ that is rapidly dismutated to H_2O_2 in the apoplasm. Thus, like EDS1, RbohD induces an increase in the levels of H_2O_2. Salicylic acid (SA), which accumulates during such stress, counteracts the effects of EDS1 and RbohD, presumably by enhancing glutathione biosynthesis. LSD1 promotes acclimation by upregulating the expression of genes that encode SOD and CAT. The mechanism is not known that flips the switch between the EDS1-NUDT7 system that tends to cause PCD and the EDS1-LSD1 system that prevents the spread of foliar lesions by inducing acclimation in cells adjacent to foliar lesions (Mullineaux and Baker 2010; Karpiński et al. 2012).

Moderate Increases in Light Intensity Can Induce the Expression of *APX2*

After the early studies on the effects of excess light on *APX2* expression (sections "Excess Light Activates Plastid-to-Nucleus Signaling" and "Plastid-to-Nucleus Signaling Contributes to Systemic-Acquired Acclimation"), fluence rates that do not cause oxidative stress were found to induce *APX2* expression. Indeed, transferring plants from 200 μmol m^{-2} s^{-1} to only 300 or 400 μmol m^{-2} s^{-1} induced *APX2* expression. Such moderate increases in fluence rate did not cause photoinhibition and did not reduce or only slightly reduced the PQ pool. These and other data are evidence that the redox state of the PQ pool does not induce the expression of *APX2*. Fluence rates that range from 300 to 650 μmol m^{-2} s^{-1} induced the expression *APX2* mostly in leaf bundle sheath cells, not in leaf mesophyll cells.

Moderate increases in fluence rate produced H_2O_2 derived from the Mehler reaction that induced the expression of *APX2*. Nonetheless, the H_2O_2 derived from this reaction is likely not sufficient to induce *APX2* expression because exogenously supplied H_2O_2 is not sufficient to rapidly induce the expression of *APX2*. In addition to H_2O_2, increases in the transpiration rate and the closing of stomata were found to increase the expression of *APX2*. These findings implicate leaf water potential, leaf turgor, or ABA as signals in the induction of *APX2*. All of these conditions may involve ABA since decreases in water potential trigger ABA signaling (Mullineaux and Baker 2010; Karpiński et al. 2012).

Subsequently, increases in the fluence rate were found to induce the biosynthesis of ABA in vascular parenchyma cells adjacent to the bundle sheath cells by increasing the transpiration rate and by lowering the water potential of leaves. The vascular parenchyma cells appear to secrete ABA, which may induce *APX2* expression in bundle sheath cells (Fig. 6). Previous studies on ABA signaling in guard cells aided in the discovery of signaling factors that contribute to the ABA-induced expression of *APX2* in bundle sheath cells. In bundle sheath cells, two antagonistic mechanisms regulated *APX2* expression. The first upregulated the expression of *APX2* and depended on a protein kinase named OPEN STOMATA1 (OST1) and two protein phosphatase 2Cs named ABA INSENSITIVE1 (ABI1) and ABI2. ABA is known to regulate the activities of these proteins. ABI1 activates

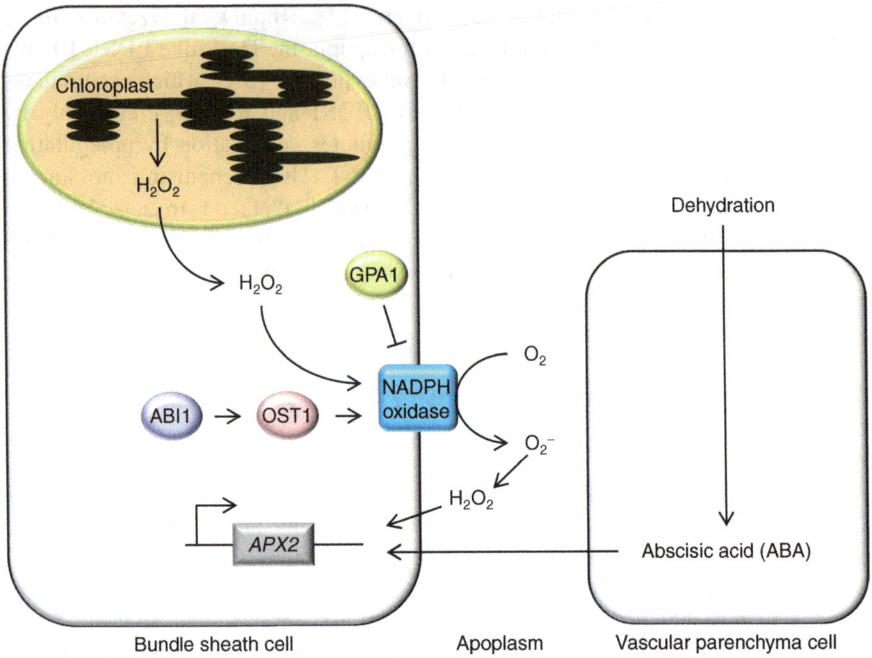

Fig. 6 Regulation of *APX2* expression by H_2O_2 and ABA. In bundle sheath cells, H_2O_2 that is derived from the Mehler reaction and the ABA-regulated factors OST1 and ABI1 induce a rise in the apoplastic H_2O_2, presumably by activating an NADPH oxidase that resides in the plasma membrane of the bundle sheath cells. Dehydration induces the biosynthesis of ABA in vascular parenchyma cells. Apoplastic H_2O_2 and ABA that is derived from vascular parenchyma cells induce the expression of *APX2* in bundle sheath cells. GPA1 attenuates the expression of *APX2*

OST1, which is thought to induce *APX2* expression by phosphorylating target proteins (Fig. 6).

Although ABA is required for the upregulation of *APX2* expression, ABA is not sufficient to induce the rapid increase in *APX2* expression associated with increases in the fluence rate. Chloroplastic H_2O_2 is thought to help induce the expression of *APX2* by activating a plasma membrane-localized NADPH oxidase. OST1 also promotes the accumulation of H_2O_2 in the apoplasm. *APX2* is unusual among stress-related genes in that its expression is induced by apoplastic H_2O_2. In summary, ABA, chloroplastic H_2O_2, and apoplastic H_2O_2 are thought to induce the expression of *APX2* (Fig. 6).

Another mechanism that depended on the Arabidopsis Gα subunit (G-PROTEIN ALPHA1 (GPA1)) of the plasma membrane-localized heterotrimeric G-protein complex was shown to downregulate the expression of *APX2*. This mechanism antagonized the ABA- and H_2O_2-based signaling that induced the expression of *APX2* by somehow attenuating the production of apoplastic H_2O_2 (Fig. 6) (Galvez-Valdivieso and Mullineaux 2010; Mullineaux and Baker 2010).

PSII Can Produce $^1O_2$

O_2 is a diradical: a molecule with a pair of electrons that can occupy two orbitals of the same energy level. In the ground state, these electrons have parallel spins and occupy different degenerate molecular orbitals. In its ground state, molecular oxygen is referred to as triplet oxygen ($^3O_2$). Transferring energy to $^3O_2$ can cause these electrons to pair with opposite spins in the same molecular orbital. This state is named the singlet state ($^1O_2$). The names singlet and triple are derived from the appearance of electron spin resonance spectra. $^1O_2$ has a short lifetime and is highly reactive. Thus, $^1O_2$ was thought to travel only short distances within the cell. However, recent findings indicate that $^1O_2$ can travel up to about 270 nm in rat nerve cells. $^1O_2$ can modify lipids, nucleic acids, and proteins. Indeed, under optimal fluence rates, $^1O_2$ causes 80 % of the nonenzymatic lipid peroxidation in leaves (Triantaphylidès and Havaux 2009; Kim and Apel 2013b).

In PSII, light excites a special pair of chlorophyll molecules named pigment 680 or P680 as though they were a single molecule (Fig. 5). P680 absorbs light with a maximum at 680 nm or it can receive energy from other chlorophylls in the photosystem. When excited, P680 transfers an electron to other electron carriers in PSII. Oxidized P680 is the strongest oxidant known in biology, as such P680 can drive the oxidation of water. Stress such as excess light, low temperature, and drought attenuates PET, which leads to the overreduction of the PQ pool and charge recombination. Under these conditions, excited P680 can transfer energy to $^3O_2$ yielding $^1O_2$, which causes oxidative damage to the D1 protein of the PSII reaction center. Because of this oxidative damage, the reaction center of PSII turns over every 20–30 min when it is exposed to optimal fluence rates. When plants are exposed to stress, the rate of oxidative damage can exceed the capacity of the protective and repair mechanisms to maintain PSII. Such stressful conditions can cause photoinhibition (Foyer et al. 2012; Kim and Apel 2013b).

The excited chlorophyll in the antennae of PSII can also produce $^1O_2$. Excited electrons of the antennae chlorophyll can convert from the singlet to the triplet state. Triplet chlorophyll exists for a few μs, which is sufficient time for the transfer of energy to $^3O_2$, yielding $^1O_2$. The antennae generate significant $^1O_2$ if the production of $^1O_2$ exceeds the $^1O_2$-scavenging capacity of $^1O_2$ quenchers, such as the xanthophylls that are bound by the antenna proteins (Kim and Apel 2013b).

$^1O_2$ Activates Plastid-to-Nucleus Signaling in the Arabidopsis *flu* Mutant

Excess light can produce diverse ROS. The Arabidopsis *fluorescent* (*flu*) mutant provides a tool to specifically test whether chloroplastic $^1O_2$ can activate plastid-to-nucleus signaling. The FLU protein localizes to the chloroplast membranes and contributes to the regulation of chlorophyll biosynthesis by ensuring that protochlorophyllide (Pchlide) does not overaccumulate during the diurnal cycle.

Pchlide is converted to chlorophyllide by a light-dependent enzyme named Pchlide oxidoreductase (POR). In the dark, POR binds Pchlide and NADPH, but does not turnover. Thus, Pchlide-POR-NADPH ternary complexes accumulate in the dark. When Pchlide accumulates to a particular threshold, the FLU protein inhibits glutamyl-tRNA reductase (GluTR), the enzyme that commits glutamyl tRNA to tetrapyrrole biosynthesis. Thus, the FLU protein prevents the overaccumulation of Pchlide in the dark (Fig. 2).

In dark-grown *flu* mutants, Pchlide accumulates to levels that exceed the capacity of POR to bind Pchlide. Free Pchlide accumulates in chloroplast membranes, and when *flu* mutants are exposed to light, the free Pchlide is not used to synthesize chlorophylls that are subsequently sequestered in the photosystems. Like other porphyrins, free Pchlide absorbs light and transfers its excitation energy to $^3O_2$, yielding $^1O_2$. *flu* mutants are indistinguishable from wild type when they grow in continuous light. When *flu* seedlings or mature *flu* plants are grown in continuous light, transferred to the dark, and reilluminated, their growth immediately stops, and necrotic lesions appear on the leaves within 2–4 h. When etiolated *flu* seedlings are transferred from the dark to the light, they bleach and die. The enhanced sensitivity of etiolated seedlings to *flu* alleles is likely explained by the four- to fivefold higher Pchlide levels in etiolated *flu* seedlings relative to mature *flu* plants and the lower ROS scavenging capacity of etioplasts relative to chloroplasts (Kim and Apel 2013b).

In Arabidopsis, transcriptome analyses indicated that increases in $^1O_2$ that occurred in *flu* mutants activated plastid-to-nucleus signaling that rapidly regulated nuclear gene expression (e.g., within 15–30 min). Increases in the levels of $^1O_2$ influenced the expression of approximately 5 % of the Arabidopsis genome and upregulated the expression of 300 genes by threefold relative to wild type. This group of genes primarily performs stress-related functions, consistent with the phenotypes of *flu* mutants. Most of the significantly regulated genes do not encode chloroplast-localized proteins.

Transcriptome analyses also reveal that the expression of several genes contributing to bacterial and fungal pathogen resistance and to herbivore tolerance were rapidly upregulated in response to increases in the levels of $^1O_2$ in *flu* (Fig. 7). One of these genes is *EDS1* (section "Excess Light Can Induce Acclimation or Cell Death"). *EDS1* attenuates both the spread of foliar lesions and the recovery from growth inhibition that occurs when continuous-light-grown *flu* plants are transferred to the dark and then reilluminated. Based on an analysis of Arabidopsis mutants, phytohormones such as ethylene, salicylic acid, and jasmonic acid that contribute to PCD induced by $O_2^{\bullet-}/H_2O_2$ during incompatible plant-pathogens interactions also contributed to PCD caused by *flu* alleles. In summary, the rise in the levels of $^1O_2$ in *flu* mutants affects a complex signaling network that can respond to other cues.

Transcriptome analyses also led to the identification of several genes that are specifically induced by $^1O_2$ in *flu* mutants and Mehler reaction-derived H_2O_2. As expected, overexpressing thylakoid-bound ascorbate peroxidase (tAPX, Fig. 5) reduced the expression of H_2O_2-inducible genes. Unexpectedly, overexpressing

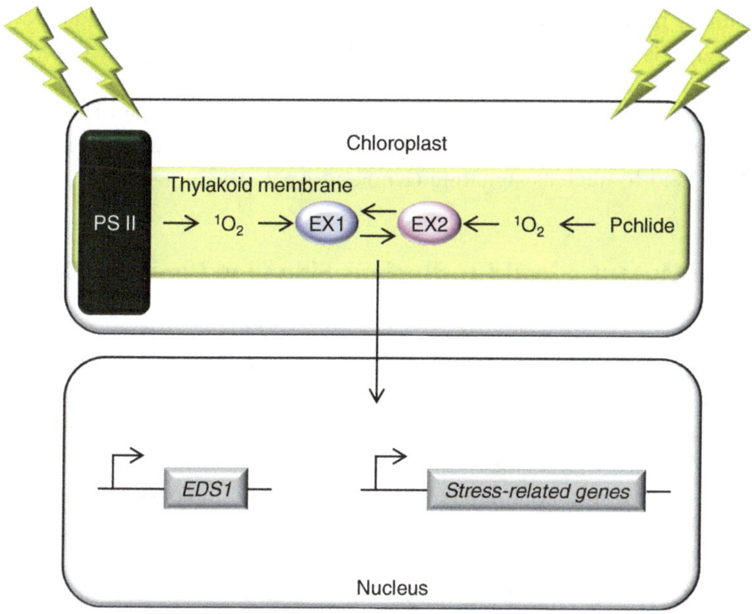

Fig. 7 Initiation of PCD by chloroplastic $^1O_2$. Light induces the production of $^1O_2$ by exciting chlorophylls in PSII and by exciting Pchlide that accumulates in the chloroplast membranes of *flu* mutants. $^1O_2$ activates a plastid-to-nucleus signaling mechanism that depends on the EXECUTER proteins (EX1 and EX2). This plastid-to-nucleus signaling mechanism induces the expression of *EDS1*, *PR1*, *PR5*, and other stress-related genes. This regulated gene expression may contribute to PCD

tAPX inhibited growth, enhanced cell death phenotypes, and caused a greater degree of regulated gene expression in *flu* mutants. Thus, the rise in the levels of $^1O_2$ in *flu* mutants and the increase in Mehler reaction-derived H_2O_2 activate distinct plastid-to-nucleus signaling mechanisms. Additionally, the Mehler reaction-derived H_2O_2 attenuated the $^1O_2$-dependent plastid-to-nucleus signaling that is activated by *flu* alleles. In *C. reinhardtii*, treatments with rose bengal – a xanthene dye that can absorb light and transfer its excitation energy to $^3O_2$, yielding $^1O_2$ – increased the potency of methyl viologen, an inhibitor that induces the production of H_2O_2 from the Mehler reaction. Therefore, $^1O_2$ can affect the signaling activated by H_2O_2. In summary, although plastidic $^1O_2$ and H_2O_2 activate distinct plastid-to-nucleus signaling mechanisms, each signaling mechanism regulates the other.

Although oxylipins produced in the chloroplast can contribute to stress responses, the increase in $^1O_2$ that occurred in *flu* mutants induced changes in nuclear gene expression prior to the accumulation of oxylipins. Thus, the plastid-to-nucleus signaling mechanism that is induced by *flu* alleles does not appear to require oxylipins. *flu* alleles led to the transient accumulation of 13S-hydroxy octadecatrienoic acid (HOT) in mature plants. This isomer is thought to accumulate only from enzymatic peroxidation of linolenic acid. Linolenic acid is the predominant fatty acid in plants and is a major target of chloroplastic ROS. The cytotoxic levels of $^1O_2$ induced in etiolated *flu* seedlings upregulated 10-hydroxy octadecadienoic acid (10-HOD), 12-HOD, 10-HOT, and 15-HOT, which are indicative of the nonenzymatic peroxidation of lipids. Thus, small quantities of $^1O_2$ that are released into chloroplast membranes are not necessarily cytotoxic and can activate plastid-to-nucleus signaling (Kim and Apel 2013b).

In the *flu* Mutant, Plastid-to-Nucleus Signaling Requires *EXECUTER1* and *EXECUTER2*

A mutant screen has provided insight into the plastid-to-nucleus signaling mechanism activated by *flu* alleles. In this screen, alleles were identified that suppress the growth inhibition and cell death in mature *flu* plants, photobleaching in *flu* seedlings, and that do not affect the rise in the levels of Pchlide and $^1O_2$ that occurs when *flu* plants are grown in continuous light, transferred to the dark for several hours, and then reilluminated. This screen yielded 15 alleles of only one gene named *EXECUTER1* (*EX1*). These data indicate that this screen specifically disrupts $^1O_2$-dependent plastid-to-nucleus signaling in *flu* mutants.

The Arabidopsis genome contains a second gene that is related to *EX1* named *EX2*. The EX1 and EX2 proteins appeared to localize to the thylakoid membranes. Loss-of-function alleles of *EX2* suppressed phenotypes of *flu* but not to the same degree as loss-of-function alleles of *EX1*. The thylakoid-membrane localization is consistent with the EX1 and EX2 proteins contributing to $^1O_2$ signaling because in the *flu* mutant, Pchlide accumulates in chloroplast membranes. The *flu*, *ex1 flu*, *ex2 flu*, and *ex1 ex2 flu* mutants accumulated similar levels of Pchlide when they are transferred to the dark and similar levels of $^1O_2$ when they are reilluminated. Thus, the suppression of *flu* phenotypes by *ex1* and *ex2* alleles is caused by defects in plastid signaling and not by a reduction in oxidative stress. The biochemical function of the EXECUTER proteins is not known.

ex1 attenuated the expression of $^1O_2$-inducible genes in a *flu* background. When the *EX2* gene is knocked out in the *flu* background, the expression of $^1O_2$ inducible genes is either enhanced or attenuated. These data provide evidence that the EX2 protein regulates the activity of the EX1 protein. Knocking out *EX1* and *EX2* in the *flu* background attenuated the expression of more $^1O_2$-inducible genes than knocking out only *EX1* in the *flu* background. These data indicated that *EX1* and *EX2* contribute to plastid-to-nucleus signaling in the *flu* mutant and that the EX1 and EX2 proteins act upstream of $^1O_2$-regulated gene expression (Fig. 7).

When continuous-light-grown *flu* plants are transferred to the dark and then reilluminated, they lost chloroplast integrity within 15 min of reillumination. *ex1* suppressed this loss in chloroplast integrity. The loss in chloroplast integrity occurred before rupture of the central vacuole and cell death. However, it is not known whether this loss in chloroplast integrity caused PCD. Thus, plastid-to-nucleus signaling that depends on *EX1* and activated by increases in chloroplastic

$^1O_2$ disrupts chloroplast integrity that occurs prior to PCD. Whether the *EX1*- and *EX2*-dependent effects on gene expression and chloroplast integrity are independent effects is not known. A loss of mitochondrial integrity at the onset of PCD is observed in numerous organisms. It is debated whether mitochondria releasing their proteins into the cytosol causes PCD in plants.

One concern about using *flu* alleles to test the effects of $^1O_2$ on plastid-to-nucleus signaling is that in wild-type plants, $^1O_2$ is produced by the reaction center and the antennae of PSII. In *flu* mutants, Pchlide accumulates in chloroplast membranes somewhat distant from PSII. To test whether the *EX* genes contribute to the plastid-to-nucleus signaling activated by PSII-derived $^1O_2$, the excitation pressure of PSII was increased by increasing the fluence rate or by increasing the fluence rate and decreasing the temperature. Although large increases in excitation pressure used in these experiments caused cell death in wild type, the cell death was suppressed by *ex1* and *ex1 ex2*. However, *ex1 ex2* did not suppress cell death brought about by more severe increases in excitation pressure that induced the production of cytotoxic levels of $^1O_2$. Thus, EX1 and EX2 contribute the PCD that is induced when PSII produces elevated levels of $^1O_2$ (Kim and Apel 2013b).

Other *flu*-Based Screens for Mutant Alleles that Disrupt Plastid-to-Nucleus Signaling

Other screens for mutant alleles that contribute to plastid-to-nucleus signaling in *flu* mutants include a screen for mutants that specifically suppress the photobleaching phenotype of Arabidopsis seedlings. Thus far, this screen has yielded single mutant alleles of three genes that contribute to diverse chloroplastic processes. These genes encode a chloroplastic sigma factor named SIG6, a chloroplastic protein that is related to the human mitochondrial transcription termination factor named mTERF, and a chloroplastic protein named CRUMPLED LEAF that contributes to chloroplast division. These data provide evidence that these screens do not specifically disrupt a particular process. These and other data provide evidence that the mutant alleles yielded by this screen likely attenuate this signaling by various indirect mechanisms.

Another screen utilized a luciferase reporter gene in which its expression is driven by the promoter of the $^1O_2$-inducible *AAA-ATPase* gene. The screen is used to identify mutant alleles that constitutively upregulate the expression of this reporter gene. Only two mutant alleles were reported from this screen. One is from a gene that encodes PLEIOTROPIC RESPONSE LOCUS1. This allele is thought to indirectly affect the expression of the *AAA-ATPase* gene. A second allele – *constitutive activator of AAA-ATPase 39* (*caa39*) – appears to directly affect ROS-regulated gene expression. *CAA39* encodes a Topoisomerase VI (Topo VI) A-subunit (AtTOP6A). Topo VI is an ATP-dependent type II topoisomerase, found in diverse organisms. Topo VI is a tetramer composed of two A and two B subunits. *CAA39/AtTOP6A* contributes to the regulation of both $^1O_2$- and H_2O_2-responsive genes. AtTOP6A binds the proximal promoter regions of *AAA-ATPase* and other $^1O_2$-inducible genes and either positively or negatively regulates a small fraction of $^1O_2$-responsive genes. The human topoisomerase II can affect transcription by interacting with the general transcription machinery. Sequence-specific transcription activators can relieve these effects. A similar mechanism is proposed to occur in plants (Kim and Apel 2013b).

$^1O_2$ and cry1 Signaling Are Required for Light-Dependent PCD

As described in section "*cry1* Mutants Are *gun* Mutants," the functional state of the chloroplast can affect *CRYPTOCHROME 1* (*CRY1*) signaling. The finding that *CRY1* is required for light-dependent PCD in *flu* mutants provides more evidence that chloroplast functions can affect *CRY1* signaling. *CRY1* does not contribute to PCD by driving the biosynthesis of photosensitizing molecules in the chloroplast. The *CRY1*-dependent component of this response is apparently a secondary response that is affected by $^1O_2$-dependent plastid-to-nucleus signaling. In plants that contain well-functioning chloroplasts, a large proportion of *CRY1*-regulated genes encode chloroplast proteins. In contrast, when a *flu* allele induces PCD, a small proportion of genes that are regulated by both $^1O_2$ and *CRY1* encode chloroplast proteins. In *flu* mutants, most of the genes that are regulated by both $^1O_2$ and *CRY1* are associated with stress and PCD. Thus, the accumulation of $^1O_2$ in the chloroplast appears to activate a plastid-to-nucleus signaling mechanism that "rewires" *CRY1* signaling to induce PCD (Larkin and Ruckle 2008).

flu-Derived $^1O_2$ Can Affect Chloroplast Biogenesis

Arabidopsis and many other plants have green embryos that become colorless as they mature. During this degreening process, the thylakoid membranes are dismantled and chloroplasts are converted to nonphotosynthetic proplastids. Results from experiments with *ex1 ex2* double mutants indicate that this dismantling of the thylakoid membranes and the resulting chlorophyll catabolism induces a rise in the levels of $^1O_2$ that is perceived by the EXECUTER proteins. Increased levels of $^1O_2$ can affect gene expression and promote the biogenesis of etioplasts and chloroplasts that occurs during germination. $^1O_2$ signaling appears to affect plastid function during embryogenesis by regulating the expression of genes that promote ABA biosynthesis. The low levels of ABA that are upregulated by $^1O_2$ signaling induced the expression of genes that encode plastid-localized proteins. Thus, $^1O_2$ signaling is thought to promote plastid function during embryogenesis by inducing ABA signaling. Although EXECUTER-dependent signaling affects etioplast and chloroplast biogenesis during germination, this signaling did not affect chloroplast biogenesis when leaves develop from the shoot apical meristem (Kim and Apel 2009).

$^1O_2$-Dependent Plastid-to-Nucleus Signaling Induces Either Acclimation or Cell Death

The Arabidopsis *nonphotochemical quenching 1* (*npq1*) *lycopene-ε-cyclase* (*lut2*) double mutant provides an alternative system for testing the effects of $^1O_2$ on plastid-to-nucleus signaling. *npq1 lut2* does not synthesize violaxanthin deepoxidase and lycopene-ε-cyclase and, thus, does not synthesize lutein and zeaxanthin, two photoprotective xanthophylls that associate with PSII. When *npq1 lut2* was transferred to 1,000 μmol m^{-2} s^{-1} and 10 °C, *npq1 lut2* accumulated similar levels of H_2O_2, $O_2^{\bullet -}$, and hydroxyl radicals as wild type and much higher levels of $^1O_2$ produced by PSII than wild type. Other factors that contribute to plastid-to-nucleus signaling such as the redox states of the PQ pool and the glutathione pool are indistinguishable between *npq1 lut2* and wild type under these conditions.

As described in section "$^1O_2$ Activates Plastid-to-Nucleus Signaling in the Arabidopsis *flu* Mutant," the rise in the levels of $^1O_2$ that occurs in *flu* mutants caused foliar lesions to form within 2–4 h of illumination. In contrast, the $^1O_2$ that is generated when *npq1 lut2* was exposed to excess light and low temperature caused photoinhibition but not PCD (Kim and Apel 2013b). Similar experiments were first conducted with the *npq1 lor1* double mutant of *C. reinhardtii*. Like the *npq1 lut2* mutant of Arabidopsis, *npq1 lor1* cannot accumulate zeaxanthin and lutein. In *npq1 lor1*, treatments with excess light led to photooxidative destruction of the thylakoids and a decline in cell viability (Ledford et al. 2004). In both systems, excess light was shown to upregulate genes that protect the cell from oxidative stress.

An alternative approach for studying $^1O_2$ signaling in *C. reinhardtii* utilizes sublethal doses of rose bengal and excess light. This treatment can induce an acclimation response to $^1O_2$ stress in *C. reinhardtii* that did not correlate with changes in the levels of small molecule antioxidants such as carotenoids and tocopherols but did correlate with changes in gene expression. Sublethal doses of rose bengal induced the expression of 14 genes in *C. reinhardtii* including the expression of genes that encode glutathione peroxidase and glutathione *S*-transferase. Constitutive overexpression of either of these two genes induced acclimation to $^1O_2$ stress in *C. reinhardtii*. This finding is consistent with the ability of glutathione *S*-transferases to detoxify xenobiotic and endogenous compounds such as lipid peroxides. However, although increases in the levels of other ROS upregulated these same genes, other ROS did not induce acclimation to $^1O_2$. Therefore, acclimation to $^1O_2$ in *C. reinhardtii* appears more complex than simply upregulating the expression of these two genes (Kim and Apel 2013b). In summary, $^1O_2$ can activate plastid-to-nucleus signaling mechanisms that induce acclimation to ROS in both Arabidopsis and *C. reinhardtii*.

A component of the signaling mechanism that helps *C. reinhardtii* to acclimatize to $^1O_2$ stress was isolated from a screen for mutants that are resistant to lethal concentrations of rose bengal. The screen yielded a mutant allele of *SOR1*, which encodes an algal-specific bZIP transcription factor. *sor1* is more tolerant of $^1O_2$, other ROS and reactive electrophile species than wild type (Chi et al. 2013). Reactive electrophile species are molecules that contain a reactive electrophilic atom such as a α,β-carbonyl groups that can react with an electron donor such as a sulfhydryl group, which are found in many biomolecules (Farmer and Mueller 2013). Based on these data, *SOR1* appears to promote a variety of stress responses (Chi et al. 2013).

β-Cyclocitral Activates Plastid-to-Nucleus Signaling

$^1O_2$ can degrade β-carotene, yielding short-chain volatile compounds. To test whether any of these short-chain volatile compounds might contribute to $^1O_2$ signaling, the volatile oxidation products of β-carotene were identified by illuminating a solution of β-carotene that contained rose bengal and analyzing these oxidation products by gas chromatography-mass spectrometry. The oxidation products of β-carotene generated in vitro were found in Arabidopsis leaves exposed to optimal fluence rates of light. The levels of these oxidation products increased within hours when leaves were exposed to 1,400 μmol m^{-2} s^{-1} of white light. This fluence rate induced the expression of $^1O_2$-inducible genes and had diverse effects on the expression of H_2O_2-inducible genes.

One of these volatile oxidation products of β-carotene – β-cyclocitral (Fig. 8) – induced the expression of several $^1O_2$-inducible genes in a dose-dependent manner and in a physiologically relevant concentration range but did not affect the expression of H_2O_2-inducible genes. Transcriptome studies indicated that β-cyclocitral regulated the expression of more than 1,000 genes in Arabidopsis. Most of the upregulated genes contribute to the interactions between plants and their environment, stress responses, and cellular transport. Most of the downregulated genes contribute to growth, development, and the biogenesis of cellular components. Thus, β-cyclocitral is a plastid signal that appears to help plants divert resources from growth to stress responses. The gene expression changes induced by both β-cyclocitral and elevated levels of $^1O_2$ in *flu* mutants are >80 % and >90 % similar for upregulated and downregulated genes, respectively. In contrast to the PCD observed in *flu* mutants, treating leaf disks and intact plants with β-cyclocitral induced tolerance to excess light and did not induce PCD. Thus, β-cyclocitral can induce an acclimation response to the $^1O_2$ that is produced by PSII (Fig. 9) (Karpiński et al. 2012; Chi et al. 2013).

Chloroplastic Ca^{2+} may Activate $^1O_2$-Dependent Plastid-to-Nucleus Signaling

Stimuli besides excess light may activate $^1O_2$-dependent plastid-to-nucleus signaling. The calcium-sensing receptor (CAS) is a calcium-binding protein that associates with the thylakoid membranes. *CAS* is required for plant immune responses that are triggered by pathogen-associated molecular patterns (PAMP) recognized by the innate immune system (PAMP-triggered immunity) and by pathogen-derived effector proteins, which trigger hypersensitive response cell death at the

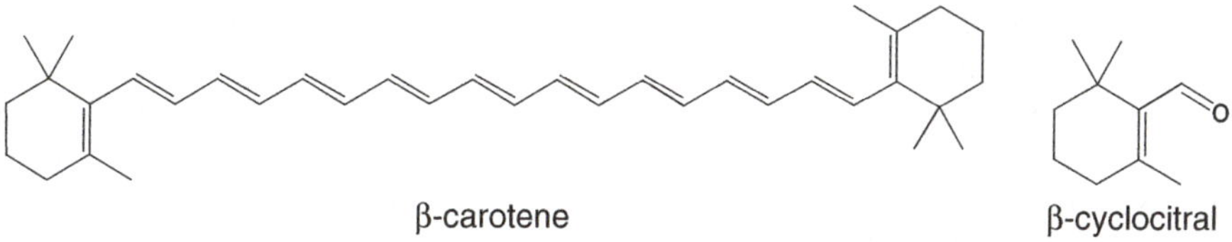

Fig. 8 Structures of β-carotene and β-cyclocitral

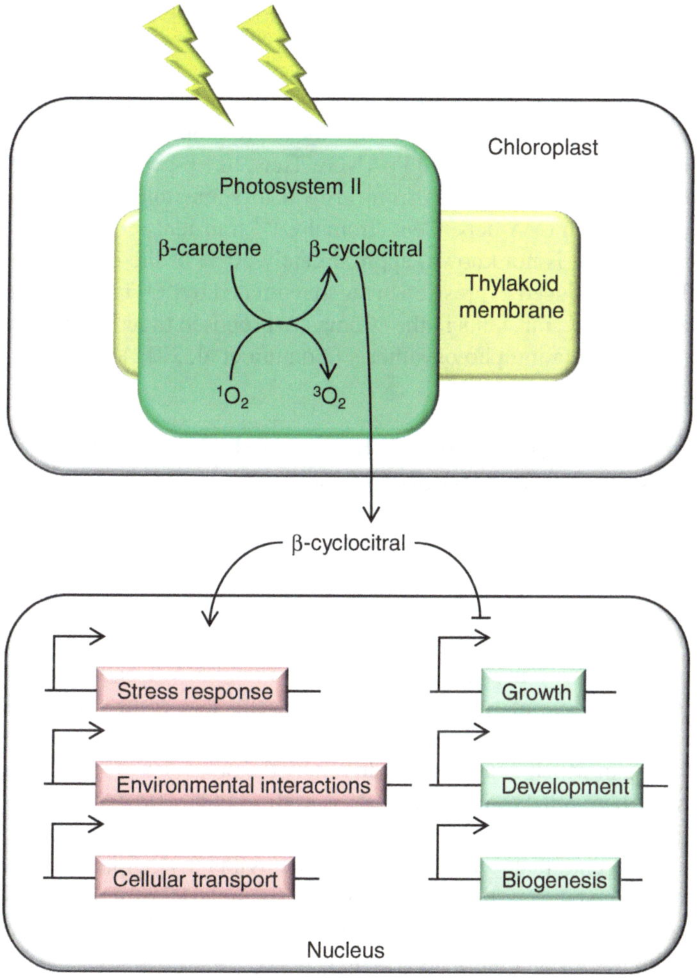

Fig. 9 β-cyclocitral-induced acclimation to $^1O_2$ stress. $^1O_2$ oxidizes β-carotene, yielding β-cyclocitral. This volatile derivative of β-carotene moves from the chloroplast to the nucleus by diffusion. β-cyclocitral regulates the expression of numerous genes. In general, β-cyclocitral *up*-regulates the expression of genes that help plants to interact with their environment, respond to stress, and promote cellular transport and *down*-regulates the expression of genes that contribute to growth, development, and biogenesis of cellular components

infection site (effector-triggered immunity). PAMP signaling activated by flagellin and chitin induced a long-lasting increase in the stromal Ca^{2+} levels and a range of basal defense responses that were attenuated in *cas-1*, a null allele of *CAS*. These defense responses include stomatal closure, callose deposition, the accumulation of defense-related phenylpropanoids, and the accumulation of salicylic acid. The hypersensitive cell death response that is activated by infection with *P. syringae* DC3000 was also attenuated in *cas-1*. Thus, the CAS protein appears to couple chloroplasts to biotic stress responses by releasing Ca^{2+} from the lumen of the thylakoid membranes into the stroma. *CAS* assumes a high position in this signaling hierarchy, acting upstream of ROS and salicylic acid.

Analysis of transcriptomes indicated a major role for *CAS* in PAMP-regulated gene expression. *CAS*-dependent plastid-to-nucleus signaling appeared to promote PAMP-triggered immunity by upregulating the expression of defense-related genes and downregulate the expression of PhANGs. Thus, like β-cyclocitral, *CAS* appears to help plants to divert resources from growth to stress tolerance. Although the mechanism by which *CAS*-dependent stromal Ca^{2+} transients contribute to plastid-to-nucleus signaling is not known, approximately 30 % of the genes that required *CAS* for flagellin-induced expression are also regulated by $^1O_2$ in *flu* mutants. Thus, *CAS*-dependent Ca^{2+} transients in the stroma are proposed to induce the production of $^1O_2$ by downregulating photosynthesis (Nomura et al. 2012).

Methylerythritol Cyclodiphosphate (MEcPP) Induces the Expression of Stress-Related Genes

Like other terpenes, carotenoids are derived from isopentenyl diphosphate (IPP) and dimethylallyl diphosphate (DMAPP). The methylerythritol phosphate (MEP) pathway drives the biosynthesis of carotenoids in chloroplasts by synthesizing IPP and DMAPP. IPP and DMAPP are also synthesized from the distinct mevalonate (MVA)-based pathway in the cytosol. The MEP and MVA pathways are largely independent. One loss-of-function allele of a gene that contributes to the MEP pathway was isolated from a screen for alleles that cause constitutive expression of a reporter gene in which expression is driven by the promoter of a nuclear gene that encodes hydroperoxide lyase (HPL). *HPL* is a stress-inducible gene that contributes to oxylipin biosynthesis. The *constitutively expressing HPL* (*cehl*) mutant has a defect in the gene that encodes hydroxymethylbutenyl diphosphate synthase (HDS), which catalyzes the conversion of methylerythritol cyclodiphosphate (MEcPP) (Fig. 10) to hydroxymethylbutenyl diphosphate (HMBPP). MEcPP accumulates to higher levels than wild type in *cehl*.

In addition to expressing elevated levels of *HPL*, the *cehl* mutant expressed elevated levels of the *ISOCHORISMATE SYNTASE 1* (*ICS1*) gene and accumulated higher levels of salicylic acid than wild type. *ICS1* contributes to the biosynthesis of salicylic acid. However, salicylic acid did not contribute to the elevated expression of *HPL* in the *cehl* mutant. Whether MEcPP affects the expression of other genes is not known.

Fig. 10 Structure of MEcPP

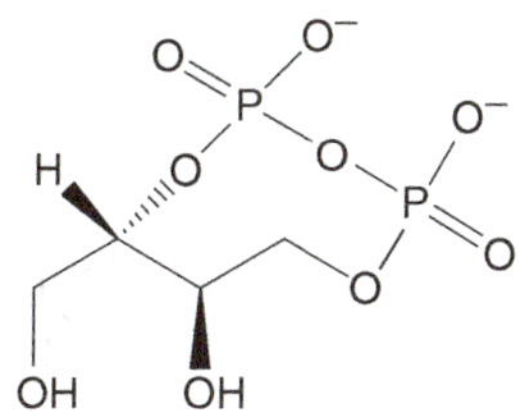

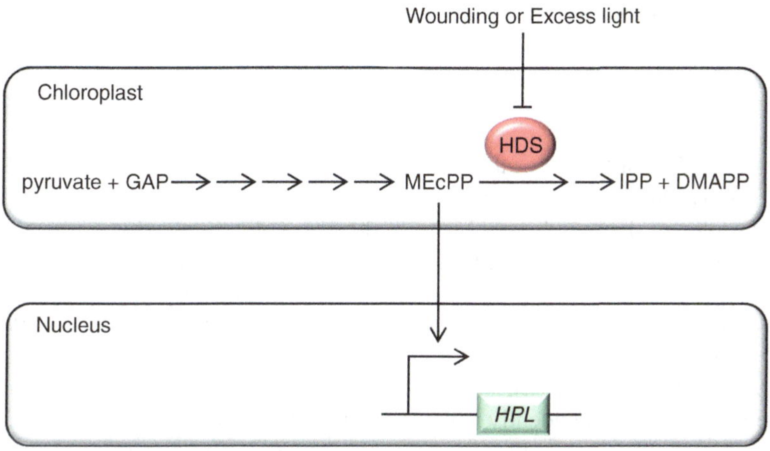

Fig. 11 MEcPP induced expression of *HPL*. Wounding and excess light can induce a rise in the levels of MEcPP, presumably by inhibiting HDS. MEcPP somehow induces the expression of HPL

Loss-of-function alleles of genes that encode enzymes that act upstream of HDS in the MEP pathway reduced the levels of MEcPP and reduced the expression of *HPL*. Exogenous applications of MEcPP induced the expression of *HPL*. Further, wounding and excess light increased the levels of MEcPP and increased the expression of *HPL*. MEcPP does not appear to affect the plastid-to-nucleus signaling that is activated by blocking chloroplast biogenesis (section "The *Genomes Uncoupled* (*gun*) Mutant Screen"). Whether MEcPP contributes to other plastid-to-nucleus signaling mechanisms is not known. MEcPP is proposed to move to the nucleus and regulate gene expression (Fig. 11) (Karpiński et al. 2012; Chi et al. 2013).

3′-Phosphoadenosine 5′-phosphate (PAP) Contributes to Excess Light and Drought Tolerance

A screen for mutant alleles that misregulate the expression of a luciferase reporter gene, driven by the *APX2* promoter, yielded a loss-of-function allele of *ALTERED APX2 EXPRESSION8 (ALX8)*. The *alx8* allele induced the expression of the endogenous *APX2* gene to higher levels than wild type regardless of whether plants are exposed to optimal or excess fluence rates. The *alx8* allele was shown to

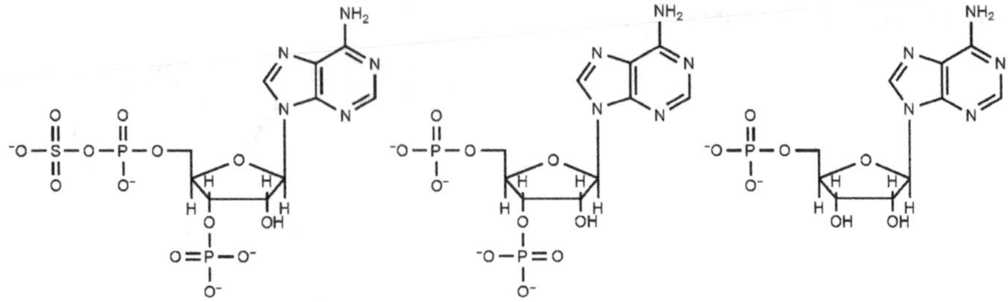

Fig. 12 Structures of PAPS, PAP, and AMP

upregulate 35 % of the genes that are induced by excess light, including genes that significantly contribute to excess-light tolerance. In addition, *alx8* was more drought tolerant than wild type. Many of the excess-light inducible genes are also induced by drought. *alx8* had a pleiotropic metabolite phenotype that stimulated the production of higher levels of putrescine and sugars than wild type. These osmoprotectants may contribute to the drought tolerance of *alx8* (Wilson et al. 2009).

ALX8 is also known as *SAL1*, *FRY1*, *RON1*, *FOU8*, and *SUPO1* because alleles of the gene were independently isolated from a number of diverse screens. Indeed, this gene contributes to other stress responses, vascular patterning, jasmonate biosynthesis, and polar auxin transport. The *SAL1/ALX8/FRY1/RON1/FOU8/SUPO1* gene (hereafter referred to as *SAL1*) encodes a phosphatase that converts 3′-phosphoadenosine 5′-phosphate (PAP) to adenosine monophosphate (AMP) (Figs. 12 and 13). PAP is produced by sulfotransferases in the cytosol. Sulfotransferases convert 3′-phosphoadenosine 5′-phosphosulfate (PAPS) (Fig. 12) to PAP upon transfer of sulfate groups from PAPS to various acceptor molecules. PAP accumulated when plants experienced drought and, to a lesser degree, when plants were exposed to excess light. Indeed, leaf water status and PAP levels were tightly correlated in both *alx8* and wild type.

The SAL1 protein was localized to both chloroplasts and mitochondria. Increasing the SAL1 activity of the chloroplast lowers the levels of PAP. Additionally, targeting SAL1 to the nucleus fully complemented the PAP, *APX2* expression, and drought-tolerance phenotypes of *sal1* mutants. These data indicate that PAP can accumulate in the chloroplast and the nucleus. At this time, PAP is the only plastid signal that is demonstrated to accumulate in both the chloroplast and the nucleus and to elicit an appropriate response in the nucleus.

The observation that PAP inhibits the activity of 5′–3′ exoribonucleases (XRNs) from *Saccharomyces cerevisiae* is consistent with PAP-based plastid-to-nucleus signaling inhibiting Arabidopsis XRNs. Indeed, the 5′–3′ exoribonucleases *XRN2* and *XRN3* appear to promote the same plastid-to-nucleus signaling mechanism as *SAL1* and PAP. For instance, both *sal1* and the *xrn2 xrn3* double mutant exhibited misregulation of 50 % of the same genes. Both *sal1* and *xrn2 xrn3* mutants

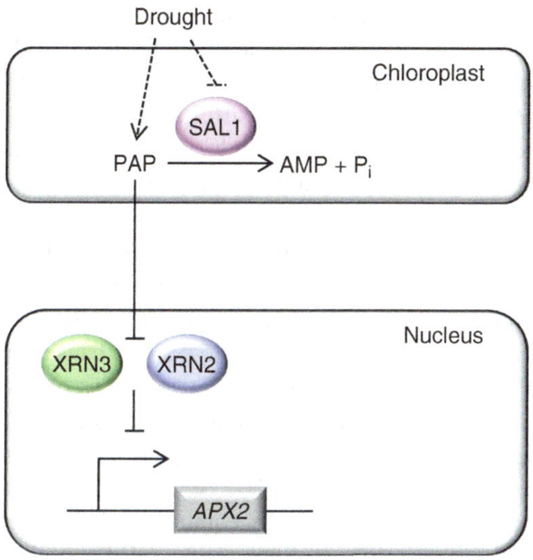

Fig. 13 PAP-induced expression of *APX2*. Drought can induce a rise in the levels of PAP by inhibiting SAL1 or by some other mechanism (*dotted arrow* and *T bar*). PAP moves from the chloroplast to the nucleus. PAP accumulates in the nucleus. In the nucleus, PAP inhibits XRN2 and XRN3. This inhibition induces the expression of *APX2* and regulates the expression of many other genes. This plastid-to-nucleus signaling mechanism promotes drought tolerance

exhibited enhanced drought tolerance, although *xrn2 xrn3* exhibited less drought tolerance than *sal1*.

In summary, chloroplastic SAL1 converts PAP to AMP when plants do not experience stress. During drought and excess-light stress, PAP accumulates and chloroplasts appear to export PAP. PAP accumulates in the nucleus and likely inhibits the activity of XRN2 and XRN3, which, in turn, downregulate the expression of stress-related genes. PAP may also regulate nuclear gene expression by mechanisms that do not involve inhibiting XRN2 and XRN3 because approximately 40 % of the gene expression changes that occur in *alx8* do not occur in *xrn2 xrn3*. The mechanism that XRN2 and XRN3 use to regulate the expression of *APX2* and other genes is not known (Fig. 13) (Chi et al. 2013).

Plastid Signals Contribute to Development

In addition to promoting metabolism, acclimation to biotic and abiotic stress, PCD, and affecting the circadian rhythm, plastid signals can promote development. The evidence that plastid-to-nucleus signaling can affect development is that inhibitors and particular mutant alleles that cause chloroplast dysfunction also cause developmental abnormalities and that mutant alleles that disrupt plastid-to-nucleus signaling affect development.

Variegated Mutants Exhibit Abnormal Leaf Development

As described in section "The Redox State of PQ Can Affect Chloroplast Biogenesis", the leaves of variegated mutants have green and yellow/white sectors because of mutations in nuclear, chloroplast, or mitochondrial genes. Analyses of a large number of variegated mutants indicated that these phenotypes are caused by loss-of-function alleles of genes that encode chloroplast proteins or mitochondrial proteins that affect chloroplast function and that these mutants often exhibit abnormal leaf morphologies. Although a variety of genetic mechanisms can cause variegated leaves in various plant species, the most well-studied variegations are caused by stable mutations in the nuclear genes of Arabidopsis. This section will consider the data from these Arabidopsis mutants that are relevant to development.

The effects of chloroplast biogenesis on leaf development are especially well studied in *immutans* (*im*) (section "The Redox State of PQ Can Affect Chloroplast Biogenesis"). Similar phenotypes are caused by mutant alleles of the orthologous gene in tomato named *ghost*. Several loss-of-function alleles that cause variegations also affect leaf development such as *atase2 deficient* (*atd2*), *chloroplast mutator* (*chm*), *cloroplastos alteradosl* (*cla1*), *pale cress* (*pac*), *rugosa2* (*rug2*), and *yellow variegated3* (*var3*). *ATD2* encodes glutamine 5-phosphoribosylpyrophosphate amidotransferase, which catalyzes the first committed step of *de novo* purine biosynthesis in the chloroplast. *CHM* encodes a mitochondrial protein that helps to maintain the integrity of mitochondrial genomes. The aberrant mitochondrial genomes of *chm* cause mitochondrial dysfunction that can attenuate chloroplast biogenesis. *CLA1* encodes 1-deoxy-D-xylulose 5-phosphate synthase, which contributes to the MEP pathway (section "Methylerythritol Cyclodiphosphate (MEcPP) Induces the Expression of Stress Related Genes"). *PAC* encodes a chloroplastic protein that contributes to RNA processing. *RUG2* is related to metazoan factors that promote the termination of mitochondrial transcription residing in both chloroplasts and mitochondria. *VAR3* encodes a chloroplastic protein that is suggested to promote carotenoid metabolism. Thus, loss-of-function alleles of diverse genes can cause variegated leaves.

Although the ultrastructures of the chloroplasts in the green sectors of variegated leaves often resembled wild type, chloroplasts had fewer thylakoid membranes and starch grains than wild type in the green sectors of *rug2-1* leaves. The plastids in the yellow and white sectors of variegated leaves were distinct from the plastids in the green leaves of wild type. The plastids in the yellow and white sectors lacked thylakoid membranes and varied in size. These plastids ranged from small plastids that resemble proplastids to large plastids that contain vacuoles and lack lamellae.

The leaf morphologies were abnormal in the yellow and white sectors of these variegated mutants. Although these abnormalities varied, in some instances they shared common features. For instance, the yellow and white sectors of these mutants all had abnormal palisade cells. They lacked palisade cells (*cla1* and *atd2*), contained fewer palisade cells (*rug2*), or exhibited no expansion of palisade cells (*atd2*, *im*, *ghost* and *var3*). In *pac*, palisade cells appeared similar to wild type during the early stages of leaf development. During the later stages of leaf development, the palisade cells of *pac* became smaller, the air spaces of *pac* leaves grew larger, and the epidermal cells of *pac* became almost 40 % larger than wild type.

Chloroplast mutator (*chm*) disrupts the developmental response of leaves to excess light. Excess light causes anticlinal elongation of palisade cells (i.e., elongation perpendicular to the surface of the leaf) and induces cell division, especially among palisade cells. These responses yield a thick palisade layer that shades the chloroplasts beneath. In the yellow and white sectors of *chm*, the anticlinal elongation response was attenuated and the cell division response was not observed. Thus, plastid-to-nucleus signaling that is activated by chloroplast dysfunction in *chm* may contribute to both of these responses. Plants may reduce their investment in leaf development without functional chloroplasts.

The green sectors of variegated leaves can appear like wild type (e.g., as in *rug2*), or they can exhibit abnormal morphologies, as in *im*. The green sectors on the leaves of *im* plants were thicker than wild type because they contain larger air spaces, mesophyll cells, and epidermal cells than wild type. The green sectors of *im* also exhibited higher rates of photosynthesis, higher levels of chlorophyll, and higher chlorophyll *a*/*b* ratios than wild type. Thus, when *im* is grown at a fluence rate of light that is optimal for wild type, the green sectors of *im* leaves resemble wild type leaves that are grown in excess light. *IM* is expressed throughout the plant, and *im* blocks not only the biogenesis of chloroplasts but also the biogenesis of amyloplasts and etioplasts. Consistent with the concept that nonphotosynthetic plastids contribute to development, *im* tended to have shorter roots than wild type.

The diverse developmental phenotypes of variegated mutants are consistent with complexity in the plastid signaling that underpins leaf development. This interpretation is supported by the finding that the alleles that cause variegations had distinct effects on PhANG expression. *Lhcb* was expressed at similar high levels in the green leaf sectors of *im* and the green leaves of wild type. *Lhcb* was expressed at much lower levels in the yellow and white sectors of *im leaves* than in the green sectors. In contrast, *Lhcb* was expressed at similar levels in green wild-type seedlings and in pale-green or albino *pac* seedlings. Additionally, *Lhcb* and *RbcS* were expressed at similar levels in green wild-type seedlings and in the yellow and white sectors of *alx13*, which is allelic to *atd2*. However, *alx13* is not a *gun* mutant (section "The *Genomes Uncoupled* (*gun*) Mutant Screen") because *Lhcb1* was expressed at wild-type levels when *alx13* was treated with norflurazon (Larkin 2014).

Plastid Translation and Proteolysis Is Required for Normal Development

Although plastid ribosome-deficient mutants of oilseed rape, maize, and barley can survive and develop normally, plastid translation is an essential process in most plant species. Thus, plastid genes that encode subunits of plastid ribosomes are essential genes in many plant species. However, plastid genes that encode subunits of the plastid ribosomes, such as *rpl36* in tobacco, are not essential. Deleting *rpl36* caused developmental phenotypes such as pale and elongated leaves and the loss of apical dominance. Perturbation in auxin biosynthesis may cause this phenotype because tryptophan-independent biosynthesis of auxin is thought to rely on plastids. Alternatively, plastid-to-nucleus signaling that is activated in *rpl36* may contribute to these phenotypes.

The development of aerial organs did not occur in tobacco plants that lack the *clpP1* gene. *clpP1* is a plastid gene that encodes a Clp protease. Clp proteases are responsible for the bulk of protein degradation in chloroplasts. Either the improper accumulation of enzymes that reside in the chloroplast or the overaccumulation of a regulatory protein in the plastid may cause these developmental abnormalities in *clp1* mutants (Larkin 2014).

Plastid Signals may Trigger Cell Expansion in Leaves

The early stages of leaf development are driven in part by cell division. Later stages in leaf development are dominated by cell expansion accompanied by the expansion of the large central vacuole. Thus, the final size of the leaf is largely determined by the time the leaf switches from cell-division-based morphogenesis to cell expansion. This transition is also marked by changes in cell shape. The epidermal cells of leaves undergoing cell-division-based morphogenesis appear square. The epidermal cells of leaves that are undergoing cell expansion resemble jigsaw puzzles pieces because they have numerous interdigitating lobes. Transcriptome analyses indicated that (1) PhANG expression is upregulated when the leaf switches from cell division to cell expansion and (2) that the expression of genes that contribute to tetrapyrrole metabolism was upregulated before the leaf switches to expansion. The upregulated expression of these genes at this transition is consistent with plastid-to-nucleus signaling contributing to this transition. Indeed, treating Arabidopsis seedlings with norflurazon (section "Early Findings") was found to inhibit the transition at the tip of the leaf. Thus, plastid-to-nucleus signaling induced by chloroplast dysfunction appears to inhibit the morphogenesis switch (Larkin 2014).

Signals from Plastids and Mitochondria may Regulate the Development of Plasmodesmata

Plasmodesmata transport a variety of molecules among adjacent cells. *INCREASED SIZE EXCLUSION*.

LIMIT1 (*ISE1*) and *ISE2* were obtained from a screen for alleles that affect plasmodesmata. Loss-of-function alleles of *ISE1* and *ISE2* increase the intercellular transport of fluorescent 10-kDa dextrans during embryogenesis. *ISE1* and *ISE2* encode RNA helicases that reside in the mitochondria and plastids, respectively. *ISE1* and *ISE2* alleles are thought to affect the processing of transcripts derived from mitochondrial and chloroplastic genes. Indeed, the expression of both mitochondrial and chloroplastic genes during embryogenesis was abnormal in *ise1* and *ise2*. The expression of nuclear genes that encode proteins that localize to chloroplasts and mitochondria and the expression of genes that contribute to both the cell wall and plasmodesmata were misregulated in *ise1* and *ise2*. Thus, chloroplast dysfunction is thought to promote intercellular trafficking by activating plastid-to-nucleus signaling that enhances the formation and function of plasmodesmata (Larkin 2014).

The finding that *ise1* affected the expression of plastid protein genes and that *ise2* affected the expression of genes that encode mitochondrial proteins is consistent with the notion that plastid signals affect mitochondria and in turn mitochondrial signals affect plastids. As described in section "Interorganellar Signaling Between Chloroplasts and Mitochondria," other findings indicate that chloroplasts can affect mitochondria and vice versa by regulating nuclear gene expression.

gun Mutants Can Exhibit Abnormal Development

Blocking chloroplast biogenesis affects light-regulated development, and null alleles of *GUN1* and *CRY1* (sections "Loss-of-Function Alleles of *GUN1* Broadly Disrupt Plastid Signaling" and "*cry1* Mutants Are *gun* Mutants") were found to attenuate these effects (Larkin 2014). Light signaling promotes the unfolding and expansion of cotyledons and inhibits the elongation of hypocotyls (Jiao et al. 2007). Plastid signals can influence these processes. Plastid signals that depend on *GUN1* attenuated the light-regulated unfolding of the cotyledons. Plastid signals that depend on *GUN1* and *CRY1* inhibited the expansion of the cotyledons. *CRY1* did not appear to affect the expansion of the cotyledons when seedlings were not treated with inhibitors of chloroplast biogenesis. Thus, the plastid signals not only convert *CRY1* from a positive to a negative regulator of *Lhcb1* expression (section "*cry1* Mutants Are *gun* Mutants"), plastid signals also convert *CRY1* into a negative regulator of cotyledon expansion.

In addition to affecting the unfolding and expansion of the cotyledons, plastid signals can affect the development of the epidermal cells of cotyledons. In untreated seedlings, epidermal pavement cells of cotyledons and leaves have interdigitating lobes. When chloroplast biogenesis was blocked, epidermal cells of wild type did not form lobes, became round, and appeared rough because of invaginations in the surface of the cotyledons. In contrast, the epidermal cells of *gun1 cry1* double mutants resembled untreated wild type in that they retained their interdigitating lobes. Thus, *GUN1*- and *CRY1*-dependent plastid signals cause abnormal development of epidermal pavement cells when chloroplast biogenesis is blocked. Like seedlings treated with inhibitors of chloroplast biogenesis, a rough appearance to the leaf surface was also observed in the variegated mutant *chm* (section "Variegated Mutants Exhibit Abnormal Leaf Development"). Abnormal epidermal cells were also observed in *pac* (section "Variegated Mutants Exhibit Abnormal Leaf Development") and *defective chloroplasts and leaves-mutable* (*dcl*), a variegated mutant of tomato. Both *PAC* and *DCL* encode chloroplastic proteins that participate in RNA processing. In addition to these phenotypes, when chloroplast biogenesis was blocked, *gun1* mutants had twice as many stomata as wild type.

The shifting of gene expression domains in the abaxial and adaxial regions of the leaf drive the expansion of the leaf lamina. *GUN1*-dependent plastid-to-nucleus signaling that is activated by chloroplast dysfunction impairs this shifting yielding narrow leaf lamina. Thus, *GUN1* is thought to prevent plants from investing in lamina expansion when they cannot develop photosynthetically efficient leaves because of chloroplast dysfunction.

Light signaling inhibits the elongation of the hypocotyl. Thus, *hy5* (section "*cry1* Mutants Are *gun* Mutants") has longer hypocotyls than wild type because like other mutant alleles that disrupt light signaling, *hy5* fails to inhibit hypocotyl elongation (Jiao et al. 2007). *gun1* alleles suppressed the long hypocotyl phenotype of *hy5* in the *gun1 hy5* double mutant when chloroplast biogenesis was blocked or when seedlings were exposed to excess light. Thus, GUN1-dependent plastid signals promote the elongation of the hypocotyl. Other findings indicate a connection between the functional state of chloroplasts and elongation of the hypocotyl. The transcription factor phytochrome interacting factor 3 (PIF3) contributes to both chloroplast biogenesis and hypocotyl elongation. Hypocotyls were shorter in the chlorophyll-deficient *chlorophyll a/b-binding protein under-expressed* (*cue*) mutants and longer in *long after far-red 6* (*laf6*), which is deficient in a chloroplastic ATP-binding-cassette protein. The interactions between plastid signals and the elongation of hypocotyls appear complex.

GUN1-dependent plastid signals affect other signaling mechanisms that regulate development. *GUN1*-dependent plastid signals interact with ABA to affect early seedling development by mechanisms that are not yet defined. Concentrations of ABA that allowed most of the cotyledons to emerge from wild-type seeds during germination permitted only a few of cotyledons to emerge from *gun1* mutant seeds. However, higher concentrations of ABA were less effective at inhibiting the germination of *gun1* mutants than wild type. Whirly1 (section "Chloroplast-Localized Transcription Factors Contribute to Plastid-to-Nucleus Signaling") was also found to affect the inhibition of germination by ABA (Larkin 2014).

alx8 Exhibits Abnormal Development

In addition to *gun* mutants, *alx8* (section "3′-Phosphoadenosine 5′-phosphate (PAP) Contributes to Excess Light and Drought Tolerance") exhibits abnormal development. The leaves of *alx8* are shorter and rounder than wild type. *alx8* has shorter petioles than wild type. *alx8* leaves have undulating surfaces. In *alx8*, the vascular bundles are disorganized, the shape of cells is abnormal, and the palisade layer is not as well defined as wild type (Wilson et al. 2009).

Interorganellar Signaling Between Chloroplasts and Mitochondria

Inhibitors and mutant alleles that specifically affect chloroplasts can affect mitochondria and vice versa. Although metabolic interactions between these two organelles are complex, there is no evidence for direct chloroplast-to-mitochondria signaling or vice versa. There is evidence that plastid-to-nucleus signaling can regulate the expression of nuclear genes that encode mitochondrial proteins and vice versa. For example, mitochondria in the white tissue of the barley mutant *albostrians* (section "Early Findings") accumulated threefold more mitochondrial DNA and expressed higher levels of mitochondrial genes than wild type. The cytoplasmic male-sterile mutant (CMSII) of *Nicotiana sylvestris* was impaired in complex I function. This mitochondrial dysfunction caused the accumulation of mRNAs that encode a chloroplastic SOD (Leister 2005). Inhibitors that specifically block chloroplast biogenesis induced diverse expression patterns in genes that encode mitochondrial proteins (Ruckle et al. 2012). A loss-of-function mutation in mitochondrial *ALTERNATIVE OXIDASE (AOX)1a* downregulated the expression of PhANGs (Carrie et al. 2013). These findings are consistent with complex interactions between chloroplasts and mitochondria.

Analyses of large sets of transcriptome data indicate that more than 25 % of the genes that were regulated by chloroplast dysfunction were also regulated by mitochondrial dysfunction. Many of these genes are also regulated by biotic and abiotic stress. Thus, these data make a case that signals from both plastids and mitochondria contribute to biotic and abiotic stress responses (Van Aken and Whelan 2012). Additionally, the signals from plastids and mitochondria can synergistically regulate the genes that encode subunits of PSI and the genes that encode subunits of PSII (Leister 2012).

ABI4 links plastid-to-nucleus signaling and mitochondria-to-nucleus signaling. ABI4 helps to downregulate the expression of *Lhcb1.2* by binding CCAC elements in the promoter of *Lhcb1.2* (section "Loss-of-Function Alleles of *GUN1* Broadly Disrupt Plastid Signaling"). ABI4 was also found to upregulate *ALTERNATIVE OXIDASE1a* (*AOX1a*) expression that is induced by mitochondria-to-nucleus signaling by binding to CACC and CCAC elements in the *AOX1a* promoter (Carrie et al. 2013).

PAP (section "3′-Phosphoadenosine 5′-phosphate (PAP) Contributes to Excess Light and Drought Tolerance") may also link plastid-to-nucleus signaling and mitochondria-to-nucleus signaling because the transcriptomes of plants that have defects in PAP signaling were more similar to the transcriptomes of plants with mitochondrial dysfunctions than plants with chloroplast dysfunctions. Based on these data, PAP was suggested to serve in both plastid-to-nucleus signaling and mitochondria-to-nucleus signaling, which is consistent with SAL1 residing in both chloroplasts and mitochondria (section "3′-Phosphoadenosine 5′-phosphate (PAP) Contributes to Excess Light and Drought Tolerance").

Although plastid signaling and mitochondria signaling regulate a common group of genes, these signaling mechanisms have distinct effects on the transcriptome. Genes that contribute to respiration and encode proteins that reside in mitochondria were overrepresented among genes that were regulated by mitochondria-to-nucleus signaling. Genes that contribute to photosynthesis and encode proteins that reside in the chloroplast were overrepresented among genes that were regulated by plastid-to-nucleus signaling (Van Aken and Whelan 2012).

The Integration of Plastid-to-Nucleus Signaling and Extraplastidic Signaling

Plastid-to-nucleus signaling mechanisms are integrated with other signaling mechanisms such as signaling mechanisms that contribute to the anterograde control of PhANG expression (section "Introduction"), ABA signaling (sections "Plastid Signals Can Regulate Expression of PhANGs at Multiple Levels," "Excess Light Activates Plastid-to-Nucleus Signaling," "Moderate Increases in Light Intensity Can Induce the Expression of *APX2*," and "*flu*-Derived $^1O_2$ Can Affect Chloroplast Biogenesis,"), light signaling (sections "Plastid Signals Can Regulate Expression of PhANGs at Multiple Levels," "*cry1* Mutants Are *gun* Mutants," and "*gun* Mutants Can Exhibit Abnormal Development"), and mitochondria-to-nucleus signaling (section "Interorganellar Signaling Between Chloroplasts and Mitochondria"). Such signal integration is thought to coordinate the expression of nuclear genes that encode chloroplast proteins. Microarray analysis of nuclear genes that mostly encode chloroplast proteins demonstrated that the expression of genes that encode chloroplast proteins is coordinated in over 100 distinct genetic and environmental conditions. Based on these data, a master switch was proposed to coordinate the expression of nuclear genes that encode chloroplast proteins. The expression of genes in two distinct regulons composed of PhANGs and genes that contribute to the expression of the chloroplast genomes is not regulated by this master switch. The master switch affects the expression of genes that contribute to the diverse chloroplast functions (Leister 2005, 2012).

Additionally, transcriptome data demonstrated coordinated expression among PhANGs, photosynthesis-associated chloroplast genes, genes that contribute to the expression of the chloroplast genome, and genes involved in chlorophyll biosynthesis. Also, chloroplast dysfunction enhanced the coordinated expression of genes that encode mitochondrial proteins. Additionally, genes that reside in the same compartment and genes that encode proteins that reside in the same organelle exhibit enhanced coexpression. Thus, intraorganellar and interorganellar signaling networks that likely include plastid-to-nucleus signaling appear to optimize the expression of genes that contribute to the production of energy and the expression of the mitochondrial and chloroplastic genomes (Leister 2012).

Chloroplast-Localized Transcription Factors Contribute to Plastid-to-Nucleus Signaling

An early proposal for a plastid-to-nucleus signaling mechanism was a protein that moves between the plastids and the nucleus (section "The First Proposals for Plastid Signals"). As described in section "Loss-of-Function Alleles of *GUN1* Broadly Disrupt Plastid Signaling," PTM is one such protein that contributes to GUN1-dependent plastid-to-nucleus signaling by moving from the plastid to the nucleus. Lagrange et al. 2003 provided the first evidence for this sort of mechanism by demonstrating that the plant-specific transcription factor IIB (TFIIB)-related protein (pBrp) moves from the chloroplast to the nucleus. TFIIB is a general transcription factor (GTF) that RNA polymerase II requires to specifically transcribe protein-encoding genes. Although pBrp is related to TFIIB, the sequence of pBrp is significantly different from canonical TFIIB, and as the name implies, pBrp is unique to plants. Canonical TFIIBs constitutively localize to nuclei. In contrast, pBrp was localized to the cytoplasmic surface of the chloroplast outer envelope. Loss-of-function alleles and inhibitors that attenuate the activity of the proteasome caused pBrp to accumulate in the nucleus. Based on these data, a plastid signal is proposed to trigger the movement of pBrp from the chloroplast to the nucleus where pBrp is degraded.

In the red alga *Cyanidioschyzon merolae*, pBrp was found to serve as a GTF for RNA polymerase I. pBrp may perform a similar function in Arabidopsis. In contrast to Arabidopsis pBrp, *C. merolae* pBrp is found in the nucleolus. There is no evidence that pBrp associates with plastids in *C. merolae*. RNA polymerase I requires a TFIIB-related GTFs only in plants and algae (Krause and Krupinska 2009). The plastid regulation of specific transcription by RNA polymerase I may generate selection pressure that retains pBrp in plants.

Other nuclear transcription factors associate with chloroplasts. Tobacco stress-induced1 (Tsi1) is an ethylene-responsive-element binding protein/APETALA2-type transcription factor that induces the transcription of stress-related genes in response to various types of biotic and abiotic stresses. Tsi1-interacting protein1 (Tsip1) is a DnaJ-type zinc-finger protein that associates with the cytosolic surface of chloroplasts. Stress induces the biosynthesis of salicylic acid in chloroplasts, which leads to the dissociation of Tsip1 from chloroplasts. After dissociating, Tsi1p associates with Tsi1 in the cytosol. This complex moves to the nucleus and induces the expression of stress-responsive genes. In tobacco, Tsi1 and Tsi1p enhance tolerance to salt and the bacterial pathogen *Pseudomonas syringae* pv. *tabaci* (Krause and Krupinska 2009).

Whirly1 is a protein that localizes to both chloroplasts and nuclei. Unlike pBrp and Tsip1, Whirly1 was found inside chloroplasts, not on the cytosolic surface of the chloroplast outer envelope. In chloroplasts, Whirly1 was found to bind and promote the stability of the nucleoids. Whirly1 also was shown to bind RNA and contribute to RNA splicing in the chloroplast stroma. The inhibition of germination by ABA was shown to depend on chloroplastic Whirly1, not on nuclear Whirly1. *GUN* genes have similar effects on the sensitivity to ABA during germination (section "*gun* Mutants Can Exhibit Abnormal Development"). In the nucleus, Whirly1 was shown to bind telomeres and promote the homeostasis of telomeres. Whirly1 was also shown to function as a transcription factor that induces the expression of the pathogen response (PR) gene *PR10α*.

Whirly1 translocates from inside the chloroplast to the nucleus by some unknown mechanism. This movement was demonstrated by inserting a gene that expresses an epitope-tagged Whirly1 into the plastid genome and testing the subcellular distribution of epitope-tagged Whirly1 in the resulting transgenic plants. This epitope-tagged Whirly1 accumulated in both the chloroplasts and nuclei. This intracellular trafficking of Whirly1 is suggested to promote pathogen resistance because Whirly1 can induce the expression of *PR10α* (Krause and Krupinska 2009; Krause et al. 2012).

Future Directions

Our understanding of plastid-to-nucleus signaling has increased significantly since the first evidence for this type of interorganellar signaling was reported by Bradbeer et al. (1979). A number of plastid-to-nucleus signaling mechanisms are known such as those that are activated by porphyrin metabolism (sections "Porphyrins Can Regulate Nuclear Gene Expression in *C. reinhardtii* and Plants," "Mg-Proto Helps to Coordinate DNA Replication in Nuclei, Chloroplasts, and Mitochondria," and "Particular *gun* Alleles Disrupt Tetrapyrrole Metabolism"), expression of the chloroplast genome (sections "Early Findings" and "Loss-of-Function Alleles of *GUN1* Broadly Disrupt Plastid Signaling"), import of proteins into the chloroplast (section "Blocking the Import of Proteins into the Plastid Activates Plastid-to-Nucleus Signaling"), PET (sections "Photosynthesis Activates Plastid-to-Nucleus Signaling in Algae," "Photosynthesis Activates Plastid-to-Nucleus Signaling in Plants," "The Plastid-to-Nucleus Signaling that Changes the Ratio of PSI to PSII may Depend on STN7," "The Redox State of PQ Can Affect Chloroplast Biogenesis," "Excess Light Activates Plastid-to-Nucleus Signaling," and "Plastid-to-Nucleus Signaling Contributes to Systemic-Acquired Acclimation"), thiols (section "Excess Light Activates Plastid-to-Nucleus Signaling"), ROS (sections "Plastid-to-Nucleus Signaling Contributes to Systemic-Acquired Acclimation," "Excess Light Can Induce Acclimation or Cell Death," "Moderate Increases in Light Intensity Can Induce the Expression of *APX2*," "PSII Can Produce $^1O_2$," "$^1O_2$ Activates Plastid-to-Nucleus Signaling in the Arabidopsis *flu* Mutant," "In the *flu* Mutant, Plastid-to-Nucleus Signaling Requires *EXECUTER1* and *EXECUTER2*," "Other *flu*-Based Screens for Mutant Alleles that Disrupt Plastid-to-Nucleus Signaling," "$^1O_2$ and cry1 Signaling Are Required for Light-Dependent PCD," "*flu*-Derived $^1O_2$ Can Affect Chloroplast Biogenesis," "$^1O_2$-Dependent Plastid-to-Nucleus Signaling Induces Either Acclimation or Cell Death," and "β-Cyclocitral Activates Plastid-to-Nucleus Signaling"), chloroplast-localized transcription factors that move to the nucleus (section "Chloroplast-Localized Transcription Factors Contribute to Plastid-to-Nucleus Signaling"), abiotic stress (sections "Plastid-to-Nucleus Signaling Contributes to Systemic-Acquired Acclimation," "Excess Light Can Induce Acclimation or Cell Death," "Moderate Increases in Light Intensity Can Induce the Expression of *APX2*," "PSII Can Produce $^1O_2$," "$^1O_2$ Activates Plastid-to-Nucleus Signaling in the Arabidopsis *flu* Mutant," "In the *flu* Mutant, Plastid-to-Nucleus Signaling Requires *EXECUTER1* and *EXECUTER2*," "Other *flu*-Based Screens for Mutant Alleles that Disrupt Plastid-to-Nucleus Signaling," "$^1O_2$ and cry1 Signaling Are Required for Light-Dependent PCD," "*flu*-Derived $^1O_2$ Can Affect Chloroplast Biogenesis," "$^1O_2$-Dependent Plastid-to-Nucleus Signaling Induces Either Acclimation or Cell Death," "β-Cyclocitral Activates Plastid-to-Nucleus Signaling," "Chloroplastic Ca^{2+} may Activate $^1O_2$-Dependent Plastid-to-Nucleus Signaling," and “Methylerythritol Cyclodiphosphate (MEcPP) Induces the Expression of Stress-Related Genes”), and biotic stress (sections “Chloroplastic Ca^{2+} may Activate $^1O_2$-Dependent Plastid-to-Nucleus Signaling” and “Chloroplast-Localized Transcription Factors Contribute to Plastid-to-Nucleus Signaling”) (Fig. 14). We know that plastid-to-nucleus signaling mechanisms are integrated with other plastid signaling mechanisms and with extraplastidic signaling mechanisms (sections “Plastid Signals Can Regulate Expression of PhANGs at Multiple Levels,” “*cry1* Mutants Are *gun* Mutants,” “Excess Light Activates Plastid-to-Nucleus Signaling,” “Moderate Increases in Light Intensity Can Induce the Expression of *APX2*,” “$^1O_2$ and cry1 Signaling Are Required for Light-Dependent PCD,” “*flu*-Derived $^1O_2$ Can Affect Chloroplast Biogenesis,” “3′-Phosphoadenosine 5′-phosphate (PAP) Contributes to Excess Light and Drought Tolerance,” “*gun* Mutants Can Exhibit Abnormal Development,” “Interorganellar Signaling Between Chloroplasts and Mitochondria,” and “The Integration of Plastid-to-Nucleus Signaling and Extraplastidic Signaling”). We are beginning to understand the mechanisms that integrate plastid signaling with light signaling (sections “*cry1* Mutants Are *gun* Mutants” and “$^1O_2$ and cry1 Signaling Are Required for Light-Dependent PCD”), mitochondria-to-nucleus signaling (section “Interorganellar Signaling Between Chloroplasts and Mitochondria”), and ABA signaling (sections “Plastid Signals Can Regulate Expression of PhANGs at Multiple Levels,” “Excess Light Activates Plastid-to-Nucleus Signaling,” “Moderate Increases in Light Intensity Can Induce the Expression of *APX2*,” “*flu*-Derived $^1O_2$ Can Affect Chloroplast Biogenesis,” “3′-Phosphoadenosine 5′-phosphate (PAP) Contributes to Excess Light and Drought Tolerance,” and “*gun* Mutants Can Exhibit Abnormal Development”). Strong evidence exists for four plastid signals: H_2O_2 (sections “Excess Light Activates Plastid-to-Nucleus Signaling,” “Plastid-to-Nucleus Signaling Contributes to Systemic-Acquired Acclimation,” “Excess Light Can Induce Acclimation or Cell Death,” and “Moderate Increases in Light Intensity Can Induce the Expression of *APX2*”), β-cyclocitral (section “β-Cyclocitral Activates Plastid-to-Nucleus Signaling”), MEcPP (section “Methylerythritol Cyclodiphosphate (MEcPP) Induces the Expression of Stress-Related Genes”), and PAP (section “3′-Phosphoadenosine 5′-phosphate (PAP) Contributes to Excess Light and Drought Tolerance”) (Fig. 15). The biological functions of plastid signals are becoming clear. Plastid signals contribute to chloroplast biogenesis (sections “*cry1* Mutants Are *gun* Mutants,” “Plastid Signals that Depend on cry1, HY5, and GUN1 Promote the Accumulation of Anthocyanins,” and “The Redox State of PQ Can Affect Chloroplast Biogenesis”), development (section “Plastid Signals Contribute to Development”), the circadian rhythm (section “Loss-of-Function Alleles of *GUN1* Broadly Disrupt Plastid Signaling”), the optimization of photosynthesis to various qualities and quantities of light (section “Photosynthetic Electron Transport Can Activate Plastid-to-Nucleus Signaling”), PCD (sections “Excess Light Can Induce Acclimation or Cell Death,” “$^1O_2$ Activates Plastid-to-Nucleus Signaling in the Arabidopsis *flu* Mutant,” “In the *flu* Mutant, Plastid-to-Nucleus Signaling Requires *EXECUTER1* and *EXECUTER2*,” “$^1O_2$ and cry1 Signaling Are Required for Light-Dependent PCD”), the response to wounding (section “Methylerythritol

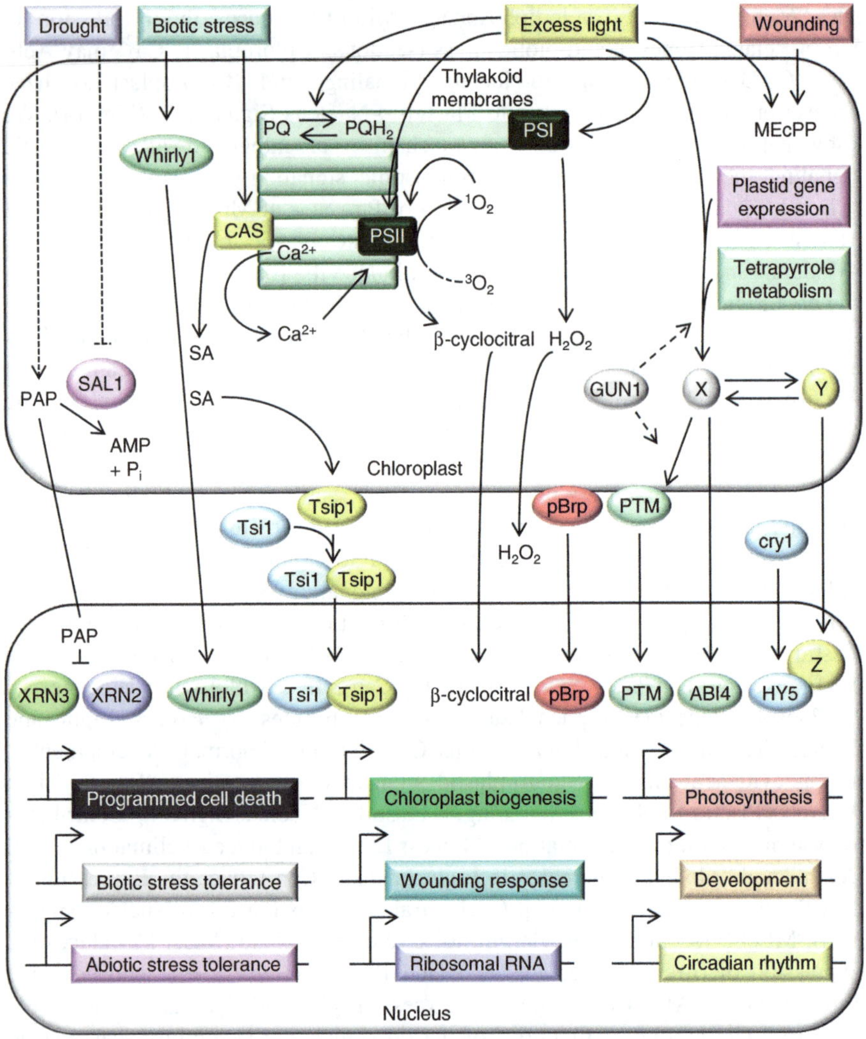

Fig. 14 Plastid-to-nucleus signaling that regulates nuclear gene expression. In the plastid, drought induces the accumulation of PAP, possibly by inhibiting SAL1. PAP moves to the nucleus and inhibits XRN2 and XRN3. Biotic stress activates plastid signaling that depends on Whirly1, CAS and Tsip1. Excess light activates plastid-to-nucleus signaling that depends on plastoquinol (PQH_2), $^1O_2$, β-cyclocitral, H_2O_2, GUN1, and MEcPP. Wounding activates plastid-to-nucleus signaling that depends on MEcPP. Plastid gene expression and tetrapyrrole metabolism affect plastid signaling that depends on GUN1, an unknown plastid signal (X), PTM, and ABI4. A distinct unknown plastid signal (Y) affects cry1 signaling that depends on HY5 by activating a factor such as a corepressor (Z) that associates with HY5. Y interacts with X by an unknown mechanism. Several plastid-localized factors are known to move to the nucleus or are thought to move to the nucleus: PAP, Whirly1, Tsip1, β-cyclocitral, pBrp, and PTM. Nuclear factors such as XRN2, XRN3, and ABI4 contribute to this signaling. These factors regulate numerous genes that contribute to programmed cell death, biotic stress tolerance, abiotic stress tolerance, chloroplast biogenesis, the wounding response, ribosomal RNA biosynthesis, photosynthesis, development, and the circadian rhythm

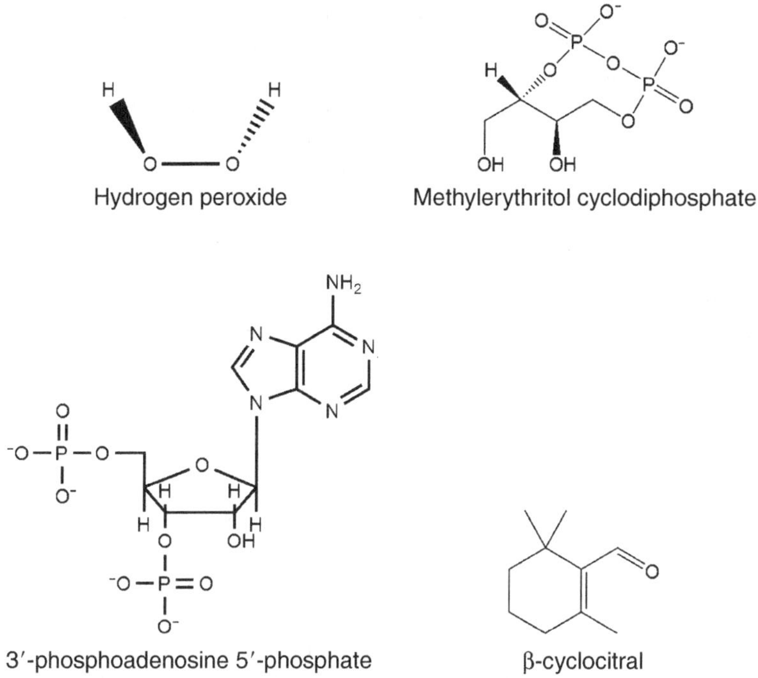

Fig. 15 Four plastid signals

Cyclodiphosphate (MEcPP) Induces the Expression of Stress-Related Genes"), biotic stress tolerance (sections "Chloroplastic Ca^{2+} may Activate $^1O_2$-Dependent Plastid-to-Nucleus Signaling" and "Chloroplast-Localized Transcription Factors Contribute to Plastid-to-Nucleus Signaling"), abiotic stress tolerance (sections "Particular *gun* Alleles Disrupt Tetrapyrrole Metabolism," "Loss-of-Function Alleles of *GUN1* Broadly Disrupt Plastid Signaling," "*cry1* Mutants Are *gun* Mutants," "Plastid Signals that Depend on cry1, HY5, and GUN1 Promote the Accumulation of Anthocyanins," "Plastid-to-Nucleus Signaling Contributes to Systemic-Acquired Acclimation," "Excess Light Can Induce Acclimation or Cell Death," "$^1O_2$-Dependent Plastid-to-Nucleus Signaling Induces Either Acclimation or Cell Death," "β-Cyclocitral Activates Plastid-to-Nucleus Signaling," "3-′-Phosphoadenosine 5′-phosphate (PAP) Contributes to Excess Light and Drought Tolerance," and "Chloroplast-Localized Transcription Factors Contribute to Plastid-to-Nucleus Signaling") (Fig. 14), and DNA replication (section "Mg-Proto Helps to Coordinate DNA Replication in Nuclei, Chloroplasts, and Mitochondria"). Our current knowledge indicates that plastids are not merely major centers of plant metabolism. Plastids are also major regulators of plant growth, development, metabolism, and plant-environment interactions.

Although the past few decades have yielded major advances to our understanding of plastid-to-nucleus signaling, there are major gaps in our knowledge of each plastid-to-nucleus signaling mechanism. Numerous plastid signals remain undefined or are poorly understood. In addition to filling these gaps in our knowledge, the apparent complexity of this signaling will undoubtedly reveal new plastid-to-nucleus signaling mechanisms in the coming years.

The accomplishments of the recent decades provide reason for optimism and challenges for the future. More research is required to understand all plastid-to-nucleus signaling mechanisms and the integration of these signaling mechanisms into networks. Unbiased and specific mutant screens should yield new alleles that disrupt plastid-to-nucleus signaling mechanisms. Such mutant screens are essential for this field because as illustrated by numerous examples in this chapter, plastid-to-nucleus signaling mechanisms are not intuitively obvious. Specific mutant screens can yield alleles of genes that encode proteins of vague or unknown functions such as *GUN1* (section "Loss-of-Function Alleles of *GUN1* Broadly Disrupt Plastid Signaling"), *EX1*, and *EX2* (section "In the *flu* Mutant, Plastid-to-Nucleus Signaling Requires *EXECUTER1* and *EXECUTER2*"). Determining biochemical functions for these proteins will provide greater insight into signaling mechanisms. Quantifying the movement of putative plastid signals within cells, between cells, and between plant organs will provide insight into signaling mechanisms and biological functions. Analyses of candidate transcription factors and *cis*-acting sequences that were identified by bioinformatics approaches may provide further insight into this signaling. Systems biology approaches may yield significant advances to our knowledge of this signaling because of the complex and interdependent nature of this signaling. Further study of plastid-to-nucleus signaling will undoubtedly provide major advances to our understanding of plant biology and information that will help plant breeders and agricultural biotechnologists develop new germplasm to meet the needs of our changing environment.

References

Bechtold U, Richard O, Zamboni A, Gapper C, Geisler M, Pogson B, Karpinski S, Mullineaux PM. Impact of chloroplastic- and extracellular-sourced ROS on high light-responsive gene expression in *Arabidopsis*. J Exp Bot. 2008;59:121–33.

Bradbeer JW, Atkinson YE, Börner T, Hagemann R. Cytoplasmic synthesis of plastid polypeptides may be controlled by plastid-synthesised RNA. Nature. 1979;279:816–7.

Brown NJ, Sullivan JA, Gray JC. Light and plastid signals regulate the expression of the pea plastocyanin gene through a common region at the 5′ end of the coding region. Plant J. 2005;43:541–52.

Carrie C, Murcha MW, Giraud E, Ng S, Zhang MF, Narsai R, Whelan J. How do plants make mitochondria? Planta. 2013;237:429–39.

Chi W, Sun X, Zhang L. Intracellular signaling from plastid to nucleus. Annu Rev Plant Biol 2013;64:559–82.

Escoubas JM, Lomas M, LaRoche J, Falkowski PG. Light intensity regulation of *cab* gene transcription is signaled by the redox state of the plastoquinone pool. Proc Natl Acad Sci U S A. 1995;92:10237–41.

Fankhauser C, Casal JJ. Phenotypic characterization of a photomorphogenic mutant. Plant J. 2004;39:747–60.

Farmer EE, Mueller MJ. ROS-mediated lipid peroxidation and RES-activated signaling. Annu Rev Plant Biol. 2013;64:429–50.

Formighieri C, Ceol M, Bonente G, Rochaix JD, Bassi R. Retrograde signaling and photoprotection in a gun4 mutant of *Chlamydomonas reinhardtii*. Mol Plant. 2012;5:1242–62.
Foudree A, Aluru M, Rodermel S. PDS activity acts as a rheostat of retrograde signaling during early chloroplast biogenesis. Plant Signal Behav. 2010;5:1629–32.
Foudree A, Putarjunan A, Kambakam S, Nolan T, Fussell J, Pogorelko G, Rodermel S. The mechanism of variegation in immutans provides insight into chloroplast biogenesis. Front Plant Sci. 2012;3:260.
Foyer CH, Neukermans J, Queval G, Noctor G, Harbinson J. Photosynthetic control of electron transport and the regulation of gene expression. J Exp Bot. 2012;63:1637–61.
Fujii S, Small I. The evolution of RNA editing and pentatricopeptide repeat genes. New Phytol. 2011;191:37–47.
Galvez-Valdivieso G, Mullineaux PM. The role of reactive oxygen species in signalling from chloroplasts to the nucleus. Physiol Plant. 2010;138:430–9.
Gordon MJ, Carmody M, Albrecht V, Pogson B. Systemic and local responses to repeated HL stress-induced retrograde signaling in *Arabidopsis*. Front Plant Sci. 2012;3:303.
Gould KS. Nature's swiss army knife: the diverse protective roles of anthocyanins in leaves. J Biomed Biotechnol. 2004;5:314–20.
Gray JC, Sullivan JA, Wang JH, Jerome CA, MacLean D. Coordination of plastid and nuclear gene expression. Philos Trans R Soc B. 2003;358:135–44.
Inaba T, Yazu F, Ito-Inaba Y, Kakizaki T, Nakayama K. Retrograde signaling pathway from plastid to nucleus. Int Rev Cell Mol Biol. 2011;290:167–204.
Jiao Y, Lau OS, Deng XW. Light-regulated transcriptional networks in higher plants. Nat Rev Genet. 2007;8:217–30.
Johanningmeier U, Howell SH. Regulation of light-harvesting chlorophyll-binding protein mRNA accumulation in *Chlamydomonas reinhardtii*. Possible involvement of chlorophyll synthesis precursors. J Biol Chem. 1984;259:13541–9.
Karpiński S, Szechyńska-Hebda M, Wituszyńska W, Burdiak P. Light acclimation, retrograde signalling, cell death and immune defences in plants. Plant Cell Environ. 2012;36:736–44.
Kim C, Apel K. $^1O_2$-mediated and EXECUTER-dependent retrograde plastid-to-nucleus signaling in norflurazon-treated seedlings of *Arabidopsis thaliana*. Mol Plant. 2013a;6:1580–91.
Kim C, Apel K. Singlet oxygen-mediated signaling in plants: moving from *flu* to wild type reveals an increasing complexity. Photosynth Res. 2013b;116:455–64.
Kim C, Lee KP, Baruah A, Nater M, Göbel C, Feussner I, Apel K. $^1O_2$-mediated retrograde signaling during late embryogenesis predetermines plastid differentiation in seedlings by recruiting abscisic acid. Proc Natl Acad Sci U S A. 2009;106:9920–4.
Kopriva S. 12-Oxo-phytodienoic acid interaction with cyclophilin CYP20-3 is a benchmark for understanding retrograde signaling in plants. Proc Natl Acad Sci U S A. 2013;110:9197–8.
Krause K, Krupinska K. Nuclear regulators with a second home in organelles. Trends Plant Sci. 2009;14:194–9.
Krause K, Oetke S, Krupinska K. Dual targeting and retrograde translocation: regulators of plant nuclear gene expression can be sequestered by plastids. Int J Mol Sci. 2012;13:11085–101.
Lagrange T, Hakimi MA, Pontier D, Courtois F, Alcaraz JP, Grunwald D, Lam E, Lerbs-Mache S. Transcription factor IIB (TFIIB)-related protein (pBrp), a plant-specific member of the TFIIB-related protein family. Mol Cell Biol. 2003;23:3274–86.
Larkin RM. The influence of plastid signals on light signaling and development. Philos Trans R Soc B Biol Sci. 2014;369:20130232.
Larkin RM, Ruckle ME. Integration of light and plastid signals. Curr Opin Plant Biol. 2008;11:593–9.
Ledford HK, Baroli I, Shin JW, Fischer BB, Eggen RI, Niyogi KK. Comparative profiling of lipid-soluble antioxidants and transcripts reveals two phases of photo-oxidative stress in a xanthophyll-deficient mutant of *Chlamydomonas reinhardtii*. Mol Genet Genomics. 2004;272:470–9.
Leister D. Genomics-based dissection of the cross-talk of chloroplasts with the nucleus and mitochondria in Arabidopsis. Gene. 2005;354:110–6.

Leister D. Retrograde signaling in plants: from simple to complex scenarios. Front Plant Sci. 2012;3:135.

Liu H, Liu B, Zhao C, Pepper M, Lin C. The action mechanisms of plant cryptochromes. Trends Plant Sci. 2011;16:684–91.

Maxwell DP, Laudenbach DE, Huner N. Redox regulation of light-harvesting complex II and *cab* mRNA abundance in *Dunaliella salina*. Plant Physiol. 1995;109:787–95.

Mayfield SP, Taylor WC. Carotenoid-deficient maize seedlings fail to accumulate light-harvesting chlorophyll *a*/*b* binding protein (LHCP) mRNA. Eur J Biochem. 1984;144:79–84.

Mullineaux PM, Baker NR. Oxidative stress: antagonistic signaling for acclimation or cell death? Plant Physiol. 2010;154:521–5.

Nomura H, Komori T, Uemura S, Kanda Y, Shimotani K, Nakai K, Furuichi T, Takebayashi K, Sugimoto T, Sano S, Suwastika IN, Fukusaki E, Yoshioka H, Nakahira Y, Shiina T. Chloroplast-mediated activation of plant immune signalling in Arabidopsis. Nat Commun. 2012;3:926.

Pfalz J, Liebers M, Hirth M, Grübler B, Holtzegel U, Schröter Y, Dietzel L, Pfannschmidt T. Environmental control of plant nuclear gene expression by chloroplast redox signals. Front Plant Sci. 2012;3:257.

Rochaix JD, Lemeille S, Shapiguzov A, Samol I, Fucile G, Willig A, Goldschmidt-Clermont M. Protein kinases and phosphatases involved in the acclimation of the photosynthetic apparatus to a changing light environment. Philos Trans R Soc Lond B Biol Sci. 2012;367:3466–74.

Rook F, Hadingham SA, Li Y, Bevan MW. Sugar and ABA response pathways and the control of gene expression. Plant Cell Environ. 2006;29:426–34.

Ruckle ME, Burgoon LD, Lawrence LA, Sinkler C, Larkin RM. Plastids are major regulators of light signaling in *Arabidopsis thaliana*. Plant Physiol. 2012;159:366–90.

Rüdiger W, Oster U. Intracellular signaling from plastids to the nucleus. In: Eaton-Rye JJ, Tripathy BC, Sharkey TD, editors. Photosynthesis advances in photosynthesis and respiration, vol. 34. Dordrecht: Springer; 2012. p. 175–90.

Saini G, Meskauskiene R, Pijacka W, Roszak P, Sjögren LL, Clarke AK, Straus M, Apel K. '*happy on norflurazon*' (*hon*) mutations implicate perturbance of plastid homeostasis with activating stress acclimatization and changing nuclear gene expression in norflurazon-treated seedlings. Plant J. 2011;65:690–702.

Schönberg A, Baginsky S. Signal integration by chloroplast phosphorylation networks: an update. Front Plant Sci. 2012;3:256.

Sullivan JA, Gray JC. Multiple plastid signals regulate the expression of the pea plastocyanin gene in pea and transgenic tobacco plants. Plant J. 2002;32:763–74.

Susek R, Chory J. A tale of two genomes: role of a chloroplast signal in coordinating nuclear and plastid genome expression. Aust J Plant Physiol. 1992;19:387–99.

Susek RE, Ausubel FM, Chory J. Signal transduction mutants of Arabidopsis uncouple nuclear *CAB* and *RBCS* gene expression from chloroplast development. Cell. 1993;74:787–99.

Tanaka K, Hanaoka M. The early days of plastid retrograde signaling with respect to replication and transcription. Front Plant Sci. 2013;3:301.

Terry MJ, Smith AG. A model for tetrapyrrole synthesis as the primary mechanism for plastid-to-nucleus signaling during chloroplast biogenesis. Front Plant Sci. 2013;4:14.

Terzaghi WB, Cashmore AR. Light-regulated transcription. Annu Rev Plant Physiol Plant Mol Biol. 1995;46:445–74.

Tikkanen M, Gollan PJ, Suorsa M, Kangasjärvi S, Aro EM. STN7 operates in retrograde signaling through controlling redox balance in the electron transfer chain. Front Plant Sci. 2012;3:277.

To JPC, Reiter W-D, Gibson SI. Chloroplast biogenesis by Arabidopsis seedlings is impaired in the presence of exogneous glucose. Physiol Plant. 2003;118:456–63.

Triantaphylidès C, Havaux M. Singlet oxygen in plants: production, detoxification and signaling. Trends Plant Sci. 2009;14:219–28.

Van Aken O, Whelan J. Comparison of transcriptional changes to chloroplast and mitochondrial perturbations reveals common and specific responses in Arabidopsis. Front Plant Sci. 2012;3:281.

Waters MT, Langdale JA. The making of a chloroplast. EMBO J. 2009;28:2861–73.
Wilson PB, Estavillo GM, Field KJ, Pornsiriwong W, Carroll AJ, Howell KA, Woo NS, Lake JA, Smith SM, Millar AH, von Caemmerer S, Pogson BJ. The nucleotidase/phosphatase SAL1 is a negative regulator of drought tolerance in Arabidopsis. Plant J. 2009;58:299–317.
Woodson JD, Chory J. Coordination of gene expression between organellar and nuclear genomes. Nat Rev Genet. 2008;9:383–95.
Woodson JD, Perez-Ruiz JM, Schmitz RJ, Ecker JR, Chory J. Sigma factor-mediated plastid retrograde signals control nuclear gene expression. Plant J. 2012;73:1–13.
Zhang H, He H, Wang X, Wang X, Yang X, Li L, Deng XW. Genome-wide mapping of the *HY5*-mediated gene networks in Arabidopsis that involve both transcriptional and post-transcriptional regulation. Plant J. 2011;65:346–58.

Further Reading

Barajas-López Jde D, Blanco NE, Strand Å. Plastid-to-nucleus communication, signals controlling the running of the plant cell. Biochim Biophys Acta. 1833;2013:425–37.
Beck CF. Signaling pathways from the chloroplast to the nucleus. Planta. 2005;222:743–56.
Chan KX, Crisp PA, Estavillo GM, Pogson BJ. Chloroplast-to-nucleus communication: current knowledge, experimental strategies and relationship to drought stress signaling. Plant Signal Behav. 2010;5:1575–82.
Estavillo GM, Chan KX, Phua SY, Pogson BJ. Reconsidering the nature and mode of action of metabolite retrograde signals from the chloroplast. Front Plant Sci. 2012;3:300.
Jung HS, Chory J. Signaling between chloroplasts and the nucleus: can a systems biology approach bring clarity to a complex and highly regulated pathway? Plant Physiol. 2010;152:453–9.
León P, Gregorio J, Cordoba E. ABI4 and its role in chloroplast retrograde communication. Front Plant Sci. 2012;3:304.
Oelmüller R. Photooxidative destruction of chloroplasts and its effects on nuclear gene expression and extraplastidic enzyme levels. Photochem Photobiol. 1989;49:229–39.
Tanaka R, Kobayashi K, Masuda T. Tetrapyrrole metabolism in *Arabidopsis thaliana*. Arabidopsis Book. 2011;9:e0145.
Taylor WC. Regulatory interactions between nuclear and plastid genomes. Annu Rev Plant Physiol Plant Mol Biol. 1989;40:211–33.
Yu F, Fu A, Aluru M, Park S, Xu Y, Liu H, Liu X, Foudree A, Nambogga M, Rodermel S. Variegation mutants and mechanisms of chloroplast biogenesis. Plant Cell Environ. 2007;30:350–65.

ER Stress Signaling in Plants

7

Stephen Howell

Contents

S. Howell (✉)
Plant Sciences Institute, Genetics, Development and Cell Biology Department, Iowa State University, Ames, IA, USA
e-mail: shh@iastate.edu

S.H. Howell (ed.), *Molecular Biology, The Plant Sciences 2*,
DOI 10.1007/978-1-4614-7570-5_9

Abstract

- The endoplasmic reticulum (ER) is the entrance and the first way station in the protein secretory pathway in plant cells. Newly synthesized proteins enter the ER in an unfolded state and are folded by the protein-folding machinery in the ER.
- The protein-folding machinery in the ER monitors the folding status and modification state of oligosaccharides on glycoproteins as they advance through the folding process.
- Protein folding is very sensitive to environmental conditions, and adverse conditions can result in protein misfolding which produces ER stress.
- Misfolded proteins are sensed by an ER quality control (ERQC) system and eliminated by the ER-associated degradation (ERAD) system, which transports misfolded proteins out of the ER and sends them for degradation to the 26S proteasome.
- When the load of misfolded proteins increases under stress conditions, it sets off an alarm called the unfolded protein response (UPR).
- The UPR involves the upregulation of stress-response genes that boost the capacity of the folding machinery and the ERAD system.
- The UPR signaling pathway has two "arms" – one arm involving ER membrane-associated transcription factors and another that involves messenger RNA splicing by the splicing factor called IRE1.
- ER stress activates the membrane-associated transcription factor arm of the UPR signaling pathway by mobilizing the factors from the ER to the nucleus where they upregulate stress-response genes.
- ER stress activates IRE1 to splice a messenger RNA encoding a transcription factor that also upregulates stress-response genes.
- Mild ER stress elicits autophagy through a signaling pathway that involves IRE1. Severe ER stress can lead to cell death.

Introduction

Abiotic stresses brought about by drought, heat, freezing, or saline conditions result in some of the greatest crop losses worldwide. The frequency of adverse conditions due to climate change presents serious challenges to global agriculture, which already suffers annual crop losses estimated at billions of dollars (Mittler and Blumwald 2010). Therefore, there is great interest in understanding how plants respond to stress and in endowing crop plants with traits for stress tolerance.

The Endoplasmic Reticulum and ER Stress

This chapter deals with a type of plant stress called endoplasmic reticulum (ER) stress. In learning about what plant cells do, the ER is sometimes overlooked – the focus is usually on chloroplasts, mitochondria, nuclei, or cell walls.

Nonetheless, the ER is important in stress management because the ER senses abiotic and biotic stresses and responds to them.

The ER plays a key role in many cellular activities because, among other things, the ER is the gateway into the protein secretion pathway. Proteins destined for secretion to the cell wall or targeted to organelles in the endomembrane system of cells (the ER, Golgi apparatus, lysosomes, exocytotic intermediate compartments, and plasma membrane) are synthesized on ribosomes on the ER membrane. Such proteins are cotranslationally inserted into the ER membrane or the ER lumen where they undergo protein folding.

Protein folding is one of the most intensely studied processes in biology. Unlike protein synthesis, which is instructed by DNA and RNA, protein folding is a self-assembly process, guided by entropic and energetic forces. The folding of proteins is also a finicky process. Because the pathway for protein folding has many detours, proteins can end up in nonnative conformations. Because protein folding can be easily perturbed, it acts as a "canary in the coal mine" serving as one of the first-line sensors and responders to adverse environmental conditions.

Under adverse conditions, misfolded proteins accumulate in the ER setting off alarms that elicit protective responses. The misfolded proteins are detected by an elaborate ER quality control system.

ER Quality Control (ERQC)

Quality control assures the safety of airplanes, cars, appliances, food, and many other products in our daily lives. Quality control is just as important in living systems. Without it, systems would be at risk – it takes only one major error or breakdown to the end the life of an organism or to deny its reproductive success. Because quality controls are important adaptive traits, organisms have employed them in many ways to insure the fidelity of DNA, RNA, and proteins. Our focus is on protein quality control in the endoplasmic reticulum (ER).

Why is quality control important for the ER? In *Arabidopsis thaliana*, somewhat over 17 % of all proteins have predicted signal peptides and 33 % have at least one transmembrane domain, many of which are likely to be associated with ER membranes or other organelles on the secretory pathway (Initiative 2000). Secreted proteins are made on ribosomes bound to the ER membrane and are cotranslationally translocated into the ER lumen or they become part of the ER membrane. Such proteins may be retained in the ER or exported from the ER to populate various organelles or the plasma membrane or to be secreted outside the cell. These proteins may play important roles such as hormone receptors or cell wall components, and quality control insures that these proteins can carry out their functions. Proteins that do not meet specifications may malfunction or be potentially toxic to cells – so quality control is of utmost importance.

The Secretory Pathway

To recap, secreted proteins are cotranslationally translocated into the ER, i.e., synthesized on ER-bound ribosomes and extruded through a translocon pore in the membrane as the growing polypeptide chain elongates (Fig. 1). Ribosomes synthesizing secreted proteins are guided from the cytoplasm to the ER membrane by a signal recognition particle (SRP), which recognizes a signal sequence on the N-terminus of a secreted protein as it emerges from the exit tunnel of the ribosome large subunit (Keenan et al. 2001). SRP binding halts further synthesis of the secreted protein until the ribosome is transferred to an SRP receptor on the ER. There the nascent chain is inserted into the sec61 translocon, the SRP dissociates and protein synthesis resumes.

Membrane proteins have one or more transmembrane domains (TMDs) that pass through a membrane and are cotranslationally inserted into the ER membrane from the translocon. A TMD is usually about 20 amino acids long, largely made up of hydrophobic residues that are able to form an α-helix. TMDs form α-helices either in the exit tunnel of the ribosome or in the translocon pore. Hydrophobic stretches of amino acid residues that form stable α-helices open a lateral gate as they enter the translocon channel, and through that gate they move passively into the lipid bilayer (Fig. 2) (Shao and Hegde 2011). Multispanning or polytopic membrane proteins

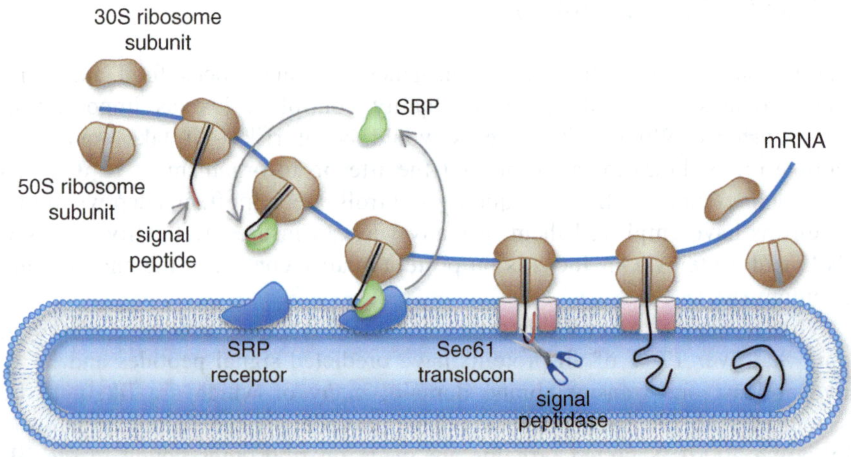

Fig. 1 Cotranslational translocation into the ER of a soluble, secreted protein. A ribosome attaches to the 5′ end of an mRNA and begins to translate the protein. If the nascent polypeptide emerging from the protein exit channel in the ribosome contains a signal peptide, then the signal recognition particle (SRP) will bind to the signal peptide. SRP blocks further translation and escorts the ribosome/mRNA complex to the SRP receptor on the ER membrane. At this point the ribosome/mRNA complex is handed off to the sec61 translocon and translation resumes. The signal peptide is proteolytically cleaved by the signal peptidase as the protein begins to enter the ER lumen. Steps involving the interaction of chaperones, glycosylation, glycan modification, and folding are described in later sections

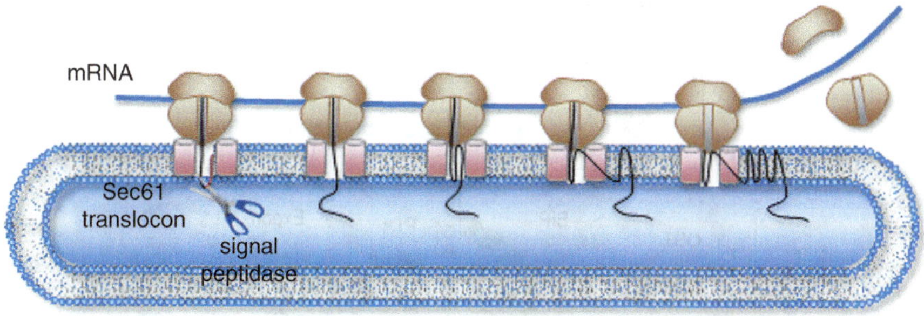

Fig. 2 Cotranslational insertion of a polytopic transmembrane membrane protein into the ER membrane. The initial steps in the synthesis of a transmembrane protein are the same as those in Fig. 1. Transmembrane proteins have transmembrane domains (TMDs), short hydrophobic a-helical domains that span the membrane. During synthesis, the entry of a TMD into the translocon channel opens a lateral gate allowing the TMD to move laterally into the lipid environment of the ER membrane. In polytopic (multispanning) membrane proteins, successive TMDs are inserted in the membrane in opposite orientations, but any TMD entering the translocon channel may sample different orientations before achieving a stable orientation

serpentine back and forth across the membrane with successive TMDs assuming alternate orientations. Multiple TMD helices appear to follow an ordered insertion pathway although there is evidence that individual helices may sample different orientations during the insertion process.

Glycosylation

Nascent proteins are immediately picked up by chaperones and other factors that modify them as they emerge in the ER lumen. Nascent soluble proteins emerging from the translocon can take one of two general paths – one involves glycosylation of glycoproteins by oligosaccharide transferase (OST). OST transfers a core glycan, a preformed, branched oligosaccharide, from a lipid carrier to asparagine residues at glycosylation sites (Asn-X-Ser/Thr) on the emerging polypeptide chain (Fig. 3). The biosynthesis of the core oligosaccharide occurs on the ER membrane in which monosaccharides are added one by one to a lipid carrier, dolichol-pyrophosphate, by monosaccharyltransferases (Kornfeld and Kornfeld 1985). Synthesis of the glycan begins on the cytosolic side of the ER membrane, and when seven sugars have been added, the oligosaccharide is flipped to the luminal side and seven more sugars are added. Flipping is catalyzed by a bidirectional flippase in an ATP-independent reaction.

OST is a heteromeric, multisubunit protein associated with the translocon complex and consists of eight subunits in yeast, five of which are essential. Mammalian cells have two OST isoforms with different catalytic subunits (STT3A and STT3B). This is also the case for Arabidopsis, which has homologs for both forms. The mammalian OST isoform STT3A is primarily involved in the cotranslational glycosylation of nascent polypeptides as they enter the ER lumen, while the STT3B

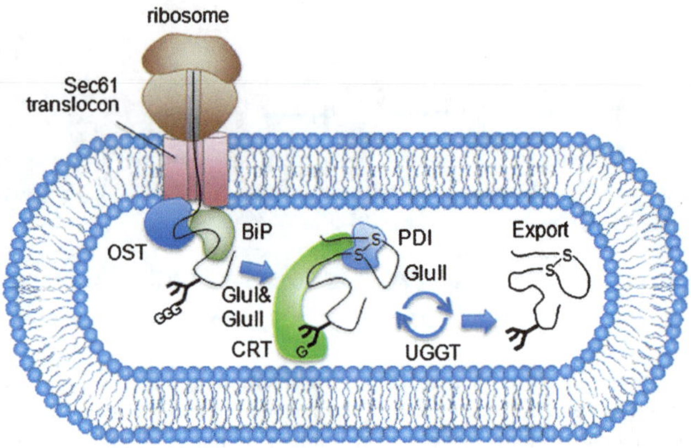

Fig. 3 Folding of a soluble, secreted protein in the ER. The newly synthesized protein emerging from the translocon channel immediately interacts with chaperones and other factors that modify the protein. Glycoproteins are glycosylated by oligosaccharide transferase (OST), which transfers core oligosaccharides to glycosylation sites on the nascent protein. Two terminal glucoses on the core oligosaccharide are quickly hydrolyzed by glucosidase I and II (GluI and GluII), leaving a monoglucosylated oligosaccharide, which is recognized by the lectin, calreticulin (CRT). CRT is a major component of the protein-folding apparatus for proteins in the ER lumen. Proteins undergo rounds of folding in which the core oligosaccharide by deglucosylated by GluII and reglucosylated by UDP-Glc/glycoprotein glucosyltransferase (UGGT). UGGT is thought to play a critical role in deciding whether the protein will be subject to additional rounds of folding. Protein folding is accompanied by disulfide bridge formation and reformation catalyzed by protein disulfide isomerase. Properly folded proteins are released from the folding apparatus and either remain in the ER (if they have an ER retention sequence) or are exported from the ER

isoform handles posttranslational glycosylation of skipped sequons. In Arabidopsis, a T-DNA insertion in STT3A results in an osmotically sensitive root phenotype and induction of UPR due to protein hypoglycosylation. The single mutants, *stt3a-1* and *stt3b-1*, are viable, but the double mutant is gametophytically lethal, suggesting that the plant OST isoforms have partially overlapping roles.

Binding Protein (BiP)

Non-glycosylated proteins generally follow another path. They are picked up by binding immunoglobulin protein (BiP), the most abundant chaperone in the ER, also known in animal systems as 78 kDa glucose-regulated protein (GRP-78) or heat shock 70 kDa protein 5 (HspA5). BiP forms a complex with the HSP40-like co-chaperones containing J domains (ERdj3) and stromal-derived factor-2 (SDF2). The BiP complex delays the folding of proteins, preventing their "hydrophobic collapse" and maintaining them in a competent state for subsequent folding and oligomerization (Braakman and Bulleid 2011). A major role of chaperones is to prevent protein aggregation. Protein folding in the ER occurs in a crowded environment in which the concentration of proteins may be as high as 100 g/L.

In such an environment, protein-folding intermediates can interact through nonspecific hydrophobic interactions leading to the formation of protein aggregates.

BiP can bind folding intermediates of a large number of the proteins because of its capacity to bind hydrophobic peptides that become buried in the interior of a protein in its native state. BiP has a nucleotide-binding site in its N-terminal domain and has ATPase activity. In the ADP-bound state, BiP binds proteins with high affinity in its C-terminal substrate-binding domain. Proteins are released from BiP by ATP binding through nucleotide exchange; therefore, cycles of nucleotide hydrolysis and exchange drive the binding and release BiP from unfolded or misfolded protein substrates, a process that terminates when the hydrophobic sequences in the protein substrate are buried (Gething 1999).

In plants, BiP binds to sites in unassembled monomers of the trimeric bean protein phaseolin. The binding and release cycles are regulated by cofactors, such as DNAJ proteins that promote ATP hydrolysis or ATP: ADP exchange. Arabidopsis has three BiP genes: BiP1, BiP2, and BiP3. BiP1 and BiP2 are nearly identical in protein sequence. Both are expressed at fairly high levels throughout the plant and are induced by stress, such as heat stress. BiP3 is normally expressed at much lower levels and because of that is highly induced (in terms of fold change in expression) by ER stress in seedlings. Plant BiP expression has also been shown to be upregulated by other environmental stresses, such as drought, cold, and insect and pathogen attack. Also, overexpression of plant BiP has been reported to confer drought tolerance in soybean and tobacco (*Nicotiana tabacum*) plants. Mutations in BiP2 make the plant more sensitive to pathogen attack resulting from the impaired induction of PR1, linking the secretory pathway to the systemic acquired resistance pathway. In addition, BiP gene expression is induced in specific cell types at developmental stages associated with high secretory activity or in plants that have expressed assembly-defective proteins, further implicating the role of BiP in protein folding.

Protein Folding of Glycoproteins

Glucosidases

The oligosaccharides (glycans) on glycoproteins are modified as they undergo protein folding, and the modifications serve as signals to summon proteins to the next stage of the process. The first modifications of the core glycans are catalyzed by glucosidase I (GluI). GluI is a type II membrane glycoprotein closely associated with OST and the translocon complex (Caramelo and Parodi 2008). GluI hydrolyzes α-1,2 glycosidic bonds, cleaving off glucose 14 (G14) from the core glycan on nascent glycoproteins (Fig. 4). Removal of the terminal glucoses generates an oligosaccharide structure that is recognized by the protein-folding machinery. Glucosidase II (GluII) is a soluble luminal enzyme that hydrolyzes α-1,3 linkages and cleaves off glucose 13 (G13) from the core oligosaccharide. GluII is composed of two glycoprotein subunits, a catalytic subunit α, and a β-subunit. The hydrolysis of the two terminal glucose residues is critical for proper protein folding and assembly. That is evidenced by the

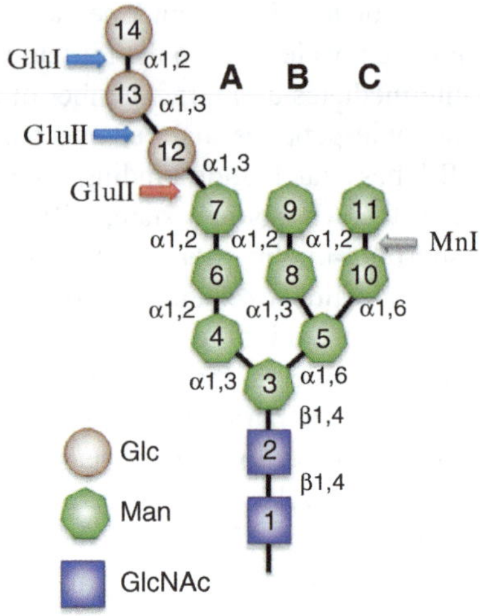

Fig. 4 Structure of the core oligosaccharide. The core oligosaccharide ($Glc_3Man_9GlcNAc_2$) contains three branches (A–C). Residues are numbered by order of addition during biosynthesis. The monoglucosylated form of the core oligosaccharide ($Glc_1Man_9GlcNAc_2$) binds to calnexin/calreticulin during protein folding (as described in Fig. 3). This form is produced by hydrolysis of the terminal a-1,2-glucose 14 by glucosidase I and the removal of the a-1,3-glucose 13 by glucosidase II (*blue arrows*). During protein-folding cycles, the terminal α-1,3-glucose 12 on the monoglucosylated form is progressively removed and readded by glucosidase II and UGGT, respectively (*red arrow*). The demannosylated form ($Man_5GlcNAc_2$) brought about by hydrolysis of the terminal a-1,2-mannose by mannosidasde I, targets glycosylated proteins to ER-associated degradation (ERAD). The terminal a-1,6-mannose 10 on the C chain is the signal recognized by the lectin OS9 for targeting to ERAD

fact that when the removal of the terminal glucose was blocked in bean (*Phaseolus vulgaris*) by glucosidase inhibitors, castanospermine and N-methyldeoxynojirimycin, the phaseolin subunits with partially trimmed glycans were unable to assemble into mature trimeric proteins (Lupattelli et al. 1997).

The CNX/CRT Folding Apparatus

The principal protein-folding machine for glycoproteins in the ER is the CNX/CRT protein-folding apparatus (Fig. 3). CNX and/or CRT are folding cages protecting nascent proteins with exposed hydrophobic surfaces from forming aggregates (Ruddock and Molinari 2006). CRT is a soluble protein, and CNX is a type I membrane protein. In Arabidopsis, there are three isoforms of CRT and two of CNX. CNX and CRT have similar structures with globular N-terminal lectin domains, extended Pro-rich arms and C domains. The major difference between CRT and CNX is that CNX has a transmembrane domain that intervenes between the arm and the C domain, which anchors CNX to the ER membrane. The N-terminal lectin regions are globular β-sandwich domains that bind the monoglucosylated oligosaccharides of nascent glycoproteins.

The CNX/CRT apparatus facilitates the folding and refolding of proteins in a protein-folding cycle (Hammond and Helenius 1994). As pointed out above, entry of a glycoprotein into the cycle begins with the successive removal of the two outer glucoses (G14 and 13) from the core glycan by the action of GluI and II (Fig. 3). Cycles of refolding involve the release and rebinding of the incompletely folded glycoprotein to the CNX/CRT apparatus. The cycles are regulated largely by the opposing actions of glucosidase II and UDP-Glc/glycoprotein glucosyltransferase (UGGT) that catalyze removal and readdition of the terminal glucose residue, respectively.

Protein Disulfide Isomerases and UGGT

Other critical components of the protein-folding apparatus are protein disulfide isomerases (PDIs) (Fig. 3). PDIs catalyze the formation of disulfide bonds that stabilize protein structures and the isomerization or reshuffling of nonnative bonds as proteins fold into their final native forms (Braakman and Bulleid 2011). In yeast, ERp57, a member of the PDI family with a thioredoxin CXXC domain, is bound to CNX/CRT folding cages and because of that specifically interacts with glycoproteins. It is not known in Arabidopsis whether one or more PDIs function as ERp57. The formation of disulfide bonds is an oxidation step and the ER, in general, is an oxidizing environment. The oxidizing character of the ER compartment is buffered by a high ratio of oxidized to reduced glutathione (GSSG/GSH), much of which is found in mixed disulfides with proteins. However, disulfide bond isomerization may involve repeated oxidation and reduction that engage different redox couples. The oxidizing equivalents for the formation of protein disulfide bonds are supplied in yeast by ER oxidoreductase1 (Ero1p). Arabidopsis has two ERo1p homologs, AERO1 and AERO2. Ero1p in yeast is a flavin-containing ER membrane-associated protein that transfers electrons directly to molecular oxygen in the cytoplasm during disulfide bond formation in the ER. It is thought that oxidizing equivalents flow from O_2 to the flavin cofactor in Ero1, then through an intercysteine relay to dithiol/disulphide sites on the surface of Ero1p. ER oxidoreductin 1 (Ero1), in turn, oxidizes members of the family of ER oxidoreductases.

UGGT plays a deciding role as a “folding sensor” in the CNX/CRT protein-folding cycles. Client glycoproteins are delivered to UGGT after the trimming of the innermost glucose residue (G12) by glucosidase II, during folding cycles (Fig. 3) (Totani et al. 2009). If a glycoprotein is incompletely folded, the enzyme will reglucosylate it using the nucleotide sugar UDP-Glc, sending it back for another round of folding. UGGT recognizes clusters of surface-exposed hydrophobic residues in molten globule-like conformers and this specificity may allow the enzyme to discriminate between folded and unfolded domains. UGGT has a 300-amino-acid C-terminal glycosyltransferase domain and an N-terminal sequence of about 1200 residues thought to be the glycoprotein recognition domain. UGGT is inactive against properly folded proteins, allowing them to proceed down the secretory pathway.

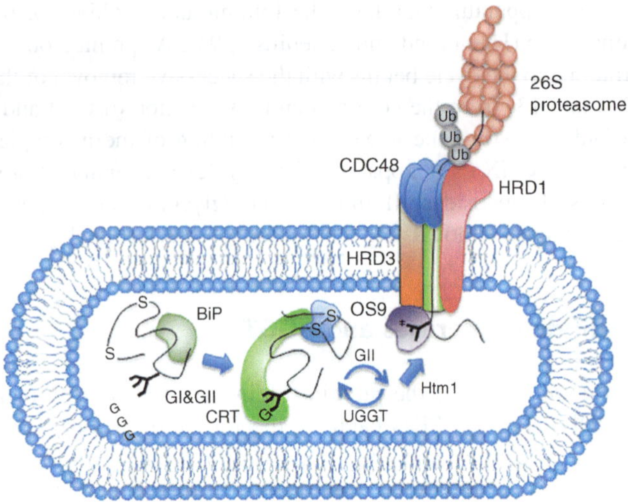

Fig. 5 Degradation of a misfolded protein by the ERAD system. In contrast to Fig. 3, which shows the export of a properly folded protein, the protein in this figure has not been properly folded and is targeted to the ERAD system for degradation. After rounds of futile folding, the protein is demannosylated by mannosidase (as described in Fig. 4) and bound by the lectin OS9. OS9 escorts the misfolded protein to the HRD3 complex, which dislocates the protein across the ER membrane, powered by the action of the motor protein, CDC48. An E3 ubiquitin ligase component of the HRD3 complex, HRD1, ubiquitinates the misfolded protein, such that it is picked up and degraded by the 26S proteasome complex

Suppressors and ERAD

In plants, the monitoring and rejection of mutant hormone or pathogen receptors during their trafficking through the secretory pathway have provided convenient readouts for the function of the ERQC system. Several components of the protein-folding and ERAD system have been identified genetically as suppressors of mutant hormone or pathogen receptors (Hong and Li 2012). Li and coworkers utilized mutant brassinosteroid receptors, such as bri1-9 (S662F) and bri1-5 (C69Y), which are functionally competent as hormone receptors, but, nonetheless, are ERAD substrates. These Arabidopsis mutants are dwarf and unresponsive to brassinosteroids because their receptors are degraded by ERAD. The mutants were then used to identify non-dwarf plant suppressors in which the mutant receptors escape destruction by the ERAD system and survive to function as brassinosteroid receptors.

One of the suppressors, *ebs1-1*, was defective in UGGT, the activity that reglucosylates the core oligosaccharides on glycoproteins in the calnexin/calreticulin protein-folding cycle (Fig. 5). UGGT plays an important role in

(continued)

ERQC because reglucosylation sends partially unfolded proteins back for additional rounds of protein folding. The defective UGGT apparently failed to detain BRI1-9 in protein-folding cycles and allowed it to leak through.

Another suppressor (*ebs2*) was a calreticulin 3 (CRT3), one of three CRTs in Arabidopsis. Again, *ebs2* is thought to act as a suppressor in that without CRT3 function, BRI1-9 is thought to escape ERQC and to survive as a functional receptor. Although BRI1 is a membrane protein, the luminal protein, CRT3, and not the membrane-associated CNX, is involved in the ER retention of BRI1-9. However, BRI1 has a large lumen-facing domain, which may explain why CRT3 was found to interact with BRI1-9 and CNX was not. It is also of interest to note that BRI1-9 is a client for CRT3 and not CRT1 and 2, two other CRT isoforms in Arabidopsis. However, CRT3 is phylogenetically distinct from CRT1 and 2 and may have a special role in ERQC that distinguishes it from CRT1 and 2.

Another system that has provided insights into the ERQC system involves the maturation of a leucine-rich receptor kinase (LRR-RK) in Arabidopsis, associated with plant innate immunity. The maturation of the pathogen receptor LRR-RK EFR that recognizes the bacterial elongation factor (EF)-Tu EFR has been shown to be dependent on the protein-folding machinery and ERQC. Arabidopsis *elfin* mutants were identified as defective in their response to (EF)-Tu surrogate peptide elf18. Several studies of elf18-insensitive mutants revealed mutations in the protein-folding machinery and in the ERQC system including mutations in CRT3, UGGT glucosidase IIα and IIβ, STT3A (a component of OST), stromal-derived factor 2 (SDF2), and the HDEL retention factor, ERD2b. The maturation of a different receptor, LRR-RK FLS2, was unaffected by most of the mutations affecting LRR-RK ELF2 indicating that the two receptors are subject to different protein-folding and ERQC systems.

Mannosidases

If UGGT is the "decider" about further rounds of folding cycles, then mannosidase is the "terminator." A key step in removing misfolded glycoproteins from the CNX/CRT folding machine is the trimming of mannose residues from the protein-bound oligosaccharide. Specific inhibition of mannose trimming by an alkaloid mannosidase inhibitor, kifunensine has been shown to delay the release of misfolded proteins from the CNX/CRT cycle and their degradation by ERAD. Misfolded polypeptides in mammalian cells are subjected to extensive demannosylation likely by several different mannosidases. In yeast, protein-bound oligosaccharides on misfolded proteins are demannosylated by two distinct α1,2 mannosidases, Mns1p (an ortholog of the mammalian ER mannosidase I, ERManI) and Htm1p (an ortholog of mammalian ER degradation enhancing α-mannosidase or EDEM). In mammalian cells, inactivation of GH family 47 mannosidases or deletion of EDEM1 substantially delays release of ERAD substrates from the CNX/CRT cycle.

In yeast, the critical demannosylation step in removing misfolded glycoproteins from the CNX/CRT cycle is the hydrolysis by Htm1p of M11 on the C branch exposing the terminal α1,6-linked mannose, M10 (Fig. 4). In mammalian cells, an additional strong signal for ERAD disposal is represented by removal of M7 from the A branch of the protein-bound oligosaccharide. This is the only residue of protein-bound oligosaccharides that can be reglucosylated by UGGT; thus, its elimination irreversibly extracts the misfolded protein from the CNX/CRT cycle, preventing futile folding attempts (Aebi et al. 2010).

Recognition of the ERAD Signal

Removal of M11 exposes a terminal α1,6-bonded mannose (M10), a signal recognized by OS-9 and the XTP3-B, ERAD lectins in mammalian cells that contain mannose 6-phosphate receptor homology domains. Although both OS-9 and XTP3-B are lectins, they can also specifically bind misfolded non-glycosylated proteins indicating that they can also directly recognize aberrant proteins (Hebert et al. 2010). In yeast, the OS-9 equivalent, Yos9p, is part of the Hrd1p complex (a ubiquitin ligase named for HMG-CoA reductase degradation) to which it is bound via Hrd3p, a transmembrane protein with a large lumen-facing domain (Fig. 5). Hrd3p recruits the misfolded proteins while Yos9p scans them for the N-glycan structure that identifies a terminally misfolded protein. ERAD substrates can be categorized based on the subcellular location of their misfolded domains (Brodsky and Wojcikiewicz 2009). Proteins with misfolded domains in the different subcellular locations are inspected by different components of the ERQC surveillance system. Misfolded soluble proteins in the ER lumen and membrane proteins with misfolded domains projecting into the ER lumen are ERAD-L substrates, and membrane proteins with misfolded domains within the membrane are ERAD-M substrates, while membrane proteins with misfolded domains projecting into the cytoplasm are ERAD-C substrates. In yeast, the Hrd1p complex generally interrogates substrates with luminal or membrane folding lesions (ERAD-L and ERAD-M substrates), while those with cytoplasmic defects (ERAD-C substrates) are acted upon by a membrane multispanning ubiquitin ligase Doa10, a ubiquitin-conjugating enzyme Ubc7p, and its membrane anchor, Cue1p (Anelli and Sitia 2008).

Defining the ERAD Network

Christianson et al. (2012) defined the human ERAD network by conducting large-scale analysis employing techniques such as proteomics, protein interaction analysis, and functional (using RNAi approaches) and coordinate gene

(continued)

expression analyses. They established an organizational framework for the principal ERAD activities consisting of functional modules involving substrate recognition, dislocation, extraction, ubiquitination, and degradation. Two principal subnetworks were defined – one involving Hrd1 and its binding partner SEL1 and another subnetwork involving the E3 ligase gp78. The authors found that most of the components of the 26S proteasome were captured with the Hrd1 and gp78 complexes (Christianson et al. 2012).

Protein Dislocation

ERAD client proteins are disposed of by the cytoplasmic 26S proteasomes, protein disposal units, located in the cytoplasm (Fig. 5). Therefore, ERAD substrates have to be exported across the ER membrane for ubiquitination and proteolysis. This process is called "protein dislocation" to distinguish it from "protein translocation," the process described above by which proteins enter the ER. ERAD proteins involved in dislocation include the Derlins. Derlins are membrane-spanning proteins that form complexes with membrane-bound components of the ERAD machinery including the E3 ubiquitin ligase Hrd1. Derlins are thought to function in the formation of an ER export channel through which ERAD substrates pass. Derlins are related to the rhomboid family of intramembrane proteases. These proteins have a membrane-embedded cavity allowing water to access the active site during hydrolysis of intramembrane substrates. This structure may provide an aqueous channel through which proteins can move during dislocation.

Chronically Misfolded Proteins

Analyzing the ERAD system has also been aided by deploying "chronically misfolded proteins" as ERAD clients. These proteins have defects that destine them for ERAD disposal, and because of that they can be used to study and test the effectiveness of the ERAD system. In mammalian systems, some of the chronically misfolded proteins that have been studied are an inactive form of carboxy peptidase Y called CPY* (G255R), a truncated form of α1-antitrypsin, the IgM heavy chain, IgG light chain, and apolipoprotein B. The latter are chronically misfolded because they fail to assemble usually because they are a subunit of a multisubunit protein expressed when the other subunits are not.

In plants, an assembly-defective form of phaseolin, the seed storage protein in the common bean *Phaseolus vulgaris*, has been studied in which a C-terminal domain (Δ363) involved in assembly was deleted. The mutant phaseolin is detained in the ER where it was found in association with the

(continued)

chaperone, BiP. Other chronically misfolded proteins that serve as reliable ERAD clients in plants are the catalytic A chain or the B chain of the heterodimeric ricin toxin when independently expressed (Di Cola et al. 2005). The catalytic A chain of the ricin toxin is degraded when it is expressed in tobacco protoplasts in a process that can be blocked by the proteasome inhibitor MG132. Yet another chronically misfolded protein that has been studied in plants is MLO, a barley powdery mildew resistance protein, an integral multispanning membrane protein. Mutated forms of MLO-1 with lesions in its various cytoplasmic loops serve as reliable ERAD clients. MLO mutants that are most unstable in barley are degraded by ERAD when expressed in Arabidopsis protoplasts, yeast, or human endothelial cells. This demonstrates that ERAD recognition signals are well conserved across the kingdoms of life. The most unstable MLO mutants were stabilized in *hrd1* mutants in yeast, but not in *doa10* mutants, indicating that the MLO mutants were likely degraded by the HRD1 ERAD pathway in yeast.

Both the ERAD-C substrates and ERAD-L and ERAD-M substrates utilize the power of Cdc48p, a protein extracting AAA-ATPase, to dislocate proteins across the ER membrane. Cdc48 associates with the 26S proteasome cap and in doing so delivers the polyubiquitinated ERAD substrates to the proteasome. In the ricin expression system (described in the box above), dislocation of the ricin A chain requires CDC48 as demonstrated by the action of a dominant negative CDC48. Di Cola et al. (2005) investigated whether ubiquitination could be dissociated from the dislocation of the ricin A chain during ERAD. Since lysine residues are a major target for ubiquitination, the authors examined the effects of eliminating lysines on the retrotranslocation and degradation of ricin A chain in tobacco protoplasts. They showed that reducing the number of lysines did not affect recognition events within the ER lumen nor the dislocation of the modified ricin A chains to the cytosol. Instead, the modifications slowed the degradation rate, uncoupling dislocation from degradation. The introduction of extra lysines had an opposite effect, tightly coupling dislocation and degradation.

Genetic Analysis of ERAD

A number of the components of the protein-folding and ERAD systems have been identified genetically as suppressors of mutant hormone or pathogen receptors by Li and coworkers (Hong and Li 2012). One of the suppressors (*ebs5-1*) encodes a homolog of yeast Hrd3, or as it is called in mammalian cells, Sel1L (Fig. 5). As described above, Hrd3 is an adapter in the Hrd1 complex that ubiquitinates misfolded proteins, thereby identifying them for ERAD disposal.

Hrd3 is an integral membrane protein with a large lumen-facing domain that senses exposed hydrophobic surfaces on misfolded proteins. It was found that EBS5 in plants binds (coimmunoprecipitates with) the misfolded BRI1 receptors (BRI1-5 and BRI1-9), but not the wild-type BRI1 receptor, demonstrating that EBS5 is capable of recognizing misfolded proteins. Two genes were also identified in Arabidopsis that were homologs of yeast Hrd1. They were found to act redundantly and a knockout of both genes suppressed the *bri1-9* phenotype.

Yet another suppressor (*ebs4*) encodes a mannosyltransferase, a putative ortholog of yeast ALG12, which is involved in the assembly of lipid-linked oligosaccharides. This mutant fails to add α1,6-linked mannose to the C-chain of the oligosaccharide and transfers incompletely assembled oligosaccharides to glycoproteins. When exposed, this α1,6 linked mannose is the critical glycan mark for an ERAD client recognized by the lectin OS9 (Fig. 5). Recently, the OS9 homolog in Arabidopsis (AtOS9) was identified in Arabidopsis through a T-DNA insertion mutation and the suppressor, *ebs6-1*. AtOS9 was found to interact biochemically and genetically with EBS5 or HRD1. Without the exposed α1,6-linked mannose or with a defect in AtOS9, the BRI1 receptors, in bri1-5 and bri1-9, escape ERAD to become functional receptors. These two suppressors emphasize the importance of oligosaccharide on glycoproteins in the functioning of the ERAD system.

Ubiquitin

Ubiquitin is a highly conserved protein consisting of ~70 amino acids. Ubiquitin is conjugated onto targets, usually other proteins, in a three-step reaction cascade E1→E2→E3. Ubiquitin is activated by the ubiquitin-activating enzyme E1 through adenylation at its C-terminal glycine to form a high-energy thiol-ester intermediate. Then ubiquitin is transferred from E1 to a ubiquitin-conjugating enzyme (Ubc) E2. E2 can transfer ubiquitin to a substrate directly or indirectly. In most cases, ubiquitin is donated to an E3 ubiquitin ligase, which then transfers the ubiquitin to the bound substrate. In other cases, E3 enzymes are not tagged by ubiquitin, rather they bind to substrates allowing E2 to transfer ubiquitin to a substrate by substrate-protein ubiquitination. A single ubiquitin or multiple ubiquitins can be conjugated onto a single target site. The conjugation of a single ubiquitin involves the conjugation of the ubiquitin C-terminal carboxyl group and a lysine ε-amino group on the substrate protein. Additional ubiquitins can be joined to one of the seven lysines present on ubiquitin itself, although the most used position is K48. In plants, nearly 5 % of the genome encodes components of the E1→E2→E3 cascade with increasing numbers of the gene families in moving through the cascade. Members of each family present different properties and cellular localization, accounting for the specificity toward the diverse substrates. In plants, the ubiquitin system is one of the most complex involving over ~1,500 E3 ligases (Vierstra 2012).

ER Stress and the Unfolded Protein Response

The ERQC and ERAD systems keep watch over the process of protein folding and eliminate those proteins that fail to fold successfully. The operation of the system usually suffices under normal conditions. However, under stress conditions the system can become overloaded with misfolded proteins creating a condition called ER stress. ER stress sets off an alarm that elicits the unfolded protein response (UPR) (Howell 2013). The UPR is a homeostatic response that upregulates genes encoding components of the protein-folding, ERQC, and ERAD systems. The response mitigates the damage done by ER stress and brings the capacity for protein folding in line with demands.

In mammalian cells there are three arms to the UPR signaling pathway (Walter and Ron 2011). One arm involves the membrane-associated transcription factors that are partially embedded in the ER membrane. When activated, these factors are released and relocate to the nucleus to upregulate UPR genes. Another arm involves IRE1, a dual protein kinase/RNA splicing enzyme that splices a messenger RNA encoding another transcription factor, which also upregulates stress-response genes. The third arm involves a membrane-associated protein kinase called protein ER kinase (PERK) that phosphorylates and inactivates a translation initiation factor, eIF2a, thereby slowing translation – allowing the protein-folding process to catch up with protein synthesis. Plants have been shown to have two arms of the UPR signaling pathway, one arm involving membrane-associated transcription factors and another arm involving IRE1 (Fig. 6). The third arm involving PERK has not yet been identified in plant system; however, there are reports that translation slows in response to ER stress.

UPR can be induced in the laboratory by treating plants with ER stress agents – agents that impede protein folding in the ER. Tunicamycin (TM) is one such agent, which interferes with N-glycosylation of secreted glycoproteins. As described in previous sections, N-glycans are recognized at various steps in the protein-folding process, and without N-glycosylation, unfolded proteins accumulate. Reducing agents, such as dithiothreitol (DTT), also produce ER stress because the proper folding of proteins containing disulfide bonds requires an oxidizing environment in the ER that promotes the formation of these bonds. In addition, inhibitors of the ER calcium pump, such as cyclopiazonic acid (CPA), function as ER stress agents because the major components of the ER protein-folding apparatus, calnexin (CNX) and calreticulin (CRT), are calcium dependent. Of course, ER stress agents are proxies for the natural conditions that elicit ER stress in plants. UPR can be induced by a variety of abiotic stresses such as heat and salt stress and by certain biotic agents.

Membrane-Associated Transcription Factors

One of the best-studied membrane-associated stress-transducing transcription factors in mammalian cells is a bZIP factor called activating transcription factor 6 (ATF6). ATF6 is a type II membrane protein with a single-pass transmembrane domain, a bZIP domain facing the cytosol, and a carboxyl terminal tail with a

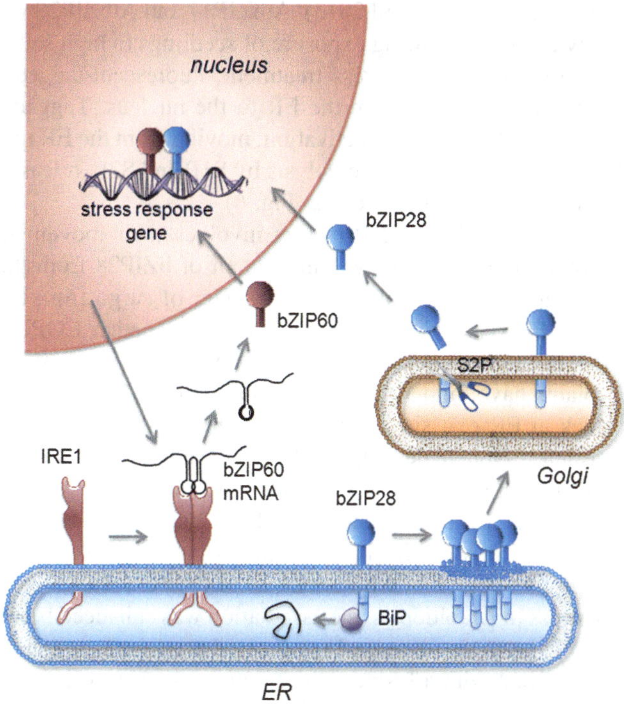

Fig. 6 The two arms of the ER stress-response signaling pathway in plants. One arm involves membrane-associated transcription factors such as bZIP28; the other involves a membrane-associated dual protein kinase/ribonuclease, IRE1, which splices the mRNA-encoding bZIP60. In response to stress, bZIP28 and IRE1 are activated by the accumulation of misfolded proteins in the ER. bZIP28 is mobilized from the ER and transported to Golgi bodies, where it is progressively processed by site 1 and site 2 protease (S1P and S2P). S2P intramembrane cleavage releases the N-terminal component of bZIP28 into the cytosol, allowing it to relocate to the nucleus. Once activated, IRE1 splices the bZIP60-encoding mRNA, creating a frameshift such that the spliced RNA now encodes a transcription factor with a nuclear targeting signal. bZIP28 and bZIP60 can heterodimerize, and it is thought that the two arms of the signaling pathway may converge in the formation of heterodimers that can upregulate stress-response genes

site-1-protease (S1P) cleavage site facing the ER lumen. Upon stress treatment, ATF6 is mobilized, exiting the ER and transported to the Golgi where it is proteolytically processed by two Golgi-associated proteases, S1P, a soluble luminal protease, and site-2-protease (S2P), a membrane-associated metalloprotease. S2P cleaves ATF6 within the membrane, releasing the cytosolic-facing component with the bZIP domain, which relocates into the nucleus to target stress-response genes.

In a search among the 75 members of the bZIP transcription factor gene family in Arabidopsis, structural homologs of ATF6 were identified as transcription factors predicted to be type II membrane proteins. There were four genes in this category, three of which, AtbZIP17, AtbZIP28, and AtbZIP49, were much like ATF6, having a bZIP domain predicted to face the cytosol and a carboxyl terminal tail with a canonical S1P cleavage site (Howell 2013). AtbZIP17 and AtbZIP28 were found to be proteolytically cleaved following exposure of seedlings to high salt stress or ER stress agents. In response to ER stress treatment, fluorescent-tagged versions of bZIP17 were observed to move from the ER to the nucleus. Tagged versions of bZIP28 followed a similar pattern of activation, moving from the ER to the Golgi in response to ER stress, followed by proteolysis by S1P and S2P, release from Golgi membranes, and relocation to the nucleus (Fig. 6).

Because activation of these bZIP factors involves their movement from one organelle to another, the steps involved in the exit of bZIP28 from the ER to the Golgi were investigated. In animal systems, transfer of cargo from the ER to the Golgi involves COPII vesicles. It is still unclear as to whether COPII vesicles are actually formed in the trafficking from the ER to Golgi in plants. It has been observed that plants have numerous, mobile Golgi, and it has been proposed that Golgi might dock at ER exit sites and pick up cargo without the involvement of intermediate vesicles. Nonetheless, ER to Golgi trafficking in plants is thought to involve the COPII machinery if only to concentrate cargo at ER exit sites.

In other systems, it has been demonstrated that the Sar1 GTPase initiates the formation of prebudding sites. Sar1 is thought to interact with cargo directly or indirectly through its interaction with Sec23/24, a COPII vesicle coat element. In response to ER stress in Arabidopsis, the association was enhanced between bZIP28 and Sar1b, one of the more abundantly expressed Sar1 isoforms in Arabidopsis. Sar1 appears to interact with a lysine-rich region on the cytosolic side of bZIP28, adjacent to the transmembrane domain. Substitution of lysine residues in this region with alanines interferes with Sar1 association and prevents the exit of GFP-bZIP28 from the ER to Golgi.

It is not known how the interaction between bZIP28 and Sar1 is enhanced in response to ER stress. Sar1's presumed binding site on bZIP28 is on the cytosolic side of the ER membrane; however, ER stress signals emanate from the opposite side of the ER membrane, in the ER lumen. In animal cells, it is thought that ATF6 is retained in the ER under unstressed conditions by binding to BiP, which is competed away or actively dissociated from ATF6 when misfolded proteins accumulate in response to stress. ATF6 is transported to the Golgi upon its release from the ER – only to be cleaved by Golgi-resident proteases and released into the cytoplasm.

Stress-Response Genes

The activation of bZIP28 leads to its relocation to the nucleus where it is involved in the upregulation of stress-response genes (Fig. 6). Genes upregulated by bZIP28 generally encode components of the ER protein-folding machinery and ERAD. Many of the upregulated genes, such as the BiP3 gene, which is highly upregulated by ER stress, have an ER stress-sensing element 1 (ERSE1) in their promoters. These elements are composed of two subelements, a CCACG subelement that binds bZIP dimers and a CCAAT subelement that binds

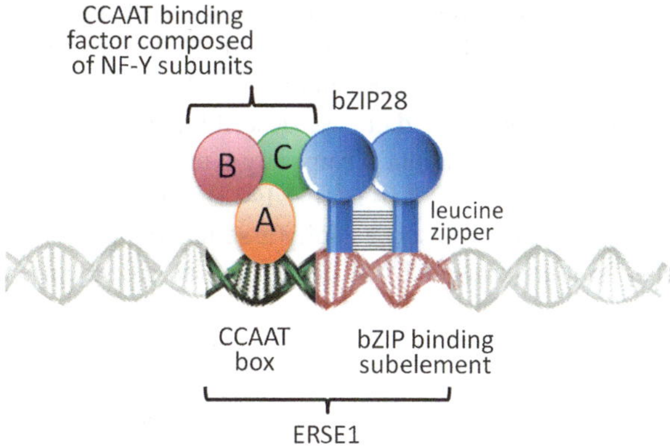

Fig. 7 Representation of the transcriptional complex associated with the ER stress promoter element (ERSE). The ERSE is composed of two subelements, a bZIP-binding subelement and a CCAAT box subelement. Shown is a dimer of bZIP28 binding to the bZIP-binding subelement and a CCAAT box factor binding to the CCAAT box. The CCAAT box binding factor is composed of NF-Y subunits A, B and C. NF-YC is transcriptionally regulated by ER stress and NF-YB is regulated by being transported into the nucleus from the cytoplasm in response to ER stress

CCAAT-box binding factors (Fig. 7). CCAAT-box binding factors are general transcription factors composed of three different NF-Y subunits, NF-YA, NF-YB, and NF-YC. The Arabidopsis genome encodes 36 NF-Y subunit genes, and because of the possible combinational complexity of subunits, it was a major challenge to sort out which NF-Y subunits were associated with bZIP28 in regulating genes by ER stress.

Yeast Three-Hybrid System to Identify Interacting Subunits

The actual composition of the NF-Y subunits in the transcriptional complex involving bZIP28 was determined using a yeast three-hybrid assay system. Such systems are used to test the ability of proteins to interact with one another. In brief, they involve a "bait" protein, which is used to identify an interacting "prey" protein. The bait and prey are set up as fusion (or hybrid) proteins such that if they interact with one another, they reconstitute a functional protein that can be recognized by a simple assay or cell selection scheme. Commonly used are the two halves of transcription factor that when joined together to form an active transcription factor that can be recognized by the gene that it activates. Typically, the assay system is a two-hybrid assay involving just the bait and the prey hybrid proteins. However, it was learned by trial and error that the bZIP28 bait did not interact with any one of the NF-Y subunits alone. Therefore, a three-hybrid system was developed allowing for an additional NF-Y subunit to provide a bridge between the bait and prey.

The three-hybrid system (described above) showed that bZIP28 as bait interacted with NF-YB3 and NF-YC2 as prey. That still left the third NF-Y subunit to be identified. A candidate for the third subunit, NF-YA4, was inferred from yeast interactome. (The interactome is a database describing a multitude of one-on-one protein interactions usually obtained from large numbers of yeast two-hybrid assays.) The three NF-Y subunits, NF-YA4, NF-YB3, and NF-YC2, together with bZIP28, all synthesized in E. coli, successfully assembled into a complex in vitro in the presence of a double-stranded DNA containing an ERSE1 promoter element (Howell 2013).

It was of interest to note that earlier studies using promoter/reporter constructs demonstrated that the genes for two of the NF-Y subunits, NF-YA4 and -YB3, but not NF-YC2, were constitutively expressed in Arabidopsis seedlings. Instead, NF-YC2 expression was induced by ER stress, and the induction was dependent, in part, on bZIP28. Furthermore, even though NF-YB3 expression was constitutive, NF-YB3 was largely located in the cytoplasm of unstressed seedlings. Following stress treatment, NF-YB3 relocated to the nucleus. Hence, the following scenario was put forward to describe the events following stress treatment leading to the assembly of a bZIP28-containing transcriptional complex. ER stress activates bZIP28 which on its own or with low levels of CCAAT-box factors modestly upregulates genes, such as BiP3 and NF-YC2. NF-YB and NF-YC are histone-fold-containing proteins that heterodimerize, and in mammalian cells they enter the nucleus as heterodimers through the importin-13 nuclear import system. In Arabidopsis it is thought, therefore, that the upregulation of NF-YC2 expression promotes the nuclear import of NF-YB3. NF-YA in mammalian cells is imported on its own by a different mechanism, likely by the importin-β system, and is then recruited to form a heterotrimeric CCAAT-box binding factor. Through its interaction with bZIP28, the CCAAT-box binding factor would reinforce the activity of bZIP28 promoting high levels of ER stress-response gene expression.

IRE1 and Its Target RNA

The second arm of the UPR signaling pathway in plants involves INOSITOL REQUIRING ENZYME 1 (IRE1) (Fig. 8a). The name of the enzyme has very little to do with its function in plants. IRE1 is a dual protein kinase/ribonuclease that plays a role as an unconventional RNA splicing factor. This arm of the ER stress-response pathway is more primal than the arm involving the membrane-associated transcription factors, because IRE1 is found in yeast, nematodes, fruit flies, and mammals. Until recently, this arm had not been described in plants, although two genes encoding IRE1, called IRE1a and IRE1b, had been identified in Arabidopsis.

In yeast, ER stress activates IRE1, which splices the mRNA for a bZIP transcription factor called Hac1p. The splice is of consequence because the intron in the unspliced RNA interferes with translation. In mammalian cells, IRE1 splices the messenger RNA for XBP1, again another bZIP transcription factor. Splicing is also consequential for XBP1 mRNA. The XBP1 coding region in the unspliced mRNA

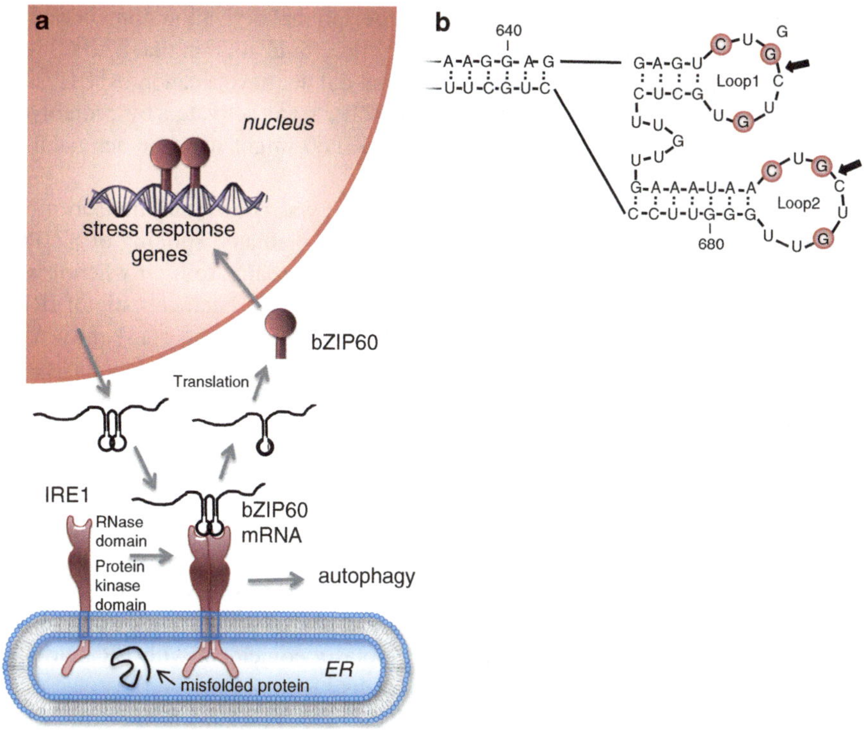

Fig. 8 The RNA splicing arm of the UPR signaling pathway. (**a**) The first step in RNA splicing is catalyzed by the membrane-associated dual protein kinase/ribonuclease, IRE1, that cuts the target RNA. The target RNA is a mRNA that encodes bZIP60, a transcription factor that upregulates ER stress response genes in response to stress. (**b**) IRE1 recognizes bZIP60 mRNA by its unusual double hairpin loop structure and cuts each of the loops (at the *arrows*) removing a 23 b intron. The removal of the intron produces a frameshift in the RNA resulting in a different reading frame downstream from the splice site. The unspliced bZIP60 RNA encodes a transcription factor with a transmembrane domain (TMD) predicted to be an ER localized protein. The spliced bZIP60 RNA encodes the same transcription factor which no longer has a TMD, but now has acquired a nuclear targeting signal. IRE1 also has an activity independent of bZIP60 that activates autophagy in response to ER stress

is composed of two open reading frames (ORFs) that are out of frame with respect to each other. Splicing produces a frameshift to bring the ORFs into alignment. In mammalian cells, it is interesting that the unspliced XBP1 protein appears to have a role in escorting the unspliced XBP1 mRNA to ER for splicing in response to stress.

In plants, the RNA splicing arm was discovered through efforts to understand the activation of another ER stress-induced, membrane-associated bZIP transcription factor, bZIP60. bZIP60 had been implicated in ER stress responses in Arabidopsis through studies demonstrating that transgenic expression of an activated form of bZIP60 upregulated the expression of UPR genes. bZIP60 was predicted to be a bZIP transcription factor, much like bZIP17 or bZIP28, a type II membrane protein with single transmembrane domain and a transcriptional activation domain facing the cytosol. However, bZIP60 differed from bZIP17 or bZIP28 in that bZIP60 was predicted to have a shorter lumen-facing tail and did not have a canonical S1P site. Based on the model by which bZIP17 and bZIP28 are activated by proteolytic cleavage, it was difficult to understand how bZIP60 might itself be activated in response to stress.

Nonetheless, the problem was resolved when it was shown that transcriptionally active forms of bZIP60 are produced by IRE1-mediated splicing of bZIP60 mRNA, not by proteolysis (Howell 2013). The matter came to light when it was recognized that bZIP60 mRNA could fold into a structure characteristic of IRE1 recognition sites, based on the structure of the splicing sites for Hac1 mRNA in yeast and XBP1 mRNA in mammalian cells. The recognition site is composed of a pair of stem-loops with conserved bases at three positions in each loop (Fig. 8b).

bZIP60 mRNA is spliced in Arabidopsis seedlings in response to ER stress agents, such as TM and DTT. Splicing excised a 23b segment of mRNA, leading to a frame shift beyond the splice site. The splice site was just upstream of the single transmembrane domain (TMD) in bZIP60, and the frame shift did away with the TMD. In the new sequence downstream from the splice site were two putative nuclear localization signals, and it was demonstrated that fluorescent-tagged forms of unspliced bZIP60 were located in the cytoplasm coincident with ER markers, while the spliced form was located in the nucleus.

Although the findings were consistent with the proposition that bZIP60 mRNA is spliced in response to stress, it was important to demonstrate whether bZIP60 mRNA splicing was, in fact, required to activate the transcriptional activity of bZIP60. This was done by substituting a conserved base in one of the twin loops of IRE1 recognition site. The substitution blocked bZIP60 mRNA splicing and inhibited the stress-induced upregulation of a bZIP60 target gene, BIP3. This demonstrated that RNA splicing, and not proteolysis, was responsible for bZIP60 activation.

Converging Pathways

Although bZIP60 and bZIP17/bZIP28 represent different arms of the UPR signaling pathway, the separate arms seem to converge in the regulation of stress-response genes (Fig. 6). The bZIP transcription factors have leucine zipper domains that function in dimer formation and binding to DNA. bZIP60 was shown in a yeast two-hybrid system to homodimerize and to heterodimerize with bZIP28 and bZIP17. Therefore, if both arms of the ER stress-response pathway are activated, it is possible that heterodimerization might take place between bZIP60 and bZIP17 or bZIP28. In addition, some genes, such as BiP3, are partially dependent on both bZIP28 and bZIP60 for upregulation in response to ER stress in Arabidopsis. From this, one might surmise that heterodimers of bZIP28 and bZIP60 are more transcriptionally active than either homodimer.

UPR Evolution and Specialization

During the course of evolution, the UPR has evolved and so have the functions of the components of UPR signaling pathway. Unicellular eukaryotes, such as yeast, have only one arm of the UPR mediated by IRE1 (Fig. 9); nonetheless, in response to ER stress, yeast upregulate a broad array of genes including ER chaperones but also genes involved in protein translocation, folding and degradation, glycosylation in the ER, lipid/inositol metabolism, ER-Golgi transport, Golgi-ER retrieval, glycosylation in the Golgi apparatus, vacuolar targeting, distal secretion, and cell wall biogenesis (Mori 2009). Despite the range of functions controlled by the UPR in yeast, IRE1 knockout mutations are not lethal; however, mutant yeast are more susceptible to ER stress.

Lower metazoans such a worms (*C. elegans*) and flies (*D. melanogaster*) have three arms of the UPR although most inducible stress-response functions are still vested in IRE1. The RNA spliced by yeast IRE1 is Hac1 mRNA, while the RNA spliced by IRE1 in worms and flies is XBP1, an mRNA similar in sequence to XBP1 in higher metazoans. Worms and flies have an ATF6 gene, which distinguishes them from yeast; however, very few UPR signaling functions have been transferred to ATF6 in lower metazoans. Instead, most all of the typical ER stress-response genes are induced by IRE1 in worms and flies. It is surprising that although ATF6 in *C. elegans* regulates few induced ER stress genes, it is responsible for the expression of many constitutively expressed genes during development and involved in homeostasis. In fact, deletion of either *ire-1* or *xbp-1* is a synthetic lethal with the deletion of ATF6, producing a developmental arrest in larvae. Worms have a third arm mediated by pek-1, a homolog of PERK in mammalian systems, and pek-1 loss-of-function mutants are normal under unstressed conditions but are more sensitive to ER stress. Nonetheless, *ire-1 pek-1* double mutants are embryonic lethals suggesting that when the IRE1/XBP1 pathway is knocked out in worms, then the regulation of translation by PEK-1 is required for viability.

In higher metazoans, the picture is more complicated, given that functions have been more widely delegated among the three arms of the UPR signaling network. In addition, higher metazoans have multiple isoforms of the various UPR signaling components. Mice have two IRE1 genes, *IRE1a* and *IRE1b*, and these genes have tissue-specific functions generally relating to the tissues in which the genes are expressed. *IRE1a* is expressed throughout the animal, but *IRE1b* is mostly expressed in the epithelial cells of the gut. *IRE1a*-knockout mice are embryonic lethals, showing widespread developmental defects, leading to embryonic death.

In mammalian cells, the membrane-associated transcription factor family has been further diversified to include ATF6α and β, OASIS, CREBH, and Luman. Tis 40 and BBF2H7. All of these factors have structural similarities, but they differ in activating stimuli, tissue distribution, and response elements to which they bind. These differences indicate specialized functions in regulating the UPR in specific organs and tissues. A growing body of evidence suggests that the processes regulated by these family members play essential roles in cell differentiation and maturation or maintenance of basal cellular homeostasis in mammals.

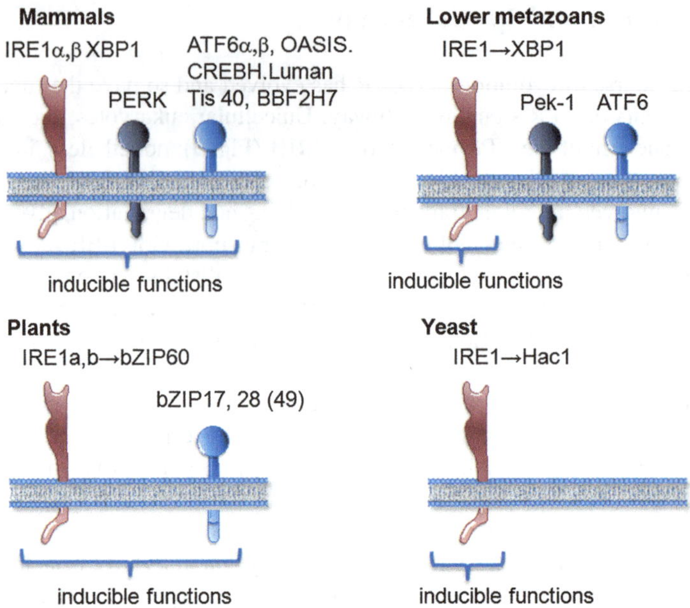

Fig. 9 The evolution of UPR signaling. Budding yeast has only a single stress sensor/transducer, IRE1 that splices Hac1 mRNA. However, IRE1 in yeast has many inducible functions, a number of which have been adopted by other factors in higher organisms. Lower metazoans, such as worms and flies, have three sensor/transducers including IRE1, Pek-1 and ATF6. IRE1 splices a mRNA that is similar to XBP1 in mammals. Although lower metazoans have three types of sensor/transducers, most of the inducible functions are vested in IRE1. Mammals also have three types of membrane sensor/transducers IRE1, PERK and a small family of proteins related to ATF6. IRE1 splices XBP1 mRNA, however, many of the ER stress inducible functions in mammals have been delegated to the other sensor/transducers. Plants, such as Arabidopsis, are known so far to have only two types of membrane sensor/transducers, IRE1 which splices bZIP60 mRNA and a small family of membrane associated transcription factors, bZIP17, bZIP28 (and bZIP49, which has not been characterized). The ER stress inducible functions in Arabidopsis are distributed between the two types of sensor/transducers such that one arm can compensate for the loss of the other

As described above, plants have only been shown to have two arms of the UPR signaling pathway, one arm involving membrane-associated transcription factors and another arm involving IRE1. The third arm involving PERK has not yet been identified in plant systems. There are three genes in the Arabidopsis genome that encode IRE1-like proteins, but only two, *IRE1a* and *IRE1b*, encode full-length proteins. In rice and maize, there only appears to a single gene encoding full-length IRE1. Studies have demonstrated that both *IRE1a* and *IRE1b* in Arabidopsis can support some level of bZIP60 mRNA splicing in seedlings in response to stress induced by ER stress agents. However, some studies indicate that IRE1b plays a major role in seedlings, while other studies suggest both *IRE1a* and *IRE1b* may contribute equally. Suffice it to say, *IRE1a* and *IRE1b* appear to have overlapping functions, but one or the other may be dominant in response to different stresses.

Recently, it was found that *IRE1a* may play an unusual role in systemic acquired resistance (SAR) response elicited by bacterial pathogen infection in Arabidopsis. It was reported that the *IRE1a* is required to support the secretion of pathogenesis-related (PR) proteins following treatment of plants with salicylic acid (SA). As a result, *ire1a* mutants show enhanced susceptibility to a bacterial pathogen, whereas *ire1b* mutants are unaffected. It was found that the immune deficiency in *ire1a* is due to a defect in SA- and pathogen-triggered, IRE1-mediated cytoplasmic splicing of the bZIP60 mRNA. How IRE1a operates in this mode and what effectors act downstream are not known.

Autophagy

Recently, it was demonstrated in Arabidopsis that ER stress induces autophagy. Autophagy literally means "self-eating," and despite its name, autophagy is thought to be a cell-sparing or cell survival activity. It is a process whereby stressed cells eliminate or turnover damaged organelles and other cellular constituents by degrading and recycling them (Liu and Bassham 2011). Autophagy is marked by characteristic cell morphology changes. When autophagy is induced, double-membrane structures called autophagosomes form around cytoplasmic components, such as organelles or pieces of organelles. The autophagosome then delivers its contents to the vacuole for degradation. The outer membrane of the autophagosome fuses with the vacuolar membrane, and autophagic bodies, consisting of the inner membrane and its cargo, are released into the vacuole where they are degraded by hydrolytic enzymes. Autophagy induced by ER stress appears to degrade the ER membrane along with other cell constituents.

The formation of autophagosomes can also be identified by visually marking proteins that make up the membrane. One such protein is called autophagy-related 8e (ATG8e) and when Arabidopsis seedlings are treated with ER stress agents, DTT or TM, autophagosomes and autophagic bodies are observed in treated cells. Another means by which it was shown that autophagy is induced by ER stress was by comparing the response of wild-type plants to those in which an essential autophagy gene, ATG18a, was knocked down by RNAi constructs. In the RNAi-ATG18a seedlings, no autophagosomes were observed in seedlings subjected to ER stress. It is possible that the appearance of autophagosomes following ER stress is due to an increase in the formation of autophagosomes or by a block in the delivery to or the degradation of autophagosomes in lysosomes. To distinguish between these, concanamycin A (concA), an inhibitor of vacuolar proton-translocating ATPase, was used to inhibit autophagic body degradation in the vacuole. ConcA raises the vacuolar pH, preventing protein degradation by vacuolar acid hydrolases, and inhibits further trafficking of proteins to the vacuole. When seedlings were treated with concA and ER stress agents, TM or DTT, a buildup of autophagic bodies in the vacuole was observed indicating that ER stress induced the formation of autophagosomes rather than preventing the breakdown of autophagic body components in lysosomes.

There are two major routes by which cellular constituents traffic to the vacuole in plants: the autophagy pathway and the biosynthetic pathway. The latter involves the Golgi apparatus, the trans-Golgi network (TGN), and the prevacuolar compartment (PVC). However, a fluorescent signal marking the ER was not found in components of the biosynthetic pathway following ER stress. Thus, the ER structures observed in the vacuole following ER stress arrive by an autophagy pathway and not via the Golgi apparatus, TGN, or PVC pathway.

How does ER stress activate autophagy? What is the link between ER stress and autophagy? Does the induction of autophagy require one or both arms of the ER stress signaling pathway? To investigate this, mutants knocking out the function of either arm were tested for the induction of autophagy following stress treatment. Mutations in the membrane-associated transcription factor arm involving bZIP28 posed some challenges in testing for their effect on ER stress-induced autophagy because bZIP28 knockout mutants showed constitutive autophagy, i.e., signs of autophagy when there was no deliberate effort to induce it. However, it was demonstrated that constitutive autophagy in the bZIP28 mutant could be suppressed by reducing oxidative stress, an independent pathway for induction of autophagy. To inactivate the RNA splicing arm of the UPR signaling pathway, two other mutants were studied – one that knocked out *IRE1a* and another that knocked out *IRE1b*. These studies demonstrated that bZIP28 and the membrane-associated transcription factor arm of the pathway were dispensable for the induction of autophagy by ER stress. However, ER stress failed to induce autophagy in mutants knocking out *IRE1b*, but not *IRE1a*. Thus, the arm of the ER stress signaling involving IRE1 plays a role in the induction of autophagy. Why *IRE1b* and not *IRE1a*? That may simply be related to the fact IRE1b is more highly expressed in vegetative tissue where the effect of ER stress on autophagy was tested.

Typically, autophagy has been studied in the laboratory by depriving plants of essential nutrients. The pathway that signals the induction of autophagy by starvation is negatively regulated by a protein kinase called TOR (target of rapamycin). The pathway by which IRE1b signals the induction of autophagy in response to ER stress in plants is not known; however, it does not appear to involve RNA splicing or bZIP60. Recall that IRE1 is a dual protein kinase/ribonuclease, and in mammalian systems it is the protein kinase activity, not the RNA splicing function of IRE1 that signals autophagy. The mammalian IRE1 protein kinase activates JNK (c-Jun N-terminal kinase) through a protein phosphorylation cascade, which activates transcription factors that upregulate stress-response genes. Whether a comparable pathway exists in plants is not known.

Cell Death

Under mild stress conditions, early ER stress responses such as autophagy and the upregulation of chaperones, protein-folding, and ERAD components are cell-sparing or cell survival responses. However, chronic or acute stress conditions lead to cell death. What then switches the response from cell survival to cell death? The answer may be that there is not an abrupt switch but that the differences may be a matter of degree.

First, it is important to point out what transpires during cell death. Unless cells die by a traumatic process, cell death is a regulated process, usually called programmed cell death. A common form of programmed cell death (PCD) in animal system is apoptosis. Apoptosis is characterized by a blebbing of the plasma membrane and engulfment of the blebs through endocytosis by phagocytes. Plant cells do not undergo apoptosis; instead, PCD is an internalization process whereby cells undergo vacuolization.

PCD has a number telltale signs including the fragmentation of DNA and chromatin condensation. DNA undergoes fragmentation by breaks in between regularly spaced nucleosomes, such that the DNA fragments form a ladder. In mammalian cells, stress signals activate the interplay of proapoptotic and antiapoptotic proteins of the Bcl-2 (B-cell lymphoma 2) family. The proapoptotic BAX and BAC proteins act on the outer mitochondrial membrane to lower the mitochondrial transmembrane potential permitting the leakage of cytochrome c. Cytochrome c complexes with other factors to form an apoptosome, a structure that recruits pro-caspase-9, and processes it into proteolytically active forms. This begins a caspase cascade resulting in apoptosis. However, plants have none of the members of the mammalian Bcl-2 gene family, so that PCD in plants does not involve a disruption in the integrity of the mitochondrial membrane or the so-called intrinsic PCD pathway in mammalian cells.

Instead PCD in plant cells involves the rupture of the tonoplast in which hydrolytic enzymes spill out in to the cytoplasm of the dying plant cell. The rupture of the tonoplast is dependent on the action of vacuolar processing enzymes (VPEs, endopeptidases of the legumain family), and metacaspases. Caspases are cysteine aspartyl proteases, which orchestrate the demolition phase of apoptosis in mammalian cells, whereas metacaspases in plant cells are cysteine proteases specific for lysine and arginine residues. In plant cells, it is not yet known how these enzymes are activated and whether they form a signaling cascade and, if so, which initiate the cascade and which are the executioners.

Historical Notes

Science builds on the discoveries of others. So what findings led to our understanding of ER stress and the unfolded protein response? Most all of the early discoveries were in yeast or animal systems, while the studies in plants occurred somewhat later. George Palade and his colleagues at the Rockefeller Institute in the 1960s laid the groundwork for our present understanding of the protein secretory pathway in animal cells. Caro and Palade (1964) published a seminal paper in which "pulse-chase" experiments were used to trace the path taken by proteins secreted by guinea pig pancreatic

(continued)

exocrine cells. Using autoradiogaphy, they demonstrated during a "pulse" of ^{3}H-leucine that the radiolabel was first incorporated into proteins in the rough ER. When the pulse of ^{3}H-leucine was followed by a "chase" of unlabeled leucine, they observed that the radiolabel moved out of the ER to the Golgi apparatus and then later to zymogen granules (storage vesicles for secreted proteins in the pancreas). These experiments demonstrated that secreted proteins are synthesized in the ER and then "trafficked" to other organelles in the secretory pathway. A multitude of papers followed the Palade study describing the protein modification, targeting, and membrane trafficking events that follow. By in large, the fundamentals of the secretory process in yeast and animals apply to plant systems (Jurgens 2004).

Peter Walter discovered the unfolded protein response in yeast in 1994. According to Walter's account, the discovery was serendipitous (Walter 2010). They realized that when there was an overload of proteins entering the ER compartment that a signal was sent to the nucleus to switch on genes to make more ER. So they wanted to know how the ER talks to nucleus. To find out, students in Walter's lab isolated yeast mutants in which the ER was defective in signaling to the nucleus. They cloned the genes responsible for the mutations and found that they encoded a protein kinase. Through a series of other experiments, they learned that when their protein kinase was activated, a messenger RNA encoding the Hac1 protein, a stress-response transcription factor, was spliced. The enzyme responsible for the splicing turned out to be their protein kinase, now called Ire1, which had two enzymatic activities – a protein kinase and a ribonuclease that cut the RNA in the first step of the splicing process. It was a remarkable finding in many ways, because no one had ever found that an RNA was involved in a signaling pathway and that an RNA splicing event occurred outside of the nucleus.

The concept that the ER could talk to nucleus by transcription factors on the ER membrane that are mobilized in response to cellular conditions grew out of very different studies on cholesterol metabolism. In a series of elegant papers in the late 1990s, Brown and Goldstein demonstrated that in response to cholesterol deficiency, an ER membrane-bound transcription factor, sterol regulatory element-binding protein (SREBP), was mobilized to the nucleus to upregulate cholesterol biosynthesis genes. The mobilization of SREBP involved its untethering from factors in the ER and its transport to the Golgi apparatus where it was proteolytically cleaved by resident proteases. Cleavage by the intramembrane S2P protease allowed the transcription factor component of SREBP to be shed from the membranes of the Golgi apparatus into the cytoplasm and then imported into the nucleus. The proteolysis was called "regulated intramembrane proteolysis," which was later found to be the way that a number of membrane-bound factors are activated, including ATF6 (Brown et al. 2000).

Future Directions

In thinking about future directions, there are many unknowns and some important questions to ask about ER stress in plants.

The ER stress response is, indeed, activated by ER stress agents, such as DTT, TM, or heat treatment, but what is the full range of conditions that elicit ER stress responses?

Do ER stress responses actually protect plants from adverse environmental conditions directly? What happens if the ER stress-response system is inactivated? Do plants become more susceptible to adverse environmental conditions?

Can plants be better protected by modifying or upregulating the expression of factors involved in ER stress responses? Certainly, one of the major goals in crop improvement is improving stress tolerance. Do ER stress responses provide an opportunity for improving plant tolerance to stress?

Do ER stress and the UPR play a role in normal development? It has been demonstrated in mammalian systems that ER stress responses help plasma cells to deal with the copious production and secretion of immunoglobulins from plasma cells. Are ER stress responses elicited in plant cells that produce and send large amounts of proteins through the secretory pathway – such as in pollen formation or seed storage protein production?

Is the basis for the induction of the UPR by various stresses the accumulation of misfolded proteins? The elicitation of UPR by biotic agents is a case in point. Do these agents cause misfolding of proteins? If so, how so? What about the activation of bZIP17 by salt stress? Is this an ER stress response? Does salt stress interfere with the unfolding of proteins?

Is there a direct way to measure the levels of unfolded or misfolded proteins in the ER? At this point, it is largely by inference that it is assumed that various treatments cause protein misfolding, but there is no direct way of measuring the load of unfolded proteins in the ER. In animals, a system has been developed to measure the immobilization of BiP in response to stress. It is argued that in the unbound state BiP diffuses more freely than when it is bound to misfolded protein. Thereby, by measuring the diffusion rates of a BiP with a fluorescent tag, one can assess relative levels of misfolded proteins. However, this method is quite demanding and not easily adapted to most laboratory settings.

Is there a third arm to the ER stress pathway in plants? Plants do not have PERK, the ER membrane-associated protein kinase that serves as the central player in the third arm in mammalian cells. In mammalian cells, PERK phosphorylates and inactivates the translation initiation factor, slowing down protein synthesis in response to ER stress. In plant cells there is evidence that translation is slowed in response to ER stress. If that is the case, then how does it happen?

Why are there two arms to the ER stress pathway? Do the arms have different functions? Are they activated with the same kinetics? Are they activated by the same stresses? Does the interplay between the two arms affect the outcome of ER stress responses?

What is the pathway by which IRE1 signals autophagy? The pathway does not appear to involve bZIP60. Is there a different mRNA target? Does signaling autophagy involve a different IRE1 function?

The ER is multifunctional. Do any of the activities of the UPR impact other ER functions? ER stress usually stimulates proliferation of the ER in other systems. What is the consequence of ER stress to ER abundance and morphology?

References

Aebi M, Bernasconi R, Clerc S, Molinari M. N-glycan structures: recognition and processing in the ER. Trends Biochem Sci. 2010;35:74–82.

Anelli T, Sitia R. Protein quality control in the early secretory pathway. EMBO J. 2008;27:315–27.

Braakman I, Bulleid NJ. Protein folding and modification in the mammalian endoplasmic reticulum. Annu Rev Biochem. 2011;80:71–99.

Brodsky JL, Wojcikiewicz RJ. Substrate-specific mediators of ER associated degradation (ERAD). Curr Opin Cell Biol. 2009;21:1–6.

Brown MS, Ye J, Rawson RB, Goldstein JL. Regulated intramembrane proteolysis: a control mechanism conserved from bacteria to humans. Cell. 2000;100:391–8.

Caramelo JJ, Parodi AJ. Getting in and out from calnexin/calreticulin cycles. J Biol Chem. 2008;283:10221–5.

Caro LG, Palade GE. Protein synthesis, storage, and discharge in the pancreatic exocrine cell. An autoradiographic study. J Cell Biol. 1964;20:473–95.

Christianson JC, Olzmann JA, Shaler TA, Sowa ME, Bennett EJ, Richter CM, Tyler RE, Greenblatt EJ, Harper JW, Kopito RR. Defining human ERAD networks through an integrative mapping strategy. Nat Cell Biol. 2012;14:93–105.

Di Cola A, Frigerio L, Lord JM, Roberts LM, Ceriotti A. Endoplasmic reticulum-associated degradation of ricin A chain has unique and plant-specific features. Plant Physiol. 2005;137:287–96.

Gething MJ. Role and regulation of the ER chaperone BiP. Semin Cell Dev Biol. 1999;10:465–72.

Hammond C, Helenius A. Folding of VSV G protein: sequential interaction with BiP and calnexin. Science. 1994;266:456–8.

Hebert DN, Bernasconi R, Molinari M. ERAD substrates: which way out? Semin Cell Dev Biol. 2010;21:526–32.

Hong Z, Li J. The protein quality control of plant receptor-like kinases in the endoplasmic reticulum. In: Tax F, Kemmerling B, editors. Receptor-like Kinases in Plants. Berlin: Springer; 2012. p. 275–307.

Howell SH. ER stress responses in plants. Annu Rev Plant Biol. 2013;64:477–99.

Initiative TA. Analysis of the genome sequence of the flowering plant *Arabidopsis thaliana*. Nature. 2000;408:796–815.

Jurgens G. Membrane trafficking in plants. Annu Rev Cell Dev Biol. 2004;20:481–504.

Keenan RJ, Freymann DM, Stroud RM, Walter P. The signal recognition particle. Annu Rev Biochem. 2001;70:755–75.

Kornfeld R, Kornfeld S. Assembly of asparagine-linked oligosaccharides. Annu Rev Biochem. 1985;54:631–64.

Liu Y, Bassham DC. Autophagy: pathways for self-eating in plant cells. Annu Rev Plant Biol. 2011;63:215–37.

Lupattelli F, Pedrazzini E, Bollini R, Vitale A, Ceriotti A. The rate of phaseolin assembly Is controlled by the glucosylation state of Its N-linked oligosaccharide chains. Plant Cell. 1997;9:597–609.

Mittler R, Blumwald E. Genetic engineering for modern agriculture: challenges and perspectives. Annu Rev Plant Biol. 2010;61:443–62.

Mori K. Signalling pathways in the unfolded protein response: development from yeast to mammals. J Biochem. 2009;146:743–50.
Ruddock LW, Molinari M. N-glycan processing in ER quality control. J Cell Sci. 2006;119:4373–80.
Shao S, Hegde RS. Membrane protein insertion at the endoplasmic reticulum. Annu Rev Cell Dev Biol. 2011;27:25–56.
Totani K, Ihara Y, Tsujimoto T, Matsuo I, Ito Y. The recognition motif of the glycoprotein-folding sensor enzyme UDP-Glc:glycoprotein glucosyltransferase. Biochemistry. 2009;48:2933–40.
Vierstra RD. The expanding universe of ubiquitin and ubiquitin-like modifiers. Plant Physiol. 2012;160(1):2–14.
Walter P. Walking along the serendipitous path of discovery. Mol Biol Cell. 2010;21:15–7.
Walter P, Ron D. The unfolded protein response: from stress pathway to homeostatic regulation. Science. 2011;334:1081–6.

Auxin Signaling in Plants

8

Catherine Perrot-Rechenmann

Contents

Abstract

- Auxin is a major hormone controlling almost every growth and developmental process in plants.
- Auxin exerts such a control through local auxin gradients, tight control of homeostasis and the regulation of gene expression.
- Auxin-dependent transcriptional regulation involves a variety of modular TIR1/AFB and AUX/IAA co-receptors and ARF downstream effectors.
- The stability versus the degradation of AUX/IAA transcriptional repressors is central for the regulation of gene expression.

C. Perrot-Rechenmann (✉)
Institut des Sciences du Végétal (ISV), UPR2355 CNRS, Saclay Plant Sciences LabEx, Gif sur Yvette Cedex, France
e-mail: catherine.rechenmann@isv.cnrs-gif.fr

S.H. Howell (ed.), *Molecular Biology, The Plant Sciences 2*,
DOI 10.1007/978-1-4939-0263-7_15

- Competitive binding of ARF transcription factors to DNA and displacement of protein–protein interactions determine downstream transcriptional activity.
- Expression of auxin-responsive genes is negatively regulated by a signaling pathway involving ABP1 (auxin binding protein 1).
- ABP1 also regulates various non-transcriptomic responses to auxin including those relying on Rho GTPase activities.

Introduction

Plant hormones are small active molecules acting in concert to control many aspects of plant growth, development, and adaptive responses to environmental stimuli. Most plant hormones are small chemical molecules acting at rather low concentrations either locally or after short- or long-distance transport within the plant. Among these, the hormone auxin is involved in a multitude of biological processes and is often considered as the major phytohormone. The word auxin originates from the Greek word *auxein*, which literally means to grow/increase, a name befitting a hormone that was initially identified as a molecule stimulating shoot organ elongation. Since then, more subtle features about auxin's regulation of the elongation response (either promoting or inhibiting the response) and many additional activities of the hormone have been demonstrated including tropisms, control of various aspects of cell division (gene expression and protein turnover of cell cycle regulators, formation and maintenance of meristems, orientation of cell plate formation, patterning, priming of cells giving rise to lateral roots, inhibition of axillary bud activity resulting in apical dominance), establishment and maintenance of axis of symmetry, cell polarity, vascular tissue differentiation, abscission, and responses to biotic and abiotic stresses. Auxin's broad repertoire of effects is largely dependent on the cellular and tissue-specific contexts in which the hormone acts as well as the developmental stage and environmental conditions of growth. The major auxin in flowering plants is the indole-3-acetic acid (IAA). This phytohormone is mainly synthesized in young shoot tissues but can also be produced in other growing organs as roots. On a whole plant scale, auxin is efficiently transported from the shoot to the root via the phloem. This long-distance transport is supplemented by a cell-to-cell directional auxin transport involving uptake and efflux carriers, the combined activity of which generates local auxin gradients that influence local cellular and developmental responses. In addition to auxin transport, plant cells regulate local concentrations of the active forms of the hormone through biosynthesis, conjugation, subcellular partitioning, and catabolism of the hormone. Not only do plants regulate local concentrations of auxin, but their capacity to respond to a variation in auxin level can vary in different tissues and at different developmental stages. These differences derive from the complexity of the mechanisms buffering and controlling auxin responses, mainly transcriptional responses, and also on cross-talk of auxin with other phytohormones acting synergistically or antagonistically with the hormone (Swarup et al. 2002; Jaillais and Chory 2010).

The present review focuses on auxin signaling mechanisms in the model plant *Arabidopsis thaliana* with particular emphasis on transcriptional regulation.

Model of Transcriptional Control of Gene Expression in Response to Auxin

Targeted proteolysis of key regulatory proteins has a central role in auxin signaling. Over the past 20 years or so, it has been revealed that the ubiquitin-dependent 26S proteasome system (UPS) controls many biological processes in all eukaryotes and is of exceptional importance in plants (Vierstra 2009). The UPS is an enzymatic cascade involving first, an E1 ubiquitin-activating enzyme that activates ubiquitin by forming a thioester bond between a cysteine and the C-terminal glycine of ubiquitin. Second, an E2 ubiquitin-conjugating enzyme transfers the activated ubiquitin from the E1 to a cysteine of the E2, and, third, an E3 ubiquitin ligase identifies a target protein and catalyzes the transfer of the activated ubiquitin from the E2 to a lysine amino group of the target (Del Pozo and Manzano 2013). Marking a protein for proteasome degradation usually requires poly-ubiquitination, rather than mono-ubiquitination at one or several sites in a protein. Poly-ubiquitination results from the reiteration of the sequence of E1, E2, and E3 enzymatic reactions to form a chain of ubiquitins. Poly-ubiquitin chains differ by the lysine residue that links the next C-terminal glycine residue. Chains of Lys11, Lys29, and especially Lys48 poly-ubiquitins ultimately promote degradation of the target protein by the 26S proteasome. Various classes of E3 ligases that differ by their subunit composition are involved in the different ubiquitination reactions associated to specific pathways. In plant hormone signaling, E3 ligases belong to multi-protein SCF (SKP1-Cullin1-F-box) complexes. The cullin1 (CUL1) acts as a scaffold protein bridging suppressor of kinetochore protein 1-like (SKP1-like), ASK1 in Arabidopsis, and an F-box protein to a ring-box1 (RBX1) protein and the ubiquitin-conjugated E2. With about 700 F-box proteins identified in Arabidopsis, these F-box proteins are the variable component of SCF complexes and their numbers reflect a tremendous enhancement of the UPS in higher plants. The F-box motif corresponds to the domain of interaction with ASK1 whereas other domains contribute to identify the target protein and recruit it as a substrate of ubiquitination within the SCF E3 ligase complex. The assembly and disassembly of the SCF complex are important for its activity. Conjugation and deconjugation of an ubiquitin-like protein, related to ubiquitin (RUB1), to CUL1 is also important for SCF activity and contributes to the dynamics and partial recycling of the system.

Elementary Module of Transcriptional Regulation in Response to Auxin

The auxin nuclear signaling pathway regulating changes in gene expression in response to auxin was identified through a combination of molecular biology and

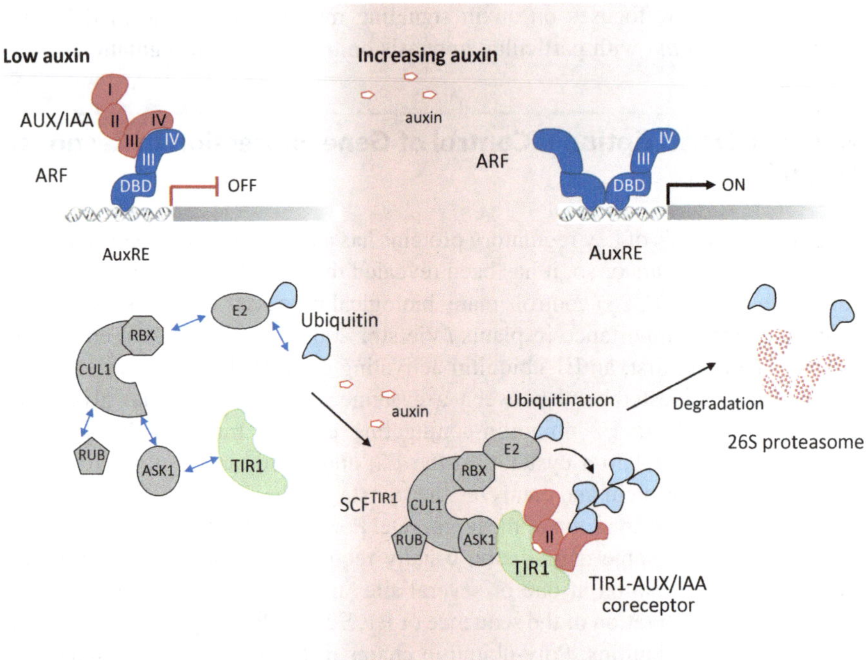

Fig. 1 Simplified model of transcriptional regulation in response to auxin. At low auxin conditions, expression of auxin response genes is repressed via AUX/IAA repressors. Assembly of the E3 ubiquitin ligase $SCF^{TIR1/AFB}$ complex is promoted by addition of RUB, a protein related to ubiquitin, to the scaffold protein cullin1. In the presence of increasing levels of auxin, interaction between TIR1/AFB F-box proteins and AUX/IAA repressors is favored. AUX/IAA proteins become substrates of $SCF^{TIR1/AFB}$ complexes, are poly-ubiquitinated, and addressed to the 26S proteasome for degradation. After degradation of AUX/IAA repressors, ARF transcriptional factors activate the transcription of auxin response genes that mediate cellular responses. Promoters of auxin response genes include auxin-responsive elements (AuxRE) that are recognized by ARF proteins

genetic approaches (Chapman and Estelle 2009). This nuclear signaling pathway is rather short and can be summarized as a module involving auxin response factor (ARF) transcriptional regulators, auxin/indole acetic acid (AUX/IAA) repressors, and transport inhibitor response 1 (TIR1) F-box proteins (Fig. 1). Most AUX/IAA proteins contain four conserved protein–protein interaction domains, among which domains III and IV are shared by most ARFs. ARF proteins contain a DNA-binding domain (DBD) toward their N-terminus, which binds to auxin regulatory elements (AuxRE) in the promoters of auxin response genes. At low auxin concentrations, interaction of AUX/IAAs with ARFs results in repression of auxin response genes. At higher auxin concentrations, auxin favors the interaction between TIR1 and AUX/IAA co-receptor and is bound at the interface between these two proteins. The immediate consequence of TIR1 and AUX/IAA interaction is that AUX/IAA becomes the substrate of the SCF^{TIR1} complex, is poly-ubiquitinated and then degraded by the 26S proteasome. The degradation of the AUX/IAA repressor

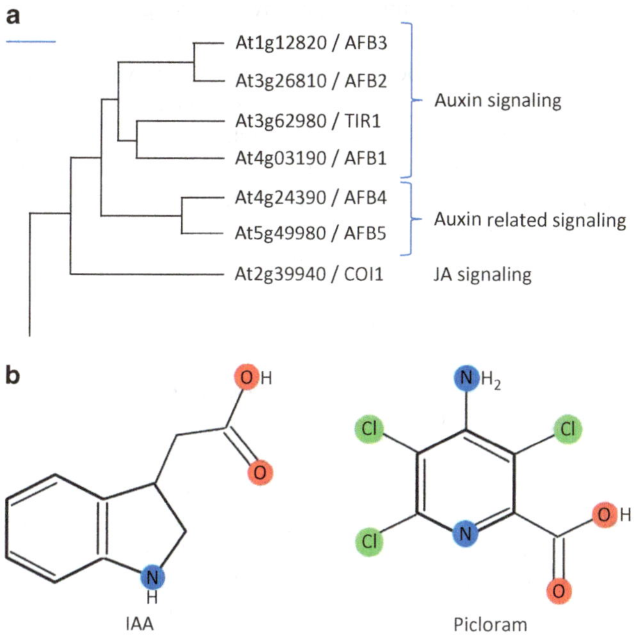

Fig. 2 TIR1/AFB family and two distant auxin compounds. (**a**) Phylogenetic tree showing the subfamily of TIR1 LRR F-box proteins. The bars represent the branch length equivalent to 0.1 amino acid changes per residue (**b**) Indole-3-acetic acid (AIA) and synthetic 4-amino-3,5,6-trichloropicolinic acid (picloram)

then allows ARF-mediated transcriptional activation of auxin response genes. Transcriptome analyses have revealed that a wide variety of genes are responsive to auxin and under the control of ARF transcription factors, which is in accordance with broad cellular and developmental effects of auxin. It is interesting that AUX/IAA genes are some of the early auxin response genes supporting the hypothesis that they are part of a feedback loop restoring transcriptional repression of auxin-induced gene expression.

Increasing Complexity Results Partly from Multigene Families

TIR1 and Auxin-Related F-Box (AFB) Proteins

Within the very large family of F-box proteins, TIR1 belongs to a small subfamily of six members among F-box proteins with leucine-rich repeats (LRR). These six F-box proteins are distributed by pairs between three clades: TIR1 and AFB1, AFB2 and AFB3, and the more divergent AFB4 and AFB5 (Fig. 2a). All these proteins are nuclear proteins and were demonstrated to interact with SKP1-like and to be part of SCF complexes. The closest relative to TIR1/AFB is coronatine-insensitive protein 1(COI1) that is involved in jasmonate signaling. TIR1 was identified from *tir1* allelic mutants isolated by screening for resistance to auxin transport inhibitors and is so far the best characterized F-box of this subfamily. *tir1* mutants were then found to be deficient in a variety of auxin-dependent developmental processes such as initiation of lateral roots, elongation of hypocotyl, or root growth inhibition by exogenous auxin. *tir1* mutants are weakly resistant to the natural auxin IAA but exhibit strong resistance to the synthetic auxin 2,4-dichlorophenoxyacetic acid (2,4-D). The phenotype of null *tir1* mutants is however not severe potentially because of partial gene redundancy with other AFBs. Similarly single *afb* mutants exhibit no or weak phenotypes but far more severe alterations were obtained in multiple *tir1afb2afb3* or *tir1afb1afb2afb3* mutants. The triple or quadruple mutants exhibit an array of phenotypes distributed into three classes (Mockaitis and Estelle 2008). A common feature is a defect in tropic responses, resistance to 2,4D and to a lesser extend to IAA. Mutants of the less severely affected class develop sufficiently well to ensure progeny, whereas most affected mutants are rootless and cannot develop further. *TIR1*, *AFB1*, *AFB2*, and *AFB3* are broadly transcribed in plant tissues but little is known about their transcriptional regulation with the notable exception of an increased expression of *TIR1* in response to phosphate deprivation. However, translational reporters (promoter: TIR1/AFB-GUS) reveal restrictions in the patterns of TIR1, AFB2, and AFB3 expression, indicative of posttranscriptional regulation. Transcripts of these three genes are targets of the microRNA miR393, a conserved microRNA in plants. miR393 mainly restricts expression of *TIR1*, *AFB2*, and *AFB3* to actively growing tissues. TIR1 and AFB2 were shown to be the major auxin-related F-box proteins in roots. Double *afb4 afb5* mutants are not significantly resistant to IAA or 2,4-D but are strongly resistant to the synthetic and auxinic herbicide 4-amino-3,5,6-trichloropicolinic acid (picloram) (Fig. 2b). The natural substrate of AFB4 and AFB5 is unknown. As reported for other F-box proteins, experimental evidence suggests that TIR1 can be itself ubiquitinated and degraded by the 26S proteasome (Stuttmann et al. 2009); it is, however, not clear whether this results from the activity of SCF^{TIR1} and co-degradation with Aux/IAA substrates or if TIR1 is the substrate of another E3 ligase.

AUX/IAAs

AUX/IAA genes were first identified in pea, soybean, or mung bean as rapidly induced genes in response to exogenous auxin in elongating epicotyl or hypocotyl tissues. In Arabidopsis, the AUX/IAA family contains 29 members distributed in 10 clades. The transcriptional response of AUX/IAA genes to auxin is highly variable in kinetics and amplitude with some genes showing no significant response. Whereas most AUX/IAA genes are induced after auxin treatment, large-scale transcriptome analyses revealed opposite responses for other members of the family. Some AUX/IAA also show transcriptional responses to other stimuli, which increases the complexity of their regulation. Most AUX/IAA proteins, but not all, contain four conserved domains I, II, III, and IV, and they are short-lived nuclear proteins. Within the whole family, AUX/IAA proteins vary in a large range regarding their size (18–31 kDa), isoelectric point (acidic 4.5 to basic 9.7), or amino acid identity (10–85 %) suggesting distinct behaviors, interactions, or functions. As already mentioned, domains III and IV are found in most AUX/IAA and ARF. They are involved in homo- or heteromerization impairing ARF-dependent transcriptional activity. Secondary structure similarity was found between domains III and IV and a protein–protein interaction domain named PB1 (for Phox and Bem1) domain (Guilfoyle and Hagen 2012). PB1 domains have acidic and/or basic motifs, and their configuration in domains III and IV leads to front-to-back interactions. Enhanced repression results also from interaction of AUX/IAA domain I with corepressors (see section "Chromatin Structure"). About ten mutants affected in developmental processes (as *short hypocotyl shy2/iaa3, bodenlos bdl/iaa12*, or *solitary root slr/iaa14*) or identified as resistant to auxin (as *auxin-resistantaxr2/iaa7, axr3/iaa17, axr5/iaa1, or massugu2 msg2/iaa19*) were found to be AUX/IAA gain-of-function mutants (Mockaitis and Estelle 2008). In each case, the mutation was a single amino acid change in domain II, which contains critical residues for interaction with TIR1/AFB F-box proteins. Mutations in these critical residues of domain II result in increased AUX/IAA protein stability because they impair the recruitment of the mutant proteins to $SCF^{TIR1/AFB}$ complexes and compromise their identification as substrates for degradation by the 26S proteasome. The elevated accumulation of mutated AUX/IAA protein enhances transcriptional repression by shifting protein interaction in favor of repressive mAUX/IAA–ARF. The precise function of each AUX/IAA is, however, difficult to determine in such unbalanced background. However, a large range of phenotypes was reported for different AUX/IAA gain-of-function mutants, potentially due to differences in their expression patterns and/or preference of interactions with other interacting proteins suggesting partial functional differences. Characterization of single *aux/iaa* loss-of-function mutants revealed no or only subtle phenotypes. Combination of double or triple mutants within a clade was even not always sufficient to reveal phenotypes, suggesting broad overlapping expression and gene redundancy.

TIR1/AFB–AUX/IAA Co-receptors

TIR1 was first claimed to be an auxin receptor but all auxin binding experiments were performed in the presence of AUX/IAA proteins or a synthetic peptide corresponding to the consensus sequence of AUX/IAA domain II. Recent analyses revealed that TIR1 alone is poorly able to bind auxin (Calderon Villalobos et al. 2012) and that at least a minimal AUX/IAA domain II is required together with TIR1 for significant auxin binding. It is now clearly established that both proteins are required for auxin binding and that pairs of TIR1/AFB and AUX/IAA form auxin co-receptors. Resolution of the 3-D structure revealed that auxin acts as a "molecular glue" facilitating or enhancing TIR1 and AUX/IAA interaction. Higher-affinity auxin binding is obtained with full-length AUX/IAA proteins (instead of their domain II), suggesting that other domains of AUX/IAA proteins are also involved in the interaction with TIR1, at least to stabilize the interaction (Pierre-Jerome et al. 2013). Yeast two-hybrid experiments showed that there are AUX/IAA substrate preferences among the different TIR1/AFB proteins. Some AUX/IAA proteins, such as IAA8, are good substrates for all TIR1/AFB proteins whatever the concentration of IAA, whereas others, as IAA3, IAA20, or IAA29, either do not interact with some TIR1/AFBs or do not interact at all. IAA binding assays performed in the presence of TIR1 and different AUX/IAA proteins revealed huge affinity differences with Kds ranging from 10 nM for TIR1-IAA14 to 217 nM or no relevant value for TIR1-IAA12 and TIR1-IAA31, respectively. For one co-receptor, there are also binding differences for various auxins. For example, the TIR1-IAA7 Kd for IAA is 17 nM, but drops down to 248 nM for 2,4-D and 3,900 nM for picloram. In other words, the auxin compound influences the interaction between the TIR1/AFB and the AUX/IAA. In this complex equation, the F-box is also an essential variable as illustrated with AFB5-IAA7 co-receptor that binds picloram 100 times better than TIR1-IAA7.

From the rather simple auxin signaling module described (Fig. 1), the number of TIR1/AFB and AUX/IAA proteins and resulting combinations of low- to high-affinity co-receptors provide a much more complex view of the auxin signaling pathway. Various combinations of TIR1/AFB and AUX/IAA proteins can be co-expressed in different tissues indicating that the relative amount of each component determines the functional modules that can be formed in a cell. The downstream response relies on the subtle balance of protein–protein interaction together with variation in auxin or auxin-related compounds.

ARFs

ARF genes also belong to a multigene family of 23 members in Arabidopsis. The first ARF, ARF1, was identified by yeast one-hybrid experiment as a transcription factor binding palindromic copies of TGTCTC auxin response element in the promoter of the *gretchen hagen3* (*GH3*) gene. ARF genes are divided into three distantly related phylogenetic branches, the largest one being subdivided into three subclasses. ARFs from the different groups are functionally divergent even if they are able to bind rather conserved AuxRE. Most ARFs exhibit a conserved structure with an N-terminal DNA-binding domain (DBD), followed by a middle region (MR) conferring preferential activation or repression activities and preceding protein–protein interaction domains III and IV. The DBD is a B3-type DNA-binding domain related to VP1/ABI3 (maize viviparous 1/Arabidopsis ABA-insensitive 3) (Guilfoyle and Hagen 2007). It was shown that ARF DBD, alone, is sufficient for the binding to AuxRE. Recent data based on the resolution of the 3-D structure of the DBD of two distant ARFs, ARF1 and ARF5, identified the residues directly interacting with DNA and revealed that they are strongly conserved within the ARF family (Boer et al. 2014). In addition, an elegant combination of surface plasmon resonance and protein binding microarrays confirmed that ARF1 and ARF5 bind the same DNA motif and revealed that binding properties rely on cooperative binding of homodimers differing from one ARF to the other by spacing preferences between inverted repeats of the AuxRE (Boer et al. 2014). The binding competition between ARF1 and ARF5 occurs for a spacing of 7 or 8 nucleotides between two binding sites, whereas ARF5 is more permissive and still binds motifs in distants from 5 to 9 nucleotides (Fig. 3a). Monomers can bind AuxRE but with reduced efficiency. A stronger binding of ARF1 and ARF5 was found for

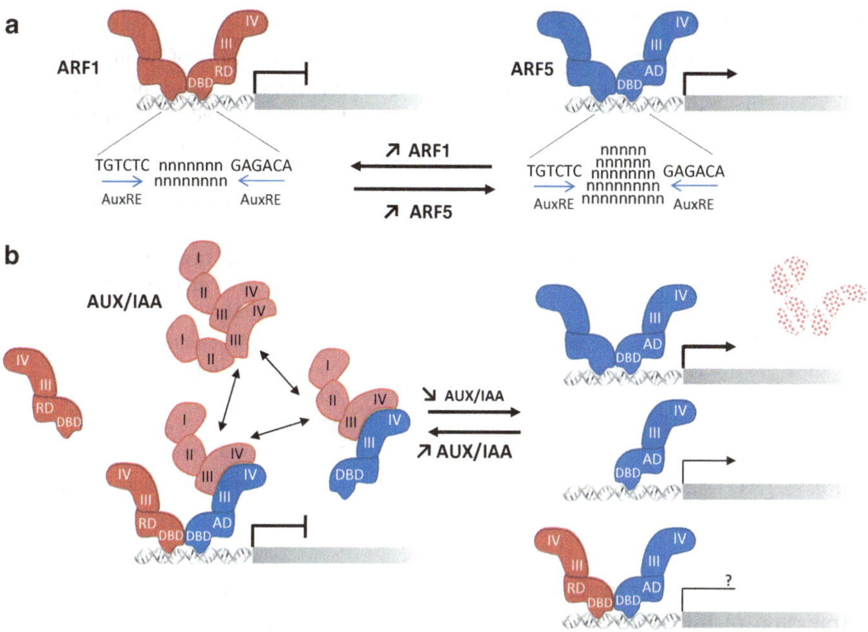

Fig. 3 Enhances complexity in the model of transcriptional regulation. (**a**) DNA-binding domains of ARF proteins promote ARF dimerization. ARF activator (in *blue*) and ARF repressor (in *red*) compete for similar auxin-responsive elements (AuxRE); the distance between neighboring AuxREs influences binding efficiency (**b**) Transcriptional regulation of auxin response genes involves subtle protein–protein interaction balances between ARFs and AUX/IAA repressors. Repression can result from various combinations of repressor proteins. Derepression results from degradation of AUX/IAA repressors and transcriptional activation mediated by ARF activators acting as monomers, dimers, or potentially heterodimers

TGTCGG instead of the core TGTCTC motif, but multiple variants of AuxRE were also found in promoters of identified ARF target genes suggesting a whole range of low- to high-affinity DNA-binding sites. Apart from the spacing between AuxRE motifs and variations in AuxRE sequences, the relative abundance of distinct ARFs co-expressed within the same cell is also critical for downstream responses.

Interestingly, a distinct set of *ARFs* is expressed in different cell types in the embryo, in the primary root, and probably in other developing tissues. Up to seven different *ARFs* were found to be co-expressed in specific embryonic cells, whereas only two were expressed in neighboring cells. The complex ARF pattern is likely to be a contributing factor to cell-type-specific responses to auxin (Wendrich and Weijers 2013). To date little is known about the transcriptional regulation of *ARFs*; expression of most of them is independent of auxin with the exception of *ARF4* and *ARF19*. In addition to differential transcriptional regulation of *ARF* genes, post-transcriptional regulation has been reported for various ARF members. miR160 and miR167 were found to target *ARF 10, 16*, and *17* and *ARF 6* and *8*, respectively. In addition, miR390 triggers the cleavage of *TAS3* (a trans-acting short-interfering RNA (tasiRNA)) precursor that further targets *ARF2, 3*, and *4* and then reduces the

Table 1 Null mutations in ARF genes

Gene	MR	Mutant	Main phenotypes of the mutant
ARF2	RD	*suppressor of hookless1(hss1) arf2*	Partial restoration of apical hook of *hsl* – agravitropic stem, partial sterility, dark green leaves, and delayed flowering time
ARF3	RD	*ettin (ett)*	Alterations of abaxial tissue identity in the gynoecium
ARF5	AD	*monopteros (mp)*	Severe defects in embryo axis formation, vascular development, rootless seedlings
ARF7	AD	*non-phototropic hypocotyl 4 (nph4)/transport inhibitor response 5(tir5)/massugu1 (msg1)*	Defects in differential growth associated to tropisms in shoot – reduced sensitivity to auxin and ethylene
ARF8	AD	*arf8*	Parthenocarpic fruits
ARF19	AD	*arf19*	Reduced sensitivity to auxin and ethylene

amount and activity of the corresponding proteins. These ARFs are required for the expression of miR390, and miR167 is induced by auxin indicating some sort of feedback mechanism. The timing and pattern of expression of these mi/tasiRNAs determine their influence on ARF targets; for example, expression of miR390 and production of tasiR-ARFs were shown to specifically downregulate ARF targets for leaf patterning or lateral root initiation (Sanan-Mishra et al. 2013).

In simplified models of transcriptional regulation in response to auxin, ARFs are usually represented as transcriptional activators. This activity is however not representative of the whole family. ARF5–8 and 19 that form one of the ARF class mainly function as activators. This class is characterized by a middle region enriched in glutamine, serine, and leucine residues behaving as an activation domain (AD). Conversely, other ARFs exhibit a middle region enriched in serine, proline, leucine, and glycine residues acting as a repression domain (RD) (Guilfoyle and Hagen 2007). For example, certain molecular and genetic studies support the view that ARF1 and ARF5 have opposite activities and act antagonistically when they are co-expressed (Fig. 3). Although other studies confirm that ARF1 is generally a repressor and ARF5 an activator, there activities are not always that contrasted. The classification of different ARFs as activators or repressors has been mainly investigated using transient expression assays in cultured cells or leaf mesophyll protoplasts, and even if it is roughly true, recent data suggest that opposite functions can occasionally occur according to specific cell types or upon environmental stimuli (Del Bianco and Kepinski 2011).

Several *arf* mutants have been identified in various genetic screens. These mutants are either hypomorphic or null alleles exhibiting an array of growth and developmental defects in agreement with the prominent role of ARFs in auxin signaling (Table 1). More systematic forward genetic screens have resulted in the identification of T-DNA insertions for most of the other ARF genes although no or only weak phenotypes were found for these single mutants under standard growth conditions. This is presumably due to partial gene redundancy at least between close relatives.

AUX/IAA–ARF Interaction

AUX/IAA and ARF form regulatory modules for transcriptional regulation of downstream genes. In theory, the conserved modular structure of AUX/IAA and ARF should promote the formation of a large number of AUX/IAA–ARF pairs. However, stage- and cell-specific expression of individual AUX/IAAs and ARFs limits the number of combinations that can be formed in vivo. In addition, sequence analyses, yeast two-hybrid experiments, and expression in protoplasts suggest no or weak interactions between ARF repressors and AUX/IAAs which reduces to five, the number of ARFs forming regulatory modules with expressed AUX/IAA (Guilfoyle and Hagen 2007). Functional regulatory modules were identified through the characterization of various mutants and analysis of specific developmental stages or localized auxin responses. For example, the *monopteros mp/arf5* null mutant and *bodenlos bdl/iaa12* gain-of-function mutant exhibit similar developmental defects with altered embryo patterning and defects in the initiation of the meristematic root pole. Further analysis showed that MP and BDL are co-expressed during early stages of embryogenesis. MP accumulates in embryonic cells and is essential for hypophysis specification in response to auxin via the regulation of transcription factors identified as target of monopteros (TMO). Another example involves the two partially redundant *ARF* genes, *NPH4/ARF7* and *ARF19*, and *SLR/IAA14* and also *MSG2/IAA19* genes that were shown to function together in the auxin-regulated initiation of lateral roots and differential hypocotyl growth upon tropic stimuli (Del Bianco and Kepinski 2011; Guilfoyle and Hagen 2012). Many other combinations still have to be identified with the resolution of cell specificity as well as their downstream targets.

Downstream Auxin Response Genes

Historically, auxin response genes covered three main gene families of genes that are rapidly induced after auxin treatment: *SAUR* (small auxin upregulated) genes, the function of which is still unclear; *GH3* genes that encode acyl adenylate-forming isoenzymes involved in hormone conjugation; and AUX/IAA repressors. Large-scale transcriptome analyses were performed on a broad range of experimental conditions varying on the nature of the auxin applied, the concentration, the kinetics, and the plant material (whole seedlings, mutants, organs, and even specific cell types). Altogether a vast number of genes were found to be differentially expressed after exogenous application of auxin, with some core genes always affected whatever the conditions and some others varying in a more subtle tissue- or even cell-type-specific manner. Transcription factors belonging to large gene families as lateral organ boundary domain containing protein (LBD), NAC domain containing protein, basic helix-loop-helix (bHLH), or MYB are found among the core genes. Expression of many cell wall remodeling genes also appears to be rapidly regulated in response to auxin. Recent spatial analysis of auxin-dependent gene expression confirmed a strong context-dependent impact of their transcriptional regulation, consistent with a cell-type-specific regulatory network (Bargmann et al. 2013).

Additional Regulators of the TIR1/AFB–AUX/IAA Pathway

Ins6P

Resolution of the crystal structure of TIR1-ASK1 complex revealed a tight association of TIR1 with inositol hexakisphosphate (InsP6, also named phytate). Positively charged residues interacting with InsP6 are strictly conserved between TIR1 and AFBs suggesting that InsP6 acts as a cofactor for the TIR1/AFB subclass of F-box proteins. Yeast 2-hybrid experiments showed that substitution of these residues results in weakening or abolishment of the interaction with ASK1 and also prevents interaction with AUX/IAA substrate even in the presence of auxin. InsP6 was thus proposed to exert a critical role in TIR1 structure. While essential it is however not clear whether InsP6 is a regulatory component modulating in vivo the TIR1/AFB-dependent pathway.

Posttranslational Modifications

Posttranslational modification of substrates of ubiquitin E3 ligases has often been reported to promote or prevent their recruitment for ubiquitination. For TIR1/AFB and AUX/IAA interaction, auxin binds these proteins together, and the question is whether posttranslational modification of AUX/IAA substrates or TIR1/AFB F-box proteins is modulating their capacity to interact in the presence or the absence of auxin. Several pieces of evidence support that posttranslational modification (s) affects TIR1/AFB–AUX/IAA interaction, at least in vitro. With short-lived proteins, validating such modification(s) in vivo is a difficult challenge.

NO

Changes in the levels of nitric oxide(NO) has been reported in response to auxin. In addition, NO was shown to enhance auxin-mediated degradation of AUX/IAA proteins by $SCF^{TIR1/AFB}$ suggesting an interference by NO in this pathway. Like InsP6, NO is known to efficiently bind metal ions but it can also act as a second messenger either by its involvement in chemical reactions or by nitration or S-nitrosylation of peptides or proteins (Gross et al. 2013). S-nitrosylation is a posttranslational protein modification resulting from the reversible binding of NO to the thiols of cysteine residues. In vitro experiments demonstrated that TIR1 or AFB2 can be S-nitrosylated and that S-nitrosylation increases their interaction with AUX/IAA targets in the presence of auxin. Interestingly, a TIR1 protein exhibiting a mutation on C140 was less efficient in complementing a *tir1-1* mutant than TIR1, supporting the possible involvement of TIR1 S-nitrosylation in vivo to modify TIR1-AUX/IAA interaction and more broadly auxin signaling (Freschi 2013).

Proline Hydroxylation

The conserved core motif of AUX/IAA domain II includes two consecutive proline residues (GWPPV), known from analysis of *aux/iaa* mutants to be critical for interaction with TIR1/AFB. In vitro immunoprecipitation assays using synthetic peptides encompassing domain II of IAA7 showed that replacement of a proline by a hydroxyproline severely reduces its interaction with TIR1; however, auxin enhanced the interaction. Proteomic analysis of plant AUX/IAA proteins has not been done to confirm or invalidate the significance of such posttranslational modification.

Phosphorylation

Posttranslational modifications of AUX/IAA by phosphorylation have also been speculated. In vitro assays suggest that various AUX/IAA proteins and truncated proteins corresponding to domains I and II can be phosphorylated by recombinant phytochrome A. The possible effect of AUX/IAA phosphorylation on protein stability or interaction with TIR1/AFB has not been investigated. As for hydroxylation, in vivo data are missing to further support this data.

By investigating the synergistic role of auxin and brassinosteroid (BR), the BR-regulated kinase BIN2 (also reported as a glycogen synthase kinase3 (GSK3)) was shown to regulate DNA-binding and transcriptional repression activity of ARF2. This effect might result from a phosphorylation of ARF2 mediated by BIN2. BIN2-dependent phosphorylation of ARF2 still needs to be confirmed in plants (Vert et al. 2008). More recently, BIN2 was shown to directly phosphorylate ARF7 and ARF19 during lateral root formation. Phosphorylation of ARF7 and ARF19 decreases their interaction with AUX/IAA repressors and then increases transcriptional activation of downstream genes. In this example, it was suggested that BIN2-mediated phosphorylation of ARF7 did not occur in response to BR but to a CLE41-CLE44 peptide/receptor signaling system (Cho et al. 2014).

RAC/ROPs

Rho-like small GTPases of plants (RAC/ROPs) are versatile proteins switching from inactive and cytosolic to active and associated to the inner face of the plasma membrane upon GDP to GTP binding, respectively (Fig. 4). They are involved in a large number of regulatory mechanisms. There are 11 members in the RAC/ROP family in Arabidopsis that exert partially overlapping or divergent functions depending on specific interacting and effector proteins (Craddock et al. 2012). Expression of a constitutively active (CA) RAC/ROP GTPase in tobacco or Arabidopsis protoplasts promotes AUX/IAA degradation in the absence of exogenous auxin, whereas expression of a dominant negative (DN) RAC/ROP GTPase prevents auxin-mediated degradation of AUX/IAA. These data suggest that RAC/ROPs are implicated in the regulation of the TIR1/AFB–AUX/IAA signaling pathway. Moreover, RAC/ROPs were shown to affect nuclear co-relocalization of AUX/IAA (IAA17), TIR1, and proteins of the UPS to nuclear protein bodies (NPB) in response to auxin. These NPBs are likely to be the sites of protein degradation.

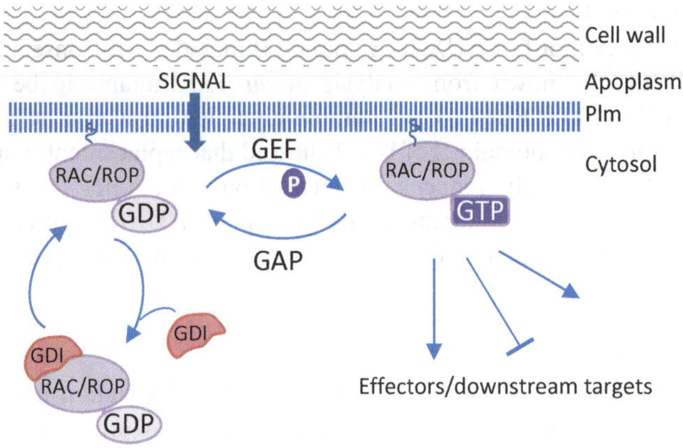

Fig. 4 **Regulatory cycle of RAC/ROP GTPase activity.** RAC/ROP GTPases shuttle between an inactive form bound to GDP and an active form bound to GTP. Posttranslational modifications anchor RAC/ROP proteins to the inner face of the plasma membrane. Association of RAC/ROP-GDP with guanine nucleotide dissociation inhibitors (GDI) results in sequestration of inactive RAC/ROP in the cytosol. RAC/ROP activation is mediated by guanine nucleotide exchange factors (GEF) in response to perception of a signal at the plasma membrane. Activated RAC/ROP-GTP binds to downstream effectors that act as positive or negative regulators of their respective targets. RAC/ROP GTPases are thus versatile actors in signaling cascades. Reciprocal conversion involves GTPase-activating proteins (GAPs) that promote GTP hydrolysis to GDP

Based on these data, inactive RAC/ROPs favor repression whereas active RAC/ROP GTPases promote AUX/IAA degradation and derepression. Molecular downstream targets of RAC/ROPs on AUX/IAA half-lives are still unknown. Auxin activates at least some RAC/ROP GTPases, thus correlating their functional activation with increased expression of auxin response genes (Nibau et al. 2006; Wu et al. 2011). This effect is mediated by the auxin binding protein 1 (ABP1), a protein identified more than 40 years ago on the basis of its capacity to bind auxin (Tromas et al. 2010; Sauer and Kleine-Vehn 2011).

ABP1 Pathway

Analysis of null mutant and plants conditional for the function of ABP1 revealed that ABP1 is involved in the control of a broad range of plant growth and developmental processes (Tromas et al. 2010; Sauer and Kleine-Vehn 2011). Moreover, recent data based on both genetic and molecular approaches established that major developmental defects observed in plants knockdowned for ABP1 result from altered transcriptional responses. Independent of its capacity to bind auxin, ABP1 protein acts as a negative regulator of the TIR1/AFB–AUX/IAA pathway, preventing AUX/IAA repressors from being degraded. Interestingly this effect is independent of ABP1's involvement in modulating endocytosis (Tromas et al. 2013). By stabilizing AUX/IAA, ABP1 buffers the responsiveness of the TIR1/AFB–AUX/IAA pathway and/or resets the system after responding to auxin. Thus, ABP1 is essential in reestablishing transcriptional repression.

Unlike TIR1/AFB, AUX/IAA, and ARFs, ABP1 does not belong to a multigene family. In Arabidopsis, there is only one gene and this gene is essential throughout development. ABP1 is not a nuclear protein and is unlikely to interact directly with AUX/IAAs or with the TIR1/AFB-dependent UPS. Therefore, the protein likely influences AUX/IAA stability through a signaling system. ABP1 was found in the ER and at the plasma membrane. The relative proportion between these two compartments varies in huge proportions between maize, from which ABP1 was first identified and which is a natural overexpressor that accumulates ABP1 in the ER, and other flowering plants which express ABP1 at a low level and show far more balanced partitioning as observed after membrane fractionation and proteomic analyses. Overexpression of ABP1 in tobacco or Arabidopsis leads to an increase in the amount of the protein in the ER where it is retained via its C-terminal KDEL sequence. Conversely, overexpression of a truncated form of ABP1 lacking a KDEL sequence or with additional amino acid residues after the leucine is not retained in the ER. No or subtle phenotype was reported for plants overexpressing the form of ABP1 retained in the ER; however, the truncated form lacking the KDEL sequence showed severe developmental defects or was lethal. This suggests that targeting ABP1 to the plasma membrane must be tightly controlled. It is interesting to note that the presence of the ER retention signal KDEL at the C-terminus of ABP1 is a recent evolutionarily acquisition as it is only found in flowering plants, whereas ABP1 itself is an ancient protein present in algae. The way in which ABP1 influences the TIR1/AFB–AUX/IAA pathway is still unknown, but it may act through intermediates such as RAC/ROP GTPases (Nibau et al. 2006; Wu et al. 2011) (see the above section).

Chromatin Structure

Domain I of AUX/IAA contains an ethylene response factor-associated amphiphilic repression motif (EAR-motif) promoting interaction with topless (TPL) or related proteins (TPR) that act as corepressors. These proteins function as adaptor proteins recruiting histone deacetylases (HDACs) to the target gene locus. By interacting with domain I of AUX/IAA, TPL mediates transcriptional repression by promoting histone deacetylation which results in a more compact chromatin structure impairing the accessibility of transcription factors to their DNA target (Fig. 5) (Ma et al. 2013).

Mutations in the chromatin remodeling factor PICKLE (PLK) suppress the phenotype of a *slr-1* mutant that expresses a mutated and stable IAA14 protein repressing the activity of ARF7 and ARF19 during lateral root initiation. PLK is a predicted ATP-dependent chromatin-remodeling chromodomain-helicase-DNA-binding protein3 (CHD3) acting either as part of the nucleosome remodeling deacetylase complex enhancing transcriptional repression or in combination with histone acetyltransferases to favor transcriptional activation. Suppression of *slr-1* by *plk* suggests the involvement of PLK in transcriptional repression during lateral root initiation; conversely PLK was reported to act antagonistically to polycomb group proteins in the control of root meristem identity and activity (Aichinger et al. 2011).

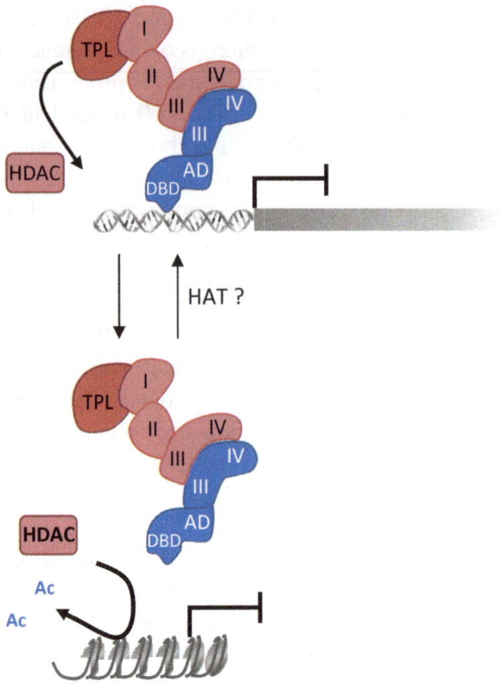

Fig. 5 Additional repression by chromatin compaction. Transcriptional repression can be enhanced by interaction of AUX/IAA with additional corepressors such as Topless (TPL). TPL recruits histone deacetylases to the locus, thus promoting histone deacetylation and increased compaction of the chromatin. After compaction, AuxREs are no longer accessible to transcription factors. Involvement of histone acetyltransferases (HAT) should be required to reverse this effect and render DNA more accessible to ARFs

Other Ubiquitin-Dependent 26S Proteasome Modules

TIR1/AFB are not the sole F-box proteins involved in the control of protein degradation in response to auxin. Auxin is a critical regulator of cell division (Perrot-Rechenmann 2010), a complex and tightly orchestrated process requiring appropriate posttranslational modifications (mainly phosphorylation), protein degradation via UPS, and the regulation of transcription at successive phases of the cell cycle. SKP2A F-box protein was shown to target E2Fc and DPB cell cycle repressors for degradation by UPS. SKP2A binds auxin and the presence of auxin increases the interaction between SKP2A and DPB. Auxin also promotes the degradation of SKP2A but it is not clear whether it results from the activity of SCF^{SKP2A} E3 ligase or from another E3 ligase targeting specifically SKP2A (Del Pozo and Manzano 2013). These findings demonstrate that modules other than TIR1/AFB–AUX/IAA are involved in auxin-dependent transcriptional responses.

Non-transcriptional Membrane and Cytosolic Responses to Auxin

In addition to transcriptional regulations, auxin also generates rapid and short-term responses occurring within seconds or minutes after an auxin stimulus. These responses correspond to electrogenic transport processes and rapid activation of RAC/ROP GTPases that function as molecular switches in signaling cascades.

Sustained modifications in these responses are more difficult to interpret as they can result also from induced changes in gene expression.

Membrane Potential

Auxin affects the activity of ion channels in various cell types in a dose-dependent manner. Changes in ion transport across the plasma membrane influence ion gradients and membrane potential. In various plant tissues, auxin induces a transient depolarization of the plasma membrane followed by a more prolonged hyperpolarization. These variations in membrane potential result from complex ion fluxes including protons, K^+, and anions. In guard cells, transient auxin-induced membrane depolarization was proposed to result from the activation of a nucleotide-dependent anion channel (Becker and Hedrich 2002). Activation of Ca^{2+} channels and remobilization of cytosolic Ca^{2+} have also been reported in response to auxin stimulus. In addition, the plasma membrane proton pump (H^+ ATPase) is activated as part of the response and precedes cell elongation at least in hypocotyls (Becker and Hedrich 2002). More recently, it was reported that auxin activation of the plasma membrane proton pump results from its phosphorylation and binding of a 14-3-3 protein (Takahashi et al. 2012). Proton extrusion modifies the local environment of cell wall proteins contributing to remodeling of the cell wall's cellulose–hemicellulose network. Paradoxically, cytosolic acidification was also observed, due to proton release from the vacuole. Potassium channels play important roles in the response to auxin. Auxin-induced activation of K^+ inward channels promotes K^+ uptake which facilitates water uptake and expansion. Conversely, high concentrations of auxin that are inhibitory for cell expansion also activate K^+ outward channels and anion channels. ABP1 was shown to be involved in the modulation of these electrogenic responses. Partial responses were obtained by adding exogenous ABP1 or synthetic peptides corresponding to the C-terminus of ABP1 indicating that modulating the relative abundance of ABP1 at the plasma membrane is sufficient to induce at least part of the responses in the absence of auxin stimulus (Tromas et al. 2010; Sauer and Kleine-Vehn 2011).

ABP1 and Interacting Protein(s)

ABP1 is associated to the outer face of the plasma membrane but it is not an intrinsic membrane protein. Biochemical analyses have demonstrated that ABP1 is localized in membrane fractions and not in soluble fractions indicating that there is no free and soluble ABP1. Its association with membrane fractions as well as its targeting from the ER to the plasma membrane probably involves protein–protein interactions. Two classes of proteins were identified as potential interactors of ABP1. The first one encodes a glycosylphosphatidylinositol (GPI)-anchored protein named C-terminal peptide-binding protein 1 (CBP1) in maize. As indicated by its name, the protein was identified as an interacting protein of a synthetic peptide corresponding to the C-terminal domain of ABP1. There is still no experimental evidence showing that ABP1 binds to CBP1 in vivo but such an interactor might be a good carrier for targeting ABP1 to the outer face of the plasma membrane. Addition of the GPI tail at the C-terminus of a protein is a posttranslational modification, occurring in the ER, by replacement of the hydrophobic C-terminus sequence by the phospholipid. First anchored in the inner face of the ER membrane, GPI-anchored proteins traffic through the Golgi and by exocytosis are exposed at the outer face of the plasma membrane. However, GPI anchors are not transmembrane moieties and cannot serve in transmitting signals from the outer to the inner faces of the membrane. Very recently, another ABP1 interacting candidate was identified among members of the receptor-like kinase (RLK) family (Xu et al. 2014). The extracellular domain of a transmembrane kinase (TMK) was reported to interact with ABP1 in an auxin-dependent manner. A subfamily of four TMK members was shown to be required for auxin-mediated activation of RAC/ROP GTPase, a response also requiring functional ABP1. Such RLKs are good candidates to form functional complexes with ABP1 because they bridge the outer and inner surface of the plasma membrane and are usually in contact with the cytosol and downstream components of signaling pathways. However, the phenotype of the quadruple *tmk* mutant differs from ABP1 loss-of-function mutants suggesting that other proteins might be involved.

RAC/ROP GTPase Auxin Signal Transducers

Auxin activates RAC/ROP GTPases in various cell systems, and the downstream effects of their activation largely depend on the combination of RAC/ROP and effectors involved (Fig. 6). There are 11 members of Rho-like small G proteins and 11 RAC/ROP-interactive CRIB motif-containing protein (RIC) effectors in *Arabidopsis* (Yang 2008). Up to now, most experimental data relative to the effect of auxin focused on ROP2 and ROP6 and their effectors RIC4 and RIC1 (Nibau et al. 2006; Wu et al. 2011). In cotyledon pavement cells, ROP2-RIC4 and ROP6-RIC1 pathways were shown to be activated within minutes after application of exogenous auxin and in a dose-dependent manner. ABP1 was required for this activation and, as mentioned above, TMK was recently reported to contribute to this response (Xu et al. 2014). Both pathways were required for coordinated expansion of pavement cells resulting in interdigitations between adjacent cells. ROP2 promotes lobe outgrowth by activating RIC4 which affects F-actin dynamics, whereas ROP6 reduces growth by promoting RIC1 interaction with cortical microtubule (CMT) ordering. RAC/ROPs are tightly associated with growth and modulation of cell polarity (Yang 2008). In root cells, auxin and ABP1-mediated activation of ROP6 and RIC1 were shown to stabilize F-actin and to inhibit clathrin-dependent endocytosis of the efflux carrier PIN2. The activation of ROP6 might occur through the action of spike (SPK1), a ROP-GEF that interacts with ROP6 and was identified in a suppressor screen of a ROP6 overexpressor. The binding of auxin to ABP1 activates RAC/ROPs and inhibits endocytosis (Sauer and Kleine-Vehn 2011).

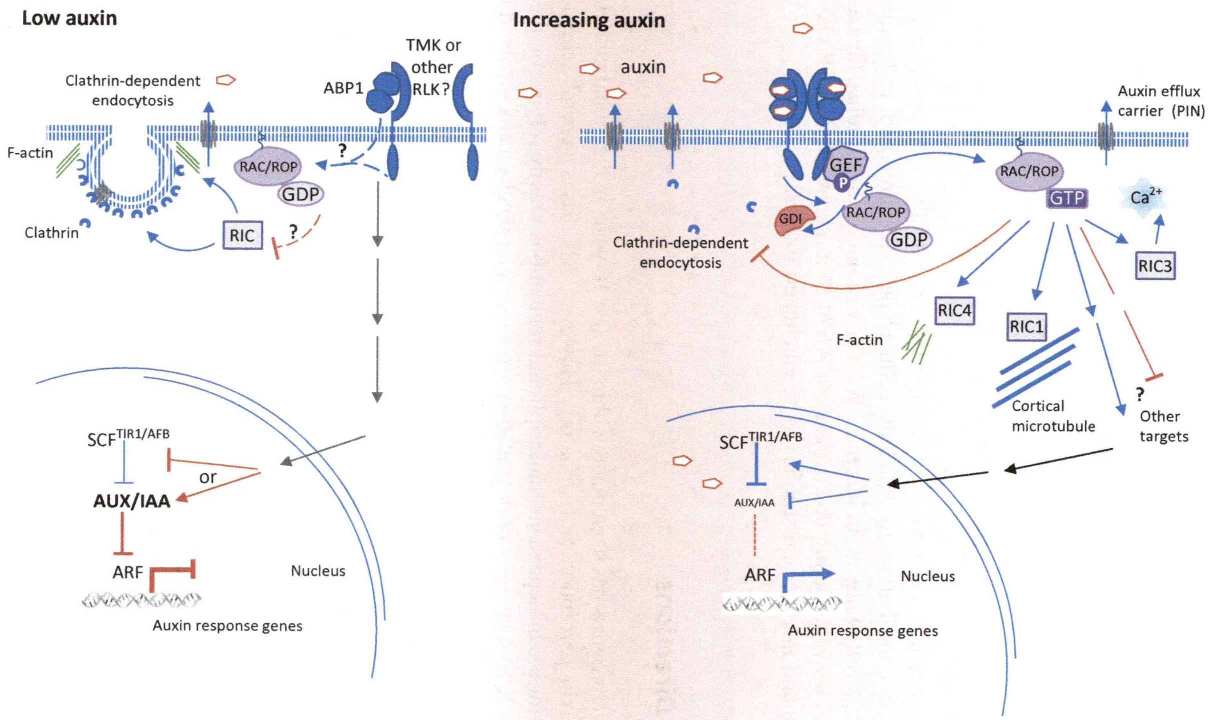

Fig. 6 Tentative model of auxin responses involving ABP1 and RAC/ROP GTPases. Under low auxin conditions, intrinsic plasma membrane proteins as auxin efflux carriers of the PIN family are recycled by clathrin-dependent endocytosis. In these conditions, RAC/ROP are predominantly found as inactive

Conversely, in the absence of auxin stimulus, ABP1 act positively on clathrin-dependent endocytosis of various plasma membrane proteins, especially PIN proteins, potentially via inactivated GDP-bound RAC/ROPs. The hypomorphic allele *abp1-5*, with a point mutation in a histidine residue in the auxin-binding pocket which is hypothesized to impair the binding of auxin to ABP1, failed to inhibit endocytosis in the presence of auxin. The weak phenotype of this mutant, compared to conditional knockdown plants, argues that ABP1 principally functions in the absence of an auxin signal. These data also suggest that ABP1-dependent modulation of RAC/ROPs influences cytoskeleton organization and dynamics and controls the recycling of at least a subset of plasma membrane proteins. The reduction in endocytosis in response to auxin leads to greater retention of PIN proteins on the plasma membrane which favors more auxin efflux. An increased efflux along with auxin conjugation, catabolism, and targeting to subcellular compartments is a mechanism contributing to efficient cellular detoxification and restoration of homeostasis following auxin increase. With or without auxin stimulus, the cascade of molecular events acting downstream of ABP1 to modulate RAC/ROP cycles of activation and inactivation still has to be elucidated further.

Future Directions

Impressive progress has been made within the last 10 years in the global understanding of auxin perception and signaling. However, whereas a general scheme has emerged for auxin-dependent transcriptional regulations mainly relying on the control of AUX/IAA stability or degradation, the molecular mechanisms involved in the regulation of this pathway in time and space still have to be elucidated. Multiple levels of regulation are reported in this review, including a subtle quantitative balance between activators and repressors, the impact of the ABP1-dependent pathway on AUX/IAA stability, post-transcriptional and post-translational modification of key regulators, and chromatin remodeling. Future work will have to identify all the actors involved in the regulation of TIR1/AFBAUX/IAA pathways and to determine the sequence of molecular events buffering and controlling these pathways in a specific

Fig. 6 (continued) proteins, bound to GDP and not interacting with their RIC effectors (RAC/ROP-interactive CRIB motif-containing proteins). ABP1 is required for clathrin-dependent endocytosis via weak interactions with the extracellular domain of receptor-like kinases (RLKs) of the TMK subfamily, but downstream events promoting endocytosis are still unclear. ABP1 also acts as a negative regulator of the $SCF^{TIR1/AFB}$ pathway, promoting AUX/IAA stability. In the presence of increasing amounts of auxin, the binding of auxin to ABP1 enhances its interaction with RLK, which results in transduction of the signal inside of the cell. RAC/ROP GTPases are thus activated which promote an array of downstream responses according to the interaction specificities between RAC/ROPs and their effectors. RAC/ROP GTPases are molecular switches that emerge as key regulators of auxin signaling downstream of ABP1

tissue or cellular context. Identification of the molecular components acting downstream of ABP1 to modulate AUX/IAA stability will be an important challenge. One of the difficulties that researchers have to face is associated with the central role of auxin signaling in an extremely broad array of responses. Spatial and temporal resolutions are thus required to dissect specific and context-dependent responses. For each cellular context, it will be necessary to identify transcription factors acting on the expression pattern of auxin-signaling components, to resolve combinatorial regulations, and to further investigate the causal relationship between modifications of the expression of target genes and cellular responses. An additional level of complexity results from crosstalks with other hormones acting either synergistically or antagonistically with auxin. Considering the increasing complexity of the regulation of auxin-dependent transcriptional responses, specific questions might be addressed using simplified experimental systems. Synthetic biology aiming at reconstituting, in a qualitative and quantitative manner, part of a pathway in a heterologous system is emerging as a promising approach to improve our understanding of combinatorial regulations. This will not resolve the overall complexity of the regulations of these modules, but should provide biochemical data on the specificity and dynamics of interaction between proteins and with DNA or the displacement of interactions under various inputs. In plants, explorations remain inescapable, and developmental processes that are limited in time and space provide favorable contexts with partially reduced *in vivo* complexity. Mathematical modeling of auxin transcriptional modules and their behavior upon challenges will also be of great help to integrate all experimental data, identify discrepancies or gaps, and raise testable hypotheses. Models that can be implemented with novel data and evolve accordingly will be of invaluable interest. Nontranscriptional responses to auxin also remain to be further investigated. Possible relations between fast electrogenic responses and more recently reported protein-trafficking modifications or cytoskeleton organizations have not been investigated yet. Whether they result from sequential processes, largely dependent on each other, or diverge at some point of a common signaling cascade is not known yet and will need to be determined. There are still question marks concerning ABP1 interacting proteins either in the control of the targeting of the protein to the plasma membrane or in the formation of a functional receptor complex anchored in the plasma membrane and connecting the apoplastic face to the cytosolic face of the membrane. RAC/ROP GTPases and their effectors appear to be promising key regulatory components; further elucidation of their role in nontranscriptomic responses to auxin as well as in the regulation of AUX/IAA stability will consolidate existing data and will lay the foundations to shed light upon auxin-signaling mechanisms.

For a good understanding of auxin signaling and downstream responses, there is still a missing critical parameter for which there is no satisfactory method of quantification at the cellular and subcellular scales: That is auxin itself. In the last 10 years, sensitive mass spectrometry–based methods were developed to analyze auxin precursors and free IAA and conjugates/metabolites in rather small amounts of tissues, which are invaluable improvements, but still far from the resolution required for investigating local auxin responses. Reporters were also developed over the years, the most broadly used being pDR5, a chimeric promoter assembling AxuRE repeats, which controls the expression of genes coding various versions of fluorescent protein(s), luciferase, or β-glucuronidase. The expression is turned on in response to an auxin stimulus and is reflecting a maximum of auxin-dependent transcriptional response mobilizing the TIR1/AFB – AUX/IAA pathway and transcriptional activation. This reporter is also activated in response to brassinolide. A few years ago, a novel auxin response reporter made of a translational fusion between the domain II of IAA28 and the fluorescent protein VENUS was developed (Vernoux et al. 2011). Expression of the reporter is constitutive, and the protein is degraded via the TIR1/AFB pathway. The advantage of this reporter compared to DR5 is the timescale, as the loss of the reporter is directly related to its degradation with no need of activation of transcription and translation. The DII:VENUS reporter does not reflect the endogenous content of auxin as it is still a reporter of the capacity of the cells to respond, and alterations of this response upon the TIR1/AFB expression pattern, post-translational modification, or any other parameter influencing the system affects its relative abundance. Development of a nanosensor that could be used *in vivo* to visualize and quantify active auxin at the cellular or even at the subcellular resolution, as developed recently for various molecules in the lab of W. Frommer (Carnegie Institution, Stanford, CA), would represent an invaluable tool allowing to discriminate between effective auxin input and any other entry point into the signaling pathway.

The road is still long before auxin-signaling mechanisms will be fully elucidated and the control exerted by auxin on the coordination of cellular responses and plant development will be understood. These are the challenges of the "auxin" research community for several decades.

Acknowledgments The author is supported by the Institut des Sciences Biologiques from the Centre National de la Recherche Scientifique. As a work of tertiary literature, most references come from edited books or book chapters explaining why only very few original research articles are cited here.

References

Aichinger E, Villar CB, Di Mambro R, Sabatini S, Kohler C. The CHD3 chromatin remodeler PICKLE and polycomb group proteins antagonistically regulate meristem activity in the *Arabidopsis* root. Plant Cell. 2011;23(3):1047–60.

Bargmann BO, Vanneste S, Krouk G, Nawy T, Efroni I, et al. A map of cell type-specific auxin responses. Mol Syst Biol. 2013;9:688.

Becker D, Hedrich R. Channelling auxin action: modulation of ion transport by indole-3- acetic acid. Plant Mol Biol. 2002;49(3–4):349–56.

Boer DR, Freire-Rios A, van den Berg WA, Saaki T, Manfield IW, et al. Structural basis for DNA binding specificity by the auxin-dependent ARF transcription factors. Cell. 2014;156 (3):577–89.

Calderon Villalobos LI, Lee S, De Oliveira C, Ivetac A, Brandt W, et al. A combinatorial TIR1/AFB-Aux/IAA co-receptor system for differential sensing of auxin. Nat Chem Biol. 2012;8 (5):477–85.

Chapman EJ, Estelle M. Mechanism of auxin-regulated gene expression in plants. Annu Rev Genet. 2009;43:265–85.

Cho H, Ryu H, Rho S, Hill K, Smith S, et al. A secreted peptide acts on BIN2-mediated phosphorylation of ARFs to potentiate auxin response during lateral root development. Nat Cell Biol. 2014;16(1):66–76.

Craddock C, Lavagi I, Yang Z. New insights into Rho signaling from plant ROP/Rac GTPases. Trends Cell Biol. 2012;22(9):492–501.

Del Bianco M, Kepinski S. Context, specificity, and self-organization in auxin response. Cold Spring Harb Perspect Biol. 2011;3(1):a001578.

Del Pozo JC, Manzano C. Auxin and the ubiquitin pathway. Two players-one target: the cell cycle in action. J Exp Bot. 2013. doi:10.1093/jxb/ert1363.

Freschi L. Nitric oxide and phytohormone interactions: current status and perspectives. Front Plant Sci. 2013;4:398.

Gross F, Durner J, Gaupels F. Nitric oxide, antioxidants and prooxidants in plant defence responses. Front Plant Sci. 2013;4:419.

Guilfoyle TJ, Hagen G. Auxin response factors. Curr Opin Plant Biol. 2007;10(5):453–60.

Guilfoyle TJ, Hagen G. Getting a grasp on domain III/IV responsible for Auxin Response Factor-IAA protein interactions. Plant Sci. 2012;190:82–8.

Jaillais Y, Chory J. Unraveling the paradoxes of plant hormone signaling integration. Nat Struct Mol Biol. 2010;17(6):642–5.

Ma X, Lv S, Zhang C, Yang C. Histone deacetylases and their functions in plants. Plant Cell Rep. 2013;32(4):465–78.

Mockaitis K, Estelle M. Auxin receptors and plant development: a new signaling paradigm. Annu Rev Cell Dev Biol. 2008;24:55–80.

Nibau C, Wu HM, Cheung AY. RAC/ROP GTPases: "hubs" for signal integration and diversification in plants. Trends Plant Sci. 2006;11(6):309–15.

Perrot-Rechenmann C. Cellular Responses to Auxin. Cold Spring Harb Perspect Biol. 2010; 2:a001446.

Pierre-Jerome E, Moss BL, Nemhauser JL. Tuning the auxin transcriptional response. J Exp Bot. 2013;64(9):2557–63.

Sanan-Mishra N, Varanasi SP, Mukherjee SK. Micro-regulators of auxin action. Plant Cell Rep. 2013;32(6):733–40.

Sauer M, Kleine-Vehn J. AUXIN BINDING PROTEIN1: the outsider. Plant Cell. 2011;23 (6):2033–43.

Stuttmann J, Lechner E, Guerois R, Parker JE, Nussaume L, et al. COP9 signalosome- and 26S proteasome-dependent regulation of SCFTIR1 accumulation in *Arabidopsis*. J Biol Chem. 2009;284(12):7920–30.

Swarup R, Parry G, Graham N, Allen T, Bennett M. Auxin cross-talk: integration of signalling pathways to control plant development. Plant Mol Biol. 2002;49(3–4):411–26.

Takahashi K, Hayashi K, Kinoshita T. Auxin activates the plasma membrane H + -ATPase by phosphorylation during hypocotyl elongation in *Arabidopsis*. Plant Physiol. 2012;159 (2):632–41.

Tromas A, Paponov I, Perrot-Rechenmann C. AUXIN BINDING PROTEIN 1: functional and evolutionary aspects. Trends Plant Sci. 2010;15(8):436–46.

Tromas A, Paque S, Stierlé V, Quettier AL, Muller P, et al. AUXIN BINDING PROTEIN 1 is a negative regulator of the SCFTIR1/AFB pathway. Nat Commun. 2013;4:2496–504. doi:2410.1038/ncomms3496.

Vernoux T, Brunoud G, Farcot E, Morin V, Van den Daele et al. The auxin signalling network translates dynamic input into robust patterning at the shoot apex. Mol Syst Biol. 2011;7:508. doi: 10.1038/msb.2011.39.

Vert G, Walcher CL, Chory J, Nemhauser JL. Integration of auxin and brassinosteroid pathways by Auxin Response Factor 2. Proc Natl Acad Sci USA. 2008;105(28):9829–34.

Vierstra RD. The ubiquitin-26S proteasome system at the nexus of plant biology. Nat Rev Mol Cell Biol. 2009;10(6):385–97.

Wendrich JR, Weijers D. The *Arabidopsis* embryo as a miniature morphogenesis model. New Phytol. 2013;199(1):14–25.
Wu HM, Hazak O, Cheung AY, Yalovsky S. RAC/ROP GTPases and auxin signaling. Plant Cell. 2011;23(4):1208–18.
Yang Z. Cell polarity signaling in Arabidopsis. Annu Rev Cell Dev Biol. 2008;24:551–575.
Xu T, Dai N, Chen J, Nagawa S, Cao M, et al. Cell surface ABP1-TMK auxin-sensing complex activates ROP GTPase signaling. Science. 2014;343(6174):1025–8.

Further Readings

Eshel A, Beeckman T, editors. Plant Roots: the hidden half. 4th ed. Boca Raton: CRC Press/Taylor and Francis Group; 2013. Chapters 3 to 20.
Estelle M, Weijers D, Ljung K, Leyser O, editors. Auxin signalling, from synthesis to systems biology. New York: Cold Spring Harbor Perspectives in Biology; 2011. p. 245.
Litwack G, editor. Plant hormones. Vitamins and hormones, vol. 72. Amsterdam: Academic Press/Elsevier; 2005. p. 535.
Perrot-Rechenmann C, Hagen G, editors. Auxin molecular biology. Dordrecht: Kluwer; 2002. p. 184.

Cytokinin Signaling in Plants

9

Chia-Yi Cheng and Joseph J. Kieber

Contents

C.-Y. Cheng (✉) • J.J. Kieber
Department of Biology, University of North Carolina, Chapel Hill, NC, USA
e-mail: chiayic@email.unc.edu; jkieber@bio.unc.edu

S.H. Howell (ed.), *Molecular Biology, The Plant Sciences 2*,
DOI 10.1007/978-1-4614-7570-5_14

Abstract

- Cytokinin is a plant hormone involved in almost every aspect of plant growth and development.
- Cytokinin signaling proceeds via a phosphorelay pathway that is similar to bacterial two-component systems.
- Cytokinin two-component elements include histidine kinase receptors, which are activated by cytokinin binding, phosphotransfer proteins, which shuttle the phosphoryl group from the receptors to the final elements of the pathway, the response regulators, which mediate the biological outputs.

Introduction

In both animals and plants, hormones generally act pleiotropically to regulate growth and development. The phytohormone cytokinin has been linked to a wide array of developmental processes since its identification in the 1950s as a factor, which, in concert with auxin, induced cell division in tobacco tissue. These processes include organ initiation, meristem maintenance, chloroplast development, vascular differentiation, and leaf senescence. Further, cytokinin also plays an important role in the responses to biotic factors, such as pathogen defense and rhizobial symbiosis, and abiotic factors such as cold, drought, and salt stress. How a signal mediates such diverse biological outputs and how these responses are intertwined with other signaling pathways remain fundamental questions in plant biology.

Molecular genetic studies in *Arabidopsis thaliana* have revealed that cytokinin signaling in this dicot model system is similar to bacterial two-component phosphotransfer signal transduction systems. Two-component phosphotransfer elements are also present in monocots, gymnosperms, as well as in lower plants such as the lycophyte *Selaginella moellendorffii* and the moss *Physcomitrella patens*. The prototypical two-component system consists of two conserved members: a sensor kinase (histidine kinase, HK) and a response regulator (RR) (Fig. 1). In most cases, in response to environmental stimuli, the HK autophosphorylates on a conserved His residue using the γ-phosphate from ATP as a donor and then transfers this phosphoryl group to a conserved Asp residue within the receiver domain of

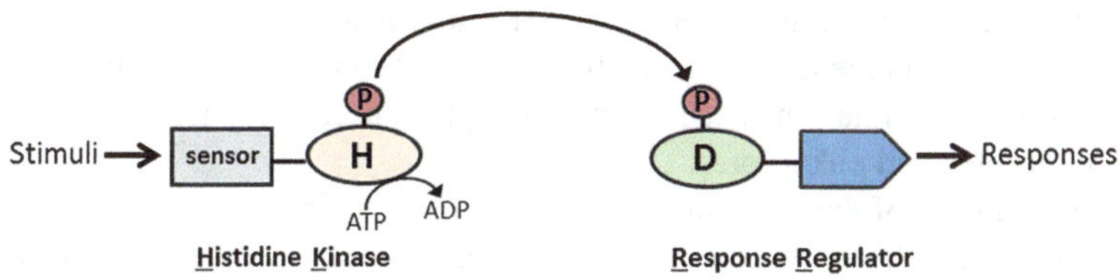

Fig. 1 Schematic diagram of two-component (TCS) system. The prototypical TCS features a phosphoryl transfer (*P*) between the conserved His residue (*H*) of the kinase to the conserved Asp residue (*D*) in the receiver domain of the response regulator

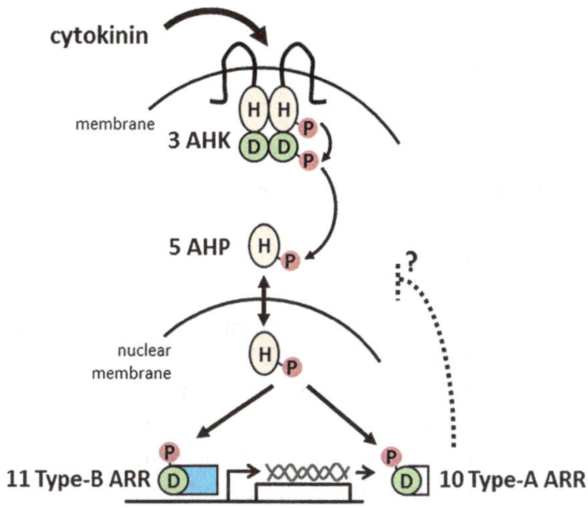

Fig. 2 Cytokinin signal transduction in*Arabidopsis*. Cytokinin binds to the membrane-bound AHK receptors, which initiates a phosphorelay through the AHPs and ultimately results in the phosphorylation of type-B and type-A ARRs. The activated type-B ARRs induce the transcription of the type-A ARRs, which in turn act to negatively feedback the pathway through an as yet undetermined mechanism that likely involved phospho-dependent protein interactions

the RR (Stock et al. 2000, 1990). In addition to the conserved N-terminal receiver domain, RRs also have a variable C-terminal output domain that confers diversity in regulatory strategies. The phosphorylation state of the RRs mediates the function of the output domain, which can participate in DNA binding and transcriptional control, perform enzymatic activities, or mediate protein-protein interactions. Thus, RRs function as phospho-mediated switches that couple environmental cues to cellular responses in a simple, direct manner.

Cytokinin signaling, as well as all other known eukaryotic two-component-like signaling, involves a more elaborate version of the two-component system known as a multicomponent phosphorelay (Fig. 2). This involves hybrid kinases containing both histidine kinase (HK) and receiver domains in a single protein,

His-containing phosphotransfer proteins (HP), and response regulators (RR). The phosphotransfer scheme occurs via a His-Asp-His-Asp phosphorelay that provides more targets for modulation. The multiple-step phosphorelay systems are present in both prokaryotes and eukaryotes.

The elucidation of the mechanism of cytokinin signaling has been hampered by the genetic redundancy of the two-component genes in plants. Nevertheless, molecular genetic studies, primarily in the model species *Arabidopsis thaliana*, have revealed the various two-component elements involved in cytokinin signaling, and the analysis of multiple loss-of-function mutants has shed light on the overlapping and distinct biological roles of these two-component proteins. However, the understanding of the underlying mechanisms by which cytokinin achieves signaling specificity in its myriad roles throughout plant growth and development, and how it integrates with other signaling pathways, is only beginning to be understood.

Two-Component Elements Are Involved in Cytokinin Signaling

Phosphotransfer Chemistry

Two-component signaling acts as a common mechanism by which bacteria sense and respond to environmental cues. Most bacteria possess numerous two-component systems to respond to a variety of environmental changes, such as temperature (thermotaxis), light (phototaxis), salinity (osmotaxis), oxygen (aerotaxis), and chemicals (chemotaxis). Two-component signal transduction has been found in all domains: eubacteria, archaea, and eukarya. However, while these systems are common in eubacteria and archaea, they are relatively rare in eukarya and have not as yet been found in animals, which instead rely on Ser/Thr/Tyr phosphorylation for much of their signaling needs (Stock et al. 2000). It is important to note that the chemistry of the phosphorelay differs substantially from the phosphoesters involved in Ser/Thr/Try phosphorylation (Stock et al. 2000).

The chemistry of the basic two-component system involves three phosphotransfer reactions:

$$\text{1. Autophosphorylation: HK} - \text{His} + \text{ATP} \leftrightarrow \text{HK} - \text{His} \sim \text{P} + \text{ADP}$$
$$\text{2. Phosphotransfer: HK} - \text{His} \sim \text{P} + \text{RR} - \text{Asp} \leftrightarrow \text{HK} - \text{His} + \text{RR} - \text{Asp} \sim \text{P}$$
$$\text{3. Dephosphorylation: RR} - \text{Asp} \sim \text{P} + \text{H}_2\text{O} \leftrightarrow \text{RR} - \text{Asp} + \text{P}_\text{i}$$

The phosphorylation at His and Asp residues is thermodynamically distinct from the phosphorylation at Ser, Thr, and Tyr residues, which are highly exothermic and thus essentially irreversible. In contrast, the free energy associated with the phosphotransfer between His and Asp residues is close to zero, which allows bidirectional flow of the phosphoryl group. That is, upon cessation of the stimulus, the phosphoryl group on the RRs is able to flow back to the HKs, effectively terminating the response and allowing rapid, adaptive responses to environmental conditions.

His-Asp Phosphorylation Pathway in *Arabidopsis thaliana*

In 1993, the identification of a candidate ethylene receptor ETR1 with putative His kinase and receiver domains in *Arabidopsis thaliana* was the first indication that eukaryotes in general, and specifically plants, harbored two-component signaling systems. In 1996, the second His kinase, CKI1, was discovered as a potential mediator of cytokinin signaling (Kakimoto 1996), and shortly thereafter response regulators (RRs) were identified as cytokinin primary response genes (Brandstatter and Kieber 1998). Similarly, two-component elements were identified in a number of fungal species. When the entire genome sequence of *Arabidopsis* was elucidated in 2000, the complete two-component element repertoire was revealed: the *Arabidopsis* genome encodes eight genes predicted to be functional HKs, five HPs, and 23 RRs. Among the HKs in *Arabidopsis*, two belong to the ethylene receptor family. There are three additional ethylene receptors in the HK family that lack residues essential for histidine kinase activity. Further, the output of the ethylene receptor HKs occurs largely through a Raf-like Ser/Thr kinase called CTR1, which does not involve a His-Asp phosphorelay. Among the six other functional HKs that are not ethylene receptors, AHK2, AHK3, and AHK4 are cytokinin receptors. CKI1 functions both in vegetative and female gametophyte development. AHK1/AtHK1 is a positive regulator of drought and salt stress responses, and CKI2/AHK5 lacks a predicted transmembrane domain and is a positive regulator of biotic and abiotic stress responses.

Cytokinin Two-Component Elements

In *Arabidopsis*, multiple two-component elements act in cytokinin signaling. Three cytokinin receptors (AHK2, AHK3, and AHK4/CRE1/WOL) contain a conserved CHASE (cyclase/histidine kinase-associated sensing extracellular) domain, which confers the ability to bind cytokinin with high affinity. These cytokinin receptors also contain a histidine kinase domain and both an authentic and a pseudo receiver domain, the latter of which lacks the Asp target of phosphorylation. AHKs are partially redundant positive elements in cytokinin signaling. The direct targets of the AHKs are the His-containing phosphotransfer proteins (AHPs), which also act redundantly as positive elements in the primary cytokinin signaling pathway. There are 23 response regulators (ARRs) in *Arabidopsis* that fall into two major classes based on phylogenetic analysis and domain structure: type-A ARRs and type-B ARRs. The eleven type-B ARRs have a conserved Myb-like GARP DNA-binding domain following the N-terminal receiver domain and act as partially redundant, positive elements in cytokinin signaling. In contrast, the ten type-A ARRs are comprised of essentially only a receiver domain, and they act as negative elements in cytokinin signaling (Schaller et al. 2011).

Cytokinin signal transduction initiates when cytokinin binds to the CHASE domain of the receptor AHKs to autophosphorylate on a conserved His residue within the histidine kinase domain (Fig. 2). This phosphoryl group is then transferred to an Asp residue within the C-terminal receiver domain of the AHKs. The AHPs then shuttle the phosphoryl group from the AHKs to the type-B and type-A ARRs. The phosphorylation activates the type-B ARRs, which regulate the expression of the primary cytokinin-responsive genes, including type-A ARRs. In turn, the type-A ARRs act as negative feedback regulators of the primary signaling pathway. Overall, the phosphorelay in cytokinin signal transduction involves four sequential phosphorylation events in the order of His-Asp-His-Asp. This more elaborate architecture of the phosphorelay provides additional opportunities for cross talk with other signaling pathways and provides a robust system for shuttling the cytokinin signal to multiple compartments within a eukaryotic cell.

Cytokinin Receptors Are Histidine Kinases

Discovery

The CKI1(cytokinin insensitive 1) HK was isolated from an activation T-DNA tagging screen as a gene that when overexpressed conferred shoot initiation in the absence of exogenous cytokinin. The CKI1 protein consists of a histidine kinase domain, a single transmembrane domain, and a receiver domain. Overexpression of *CKI1* induces typical cytokinin responses in cultured cells, including rapid proliferation, greening, shoot formation, and inhibition of root formation. The activation of cytokinin responses by *CKI1* implicates it as a cytokinin signaling element, but the gain-of-function nature of the allele complicates this conclusion. Subsequent studies identified *cre1* (*cytokinin response 1*) as a mutant that showed reduced cytokinin sensitivity. *CRE1* also encodes a histidine kinase. However, in contrast to CKI1, CRE1 binds cytokinin with high affinity and specificity and complements yeast and bacterial HK mutants in a cytokinin-dependent manner. This Arabidopsis histidine kinase (AHK4/CRE1/WOL) and its paralogues (AHK2 and AHK3) contain a CHASE domain that binds cytokinin. How binding of cytokinin to this CHASE domain transduces the signal across the membrane is not known. CKI1 and other hybrid HKs that do not include the CHASE domain do not bind cytokinin; nevertheless, they could feed into cytokinin signaling through phosphorylation of the AHPs in response to the sensing of signals other than cytokinin and/or by forming heterodimers with the cytokinin receptor AHKs.

Cytokinin Receptors Have Distinct Biochemical Properties

Naturally occurring cytokinins are adenine derivatives with an N^6-side chain and are classified as isoprenoid or aromatic depending on the nature of the side chain. The binding preferences for the different AHKs have been studied by expression in *Escherichia coli*. AHK3 or AHK4 are most sensitive to the isoprenoid-type *trans*-zeatin (*t*Z) and isopentenyladenine (iP), but differ significantly in the recognition of other cytokinin compounds. Interestingly, the maize HKs recognize *cis*-zeatin (*c*Z) with high affinity, but the Arabidopsis AHK cytokinin receptors do not. Results from in planta experiments confirm the in vitro data and further reveal different affinities of various AHKs toward *t*Z and iP; AHK2 and AHK4 show comparable activity in response to *t*Z and iP, while AHK3 displayed tenfold higher sensitivity to *t*Z as compared to iP. A higher functional similarity between AHK2 and AHK4 is supported by promoter-swap experiments in which AHK4 expressed under the control of the *AHK2* promoter (but not the *AHK3* promoter) is sufficient to complement the *ahk2 ahk3* loss-of-function phenotype. A chimeric protein that includes the CHASE-TM (transmembrane) from AHK3 and the cytoplasmic domain from AHK4 could partially complement an *ahk2 ahk3* loss-of-function mutant. The partial complementation suggests that the nature of the CHASE-TM domain is critical, but not the sole feature required for proper AHK3 function.

Hypomorphic mutations in *AHK4* do not have a substantial effect on plant growth and development. However, certain point mutations within the AHK4 coding region produce what are known as the *wooden leg* (*wol*) alleles, which have fewer vascular initials in the root during embryogenesis and, as a result, cause defects in root vascular morphogenesis postembryonically. One of these point mutations (T278I) in the CHASE domain blocks the ability of AHK4 to bind cytokinin in vitro. The recently resolved crystal structure of the CHASE domain of AHK4 has revealed that the T278I mutation likely restricts the overall size of the binding pocket and thus affects the binding capacity for cytokinin. The dominant-negative nature of wol^{T278I} may result from the phosphatase activity of AHK4 that is unbound to cytokinin, similar to some prokaryotic histidine kinases that possess both kinase and phosphatase activity. Biochemical analyses show that CRE1 can dephosphorylate multiple AHPs, and this phosphatase activity requires the conserved Asp residue in its receiver domain. These results suggest that AHK4 can act as either a kinase or a phosphatase in a bidirectional phosphorelay. Thus, in the absence of cytokinin, phosphate from the ARRs could be removed via flow back to AHK4, through the AHPs, reflecting the reversibility of the various phosphorylation events in this pathway.

A subset of *wol* alleles of *AHK4* display intragenic complementation, implying that the signal transduction involves dimerization or higher-order oligomerization. *Trans*-hetergozygous plants combining certain alleles (*wol-1*/*wol-2*) have a wild-type root vascular phenotype, though they remain insensitive to exogenous cytokinin. It is puzzling how these recessive mutations cause dominant-negative effects on procambial cell proliferation and display intragenic complementation exclusively in vascular development. One explanation is that AHK4 represses, perhaps via its phosphatase activity, cambium morphogenesis as a monomer. Binding of cytokinin triggers the dimerization (or higher-order oligomerization) and subsequently derepresses the negative effect on vascular initiation. Homozygous *wol* plants lack the ability either to bind cytokinin (*wol-1*) or to dimerize (*wol-2*) and thus repress procambial development. In this model, the *trans*-heterozygotes *wol-1*/*wol-2* would only have compromised receptors including one copy of WOL-2 that can bind cytokinin but not dimerize and one copy of WOL-1 that can dimerize but not bind cytokinin. This would be sufficient to release the repressing effect on vascular initiation but not for the response to elevated levels of exogenous cytokinin.

Functions of the AHK Receptors

Since their discovery, cytokinins have been shown to positively regulate shoot growth and negatively regulate root growth. Much of the work, however, has been based on experiments utilizing overexpression of cytokinin biosynthetic genes and exogenous cytokinin treatment. Mutations in the genes encoding various cytokinin two-component elements have provided novel ways to explore the functions of cytokinin in plant growth and development.

Primary root elongation and lateral root formation are inhibited in the *ahk2 ahk3 ahk4* mutant, which is associated with cell cycle arrest as the transition from $G_2 \rightarrow M$ phase is delayed. Moreover, the *wol* mutant displays a lack of the periclinal cell divisions that occur during vascular morphogenesis. However, a single mutation in *AHK3* and multiple mutations in isopentenyltransferases (*IPTs*), which are essential to cytokinin biosynthesis, result in a longer primary root and a larger root meristem. These results led to a model in which the cytokinin response curve, at least in the root, is bell-shaped rather than linear (Ferreira and Kieber 2005). If this were the case, a minor reduction of cytokinin signaling would induce root growth, while reduction beyond the threshold would abolish growth. This is similar to the bell-shape response curve observed for cytokinin in shoot initiation assays in cultured cells.

Disruption of the cytokinin receptors also perturbs shoot and floral development. Shoot meristem size and leaf cell numbers are reduced in *ahk2 ahk3 ahk4* mutants, consistent with a role for the AHKs as positive regulator of cell division. The *ahk2 ahk3 ahk4* mutants only occasionally form an inflorescence stem and produce few sterile flowers. These results suggest the transition from vegetative to inflorescence meristem is defective and that the floral meristem activity is depleted in *ahk2 ahk3 ahk4* mutants. The cytokinin receptors AHKs are also required in gametophyte development as strong triple mutant combinations result in complete male and female sterility. Interestingly, the weak *ahk* triple mutants produce a few flowers with reduced fertility, but are capable of producing a few seeds. These findings indicate that cytokinin is essential in floral development; however, the dosage of cytokinin signaling required differs for different developmental stages.

Cytokinin Receptors Have Overlapping and Specific Expression Pattern

The expression of the *AHK* genes overlap, but at different levels in various tissues. In roots, *AHK4* is expressed at a higher level than either *AHK2* or *AHK3*; in rosette leaves, *AHK2* and *AHK3* are more highly expressed and *AHK4* is barely detectable. Consistent with the expression patterns, *ahk2 ahk3* mutants have smaller rosettes, while *ahk3 ahk4* and *ahk2 ahk4* are similar in size to the wild type; single *ahk4* mutants have reduced cytokinin sensitivity in the root, while the *ahk2* and *ahk3* mutants exhibit normal sensitivity. The cytokinin-dependent induction of *ARR15* and *ARR16* is compromised in the roots of the *ahk4* mutants, but not in leaves. This suggests that the transcriptional induction of a subset of type-A *ARRs* in roots by exogenous cytokinin treatment is dependent on AHK4. Although *AHK2* and *AHK3* both have high expression levels in leaves, a specific role in regulating senescence is mediated exclusively by *AHK3*. A gain-of-function mutation in *AHK3* causes delayed leaf senescence, whereas a loss-of-function *ahk3* mutant (but not *ahk2* nor *ahk4* single mutants) confers reduced cytokinin sensitivity in leaf senescence. Chlorophyll retention is impaired in the *ahk3* mutant, and addition of an *ahk2* mutation further magnifies this effect. The involvement of *AHK4* only becomes significant in the triple mutant, consistent with its weak expression level in rosette leaves.

The Majority of Cytokinin Receptors Localize in the ER

In addition to the redundant roles of the receptors in some contexts, genetic studies have also reported specific roles for individual AHKs. These unique receptor functions could be the result of differences in expression patterns, ligand binding affinity, interacting targets, or subcellular localizations. Hydrophobicity analysis indicated that there are putative transmembrane segments in the N-termini of AHK2, AHK3, and AHK4. Bioinformatic analysis using the PSORT (prediction of protein sorting signals and localization sites) program suggested that they are localized in the plasma membrane (PM). However, endomembranes have higher saturable cytokinin binding than the PM. Biochemical and cell biological assays show that at least in *Arabidopsis* and maize, the majority of cytokinin receptors are localized in the endoplasmic reticulum (ER) and the cytokinin-binding CHASE domain is exposed to the ER lumen. Studies of transiently expressed AHK fluorescent fusion proteins support the predominant localization to the ER for all three cytokinin receptors. Nevertheless, there is a minor, but perhaps functional relevant fraction of the receptors at the PM. The canonical model for cytokinin signaling has assumed these receptor sense extracellular cytokinins. The finding that their location is predominantly in the ER suggests that active cytokinins must cross the plasma membrane and ER membrane in order to bind to the lumen-localized CHASE domain. Purine permeases (PUP) have been demonstrated to transport cytokinin and adenine into the cytosol in *Arabidopsis* cell culture and in yeast. Genetic evidence for the role of PUPs in cytokinin *in planta* is still lacking due to the large number of *PUP* genes present in the *Arabidopsis* genome.

Non-receptor Kinases also Feed into Two-Component Signal Transduction

CKI1

As noted above, overexpression of CKI1 induced cytokinin-independent callus formation in cultured *Arabidopsis* cells. However, the lack of a CHASE domain in CKI strongly suggests that this HK is not a cytokinin receptor per se. Recent studies have shown that CKI1 can feed in downstream into two-component signal transduction via the AHP phosphotransfer proteins and the type-B ARRs (Deng et al. 2010). The receiver domains of CKI1 can interact with AHP2, AHP3, and AHP5 in yeast and plant protoplasts. The phenotype induced by overexpression of CKI1 is eliminated in an *ahp1,2,3,4,5* mutant, suggesting that *AHPs* act epistatically to *CKI1*. Further, in high-order *ahp* and type-B *arr* mutants, a subset of female gametophytes have phenotypes similar to those observed in *cki1* loss-of-function alleles, which supports a role for AHPs and type-B ARRs acting downstream of *CKI1*. Expression of ARR1, one of the type-B ARRs, under the control of *CKI1* promoter is able to partially rescue the *cki1* phenotype, further suggesting *ARR1* is epistatic to *CKI1*.

Despite the lack of the cytokinin-binding CHASE domain, overexpression of CKI1 was found to partially rescue multiple phenotypes in a *wol* mutant, including the shortened primary root, defects in xylem development, and cytokinin insensitivity in shoot regeneration assay. In addition, ectopic expression of cytokinin biosynthetic isopentenyltransferase *IPT8* under the control of the *CKI1* promoter is able to partially rescue *cki1* phenotypes. The mechanisms by which elevated cytokinin levels complement the *cki1* phenotype is not clear, though likely involve increased activity of the downstream phosphorelay.

Other Non-receptor Histidine Kinases

The role of AHK1 in plant growth and development is only obvious in a triple *ahk1 ahk2 ahk3* mutant background. Addition of *ahk1* to an *ahk2 ahk3* mutant significantly reduces the plant size and retards growth. In addition, the *ahk1* mutant is more sensitive to drought stress, while *ahk2* and *ahk3* are more tolerant. This suggests that cytokinin receptors act in stress responses in a manner opposite to that of *AHK1*. Comprehensive coexpression analysis reveals that *AHK1* is coexpressed with a set of type-A ARRs (*ARR4*, *ARR5*, *ARR6*, *ARR8*, and *ARR9*) under abiotic stress conditions and cytokinin treatment. These results suggest a potential interaction between AHK1 and cytokinin signaling in abiotic responses, though further studies are needed to confirm this.

CKI2/AHK5 is a histidine kinase that lacks a transmembrane domain. The role of *AHK5* in the cytokinin two-component phosphorelay is not clear. AHK5 interacts with multiple phosphotransfer proteins, except AHP4, though the biological significance of these interactions is unknown. AHK5 is a biochemically active histidine kinase, but this activity is not dependent on cytokinin. Disruption of *AHK5* results in a wild-type shoot; the roots displayed wild-type sensitivity to cytokinin, but were hypersensitive to abscisic acid and ethylene, suggesting a potential role as a negative regulator of these signaling pathways.

Are AHKs the Only Cytokinin Receptors?

Cytokinins were long thought to be essential for plant growth and development as they regulated essential processes such as cell division and organogenesis. Surprisingly, three independent *ahk* triple mutants harboring nonoverlapping T-DNA alleles are seedling viable, albeit quite stunted. One allelic combination has marginally reduced fertility, while the other two mutants are completely male and female sterile. Recent studies have revealed that even in the strongest *ahk* triple mutant, there is residual full-length *AHK3* transcript (~0.8 % compared to wild type), indicating that none of these three triple mutants completely lack AHK cytokinin receptor activity. This raises the question as to whether or not cytokinin is essential for plant growth and development. One possibility is that cytokinin is not essential for viability, notwithstanding the residual AHK3 transcript. A second possibility to explain the viability of the *ahk2 ahk3 ahk4* mutants is that the residual *AHK3* transcript, although insufficient for male and female gametophyte development, is sufficient to support some vegetative development. A final possibility is that there are additional cytokinin receptors in addition to AHK2, AHK3, and AHK4. Several cytokinin-binding proteins have been isolated from various plant species including barley, maize, oat, and tobacco. However, the evidence linking these to a physiological function in cytokinin signaling is lacking. Another candidate for a novel cytokinin receptor is *CHARK*, a gene found in rice that encodes a protein containing a CHASE domain at the N-terminus followed by a serine/threonine kinase domain. The CHASE domain of CHARK is 49–67 % identical to the cytokinin HK receptors in rice, maize, and *Arabidopsis*. CHARK may be a cytokinin-binding element unique to rice or monocots, although the cytokinin-binding activity of the encoded product CHARK has not been verified. Further analysis is needed to resolve this important question.

The Phosphotransfer Proteins in *Arabidopsis*

AHPs Shuttle Phosphate Between the AHKs and ARRs

There are five genes in *Arabidopsis* that are predicted to be functional histidine-containing phosphotransfer (Hpts) proteins. These *Arabidopsis* Hpts (AHPs) complement a YPD1 loss-of-function mutation in yeast, which encodes an Hpt protein involved in osmosensing, indicating that they can act as Hpt proteins. The AHPs were shown to rapidly transfer phosphoryl groups from their own conserved His residue to the Asp residue of response regulators. In addition to these functional AHPs, there is a pseudo-AHP, AHP6, that lacks the conserved His residue that acts as the phosphorylation site (see below). The AHPs directly interact with both the upstream cytokinin AHK receptors and the downstream type-A and type-B ARRs. The AHPs act to shuttle a phosphoryl group from the Asp residue of the receiver domains in the cytokinin AHK receptors to the Asp residue in the receiver domains of the ARRs, thus transducing the signal from the site of perception to the nucleus to regulate gene expression and to the type-A ARRs, some of which are in the nucleus and some of which are cytoplasmic.

AHPs Act Downstream of AHKs and CKI1

Reverse genetic experiments provided direct, compelling evidence that the *AHPs* act as positive regulators in cytokinin signaling. Various combinations of T-DNA insertion alleles in the five *AHP* loci were analyzed. *AHP1*, *AHP2*, *AHP3*, and *AHP5* were found to have overlapping roles as positive elements in cytokinin signaling using different cytokinin response assays, including the induction of primary response genes. The quintuple *ahp1ahp2-1 ahp3 ahp4 ahp5* mutant displayed phenotypes similar to *ahk2 ahk3 ahk4* triple receptor mutants, including inhibition of primary root growth and loss of metaxylem development. Interestingly, shoot development in the *ahp1ahp2-1 ahp3 ahp4 ahp5* mutant was not as severely affected compared to the *ahk2 ahk3 ahk4* mutant, which is likely due to the residual full-length *AHP2* transcript from the *ahp2-1* allele used. A quintuple *ahp1ahp2-2 ahp3 ahp4 ahp5* mutant that incorporates the null *ahp2-2* allele is severely delayed in leaf formation and dies at the seedling stage, which has not been reported even in the strongest *ahk2-7 ahk3-3 cre1-12* triple receptor mutant. As noted above, the discrepancy in the phenotypic strength of these lines as compared to the null *AHP* mutant is likely due to the fact that none of the *ahk* triple mutant combinations represent complete receptor nulls (see section "Are AHKs the Only Cytokinin Receptors"). Alternatively, as the AHKs are likely not the sole upstream regulators of the AHPs (see section on "CKI1" earlier), the stronger phenotype of the *ahp1ahp2-1 ahp3 ahp4 ahp5* mutant may reflect disruption of other HK signaling pathways.

The Subcellular Localization of AHPs Suggests a Function in a Bidirectional Phosphorelay

Phospho-His and phospho-Asp residues are high energy molecules. As noted above (see section "Phosphotransfer Chemistry"), the free energy associated with the various phosphorylation reactions that occur in the phosphorelay is close to zero, which allows HP domains to act both as phosphodonors and phosphoreceivers, and so to shuttle a phosphoryl group between two or more receiver domains. That is, the high energy cytosolic phospho-AHP is capable of donating the phosphoryl group to both the type-A and type-B response regulators or to the upstream AHKs. Early studies suggested that the AHP-GFP fusion proteins moved into the nucleus in response to cytokinin treatment, but a more quantitative analysis demonstrated that while the AHPs do shuttle between the cytoplasm and the nucleus via an active transport mechanism, this movement is not responsive to cytokinin or phosphorylation, and thus the AHPs appear to continuously move in and out of the nucleus. Together, this suggests that the AHPs can mediate the phosphorelay from the membrane-bound receptors to the mainly nuclear-localized response regulators. In addition, the flow of phosphate can proceed from the ARRs back to the AHKs via the AHPs upon cessation of the cytokinin signal.

AHP6 Is a Negative Element in Cytokinin Two-Component Signaling

AHP6 was isolated in a genetic screen for suppressors of the determinate root phenotype of *wol*. AHP6 lacks a conserved His residue at the site of phosphorylation and thus is predicted to be a nonfunctional Hpt protein. AHP6 inhibits the phosphotransfer from the His kinase domain of SLN1 to its fused receiver domain in vitro. SLN1 is a hybrid HK involved in the *Saccharomyces cerevisiae* osmosensing pathway. AHP6 also inhibits the phosphotransfer from phosphorylated AHP1 to ARR1 in vitro, which suggests that AHP6 acts as an inhibitor of phosphotransfer, likely through a dominant-negative mechanism. The role of AHP6 as a negative regulator of cytokinin signaling is also supported by functional analysis in vivo. The *ahp6-1* loss-of-function mutant has elevated basal expression of the cytokinin primary response gene *ARR15* and is hypersensitive to the effects of exogenous cytokinin on adventitious root formation and protoxylem differentiation. Interestingly, cytokinin also negatively regulates *AHP6* expression, forming a mutual regulatory circuit in regulating root development.

Nitric Oxide Regulates Phosphotransfer Proteins Through *S*-Nitrosylation

In cells, nitric oxide (NO) can directly modify the cysteine thiol of proteins as a redox-based posttranslational modification, which is known as *S*-nitrosylation. Most AHP proteins (AHP1, AHP2, AHP3, and AHP5) have a conserved cysteine residue that is *S*-nitrosylated by NO as shown by in vitro and in planta experiments. As a result, NO negatively regulates cytokinin signaling as shown by decreased expression of the cytokinin reporter TCS-GFP and multiple cytokinin primary response genes in the *nox1* and *gsnor1*-3 mutants, which have elevated levels of endogenous NO. Further, these NO-overexpressing lines are less sensitive to cytokinin in root and hypocotyl elongation, root apical meristem size, and in the induction of cytokinin primary response genes, consistent with NO acting as a negative regulator of cytokinin signaling. *S*-nitrosylation of the AHP proteins reduces their ability to act as phosphotransfer proteins. AHP1^{C115S}, a non-nitrosylatable mutant protein, was resistant to the inhibitory effect of NO donors and was able to complement the cytokinin insensitivity of high-order *ahp* mutants. In contrast, an AHP1^{C115W} mutant protein that mimics the *S*-nitrosylation modification, displayed reduced phosphorylation even in the absence of an NO donor and did not complement *ahp* mutants. In vitro *S*-nitrosylation of AHP1 repressed its phospho-receiving ability from the histidine kinase and phosphotransfer activity to ARR1, a type-B ARR, demonstrating that *S*-nitrosylation compromised its function. This represents a novel mechanism by which environmental stimuli can intertwine with endogenous signal transduction pathways. Furthermore, *S*-nitrosylation by NO may not be exclusive to AHPs as putative *S*-nitrosylated cysteine residues are also present in AHKs and ARRs, although these have not been demonstrated to be *S*-nitrosylated.

Response Regulators in *Arabidopsis*

Response regulators (RRs) were first implicated in cytokinin signaling when they were identified as primary cytokinin response genes in *Arabidopsis* and maize. Response regulators contain a conserved receiver domain with a conserved phospho-receiving Asp residue that generally regulates their output activities. The *Arabidopsis* response regulators (ARRs) fall into four major classes based on their domain structure and the similarity of the amino acid sequences of the receiver domains: type-A, type-B, type-C ARRs, and the *Arabidopsis* pseudoresponse regulators (APRRs) (Fig. 3). The ten type-A ARRs are primary transcriptional targets of cytokinin signaling and contain short C-terminal extensions following the conserved receiver domain. The eleven type-B ARRs, which are not transcriptionally induced by cytokinin, contain a receiver domain followed by an output domain that has DNA-binding activity. The two type-C ARRs are structurally similar to type-A ARRs as they contain only the receiver domains; however, they are not transcriptionally induced by cytokinin. The role of type-C ARRs in cytokinin signaling, if any, remains unclear, although overexpression of one type-C ARR confers reduced cytokinin sensitivity. The receiver domains of the APRRs lack the conserved Asp residue for phosphorylation, although many appear to be phosphorylated on Ser/Thr residues. A subset of the APRRs plays a role in modulating circadian rhythms, and their phosphorylation status oscillates throughout the day.

Biochemical and genetic analyses have demonstrated that bacterial RRs function as phosphorylation-mediated switches. Phosphorylation of the highly conserved Asp residue in the receiver domain inactivates the protein in some RRs and activates it in others. In agreement with a regulatory role of phosphorylation, phosphorylation of the conserved Asp residue of type-A ARRs is critical for proper function. A subset of type-A ARR proteins is stabilized by cytokinin via phosphorylation of the Asp residue (see the section on "Type-A Response Regulators Negatively Regulate Cytokinin Signaling"). Similarly, mutation of the phospho-receiving Asp to phospho-insensitive Asn in type-B ARRs abolished its activity to transactivate a target type-A *ARR6* (see the section on "The Receiver Domains of Type-B ARRs Have Inhibitory Effect on DNA-Binding Domains"). Together, these data suggest that phosphorylation is a common strategy utilized by the ARRs to modulate their output response.

Type-A Response Regulators

Type-A ARRs Are Primary Response Genes in Cytokinin Signaling

The *Arabidopsis* type-A ARRs are a family of ten genes that fall into five distinct pairs which, based on the analysis of the locations of the genes within the genome, likely arose from the most recent genome duplication event in the evolution of *Arabidopsis*. The amino acid sequences of type-A ARRs are somewhat similar to that of the bacterial single-domain response regulator CheY, which is

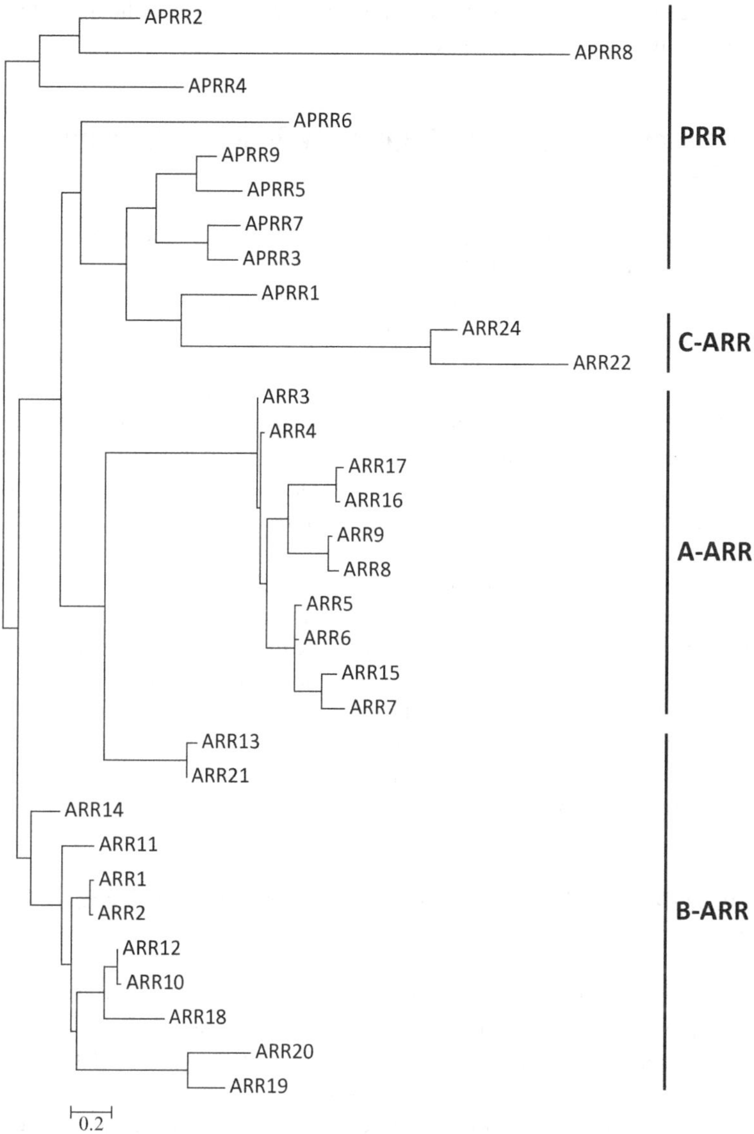

Fig. 3 **Phylogenetic analysis of the Arabidopsis response regulators (ARR)**. The amino acid sequences of the receiver domains were aligned using MUSCLE. Gaped regions that were poorly conserved were first removed from each alignment. The phylogenetic trees were generated with MEGA5.1. *PRR* pseudoresponse regulator. Scale bar indicates substitutions per site

comprised of essentially of only a receiver domain. The transcripts of type-A ARRs are rapidly and specifically induced by cytokinin, and this induction is insensitive to inhibition of protein synthesis, and thus they are cytokinin primary response genes. The cytokinin receptor AHKs, the AHPs, and the type-B

ARRs are required for the rapid induction of type-A *ARRs*, indicating that their induction requires an intact phosphorelay.

Type-A Response Regulators Negatively Regulate Cytokinin Signaling

Analysis of loss-of-function and gain-of-function type-A *arr* mutants shows that at least eight of the ten type-A ARRs are negative regulators of cytokinin signaling in multiple cytokinin response assays (To et al. 2007). Single loss-of-function type-A *arr* mutants show no significant difference from the wild type in the response to exogenous cytokinin treatment in root elongation assays, while double and higher-order *arr* mutants show increasing hypersensitivity to cytokinin. In monocots and dicots, disruption of type-A response regulators produces enlarged shoot apical meristems, presumably through increased cytokinin signaling output. Consistently, overexpression of the rice type-A response regulator *OsRR6* leads to repression of shoot regeneration in tissue culture and an aberrant dwarf phenotype in transgenic plants.

Despite the well-described role of type-A ARRs as negative regulators in cytokinin signaling, mutations in eight out of ten type-A *ARRs* do not cause dramatic morphological phenotypes. It is possible that additional negative feedback loops might compensate for the cytokinin hypersensitivity in this octuple mutant by decreasing cytokinin levels, as several cytokinin degrading oxidases (*CKX*) are induced upon cytokinin treatment. An alternative, but not a mutually exclusive possibility, is that type-A ARRs might functionally overlap with other signaling elements to regulate plant growth and development.

Phosphorylation of Receiver Domain Is Essential for Type-A ARR Function

The type-A ARR proteins exhibit in vitro activity typical of bacterial receivers as they can be phosphorylated on the conserved Asp residue using a phospho-HP domain as a phosphodonor. In *Arabidopsis*, a subset of type-A ARR proteins is stabilized by cytokinin via phosphorylation of the Asp residue. Aspartate (D) to glutamate (E) substitutions in the conserved phospho-receiving Asp residue of RR proteins can partially mimic the active, phosphorylated state in some but not all RRs. Type-A ARR5^{D87A}, a non-phosphorylatable mutant protein, fails to complement the cytokinin hypersensitivity of an *arr3 arr4 arr5 arr6* quadruple mutant, indicating that phosphorylation of the conserved Asp residue is required for its proper function. In contrast, ARR5^{D87E}, a phosphomimic version of the protein, can partially rescue the cytokinin hypersensitivity of *arr3 arr4 arr5 arr6* mutant. As this ARR5^{D87E} protein cannot be phosphorylated by the phosphorelay system, it leads to the hypothesis that the type-A ARRs act via phospho-dependent interactions with other proteins to negatively feedback cytokinin signaling rather than via competition with the type-B ARRs for phosphorylation by the upstream AHPs. Such phospho-dependent interactions are a common strategy used by bacterial single-domain response regulators that lack a distinguishable output domain.

Other Signals Cross Talk with Cytokinin Signaling via Type-A ARRs

In addition to cytokinin, type-A *ARRs* are regulated by other inputs, presumably as a means to locally dampen cytokinin signaling (El-Showk et al. 2013). For example, WUSCHEL (WUS), a homeodomain transcription factor, specifically represses the expression of *ARR5*, *ARR6*, *ARR7*, and *ARR15* in the shoot apical meristem presumably to reduce their attenuation of cytokinin signaling, thus maintaining stem cell fate. Overexpression of the phosphomimic form of ARR7 (35S:: ARR7^{D85E}) results in an arrested shoot apical meristem that is similar to the phenotype observed in a *wus* mutant. Similarly, auxin, also cross talks with cytokinin by regulating two type-A *ARRs*, *ARR7* and *ARR15*. Whereas cytokinin induces *ARR7* and *ARR15* in shoot apical meristem, auxin represses the expression of these two type-A *ARRs* via auxin response factor 5 (ARF5). Mutations of the ARF-binding site in the promoter region of *ARR15* result in ectopic expression of *ARR15*. Mutations of *ARR7* and *ARR15* in *arf5* background partially rescue an *arf5* phenotype. Further, the CIN-TCP transcription factors that promote leaf maturation trigger *ARR16* to locally reduce leaf sensitivity to cytokinin, which has an inhibitive effect on shoot differentiation. In leaves with reduced CIN-TCP activity, reconstitution of *ARR16* levels partially restored growth defects. Overall, these experiments provide examples in which endogenous signals cross talk with cytokinin signaling via modulation of the expression of type-A *ARRs*.

Type-B Response Regulators

The type-B *ARR* gene family is comprised of eleven members that belong to three subfamilies based on amino acid sequence similarities in their receiver domains. The subfamily I type-B *ARRs* have been the most thoroughly studied as they seem to play the predominant role in cytokinin-mediated outputs. The type-B ARRs contain a receiver domain at the N-terminus followed by a conserved plant-specific Myb-related GARP DNA-binding domain.

Type-B ARRs Are Positive Regulators in Cytokinin Signaling

Type-B *ARRs* play overlapping roles in cytokinin signaling and plant development. Analysis of loss-of-function subfamily I type-B *arr* mutants has elucidated the functions of these genes. Single type-B *arr* mutants display slight cytokinin insensitivity in seedlings, while multiple mutations in the subfamily I ARRs *ARR1*, *ARR10*, *ARR12* confer additive cytokinin insensitivity in primary root growth, hypocotyl elongation, and shoot formation in tissue culture. These results suggest that subfamily I type-B ARRs act as redundant positive elements in cytokinin signaling. Mutations in type-B *ARRs* also compromise the induction of cytokinin primary response genes, indicating that they mediate the immediate transcriptional response to cytokinin.

Subfamily I type-B *ARR*s are broadly detected in vegetative and reproductive tissues by RT-PCR, with the highest expression in young developing leaves and meristems, which is consistent with the phenotypes resulting from disruption of these genes. Subfamily II and III type-B *ARR*s have more restricted expression patterns but still overlap to some extent with those of subfamily I members. The restricted expression profile suggests that subfamily II and III type-B ARRs may function in specific developmental processes. Consistent with this, *ARR20*, which belongs to subfamily III type-B *ARR*, is a positive regulator of cytokinin signaling in regulating pavement cell morphogenesis. Similar to the cytokinin double receptor mutant *ahk3-3 cre1-12*, the single *arr20* mutant displays a modest enhancement of pavement cell interdigitation. This phenotype is not observed even in *arr1-3 arr10-5 arr12-1* mutant, which shows almost complete insensitivity to exogenous cytokinin in multiple assays.

The Receiver Domains of Type-B ARRs Have Inhibitory Effect on DNA-Binding Domains

Cytokinin signaling activates type-B *ARRs* via phosphorylation at the conserved Asp residue in the receiver domain. The N-terminal receiver domains of type-B ARRs interact with the AHP phosphotransfer proteins in both in vitro and in vivo assays. Mutation of the phospho-receiving Asp in ARR2 to a phospho-insensitive Asn abolishes the ability of this protein to be phosphorylated via phosphotransfer activity. This ARR2^{D80N} mutant protein lacks the ability to transactivate type-A *ARR6* promoters in response to cytokinin, indicating that phosphorylation of the receiver domain is required for type-B ARR function.

Transgenic plants overexpressing full-length type-B ARRs display wild-type morphology; however, deletion of the receiver domain leads to constitutive activation of this transcription factor. Overexpressing several N-terminal truncated type-B ARRs (ARR2, ARR11, ARR18, ARR19, ARR20, and ARR21), which contain only the C-terminal DNA-binding domain, results in their constitutive activation and pleiotropic phenotypes. These results indicate potential novel roles of type-B ARRs in plant growth and development, although the ectopic and overexpression of these activated type-B ARRs may not faithfully reflect their endogenous functions.

One simple model for the autoinhibitory effect of the type B N-terminal domains is that the receiver domain blocks, through steric hindrance or direct interaction, the activity of the DNA-binding domain. This is similar to the mechanism in many bacterial RRs. Phosphorylation of the receiver domain triggers conformational changes that derepress the inhibitory effect on the DNA-binding domain.

Type-B ARRs Function as Transcription Factors

In vitro assays have shown that the GARP motifs of at least four type-B ARRs bind to DNA in a sequence-specific manner; the core sequences that ARR1/ARR2, ARR10, and ARR11 preferentially bind are GAT(T/C), AGATT, and GGATT, respectively. These core sequences are too short to specify direct targets as they appear too frequently by random in genomes. A study analyzing the *cis*-elements of the target genes induced by ARR1 extended the ARR1-binding sequences to AAGAT(C/T), GAT(C/T)TT, and AAGAT(C/T)TT, which were found to be tandemly enriched in target promoters. A meta-analysis of cytokinin-regulated genes further showed that the ARR1 consensus binding site AAGAT(C/T)TT was substantially overrepresented in the regulatory regions of robustly cytokinin-responsive genes. Single mutations in the ARR1 binding sites AGATT to ACATT in the promoter region of type-A *ARR15* were sufficient to eliminate its cytokinin responsiveness in planta, demonstrating that this regulatory element is required for cytokinin responsiveness.

Despite sharing a core binding sequence (GAT), ARR11, unlike other type-B ARRs, does not bind to the cytokinin-responsive element AGATT, suggesting nonoverlapping targets and hence function of type-B ARRs. Indeed, seven of the type-B ARRs (ARR11, ARR14, and ARR18 of subfamily I; ARR13 of subfamily II; ARR19 and ARR20 of subfamily III) under the control of the *ARR1* native promoter were not able to complement the root growth phenotype of *arr1 arr12*, suggesting these type-B ARR proteins may have distinct functions.

The Direct Targets of Type-B ARRs

Type-B ARRs-regulated genes have been identified by microarray analyses of wild-type and multiple type-B *arr* mutants in response to cytokinin treatment. These studies indicate that type-B ARRs are essential for nearly all cytokinin-regulated gene expression. Further, the endogenous levels of many genes not identified as cytokinin-responsive are differentially expressed in the type-B *arr* mutants compared with wild-type seedlings, suggesting the responsiveness of these genes is saturated by endogenous levels of cytokinin.

The different experimental conditions and various combinations of high-order mutants used in these experiments make it difficult to decipher the target specificity for individual type-B ARRs. Transgenic plants expressing ARR1ΔDDK–GR, a glucocorticoid (DEX) inducible chimeric transcription factor fused to ARR1 lacking its receiver domain, have been used to identify the genes rapidly activated by ARR1ΔDDK upon DEX treatment. These genes potentially represent direct targets of ARR1 and include genes such as cytokinin oxidase, cytokinin hydroxylase, putative disease resistance response proteins, and IAA3/SHY2. Chromatin immunoprecipitation and gel mobility shift analyses confirmed that ARR1 directly associates with the promoter region of *IAA3/SHY2*.

Cytokinin Response Factors

Multiple *cytokinin response factors* (*CRFs*), a subset of the large, plant-specific AP2/ERF superfamily of transcription factors, are upregulated in response to exogenous cytokinin (Rashotte et al. 2006). The CRFs include a group of six core members that contain an AP2/ERF domain and a CRF motif. In *Arabidopsis* seedlings, *CRF2* and *CRF5* show rapid (<30 min) induction upon cytokinin treatment, while *CRF6* induction does not peak until later (>8 h), indicating different kinetics in their response to cytokinin. Despite the name, not every *CRF* has been found to be cytokinin responsive. Similarly, cytokinin upregulates only a subset of the *CRFs* in tomato, which also display distinct kinetics in response to cytokinin.

The rapid induction of the *CRFs* by cytokinin is compromised in type-B *arr* mutants, leading to the hypothesis that CRFs regulate part of the transcription network downstream of type-B ARRs. Indeed, multiple cytokinin-regulated genes exhibit reduced responsiveness in the loss-of-function *crf* mutants, suggesting CRFs are responsible for a subset of cytokinin responses. The induction of type-A *ARRs*, however, is not dependent on CRFs as they retain the wild-type level of induction in *crf* mutants. Overall, *CRFs* are hypothesized to form a side branch of the cytokinin response downstream of type-B *ARRs*.

Future Directions

Since the initial discovery of a cytokinin receptor, remarkable progress has been made in our understanding of cytokinin signaling. Two-component elements modulate, via sequential phosphorelay events, cytokinin signal transduction. Loss-of-function and gain-of-function mutants in these elements have helped define the roles of cytokinin signaling in plant growth and development. Meanwhile, the elucidation of the pathway has raised many fundamental questions that range from the perception of cytokinins to their outputs. For example, the subcellular localization of cytokinin receptors in ER raises the question as to how cytokinins are transported across the plasma membrane and ER membrane to reach the receptors in the ER lumen. Are there functionally relevant cytokinin receptors at the cell surface, and are the AHKs the only receptors? How is specificity in outputs achieved in the pathway, and how is this pathway integrated with other signals to achieve appropriate growth and development? How does the non-receptor histidine kinase CKI1 sense stimuli and feed into cytokinin signaling?

The phosphotransfer proteins that shuttle the phosphoryl group between the receptors and the response regulators provide another layer of regulation in this system. At the cellular level, studies are needed understand the mechanism underlying the transport of AHPs in and out of the nucleus, which allows the phospho-AHPs to maintain the appropriate phosphorylation status of the system elements. As for the ARRs, the mechanism by which type-A ARRs negatively feedback on cytokinin signaling is not yet clear. Likewise, how phosphorylation derepresses the inhibitory effect of the receiver domain on the DNA-binding domain of type-B ARR proteins has not been explored.

Disruption of cytokinin signal transduction, at multiple branch points, implicates the requirement of cytokinin in many aspects of plant growth and development. The next step is to illuminate the mechanisms by which cytokinin signaling elicits multiple, diverse biological outputs and interacts with endogenous and environmental stimuli.

References

Brandstatter I, Kieber JJ. Two genes with similarity to bacterial response regulators are rapidly and specifically induced by cytokinin in *Arabidopsis*. Plant Cell. 1998;10:1009–19.

Deng Y, Dong H, Mu J, Ren B, Zheng B, Ji Z, Yang WC, Liang Y, Zuo J. Arabidopsis histidine kinase CKI1 acts upstream of histidine phosphotransfer proteins to regulate female gametophyte development and vegetative growth. Plant Cell. 2010;22:1232–48.

El-Showk S, Ruonala R, Helariutta Y. Crossing paths: cytokinin signalling and crosstalk. Development. 2013;140:1373–83.

Ferreira FJ, Kieber JJ. Cytokinin signaling. Curr Opin Plant Biol. 2005;8:518–25.

Kakimoto T. CKI1, a histidine kinase homolog implicated in cytokinin signal transduction. Science. 1996;274:982–5.

Rashotte AM, Mason MG, Hutchison CE, Ferreira FJ, Schaller GE, Kieber JJ. A subset of Arabidopsis AP2 transcription factors mediates cytokinin responses in concert with a two-component pathway. Proc Natl Acad Sci U S A. 2006;103:11081–5.

Schaller GE, Shiu SH, Armitage JP. Two-component systems and their co-option for eukaryotic signal transduction. Curr Biol. 2011;21:R320–30.

Stock JB, Stock AM, Mottonen JM. Signal transduction in bacteria. Nature. 1990;344:395–400.

Stock AM, Robinson VL, Goudreau PN. Two-component signal transduction. Annu Rev Biochem. 2000;69:183–215.

To JP, Deruère J, Maxwell BB, Morris VF, Hutchison CE, Ferreira FJ, Schaller GE, Kieber JJ. Cytokinin regulates type-A Arabidopsis response regulator activity and protein stability via two-component phosphorelay. Plant Cell. 2007;19:3901–14.

Future Reading

Bhargava A, Clabaugh I, To JP, Maxwell BB, Chiang YH, Schaller EG, Loraine A, Kieber JJ. Identification of cytokinin responsive genes using microarray meta-analysis and RNA-seq in *Arabidopsis thaliana*. Plant Physiol. 2013;162:272–94.

Efroni I, Han SK, Kim HJ, Wu MF, Steiner E, Birnbaum KD, Hong JC, Eshed Y, Wagner D. Regulation of leaf maturation by chromatin-mediated modulation of cytokinin responses. Dev Cell. 2013;24:438–45.

Feng J, Wang C, Chen Q, Chen H, Ren B, Li X, Zuo J. *S*-nitrosylation of phosphotransfer proteins represses cytokinin signaling. Nat Commun. 2013;4:1529.

Hwang I, Sheen J, Müller B. Cytokinin signaling networks. Annu Rev Plant Biol. 2012;63:353–80.

Kim HJ, Chiang YH, Kieber JJ, Schaller GE. SCFKMD controls cytokinin signaling by regulating the degradation of type-B response regulators. Proc Natl Acad Sci U S A. 2013;110:10028–33.

Tsai YC, Weir NR, Hill K, Zhang W, Kim HJ, Shiu SH, Schaller GE, Kieber JJ. Characterization of genes involved in cytokinin signaling and metabolism from rice. Plant Physiol. 2012;158:1666–84.

Brassinosteroid Signaling in Plants

10

Steven Clouse

Contents

Abstract

- Brassinosteroids (BRs) are a class of plant hormones comprising more than 70 polyhydroxylated sterol derivatives with structural similarity to animal steroid hormones.
- BRs are widely distributed across the plant kingdom and play essential roles in regulating multiple physiological processes and developmental programs including cell and organ elongation, cell division and differentiation, vegetative and reproductive development, and responses to the environment.

S. Clouse (✉)
Department of Horticultural Science, College of Agriculture and Life Sciences, North Carolina State University, Raleigh, NC, USA
e-mail: steve_clouse@ncsu.edu

S.H. Howell (ed.), *Molecular Biology, The Plant Sciences 2*,
DOI 10.1007/978-1-4614-7570-5_12

- BR signaling is initiated by binding of the BR ligand to the extracellular domain of a membrane-bound receptor kinase, BRASSINOSTEROID-INSENSITIVE 1 (BRI1).
- BRI1 forms heterooligomers with its co-receptor, BRI1-ASSOCIATED RECEPTOR KINASE 1 (BAK1), and its close homologs to initiate a phosphorelay that ultimately results in changes in nuclear gene expression.
- Specific phosphorylation sites of BRI1 and BAK1 have been analyzed by biochemical and genetic approaches, and phosphorylation at individual residues has been shown to have profound effects on BR signaling and overall plant growth.
- Downstream substrates of BRI1 have been identified and characterized as well as intermediate signaling components including kinases, phosphatases, and transcription factors.
- BR signaling generates pleiotropic responses by regulating transcription factors that target genes involved in cell elongation and cell wall metabolism, cell division and differentiation, development of numerous plant organs, environmental responses to light and abiotic and biotic stress, and genes involved in the biosynthesis, transport, and signaling of other hormones.
- Specific components of BR, gibberellin (GA), and light signaling pathways directly interact to form a core transcription module that coordinately regulates seedling morphogenesis.

Introduction

Plants respond to internal developmental cues as well as external environmental factors by activating signal transduction pathways that regulate growth and development, metabolism, and homeostasis. Protein receptors perceive specific ligands resulting in a cascade of biochemical events that often culminates with changes in gene expression. The newly synthesized gene products then act to alter cell size, number, shape, and function, which ultimately can lead to organ initiation, patterning, and morphogenesis. Thus, specific ligand/receptor interactions, often at the cell surface, regulate physiological processes and developmental programs through signaling pathways that alter nuclear gene expression. Intermediate components of the signaling pathway are required to relay information from signal perception to changes in gene expression. One of the most common signaling mechanisms in both plants and animals is reversible phosphorylation of Ser, Thr, and/or Tyr residues in signaling proteins. Reversible protein phosphorylation by a kinase or dephosphorylation by a phosphatase can dramatically change protein function. For example, a change in phosphorylation status at a specific residue may alter enzyme activity, affect DNA-binding properties of a transcription factor, determine the protein partners in a complex, and regulate protein stability and subcellular localization. Because of the importance of protein phosphorylation in signal transduction mechanisms, methods to determine phosphorylation sites in a protein by liquid chromatography-tandem mass spectrometry (LC/MS/MS), as well as biochemical and genetic approaches to determine the functional significance of phosphorylated residues, are critical in research characterizing signaling pathways.

In mammals, proteins called receptor kinases that combine the ability to bind a specific ligand with the capacity to phosphorylate the same protein (autophosphorylation) and/or an independent substrate (transphosphorylation) have been thoroughly studied for several decades and are known to be key regulators of cellular function and organ development. Plants also contain very large families of receptor-like kinases (RLKs) with more than 600 RLKs in Arabidopsis (*Arabidopsis thaliana*) and over 1,000 in rice (*Oryza sativa*). The overall domain structure of both plant and animal receptor kinases is conserved, although their evolutionary origin may not be. Both plant and animal receptor kinases have an ectodomain that may be responsible for binding a ligand followed by a single-pass transmembrane domain that anchors the protein within a membrane and finally a kinase domain that phosphorylates target proteins in response to ligand binding by the ectodomain. Receptor kinases are often localized in the plasma membrane with the ectodomain in the extracellular space and the kinase domain within the cytoplasm. Another key feature of receptor kinase mechanisms is the ability to form complexes of homo- or heterooligomers, often in response to ligand binding. The formation of such complexes can lead to transphosphorylation and activation of the kinase domains, which in turn generate docking sites for specific downstream substrates, resulting in their phosphorylation and the initiation of a ligand-dependent phosphorelay.

Plant RLKs are classified based on the nature of functional moieties in the ectodomain, with the most common category being leucine-rich repeat (LRR) RLKs. Plant LRRs are on average 24 amino acids in length with the consensus sequence, LxxLxxLxLxxNxLSGxIPxxLGx (where x is any amino acid). Ligands for several plant LRR RLKs have been identified, and their role in developmental processes such as meristem and abscission zone formation, vascular differentiation, and generation of stomata has been studied, along with LRR RLKs involved in responses to pathogen-associated molecular patterns (PAMPs). Perhaps the best-characterized LRR RLKs in plants are the BR receptor, BRI1, and its co-receptor BAK1.

Early Events in BR Signaling

BRI1 Structure

Loss-of-function, BR-insensitive mutant alleles of BRI1 were first identified in the early 1990s with genetic screens for mutants capable of elongating primary roots in the presence of BR concentrations that were inhibitory to wild-type Arabidopsis. Subsequent genetic screens in Arabidopsis have resulted in a collection of more than two dozen *bri1* mutant alleles, which have been important tools for studies of BRI1 protein function. Null *bri1* alleles with mutations that critically affect either the ectodomain or the kinase domain show an extreme dwarf phenotype with altered leaf morphology, delayed flowering, altered vascular development, and male infertility, suggesting that full BRI1 function is essential for multiple aspects of normal plant development. Mutant alleles of *bri1* have also been studied in several crops including tomato, pea, barley, and rice, with similar dwarfism and other developmental defects being observed. The *BRI1* locus was mapped to the bottom of Arabidopsis chromosome IV in 1996 and positionally cloned in 1997, revealing that *BRI1* encoded an LRR RLK consisting of 1196 amino acids (Clouse 2011b).

The Arabidopsis BRI1 ectodomain is comprised of 25 tandemly arrayed 24-amino acid LRRs with a novel 70 amino acid island domain inserted between LRRs 21 and 22 (Fig. 1a). Mutational analysis of the island domain region suggested its critical importance in BRI1 function, and biochemical experiments using labeled BR and recombinant BRI1 protein confirmed that BR bound directly to the island domain in conjunction with LRR 22, thus uncovering a novel plant steroid binding motif and defining BRI1 as the BR receptor (Kim and Wang 2010). Interestingly, the BR steroid binding motif in plants is structurally distinct from the nuclear receptors in animals that bind a range of steroid hormones. The three-dimensional structure of the BRI1 ectodomain was recently solved by x-ray diffraction which further refined the role of the island domain in BR binding. The BRI1 ectodomain forms a superhelix of twisted LRRs with the island domain folding into the interior where it interacts with numerous amino acids in LRRs 13–25 to generate a hydrophobic pocket that binds one molecule of brassinolide (BL), the most active naturally occurring BR. Steroid binding to BRI1 results in a conformational change in the island domain that may lead to co-receptor binding and kinase domain activation by transphosphorylation. Loss-of-function mutations that map to this region interfere with BL binding and thus disrupt BR signaling (Jiang et al. 2013).

Immediately downstream of the BRI1 ectodomain lies a hydrophobic membrane-spanning region from amino acids 792–814 which anchors the protein in the membrane and separates the ectodomain from the intracellular cytoplasmic domain. The cytoplasmic domain begins with the juxtamembrane region (amino acids 815–882) followed by the catalytic kinase domain from residues 883–1,155 and ending with the carboxy-terminal domain from residues 1,156–1,196. Biochemical experiments using recombinant BRI1 cytoplasmic domain and radiolabeled ATP confirmed that BRI1 was an active kinase capable of both auto- and transphosphorylation. Similar to the ectodomain, mutations mapping to regions of the cytoplasmic domain of BRI1 that result in loss of kinase activity also lead to the typical dwarf stature and other characteristic phenotypes of *bri1* null alleles, suggesting that both BR binding to the ectodomain and kinase activity in the cytoplasmic domain are required for normal BRI1 action and BR signaling. Because of the importance of phosphorylation to BRI1 function, numerous experiments involving mass spectrometry were undertaken to identify individual BRI1 phosphorylation sites in the juxtamembrane, kinase, and carboxy-terminal domains and to characterize their function through biochemical and molecular genetic

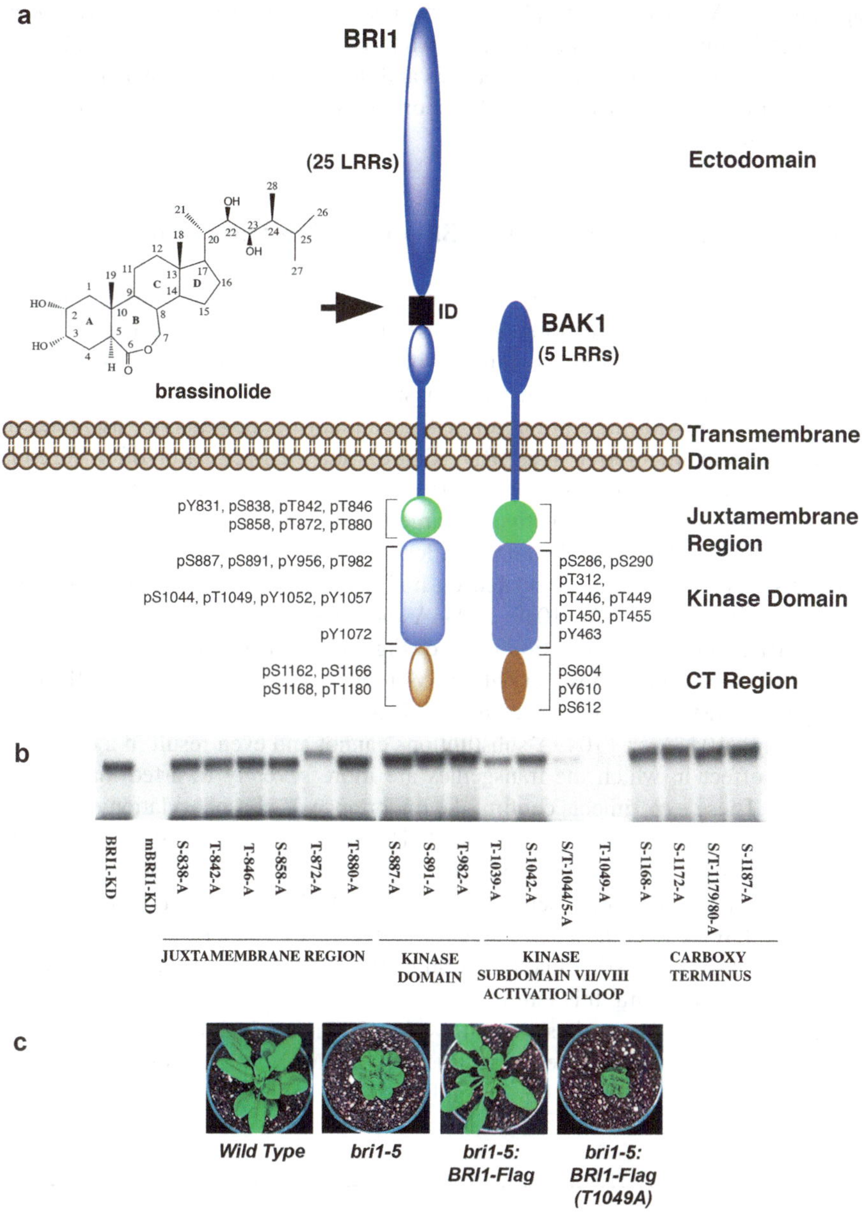

Fig. 1 Structure and function of the BRI1 and BAK1 receptor kinases. (**a**) The organization of structural domains in BRI1 and BAK1 is shown. The structure of brassinolide, the most active BR, is enlarged relative to the BRI1 ectodomain for greater legibility. Brassinolide binds to the island domain (*ID*) between LRR 21 and 22 in the BRI1 ectodomain. Identified phosphorylation sites in the BRI1 and BAK1 cytoplasmic domains are indicated. (**b**) Kinase autophosphorylation assays of wild-type Flag-BRI1 kinase domain and a series of mutants in which Ser and Thr residues were substituted with Ala to prevent phosphorylation at that residue. Flag-mBRI1-KD is a

approaches. At least 12 phosphorylation sites in Arabidopsis BRI1 have been shown to be important for kinase activity and BR signaling (Clouse 2011a). This was demonstrated by LC/MS/MS analysis and in vitro mutagenesis experiments in which Ser or Thr residues at these phosphorylation sites were substituted with Ala (or Tyr with Phe).

The Role of Phosphorylation in BRI1 Function and BR Signaling

In many mammalian receptor kinases, multiple phosphorylation sites are distributed throughout the juxtamembrane, kinase, and carboxy-terminal domains, and a hierarchy of phosphorylation beginning with general kinase activation followed by generation of phosphorylation-dependent specific substrate binding motifs often occurs. Kinases with an arginine directly upstream of the catalytic aspartate in subdomain VIb are referred to as "RD" kinases, and their activation often depends on phosphorylation of one to three residues in the activation loop of subdomains VII and VIII, which allows the negatively charged phosphate groups to interact with the positively charged R in subdomain VIb, leading to kinase activation. BRI1 is also an RD kinase and is phosphorylated on multiple residues in the activation loop, including Ser-1044 and Thr-1049. S1044A and T1049A substitutions dramatically reduce BRI1 autophosphorylation in vitro (Fig. 1b) compared to wild type, as well as a loss of substrate phosphorylation. Moreover, while a wild-type BRI1-Flag transgenic construct can rescue the dwarf phenotype of a *bri1-5* mutant, constructs containing S1044A or T1049A substitutions cannot and even result in a dominant negative effect in which the transgenics are more severely dwarfed than *bri1-5* (Fig. 1c). These experiments confirm the importance of phosphorylation on activation loop residues S1044 and T1049 for BRI1 kinase function and also demonstrate the critical importance of BRI1 phosphorylation for BR signaling and normal plant growth and development. When the sequence of Arabidopsis RD-type LRR RLKs are aligned, more than 99 % have a Ser or Thr residue at the position equivalent to BRI1 Thr-1049, suggesting that this residue may be of fundamental importance in kinase activation within this family.

BRI1 phosphorylation sites mapping within the kinase domain, but outside of the activation loop, have also been identified by LC/MS/MS analysis and by using phospho-specific antibodies. For example, Ser-891 in the ATP-binding domain is

Fig. 1 (continued) kinase-inactive mutant generated by a K911E substitution (Reproduced with permission from Fig. 6 of Wang et al. Plant Cell 17:1685–1703, 2005). (**c**) The activation loop residue Thr-1049 is critical for BRI1 function and BR signaling in vivo. The weak *bri1-5* mutant allele exhibits an intermediate dwarf phenotype and is rescued by expression of a wild-type BRI1-Flag construct. A similar construct in which Thr-1049 is substituted with Ala (which prevents phosphorylation at that residue) fails to rescue and leads to a dominant negative effect with phenotype similar to the extreme dwarf phenotype of *bri1* null alleles (Reproduced with permission from Fig. 1 of Clouse 2011a)

phosphorylated, but in contrast to Ser-1044 and Thr-1049, the result of Ser-891 phosphorylation is inactivation of BRI1 kinase activity rather than activation. BRI1-Flag with a S891A substitution yields plants larger than wild-type BRI1-Flag when transferred into *bri1-5*, while an S891D substitution, which mimics constitutive phosphorylation at that residue due to the negative charge of Asp, is an extreme dwarf and has reduced kinase activity in vitro. These data suggest that Ser-891 phosphorylation inhibits BRI1 kinase activity, most likely by interfering with ATP binding, and also attenuates BR signaling in vivo. Thus, phosphorylation can have either a positive or negative effect on kinase function and signaling depending on the specific residue that is phosphorylated.

In mammalian receptor kinases, phosphorylation of the juxtamembrane and carboxy-terminal regions that flank the kinase domain can also have a general regulatory effect on kinase activation, and BRI1 also exhibits some of these same properties. Deletion of the BRI1 carboxy-terminal region results in larger plants and higher in vivo phosphorylation levels when the deletion construct is compared to wild-type BRI1 expressed in the *bri1-5* mutant. Several BRI1 carboxy-terminal phosphorylation sites have been identified, and when these are substituted with Asp to generate phosphomimetic mutants, BRI1 kinase activity towards peptide substrates is increased, and transgenic plants have longer hypocotyls than wild-type BRI1. Taken together, these data suggest that the carboxy-terminal region negatively regulates BRI1 function and BR signaling, and this negative regulation is released by phosphorylation of multiple residues within this domain. In contrast to the carboxy-terminal region, deletion of the BRI1 juxtamembrane domain results in reduction of BR signaling in vivo and loss of autophosphorylation and peptide substrate phosphorylation in vitro, suggesting that the juxtamembrane domain is an activator of BRI1 kinase function.

In addition to the general regulatory effects of the juxtamembrane and carboxy-terminal regions on kinase activity discussed above, phosphorylation of specific residues within the juxtamembrane and carboxy-terminal regions in many mammalian receptor kinases generates recognition sites for downstream substrate binding and subsequent phosphorylation of these substrates by the activated receptor kinase. BRI1 again is similar to mammalian receptor kinases in this respect. LC/MS/MS analysis identified the BRI1 juxtamembrane residues Ser-838, Thr-842, Thr-846, and Ser-858 as phosphorylation sites in vivo and in vitro, and functional analysis of constructs in which these residues were substituted with Ala showed that elimination of phosphorylation at these residues had no detectable effect on BRI1 autophosphorylation, but dramatically decreased the phosphorylation of a peptide substrate. Consistent with these results, the phosphomimetic mutants S838D, T842D, T846D, and S858D showed enhanced phosphorylation of the peptide substrate. Moreover, S838A, T846A, and S858A mutants all showed reduced BR signaling in vivo when hypocotyl elongation was used as a readout of BR action. Taken together, these results suggest that these four BRI1 juxtamembrane sites have a positive effect on BR signaling by enhancing the ability of BRI1 to phosphorylate downstream substrates.

Besides the positive effects of juxtamembrane phosphorylation discussed above, phosphorylation of specific sites in the BRI1 juxtamembrane region can also have negative effects on kinase function and BR signaling. Thr-872 was demonstrated by LC/MS/MS analysis to be a juxtamembrane site in vitro and in vivo. T872A substitutions show a tenfold increase in BRI1 peptide substrate phosphorylation, indicating that Thr-872 is a negative regulatory site when phosphorylated and T872A mutations release this inhibition. The role of Thr-872 phosphorylation in vivo is currently unknown. Tyr831 is another BRI1 juxtamembrane phosphorylation site, and Y831F substitutions expressed in *bri1-5* had larger leaves and earlier flowering time than *bri1-5* expressing wild-type BRI1, suggesting that phosphorylation at Tyr-831 has a negative effect on leaf development and time of flowering (Oh et al. 2009). Further analysis showed that Y831F mutants had higher rates of photosynthesis and enhanced accumulation of starch, sucrose, and amino acids, demonstrating that phosphorylation status at a specific BRI1 residue can have effects on overall leaf growth, photosynthetic rates, and carbon assimilation in the whole plant.

BRI1 Interacting Partners: BAK1 and Other SERKs

Oligomerization is a common feature of mammalian receptor kinase mechanisms. Homo- or heterodimers and/or higher-order oligomers may form in response to ligand binding, or preformed oligomers may exist in the absence of ligand, which are stabilized and activated in response to ligand binding. Transphosphorylation between members of receptor kinase complexes in response to ligand binding is also commonly observed, and the kinetics, sequence, and symmetry of phosphorylation of components vary with specific receptor complexes. In the case of BRI1, association with members of another LRR RLK family, the SOMATIC EMBRYOGENESIS RECEPTOR KINASEs (SERKs), is required for full activation of BR signaling in vivo. SERKs are smaller LRR RLKs than BRI1 with only five LRRs in the ectodomain, followed by a unique Ser-Pro-Pro repeat before the transmembrane domain (Chinchilla et al. 2009). SERK3 was initially shown to interact with BRI1 both in vitro and in vivo by a variety of biochemical and genetic approaches, and it was shown to have a positive effect on BR signaling and BRI1 function. Because of this, SERK3 has been renamed as BRI1-ASSOCIATED RECEPTOR KINASE 1 (BAK1). SERK1 and SERK4, alternatively named BAK1-LIKE 1 (BKK1), also interact with BRI1 in vivo and play a role in BR signaling. Independent positive and negative regulatory roles for BAK1 and BKK1 in BR signaling and BR-independent cell death pathways, respectively, have been demonstrated. Moreover, BAK1 heterodimerizes with other LRR RLKs besides BRI1 and promotes their function in plant defense responses. Thus, BAK1 operates in independent pathways by enhancing the signaling output of distinct LRR RLK partners that bind different ligands, suggesting that SERKs in general are co-receptors that regulate multiple independent pathways by heterooligomerization with different LRR RLKs (Li 2010a).

The binding of BRI1 to multiple SERK family members serves as a model for the interactions of other LRR RLKs, and a great deal of effort has been expended to examine the mechanisms of the BRI1/BAK1 interaction and its role in early events in BR signaling. BAK1 was discovered independently in 2002 by yeast two-hybrid analysis and genetic screens for suppressors of the *bri1-5* weak mutant allele. These early studies also showed that BRI1 and BAK1 transphosphorylated each other when expressed in yeast and bacterial cells. Subsequent studies expressing different combinations of kinase-inactive and wild-type epitope-tagged versions of BRI1 and BAK1 in the same transgenic Arabidopsis plant suggested that an active BRI1 kinase, but not BAK1 kinase, was required for BR-dependent association of the pair, as determined by co-immunoprecipitation. Moreover, BRI1 overexpression increased phosphorylation of BAK1, while expressing BAK1-GFP in the *bri1-1* null mutant background reduced BAK1-GFP phosphorylation levels dramatically. Immunoblot analysis further showed that BR treatment increased phosphorylation of both BRI1 and BAK1 in vivo. As with BRI1, combining LC/MS/MS analysis, functional characterization in mutant backgrounds, and in vitro biochemical studies also identified multiple in vivo and in vitro phosphorylation sites for BAK1 (Fig. 1). Interestingly, phosphorylation of the activation loop residue Thr-455 appears essential for BAK1 function just as the corresponding BRI1 residue Thr-1049 was also required for kinase function and BR signaling in planta, as discussed above. Solving the crystal structure of the BAK1 cytoplasmic domain also confirmed that Thr-450 and Thr-455 phosphorylation played a key role in stabilizing key BAK1 functional motifs. A range of in vitro kinase assays also showed that BAK1 interaction stimulates BRI1 activity towards peptide substrates and that both BRI1 and BAK1 can transphosphorylate each other on specific residues. BRI1 primarily transphosphorylates BAK1 on kinase domain residues, including the essential Thr-455, while BAK1 transphosphorylates BRI1 on the juxtamembrane and carboxy-terminal residues discussed above that appear to increase BRI1 activity on downstream substrates, leading to enhanced BR signaling. Like BRI1, BAK1 can also autophosphorylate on Tyr residues, and Tyr-610 appears to be an important residue for the interaction with BRI1 since Y610F BAK1 mutants have reduced BR signaling in vivo and reduced transphosphorylation of BRI1 by BAK1 in vitro.

The combined input of the studies described above allowed the development of an initial sequential transphosphorylation model of BRI1/BAK1 interaction (Fig. 2) that suggests plant receptor kinases share some of the properties of mammalian receptor kinase mechanisms while retaining unique plant-specific features. In this model, activated BRI1 can signal independently of BAK1 at a basal level to promote plant growth or oligomerize with inactive BAK1 and transphosphorylate it on kinase domain residues. BRI1-activated BAK1 then transphosphorylates BRI1 on juxtamembrane and carboxy-terminal residues, thus quantitatively enhancing BRI1 phosphorylation of downstream substrates and increasing BR-signaling output. In this model, the BAK1-independent function of BRI1 was based on observations that BRI1 can form homodimers in vivo and that BRI1 overexpression increases hypocotyl elongation even in a *bak1/bkk1* double mutant background. Therefore, it was proposed that BRI1 homooligomers might bind BR and lead to

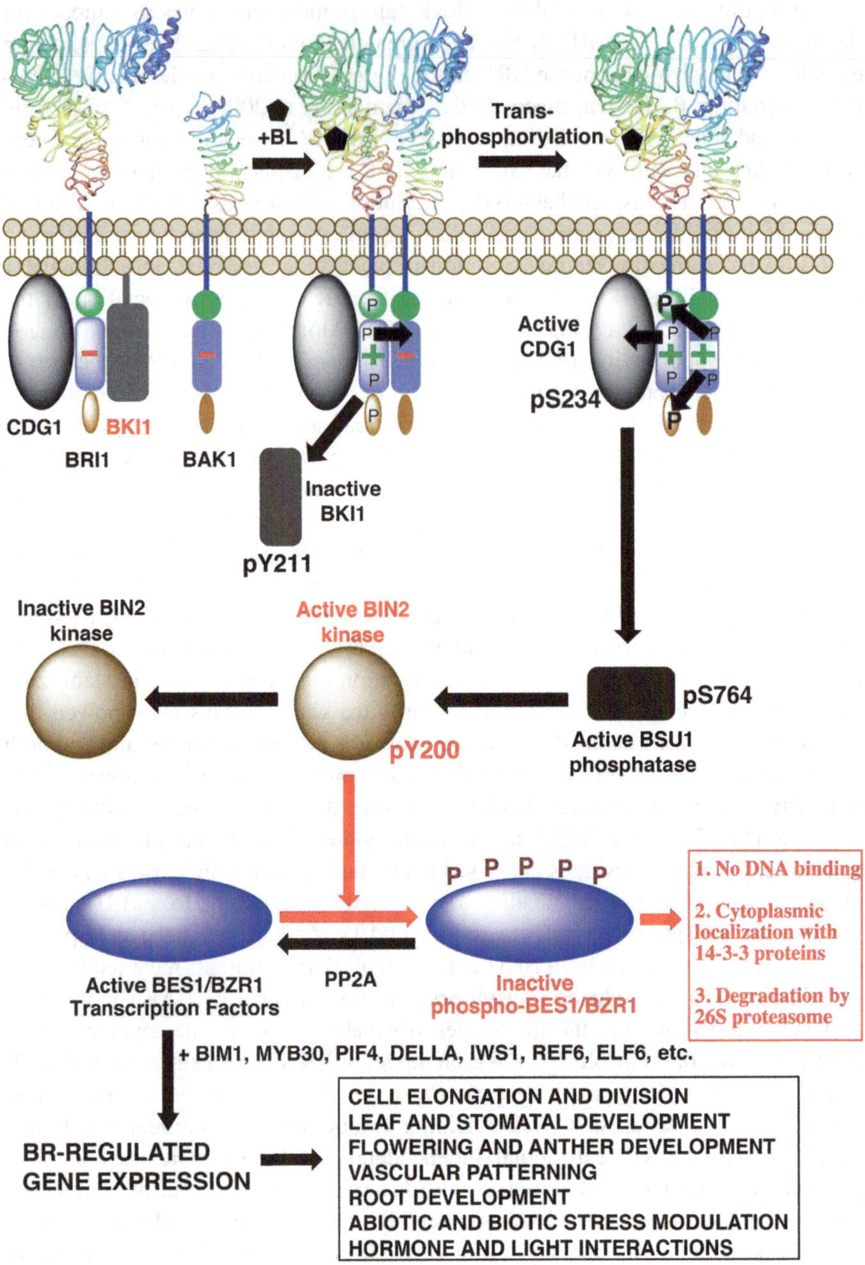

Fig. 2 Overview of the BR-signaling pathway. The three-dimensional structure of the BRI1 and BAK1 ectodomains is shown. BL binding to BRI1 leads to release of the BKI1 inhibitor and transphosphorylation with the co-receptor BAK1 (or SERK1 and BKK1, not shown) which activates the signaling pathway. Activated BRI1 phosphorylates CDG1 on Ser-234 (and BSK1, not shown) which phosphorylates the BSU1 phosphatase on Ser-764. The activated BSU1 then

initial activation of the BRI1 kinase in the absence of BAK1. However, recent structural studies of the BRI1 ectodomain indicated that it was unlikely that BRI1 homodimers could bind BR, and the role of these BRI1 oligomers in vivo remains uncertain. Moreover, while BRI1 was demonstrated to function in a *bak1/bkk1* double mutant background, subsequent studies with *serk1/bak1/bkk1* triple mutants showed that some SERK activity was essential for BRI1 function in BR signaling. Finally, the initial sequential phosphorylation model was derived from observations of BRI1/BAK1 interaction by co-immunoprecipitation which requires a fairly stable interaction to survive tissue extraction and membrane purification protocols. Very recent work using fluorescence lifetime imaging microscopy in intact Arabidopsis root tissue verified that BR treatment can increase BRI1/BAK1 association, but also demonstrated that a substantial number of preformed BRI1/BAK1 heterooligomers exist in the absence of BR ligand. Thus, while transphosphorylation within the BRI1/BAK1 complex appears with certainty to be essential for BRI1 signaling, the early events leading to ligand-dependent activation of BRI1 and BAK1 are still uncertain and modification of the sequential transphosphorylation model will be required.

BRI1 Interacting Partners: BKI1

Further complicating the story of BRI1/BAK1 interaction in early BR signaling was the discovery of BRI1 KINASE INHIBITOR1 (BKI1), a negative regulator of BR signaling which is membrane bound and interacts with BRI1 in the absence of BR. BR activation of BRI1 causes release of BKI1 from the membrane, most likely by phosphorylation of BKI1 by BRI1 on Tyr-211 within a membrane-binding motif. Consistent with this view, overexpression of a BKI1-Y211F mutant that cannot be phosphorylated on this residue results in severely dwarfed plants and constitutive membrane localization of the Y211F mutant protein even after BR treatment. Further work has shown that a C-terminal 20 amino acid residue of BKI1 is responsible for binding BRI1 and that a synthetic peptide representing this domain interferes with the interaction of BRI1 and BAK1 cytoplasmic domains in vitro. This suggests that the negative effect of BKI1 on BR signaling results from inhibition of BRI1 interaction with its essential SERK co-receptors, which is released by BR-dependent phosphorylation of BKI1. Structural studies of the SERK1 ectodomain along with mutational analysis of the BAK1 LRR domain have shown that the ectodomain of SERKs can interact directly with the

Fig. 2 (continued) dephosphorylates the BIN2 kinase on Tyr-200, leading to its inactivation. In the absence of BIN2 and the presence of PP2A, the unphosphorylated and active forms of the BZR1 and BES1 transcription factors then accumulate in the nucleus where they heterodimerize with a variety of other transcription factors, leading to the regulation of over 1,000 genes involved in various physiological responses modulated by BRs. Positive and negative regulatory inputs are indicated with *black* and *red arrows* and text, respectively

ectodomain of BRI1 and that this interaction is enhanced by BR. This suggests that even though BKI1 may interfere with the interaction of BRI1 and SERKs via the cytoplasmic domains, the ectodomains could interact in response to BR and activate BRI1 sufficiently to phosphorylate BKI1, releasing the inhibitor and allowing interaction and transphosphorylation of BRI1 cytoplasmic domains and those of its SERK co-receptors. Additional work is required to understand the order and kinetics of BRI1, SERK, and BKI1 interaction during early events of BR signaling at the membrane.

Downstream Intermediates in BR Signaling

After initial activation of BRI1 and BAK1, phosphorylation of downstream components takes place, which initiates the BR-dependent phosphorelay (Gruszka 2013; Zhu et al. 2013). Approximately 25 % of the 610 members of the Arabidopsis RLK family lack an ectodomain and are therefore referred to as the receptor-like cytoplasmic kinase (RLCK) subfamily. Several members of this group have been demonstrated to be substrates of the BRI1 kinase domain and positive regulators of BR signaling. BR-SIGNALING KINASE 1 (BSK1), a member of the RLCK-XII subfamily, interacts with BRI1 in vivo in the absence of ligand and is phosphorylated by BRI1 on Ser-230 in response to BR, which activates BSK1 and releases it from the receptor complex. Similarly, CONSTITUTIVE DIFFERENTIAL GROWTH 1 (CDG1), a member of the RLCK-VIIc subfamily, interacts with BRI1 and is activated by phosphorylation on Ser-234 (equivalent to BSK1 Ser-230), which enhances CDG1 kinase activity. Furthermore, S234A mutants of CDG1 have a reduced positive effect on BR signaling in vivo, showing that this residue is critical for CDG1 function. Phosphorylation of BSK1 and CDG1 by BRI1 promotes their interaction with and activation of another downstream positive regulator of BR signaling, BRI1 SUPPRESSOR 1 (BSU1) phosphatase. The exact mechanism of how BSK1 activates BSU1 is not known, but it has been demonstrated that CDG1 phosphorylates BSU1 on Ser-764 in the C-terminal region of BSU1, which increases BSU1 binding to the key negative regulator of BR signaling, BR-INSENSITIVE 2 (BIN2) kinase, resulting in dephosphorylation of Tyr-200 in BIN2 which renders the kinase inactive.

BIN2 shows sequence similarity to Drosophila shaggy kinase and mammalian glycogen synthase kinase 3 (GSK3), both of which often function as negative regulators of signaling pathways involved in metabolism, cell fate determination, and pattern formation. In Arabidopsis, BIN2 belongs to a small multigene family of 10 members, several of which have been implicated in BR signaling. BIN2 phosphorylates two closely related transcription factors, BRASSINAZOLE-RESISTANT1 (BZR1) and BRI1-EMS SUPPRESSOR1 (BES1), also known as BZR2, which play essential roles in BR signaling by regulating the expression of a network of more than 1,000 genes involved in multiple aspects of the BR response. Phosphorylation of BZR1 and BES1 by BIN2 inactivates these transcription factors by a variety of mechanisms including loss of their DNA-binding capacity, their nuclear export and retention in the cytoplasm in a complex with 14-3-3 proteins, and their proteasome-mediated degradation. In contrast to animal GSKs, which require priming phosphorylation and/or scaffolding proteins to phosphorylate their substrates, BIN2 phosphorylates BZR1 by direct binding through a 12 amino acid docking motif. Deletion of this motif results in disruption of the BIN2-BZR1 interaction and loss of in vivo phosphorylation of BZR1, demonstrating its importance.

Phosphorylation of BIN2 itself is also critical for its function. Tyr-200 is a highly conserved residue in GSK3 kinases, and its phosphorylation is essential for kinase activity of BIN2 as well as mammalian kinases. A BIN2 mutant with a Y200F substitution losses its ability to negatively regulate BR signaling in vivo, which confirms the importance of phosphorylation at this residue. When BSU1 dephosphorylates BIN2 Tyr-200 in response to activation of BR signaling, BIN2 loses kinase activity and can no longer phosphorylate BZR1 and BES1. The hypophosphorylated forms of these transcription factors then accumulate in the nucleus, which is aided by the action of a protein phosphatase 2A (PP2A), which dephosphorylates BZR1 and BES1 and is thus a positive regulator of BR signaling. The B' subunit of PP2A binds directly to the Pro-, Glu-, Ser-, and Thr-rich (PEST) domain of BZR1, which includes Ser-173, a residue whose phosphorylation promotes 14-3-3 binding and cytoplasmic retention of BZR1. Dephosphorylation of Ser-173 by PP2A interferes with 14-3-3 binding and so increasing the accumulation of the active, dephosphorylated form of BZR1 in the nucleus. Thus, PP2A and BIN2 have directly antagonistic roles on BZR1 phosphorylation and BR signaling.

The BZR1 and BES1 Transcription Factors and Gene Regulatory Networks

The BZR1 and BES1 transcription factors distribute the BR signal from the initial single binding event of BR to BRI1 at the cell surface to a multitude of gene responses that affect numerous physiological processes (Gudesblat and Russinova 2011). Among the BZR1 and BES1 target genes are genes for additional transcription factors, which promote amplification of the signal, and both BZR1 and BES1 can heterodimerize with additional protein partners to partition the signal to specific response pathways. BZR1 and BES1 share 88 % sequence identity, and both have an N-terminal DNA-binding domain and a C-terminal atypical basic helix-loop-helix (bHLH) domain, which may be involved in dimerization with other transcription factors. Both BZR1 and BES1 recognize two different promoter binding motifs, the BR-response element (BRRE, CGTG(T/C)G) and the E box element (CANNTG). Both BZR1 and BES1 can either activate or repress gene transcription depending on promoter structure and dimerization partners. A dominant mutant, *bzr1-1D*, contains a P234L substitution that stabilizes the protein in the active form and thus leads to constitutive BR signaling even in the absence of ligand (Li 2010b).

Experiments using chromatin immunoprecipitation microarray (ChIP-chip) analysis to identify direct binding targets of BZR1 and BES1 in BR-regulated genes have uncovered an extensive regulatory network that modulates growth and development in Arabidopsis. ChIP-chip analysis showed that 953 BR-regulated genes bind BZR1 in their promoter, resulting in 450 genes being activated, 462 repressed, and 41 showing complex regulation. Similarly, there are 250 direct targets of BES1 that are also BR-regulated with 165 of these showing activation by BES1 binding and 85 showing repression. A total of 120 genes appear in both BZR1 and BES1 lists indicating BZR1 and BES1 can regulate the same gene in many cases. Looking at the confirmed or predicted function of many of these target genes, it is clear why BR signaling can generate so many pleiotropic responses. BZR1 and BES1 target genes include those involved in cell elongation and cell wall metabolism; cell division and differentiation; development of numerous organs including flowers, leaves, shoots, and roots; environmental responses to light and abiotic and biotic stress; and genes involved in the biosynthesis, transport, and signaling of other hormones, including auxin, cytokinin, GA, abscisic acid (ABA), ethylene, and jasmonate (Wang et al. 2012). This diversity of response is likely facilitated by the interaction of BZR1 and BES1 with other transcription factors including those in the bHLH, MYB, and Jumonji N/C families of nuclear proteins. The current status of the BR-signaling pathway is summarized in Fig. 2.

BR Signaling and Plant Growth and Development

Vegetative Growth and Development

Since their discovery more than 30 years ago, it has been clear that BRs have a pronounced positive effect on the elongation of hypocotyls, epicotyls, and peduncles in dicots and coleoptiles and mesocotyls in monocots. Examination of cell files in wild-type plants versus dwarf mutants defective in BR biosynthesis or signaling provides convincing evidence that active BR signal transduction components such as BRI1 are essential for normal cell expansion. Turgor-driven cell expansion requires extensive modification and remodeling of the cell wall through the action of wall-modifying enzymes including xyloglucan endotransglycosylase/hydrolases (XTHs), glucanases, pectinases, expansins, sucrose synthase, and cellulose synthase. BR regulation of genes encoding XTHs and expansins was demonstrated in soybean, tomato, and Arabidopsis as early as 1994, and BRs are known to promote wall loosening in soybean epicotyls and hypocotyls of *Brassica chinensis* and *Cucurbita maxima*. Studies of individual genes as well as global gene expression analyses have shown that many genes involved in cell wall modifications, ion and water transport, and cytoskeleton rearrangements are BR regulated. ChIP-chip analysis further shows that many of these expansion-related genes are direct targets of the BZR1 and/or BES1 transcription factors, thus directly linking BR signaling to altered expression of genes essential for cell elongation. Changes in membrane polarization and H^+-ATPase activity are also involved in cell expansion, and recent studies utilizing microscopy and electrophysiology suggest that BR promotes plasma membrane hyperpolarization and rapid cell elongation through the direct interaction of BRI1 with an ATPase (Clouse 2011b).

Besides their role in cell elongation, BRs also promote cell division and differentiation. BRs (in the presence of auxin and cytokinin) stimulate cell division in a variety of protoplast and cell culture systems as well as in intact plants. Auxin and cytokinin are generally considered essential for continued cell division in cultured plant cells. However, BR could substitute for cytokinin in the growth of certain Arabidopsis cell cultures, and, interestingly, BR treatment increased the transcript levels of a gene encoding cyclinD3 (CycD3), a key cell cycle protein involved in regulation of the G1/S transition at the onset of cell division. The importance of BR regulation of CycD3 and progression of the cell cycle was also demonstrated in planta when it was found that BR mutants have reduced root meristem size and altered expression of cell cycle markers, which was rescued by overexpression of CycD3 (Fridman and Savaldi-Goldstein 2013). Arabidopsis plants overexpressing BRI1 have enlarged leaves, and this increase in leaf size was found to be predominately due to increased cell numbers, suggesting another link between BR signaling and the promotion of cell division. With respect to cell differentiation, BRs are known to promote vascular development, particularly xylem elements. BR signaling or biosynthetic mutants have altered vascular development, and BRs appear to regulate the differentiation of procambial cells into xylem elements in part through transcriptional regulation of genes necessary for xylem differentiation. BRI1, as well as two closely related LRR RLKs, BRI1-LIKE 1 (BRL1) and BRL3, which bind BR and are expressed in vascular tissue, is required for normal vascular development.

Besides their role in cellular functions, BRs also contribute to the regulation of morphogenesis and the formation of entire plant organs. The generation of boundary layers between the undifferentiated meristem and the developing primordia for leaves and other lateral organs is essential for proper plant architecture. Recent work has shown that the LATERAL ORGAN BOUNDARIES (LOB) transcription factor reduces BR signaling in boundary layer cells, which allows other transcription factors necessary for boundary layer development (which are repressed by the BZR1 transcription factor), to be expressed (Arnaud and Laufs 2013). Active BR signaling in primordia results in BZR1 repressing these boundary layer genes, thus establishing a spatial distribution of cellular differentiation in response to different levels of BR signaling (Cano-Delgado and Blazquez 2013). Spatial distribution of BR signaling is also important in root development, as BR signaling in the root epidermis is required to control the size of the root meristem in inner tissue. Finally the development of stomata in hypocotyls and leaves is also influenced by BRs. Formation of stomata is regulated by a pathway involving peptide ligands which bind to the ERECTA family of LRR RLKs and include mitogen-activated protein kinases (MAPK) such as the MAPK kinase kinase YODA as well as the bHLH transcription factor SPEECHLESS (SPCH). BIN2 phosphorylates and inactivates YODA in leaves, allowing the unphosphorylated form of SPCH to initiate development of stomata. High levels of BR inactivate BIN2, which allows YODA to initiate phosphorylation of SPCH via several intermediate steps, thereby preventing stomata formation. In hypocotyls, BIN2 directly phosphorylates SPCH, preventing formation of stomata, while BR inactivates BIN2, allowing SPCH to initiate formation of stomata (Fig. 3). Thus, BR signaling

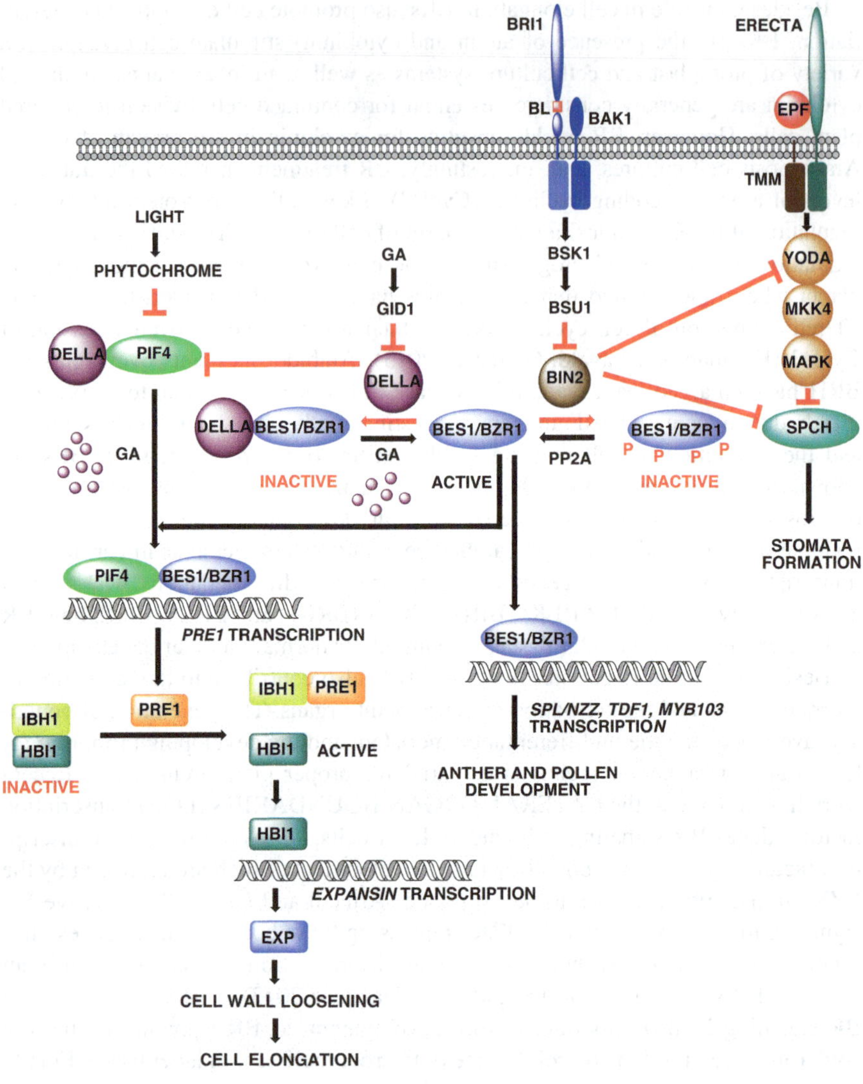

Fig. 3 Interaction of BR signal transduction components with other signaling pathways regulating plant growth and development. As detailed in Fig. 2, BL binding to BRI1/BAK1 results in inactivation of the BIN2 kinase and accumulation of the unphosphorylated, active form of BES1 and BZR1 transcription factors. BIN2 can also phosphorylate components of the ERECTA signaling pathway and thus influence formation of stomata. BES1/BZR1 bind to the promoters of approximately 1,000 genes including many transcription factors associated with different development responses, including elongation and anther and pollen development. DELLA, a negative regulator of GA signaling, binds directly to BES1/BZR1 and PIF4 transcription factors, causing their inactivation. When GA binds to its receptor, GID1, DELLA is degraded (indicated by *small purple circles*), and the active PIF4 and BES1 or BZR1 form dimers that bind to target promoters, including *PRE1*, encoding an atypical bHLH. PRE1 protein binds to and sequesters another bHLH protein, IBH1, which allows a third bHLH, HBI1, to bind to and activate promoters

promotes stomata formation in hypocotyls and reduces it in leaves (Gudesblat et al. 2012; Kong et al. 2012; Serna 2013).

Reproductive Growth and Development

The vegetative to reproductive transition during plant development is regulated by several integrated signaling pathways responding to photoperiod, cold treatment, and GAs. It now appears that BR signaling is also involved in several reproductive processes. Mutants defective in BR signaling flower later than wild-type Arabidopsis, in part because FLOWERING LOCUS C (FLC), which quantitatively represses flowering, is expressed at higher levels in these mutants. Moreover, BES1 interacts directly with RELATIVE OF EARLY FLOWERING6 (REF6), a Jumonji N/C domain-containing transcriptional regulator that functions in chromatin modification. REF6 is also a repressor of FLC, and ref6 mutants accumulate FLC transcripts, resulting in late flowering. BRs also interact with GAs to regulate time of flowering. With respect to floral organ development, microscopy and gene expression studies have shown that many genes essential for anther and pollen development have altered expression in BR mutants, resulting in defects in anther and pollen morphology. Moreover, several of these critical genes are regulated by BES1 binding to their promoters, providing a direct connection between BR signaling and anther and pollen development (Fig. 3). BRs also affect gynoecium and ovule development through a pathway involving the Arabidopsis SEUSS gene, which encodes a transcriptional adaptor protein involved in floral organ identity and carpel development. Seed germination is also affected by BRs. ABA and GA play antagonistic roles in establishing and breaking dormancy during seed development and germination. BRs rescue the germination defect in GA biosynthetic and insensitive mutants, and BR signaling may be required to reverse ABA-induced dormancy and stimulate germination.

Interaction with Other Signaling Pathways

Regulation of cell and organ elongation during plant development requires integration of multiple signals including hormones, light, and temperature. Auxin, BRs, and GAs all promote cell expansion, and this apparent redundancy in signals modulating the same physiological response may allow fine tuning of the regulation of cell expansion. Many genes involved in cell elongation are regulated by both

Fig. 3 (continued) for expansin genes. Expansins then cause cell wall loosening, which initiates cell elongation. TMM, TOO MANY MOUTHS; SPL/NZZ, SPOROCYTELESS/NOZZLE; TDF1, DEFECTIVE IN TAPEL DEVELOPMENT AND FUNCTION 1. Other abbreviations are defined in the main text

auxin and BRs, and it is now known that there is also a direct interaction between major regulatory components of BR and GA signal transduction. DELLA proteins are GRAS family transcription factors that negatively regulate genes required for GA responses, which include cell elongation, seed germination, and flowering. When GA binds to its receptor, DELLA proteins are degraded through a pathway involving ubiquitination and the 26S proteasome complex, allowing GA-responsive genes to be expressed and GA-promoted growth to occur. It is now known that DELLA proteins directly interact with BZR1 and BES1 transcription factors and inhibit their activity. Thus, GA may function in part by releasing the negative effects of DELLA proteins on BR signaling (Li and He 2013).

Light has a dramatic effect on plant morphology and activates the developmental switch from the etiolated (skotomorphogenesis) to the de-etiolated (photomorphogenesis) state. Dark-grown seedlings have rapidly elongating hypocotyls, a pronounced apical hook, and undeveloped cotyledons. Upon exposure to light, stem elongation is inhibited, cotyledons open, and chloroplasts differentiate. BR biosynthetic and signaling mutants both show de-etiolated phenotypes in the dark, suggesting that BRs are negative regulators of photomorphogenesis. The mechanism for this regulation likely involves BR signaling rather than changes in BR levels, and the BZR1 transcription factor has been shown to directly affect genes in light signaling pathways. BZR1 represses several positive regulators of photomorphogenesis including the GATA2 transcription factor and BES1 represses the GOLDEN2-LIKE (GLK) transcription factors required for chloroplast development. Furthermore, more than 750 genes were regulated in a similar manner by loss of BR signaling in the *bri1-116* mutant or by red light treatment of wild-type plants, suggesting that BR signaling and red light have antagonistic effects on gene expression involved in regulating photomorphogenesis (Wang et al. 2012).

The phytochrome family of light receptors positively regulates photomorphogenesis in part by light-dependent degradation of the phytochrome-interacting factor (PIF) family of transcription factors that promote cell elongation. Recently, BZR1 and PIF4 have been shown to form heterodimers and bind to the promoters of many genes involved in cell elongation. Furthermore, DELLA proteins also bind to PIF4 and the BZR1/PIF4 heterodimer. Thus, BZR1/PIF4/DELLA forms a core transcription module that regulates seedling morphogenesis through coordination of the BR, GA, and light signaling pathways (Jaillais and Vert 2012).

The BZR1/PIF4 module promotes elongation by targeting another set of transcription factors, the small family of atypical bHLH proteins containing six proteins named PACLOBUTRAZOL-RESISTANT 1 (PRE1) through PRE6. PREs were originally identified as positive regulators of GA signaling and have now been shown to be positive regulators of BR signaling as well. Classical bHLH proteins have a basic region potentially involved in DNA binding and the HLH region responsible for dimerization with other transcription factors. Atypical bHLH proteins contain the HLH domain but lack features of the basic domain and thus do not bind DNA. They often function by heterodimerizing with classical bHLH proteins, which inhibits DNA binding of the partner. PRE1 is a positive regulator of BR signaling that is thought to act by dimerizing with a classical bHLH protein, IBH1 (ILI1 BINDING BHLH PROTEIN1), which is a negative regulator of BR signaling. PRE1 association with IBH1 likely releases the inhibition of BR signaling conferred by IBH1. PRE1 is expressed at higher levels in young growing tissue, while IBH1 is more highly expressed in mature organs, suggesting an antagonistic regulatory mechanism that enhances BR-promoted cell elongation in immature tissue that is highly responsive to BR while arresting growth in mature tissue (Zhu et al. 2013). Recently, it was found that IBH1 interacts with another bHLH protein, HBI1 (HOMOLOG OF BEE2 INTERACTING WITH IBH1). Expansins (EXP) are wall-modifying proteins essential for cell elongation. HBI1 binds to the promoters of the EXP1 and EXP8 genes, activating their transcription and promoting cell expansion. IBH1 binding to HBI1 inhibits transcription of EXP1 and EXP8, while BR or GA activation of PRE1 promotes PRE1/IBH1 binding, releasing IBH1 inhibition of HBI1 and allowing it to activate EXP1 and EXP8 transcription and cell elongation. Thus, in this particular example of cell elongation, a complete signaling pathway with every step from BR binding at the cell surface to activation of a specific gene required for cell expansion has been defined (Fig. 3).

Future Directions

Since the discovery of the BRI1 receptor 16 years ago, the field has progressed to the point that all major components of BR signal transduction and their primary function in BR action have been uncovered. These primary components are summarized in Table 1. However, much remains to be done to refine the precise mechanisms of action at each step in the pathway. The three-dimensional structure of the BRI1 ectodomain has been solved, but further research is needed to determine the precise order of how BR binding to the ectodomain leads to initial activation of BRI1 kinase function, the release of the BKI1 inhibitor, and the association and activation of BAK1 by transphosphorylation. The use of advanced techniques in microscopy holds promise in clarifying early events at the plasma membrane including states of oligomerization of the BRI1 and SERK family receptor kinases. The combined approaches of mass spectrometry, biochemical analysis of kinase function, and molecular genetic studies of signaling in planta have identified and characterized many specific phosphorylation sites in BRI1, BAK1, BKI1, BSK1, CDG1, BSU1, and BIN2. Differential phosphorylation in these BR-signaling components may lead to alternative downstream signaling responses by activation of different gene sets involved in specific physiological responses. The use of mass spectrometry approaches that quantitatively monitor phosphorylation levels at specific residues in response to BR treatment may help to elucidate the phosphorelay crucial to BR signaling.

It is now evident that many of the pleiotropic effects of BRs on plant physiology and development are due to the regulatory gene networks responding to BZR1 and BES1 transcription factors. While recent publications have begun to characterize these BZR1/BES1 targets and their roles in plant development, continued analysis

Table 1 Primary components of the BR-signaling pathway

Gene ID	Name	Classification	Interactors	Regulation	Other names
AT4G39400	BRI1	LRR RLK	BAK1, BKI1, CDG1, BSK1	Positive	BIN1, CBB2, DWF2
AT1G55610	BRL1	LRR RLK	BAK1	Positive	
AT3G13380	BRL3	LRR RLK	BAK1	Positive	
AT4G33430	BAK1	LRR RLK	BRI1	Positive	SERK3
AT1G71830	SERK1	LRR RLK	BRI1	Positive	
AT2G13790	BKK1	LRR RLK	BRI1	Positive	SERK4
AT5G42750	BKI1	Inhibitor	BRI1	Negative	
AT4G35230	BSK1	Cytoplasmic RLK	BRI1, BSU1	Positive	
AT3G26940	CDG1	Cytoplasmic RLK	BRI1, BSU1	Positive	
AT1G03445	BSU1	Phosphatase	BSK1, CDG1, BIN2	Positive	
AT4G18710	BIN2	GSK3 kinase	BSU1, BZR1, BES1	Negative	ATSK21, DWF12, UCU1
AT1G75080	BZR1	Transcription factor	BIN2, 14-3-3, BIM1, PP2A, PIF4, DELLA	Positive	
AT1G19350	BES1	Transcription factor	BIN2, 14-3-3, BIM1, MYB30, REF1, ELF1, IWS1, PIF4, DELLA	Positive	BZR2
AT5G03470	PP2AB'a	Phosphatase	BZR1, BES1	Positive	
AT5G39860	PRE1	Atypical bHLH	IBH1	Positive	BNQ1
AT2G43060	IBH1	bHLH	PRE1, HBI1	Negative	
AT2G18300	HBI1	bHLH	IBH1	Positive	

of the BR gene network will likely uncover genes with previously undefined functions, which may reveal new physiological processes regulated by BR signaling. The spatial regulation of BR signaling in cell differentiation and organ development is just beginning to be examined, and the findings that BRs affect development of stomata, root organization, and cell boundaries in leaf primordia suggest that this line of research will be very productive in the next several years. The discovery of the BZR1/PIF4/DELLA transcriptional module clearly showed that BR signaling is integrated with other signaling networks to regulate seedling morphogenesis. Continued analysis of the interaction of BR signaling with other hormone and environmental signaling pathways will help to clarify the central role of BRs in regulating overall plant growth and development. Finally, BRs are known to affect many developmental and physiological events, such as plant architecture, seed development, flowering, and fruit ripening, that are critical to yield in agricultural crops. Field studies have shown that altering BRI1 expression can enhance rice yields by up to 30 %. Studies of BR signaling in Arabidopsis have now uncovered the importance of BRs in basic plant development. Expanding these types of detailed studies of BR signaling to important agronomic and horticultural crops may also have practical agricultural applications in the future.

References

Arnaud N, Laufs P. Plant development: brassinosteroids go out of bounds. Curr Biol. 2013;23: R152–4.

Cano-Delgado AI, Blazquez MA. Spatial control of plant steroid signaling. Trends Plant Sci. 2013;18:235–6.

Chinchilla D, Shan L, He P, de Vries S, Kemmerling B. One for all: the receptor-associated kinase BAK1. Trends Plant Sci. 2009;14:535–41.

Clouse SD. Brassinosteroid signal transduction: from receptor kinase activation to transcriptional networks regulating plant development. Plant Cell. 2011a;23:1219–30.

Clouse SD. Brassinosteroids. Arabidopsis Book. 2011b;9:e0151.

Fridman Y, Savaldi-Goldstein S. Brassinosteroids in growth control: how, when and where. Plant Sci. 2013;209:24–31.

Gruszka D. The brassinosteroid signaling pathway-new key players and interconnections with other signaling networks crucial for plant development and stress tolerance. Int J Mol Sci. 2013;14:8740–74.

Gudesblat GE, Russinova E. Plants grow on brassinosteroids. Curr Opin Plant Biol. 2011;14:530–7.

Gudesblat GE, Betti C, Russinova E. Brassinosteroids tailor stomatal production to different environments. Trends Plant Sci. 2012;17:685–7.

Jaillais Y, Vert G. Brassinosteroids, gibberellins and light-mediated signalling are the three-way controls of plant sprouting. Nat Cell Biol. 2012;14:788–90.

Jiang J, Zhang C, Wang X. Ligand perception, activation, and early signaling of plant steroid receptor BRI1. J Integr Plant Biol. 2013;55:1198–211.

Kim TW, Wang ZY. Brassinosteroid signal transduction from receptor kinases to transcription factors. Annu Rev Plant Biol. 2010;61:681–704.

Kong X, Pan J, Cai G, Li D. Recent insights into brassinosteroid signaling in plants: its dual control of plant immunity and stomatal development. Mol Plant. 2012;5:1179–81.

Li J. Multi-tasking of somatic embryogenesis receptor-like protein kinases. Curr Opin Plant Biol. 2010a;13:509–14.

Li J. Regulation of the nuclear activities of brassinosteroid signaling. Curr Opin Plant Biol. 2010b;13:540–7.

Li QF, He JX. Mechanisms of signaling crosstalk between brassinosteroids and gibberellins. Plant Signal Behav. 2013;8:e24686.

Oh MH, Clouse SD, Huber SC. Tyrosine phosphorylation in brassinosteroid signaling. Plant Signal Behav. 2009;4:1182–5.

Serna L. What causes opposing actions of brassinosteroids on stomatal development? Plant Physiol. 2013;162:3–8.

Wang ZY, Bai MY, Oh E, Zhu JY. Brassinosteroid signaling network and regulation of photomorphogenesis. Annu Rev Genet. 2012;46:701–24.

Zhu JY, Sae-Seaw J, Wang ZY. Brassinosteroid signalling. Development. 2013;140:1615–20.

Further Reading

Bucherl CA, van Esse GW, Kruis A, Luchtenberg J, Westphal AH, Aker J, van Hoek A, Albrecht C, Borst JW, de Vries SC. Visualization of BRI1 and BAK1(SERK3) membrane receptor heterooligomers during brassinosteroid signaling. Plant Physiol. 2013;162:1911–25.

Choudhary SP, Yu JQ, Yamaguchi-Shinozaki K, Shinozaki K, Tran LS. Benefits of brassinosteroid crosstalk. Trends Plant Sci. 2012;17:594–605.

Cutler HG, Yokota T, Adam G. Brassinosteroids: chemistry, bioactivity, & applications, vol. 474. Washington, DC: American chemical Society; 1991.

Hao J, Yin Y, Fei SZ. Brassinosteroid signaling network: implications on yield and stress tolerance. Plant Cell Rep. 2013;32:1017–30.

Hayat S, Ahmad A. Brassinosteroids: bioactivity and crop productivity. Dordrecht: Kluwer; 2003.

Hayat S, Ahmad A. Brassinosteroids: a class of plant hormone. Dordrecht: Springer; 2011.

Kutschera U, Wang ZY. Brassinosteroid action in flowering plants: a Darwinian perspective. J Exp Bot. 2012;63:3511–22.

Li J, Li Y, Chen S, An L. Involvement of brassinosteroid signals in the floral-induction network of Arabidopsis. J Exp Bot. 2010;61:4221–30.

Sakurai A, Yokota T, Clouse SD. Brassinosteroids: steroidal plant hormones. Tokyo: Springer; 1999.

Santiago J, Henzler C, Hothorn M. Molecular mechanism for plant steroid receptor activation by somatic embryogenesis co-receptor kinases. Science. 2013;341:889–92.

Vriet C, Russinova E, Reuzeau C. Boosting crop yields with plant steroids. Plant Cell. 2012;24:842–57.

Vriet C, Russinova E, Reuzeau C. From squalene to brassinolide: the steroid metabolic and signaling pathways across the plant kingdom. Mol Plant. 2013;6:1738–57.

Zhao B, Li J. Regulation of brassinosteroid biosynthesis and inactivation. J Integr Plant Biol. 2012;54:746–59.

COP9 Signalosome Network

11

Anna Franciosini, Giovanna Serino, and Xing-Wang Deng

Contents

A. Franciosini (✉) • G. Serino
"Charles Darwin" Department of Biology and Biotechnology, Sapienza Università di Roma, Rome, Italy
e-mail: anna.franciosini@uniroma1.it; giovanna.serino@uniroma1.it; giovanna.serino@gmail.com

X.-W. Deng
Department of Molecular, Cellular and Developmental Biology, Yale University, New Haven CT, USA
e-mail: xingwang.deng@yale.edu

S.H. Howell (ed.), *Molecular Biology, The Plant Sciences 2*,
DOI 10.1007/978-1-4614-7570-5_13

Abstract

- The COP9 signalosome (CSN) is an evolutionary conserved multiprotein complex that functions in the ubiquitin–proteasome pathway.
- The CSN has an isopeptidase activity which removes the NEDD8 peptide from the *c*ullin–*R*ING ubiquitin *l*igases (CRLs). This reaction ensures the proper function of the CRLs.
- The CSN is involved in the control of multiple signaling processes in virtually all eukaryotes. In plants, the CSN takes part in a plethora of developmental processes and environmental responses, including photomorphogenesis, flower development, hormone signaling, and plant pathogen response.
- The CSN is also required for the regulation of cellular pathways, such as cell cycle, DNA repair, and regulation of gene expression.

Introduction

The COP9 signalosome is a protein complex conserved in all eukaryotic organisms and initially isolated in plants as a repressor of light-dependent development.

Light plays an essential role in multiple aspects of plant development. Immediately after having emerged from the seed coat, seedlings must choose between two different developmental pathways, depending on the presence or absence of light: photomorhogenesis in the light or skotomorphogenesis in darkness. During photomorphogenesis, seedlings develop a short hypocotyl, and open, green, and enlarged cotyledons, in order to optimize light absorption. During skotomorphogenesis, seedlings develop an elongated hypocotyl, while cotyledons stay closed and unexpanded with an apical hook. In the late 1980s and the early 1990s, two genetic screens aimed at identifying novel components of light signal transduction and photomorphogenesis led to identification of a group of nine *Arabidopsis thaliana* mutant loci that displayed a complete constitutive photomorphogenic phenotype, regardless of the presence or absence of light (Wei and Deng 1996). These mutants, called pleiotropic *cop/det* (*constitutive photomorphogenic/de-etiolated*), when grown in darkness, mimicked the appearance of light-grown seedlings, with a short hypocotyl, opened cotyledons, plastid differentiation, and expression of light regulated genes. If grown in the light, the *cop/det* mutants exhibited features of light stress and accumulated a purple pigment (anthocyanin) in the mature seed and young seedling. Indeed, same *cop/det* loci were also isolated from other genetic screens aimed at identifying mutants with high accumulation of anthocyanin pigments (*fusca*). To reflect this finding, these mutants were then collectively named *cop/det/fus* (Wei and Deng 1996). Based on the recessive nature of all *cop/det/fus* mutations, it was then hypothesized that the *COP/DET/FUS* genes were negative regulators of photomorphogenesis. In addition, the fact that all *cop/det/fus* mutants share a common phenotype suggested that their corresponding loci might encode proteins that acted together in the same pathway. Indeed, the cloning of the corresponding genes, together with biochemical analysis, confirmed that six out of the nine loci encoded subunits of a protein complex that was initially named the COP9 complex after the name of the first gene, *COP9*, of the complex. Another COP/DET/FUS protein (COP1) was not integral part of the complex but was discovered to work together with the COP9 complex to repress photomorphogenesis in the dark. The two remaining proteins, COP10 and DET1, are part of the COP10–DDB1–DET1 (CDD) complex (Lau and Deng 2012).

The finding that all null mutations in *COP/DET/FUS* genes are lethal after the seedling stage suggested from the beginning that these genes were controlling other processes, in addition to photomorphogenesis. As expected, in the following years the COP9 complex was associated with other plant developmental pathways, ranging from auxin signaling to pathogen response. The discovery, made in the year 1998, that this complex is also conserved in animals including human provided the final evidence that the COP9 complex played an essential role in the development not only of plants but of all higher eukaryotes. Following the revelation of its conservation in other eukaryotic organisms, the complex was renamed the COP9 signalosome (abbreviated as CSN), and the COP9 subunit is now known as CSN8 (Deng et al. 2000).

CSN Structure and Function

CSN Architecture

It is now known that the canonical CSN is composed of eight core subunits, designated as CSN1 to CSN8 (Table 1). This composition is conserved among all eukaryotic organisms, with the exceptions of some unicellular fungi and of *Caenorhabditis elegans*, which possess smaller versions of the complex (Wei et al. 2008) (Table 1). Genes coding for CSN subunits 1–4 and 7–8 were found in the initial genetics screens, while CSN5 and CSN6 were identified as subunits after the biochemical purification of the complex. The reason why CSN5 and CSN6 were not isolated in the genetic screens is likely due to the fact that they are both encoded by two genes in *Arabidopsis* (Stratmann and Gusmaroli 2012).

Each CSN subunit contains one of two conserved domains: subunits CSN1–CSN4 and CSN7–CSN8 contain a PCI (*p*roteasome, *C*OP9 Signalosome and *i*nitiation factor 3) domain, while CSN5 and CSN6 have a MPN domain (*M*pr1 and *P*ad1 *N*-terminal) (Table 1) (Pick et al. 2009).

Table 1 CSN subunit composition in different organisms

Domains	*S. cerevisiae*	*S. pombe*	*C. elegans*	*Drosophila*	*Human*	*A. thaliana*
PCI	Csn11	Csn1	CSN1	CSN1	CSN1	CSN1/FUS6
PCI	Csn10	Csn2	CSN2	CSN2	CSN2	CSN2/FUS12
PCI	–	Csn3	CSN3	CSN3	CSN3	CSN3/FUS11
PCI	Rpn5[a]	Csn4	CSN4	CSN4	CSN4	CSN4/COP8
MPN+	Csn5	Csn5	CSN5	CSN5	CSN5	CSN5
MPN–	Csi1	–	CSN6	CSN6	CSN6	CSN6
PCI	Csn9	Csn7	CIF-1	CSN7	CSN7	CSN7/FUS5
PCI	–	–	–	CSN8	CSN8	CSN8/COP9

[a]Also a proteasome subunit (Serino and pick 2013)

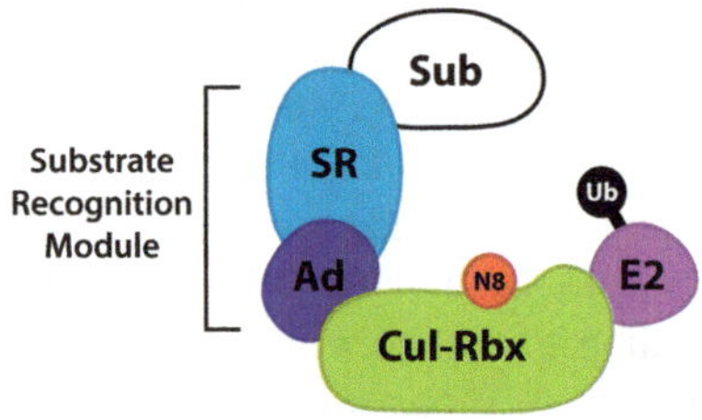

Fig. 1 **Model of a CRL**. The CRLs are composed by an enzymatic core that contains a cullin scaffold subunit and Rbx1 (Cul-Rbx) that interacts with the E2 Ub-conjugating enzyme. Specific substrates are recruited to the core by a substrate recognition module consisting of an adaptor protein (*Ad*) and a substrate receptor (*SR*). In the CRL active form, cullin subunit is modified by the attachment of the NEDD8 (*N8*) peptide

The PCI domain displays a bipartite fold consisting of a N-terminal helical bundle and a C-terminal winged helix, which are connected through a central helix. This domain mediates protein–protein interaction within multiprotein complexes and is therefore essential for the maintenance of the structural integrity of the complexes. The MPN domain contains a beta-sheet motif with nine beta-strands, surrounded by three alpha-helices. Structural and biochemical studies have indicated that the MPN domain is present in two distinct versions. The first one, known as MPN+/JAMM (for Jab1/MPN/Mov34), is found in the CSN5 subunit and harbors a metalloprotease motif that is responsible for the catalytic activity of the complex. This activity is essential for the removal of an ubiquitin-like peptide, NEDD8 (*n*eural precursor cell *e*xpressed, *d*evelopmentally *d*ownregulated 8) (called RUB1 in plants), from the cullin subunit of the *c*ullin–*R*ING type of ubiquitin *l*igases (CRLs) (Fig. 1). The second version, which has recently been renamed MPN-, is located in CSN6, lacks the metal coordinating residues and is therefore biochemically inactive but likely plays a structural or regulatory function (Nezames and Deng 2012; Serino and Pick 2013).

Because pure crystals of the entire complex have not been obtained to date, a tridimensional analysis of CSN architecture based on X-ray crystallography is not available. However, several other approaches have been used to circumvent

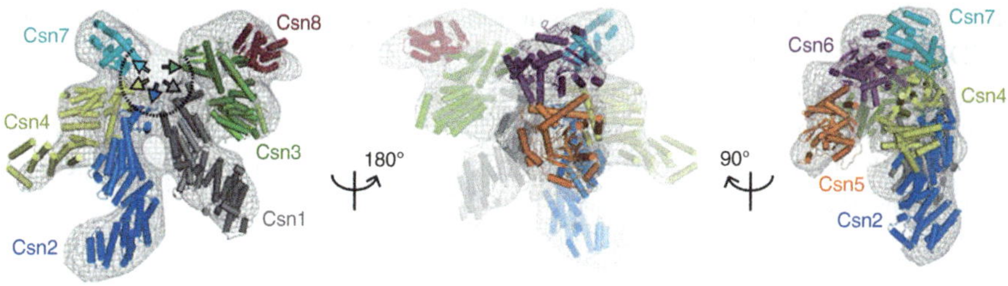

Fig. 2 Model of the CSN structure. *Left*, PCI cluster side view. A *dotted arc* and *color-coded arrows* indicate the approximately coplanar positions of the winged-helix domains. MPN subunits are omitted for clarity. *Center*, opposite side, characterized by a protrusion formed by the two MPN domain subunits, Csn5 and Csn6. *Right* view showing the edge of the coplanar PCI cluster. The protrusion formed by the Csn5 and Csn6 MPN subunits is *left* of the PCI cluster (Reprinted from Enchev et al. 2012 with permission from Elsevier)

this problem and to allow a preliminary characterization of CSN structure. Mass spectrometry analysis and electron microscopy studies have contributed to elucidate the topology and the structure of the CSN. These data have shown that the human CSN is composed by two symmetrical modules, CSN1/2/3/8 and CSN4/5/6/7, connected by the interaction between CSN1 and CSN6 (Fig. 2) (Nezames and Deng 2012). High-resolution electron microscopy has provided the latest structure model depicting the refined subunit organization of the complex (Enchev et al. 2012).

A Common Molecular Structure for the CSN, the Proteasome Lid, and the eIF3 Complexes

As soon as the composition of the CSN was determine, it became clear that it has striking similarities, in terms of both structure and subunit composition, to two other eukaryotic protein complexes: the "lid" subcomplex of the proteasome and the *e*ukaryotic translation *i*nitiation *f*actor 3 (eIF3). All three complexes contain PCI and MPN subunits and collectively are known as PCI complexes (Pick et al. 2009).

Despite being all related to the control of protein turnover and sharing similar features, the three PCI complexes perform different cellular functions. The eIF3 facilitates the assembly of the initiation complex during protein synthesis and promotes dissociation of the 40S and 60S ribosomal subunits. The eIF3 complex contains both PCI and MPN proteins, but its structure and composition are more divergent from those of the CSN and the lid (Pick et al. 2009).

The lid is the distal cap of the proteasome, a proteolytic complex responsible for the degradation of ubiquitinated proteins (see review "Ubiquitin/Proteasome" http://www.springerreference.com/docs/html/chapterdbid/138039.html). Interestingly, the CSN and the lid of the proteasome share a very common architecture, both showing a 6:2 PCI:MPN subunit ratio, with each subunit of a complex having its counterpart, or paralog, in the other. Moreover, these two complexes perform different but related biochemical functions. In fact, while the CSN removes NEDD8 from CRL ubiquitin ligases through its CSN5 subunit, the lid removes ubiquitin from proteasome substrates before the degradation. This deubiquitination activity resides in the MPN + subunit of the lid (Rpn11). This suggests that the CSN and the lid of the proteasome might have evolved from a common ancestor (Serino and Pick 2013).

CSN Regulates CRLs

A major breakthrough which led to the discovery of the biochemical activity of the CSN came from the laboratories of Raymond Deshaies and Xing-Wang Deng. Using human and *Arabidopsis*, these laboratories found a physical direct and functional interaction between CSN and CRLs, one of largest class of E3 ubiquitin ligases.

As shown in Fig. 1, all CRLs enzymes are composed of a backbone cullin subunit (CUL1, CUL3, and CUL4 in *Arabidopsis*) that interacts via its C-terminus with a *R*ING-*box* protein 1 (RBX1) subunit, which functions in turn as a platform for the E2 (ubiquitin-conjugating enzyme) charged with ubiquitin. At its N-terminus, the cullin subunit binds specific substrate recognition modules that recognize and deliver appropriate substrates for ubiquitination. Different subclasses of CRLs exist, depending on the different assembly based on the type of cullin subunit, and each cullin interacts with a different class of substrate recognition modules (Hua and Vierstra 2011). In CUL1-based CRLs, the substrate recognition module is composed of a F-box protein that is responsible for the interaction with the substrate and which is anchored to CUL1 through the SKP1 (*S*-phase *k*inase-associated *p*rotein 1) adaptor subunit, called ASK1 (*A*rabidopsis *SK*P1-like1) in *Arabidopsis*. CUL1-containing CRL complexes are named SCFs (*S*kp1/*C*ul1/*F*-box) (Hua and Vierstra 2011). CUL3-based ubiquitin ligases contain a BTB/POZ (*b*road complex/*t*ramtrack/*b*ric-a-brac and *Po*x virus and *z*inc finger) subunit that serves as the substrate recognition module. CUL4-based CRLs contain a substrate recognition module composed of a DWD (*D*DB1-binding/*WD*-40 domain) protein, which interacts with the substrate, and of a DDB1 (*D*NA *d*amage *b*inding protein 1) adaptor that connects DWD proteins to and CUL4 (Hua and Vierstra 2011).

The *Arabidopsis* genome, unlike the genome of other organisms, encodes a remarkably large number of genes related to protein degradation. In *Arabidopsis* there are more than 700F-box proteins, 85 DWD and 80 BTB/POZ substrate receptors; thus, an incredible number of CRLs can be assembled. This suggests that, in this sessile organism, protein degradation has a crucial role in cellular and developmental processes. Because the CSN has been shown to interact with all three types of CRLs in *Arabidopsis*, this complex might be required to ensure the proper life span of hundreds, if not thousands, specific proteins (Hua and Vierstra 2011). This might help explaining the highly pleiotropic phenotype of *csn* mutants, because lack of CSN would cause malfunction of multiple CRLs (Table 2).

Table 2 Representative plant processes controlled by the CSN

Plant processes	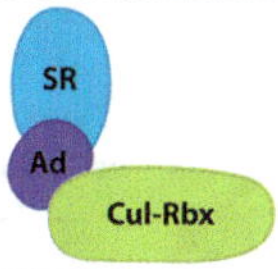 CRL	 Substrate	References
Photomorphogenesis	COP1, CDD-CUL4	phyA, phyB, HYS, HYH, LAF1, HFR1	Chen et al. 2006
Flower development	SCF^{UFO}	Unknown	Wang et al. 2003
Auxin perception	SCF^{TIR1}	AUX/IAA Proteins	Schwechheimer et al. 2001
Jasmonate response and plant defense	SCF^{COi1}	JAZ Proteins	Feng et al. 2003; Hind et al. 2011
Gibberellin signaling	SCF^{SLY}	DELLA Proteins	Dohmann et al. 2010
R-mediated resistance	SGT1-associated SCF	Unknown	Liu et al. 2002
SA-mediated defense	CUL3-based Ubiquitin Ligases	NPR1	Spoel et al. 2009

The CSN and CAND1-Regulated CRL Cycle

CRL function is regulated by the covalent attachment (neddylation) or removal (deneddylation) of the NEDD8 peptide on their cullin subunit. NEDD8 is conjugated to CRLs by a three-step cascade reaction similar to ubiquitination cascade (see review "Ubiquitin/Proteasome" http://www.springerreference.com/docs/html/chapterdbid/138039.html). NEDD8 binding promotes substrate ubiquitination, because it leads to a conformational change on the CRL that allows for the transfer of the ubiquitin peptide from the E2 enzyme to the substrate (Fig. 3). CSN removes NEDD8 from the CRLs, through its catalytic CSN5 subunit leading to CRL inactivation.

Evidence from plants and other organism show that CRL neddylation and deneddylation are required for the proper functioning of the CRLs (Wei et al. 2008). Understanding the precise mechanism of the CRL neddylation and deneddylation cycle is still a focus of active research. In the current model, the substrate itself plays a role in the regulation of the CRL cycle. In the presence of a substrate, the substrate receptor subunit can inhibit deneddylation of its cullin partner and prevent CSN activity in order to ensure substrate degradation (Fig. 3) (Stratmann and Gusmaroli 2012). Thus, substrate availability promotes accumulation of active, NEDD8-conjugated CRL complexes. After substrate ubiquitination and degradation, the CRL recruits the CSN and is deneddylated or, alternatively, can undergo auto-ubiquitination of its own substrate receptor subunit, followed by release of the cullin–RBX1 core (see Fig. 3). CSN has high affinity for its reaction products and might remain bound to the cullin–Rbx1 core. When substrates are again available, CSN is displaced and CRLs become neddylated, returning to their activated form (Emberley et al. 2012).

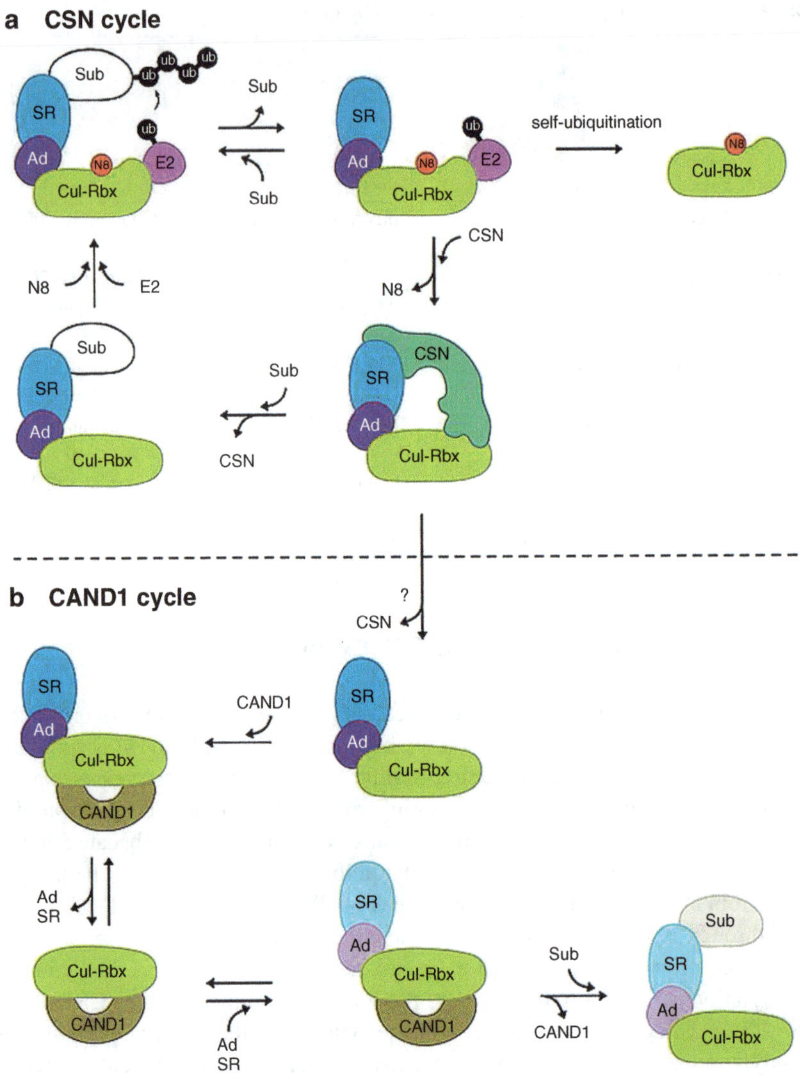

Fig. 3 Proposed regulatory CRL cycles involving CSN and CAND1. A completely assembled CRL binds a E2 Ub-conjugating enzyme and poly-ubiquitinates a substrate (*upper left*). Upon consumption of the substrate, the CRL complex could either be subjected to auto-ubiquitination and degradation of the substrate receptor (*SR*) or recruit the CSN and be deneddylated. The reappearance of the substrate leads to displacement of the CSN and to the reformation of neddylated active CRL complex, completing the CSN cycle. In addition, following the dissociation of the CSN that occurs in an unknown manner, CRL can enter in the CAND1 cycle. The protein CAND1 binds the cullin–Rbx1 scaffold (Cul-Rbx), promoting the dissociation of the adaptor (*Ad*) and the substrate receptor from Cul-Rbx. The availability of a second Ad–SR module leads to the dissociation of CAND1 and to the formation of a new CRL complex

CSN regulation of CRLs works in concert with another regulation module, centered around the protein CAND1 (*c*ullin-*a*ssociated and *n*eddylation *d*issociated 1). CAND1 binds the unmodified cullin–RBX core, preventing it from associating with the other subunits of the CRL complex. This results in an inactive CRL, and the active state is restored only when the levels of its respective substrate recognition module increase. The substrate recognition module then displaces CAND1, allowing cullin neddylation and substrate ubiquitination. CAND1 might play the role of a substrate receptor exchange factor and stimulate CRL activity, thus allowing the exchange of different substrate recognition modules on the same cullin–Rbx1 scaffold (see CAND1 cycle in Fig. 3). In other words, after substrate degradation and CSN-mediated deneddylation, the cullin–Rbx1 core can follow two different routes: (1) it can interact with another substrate, promote CSN displacement, and become neddylated or (2) can bind CAND1, enter in an "exchange regime", and as a result associate with a newly available substrate recognition module. This triggers the dissociation of CAND1. A neddylation event completes the cycle, and CRL is now back in its active form again (Pierce et al. 2013).

CSN regulates the CRL cycle also in a non-catalytic fashion. Results obtained for SCF and CUL4-based CRL suggest that CSN occludes two CRL functional sites: the Cul1–Rbx1 C-terminal domain and the substrate receptor, thus maintaining CRL assembly. This indicates that this CSN-mediated inhibition by steric hindrance could be conserved among all CRLs. Thus, CSN works as a CRL inhibitor by deneddylation and protein interactions, with a result that it promotes sustained functions of CRLs by maintaining its stability by facilitating a rapid and efficient substrate turnover in vivo (Enchev et al. 2012).

CSN Holocomplex and Subcomplexes

In *Arabidopsis*, as well as in other organisms, mutation of a single CSN subunit leads to instability of the entire complex, and CSN catalytic activity is also dependent on the integrity of the complex (Wei et al. 2008). There are exceptions: in *Arabidopsis* and *Drosophila* loss of CSN5 does not prevent formation of a CSN. Likewise, subcomplexes containing only a subset of subunits have been identified (Chamovitz 2009). The biological role of these "mini-complexes" remains unclear, but it is possible that the formation of smaller subcomplex may contribute to explain the multi-functionality of the CSN in vivo (Nezames and Deng 2012).

CSN Roles in Plant Development

As described in the "Introduction" section, *Arabidopsis csn* mutants have a pleiotropic phenotype, which includes a deregulated activation of light-induced development, as well as defect in other multiple cellular pathways, which culminates in growth arrest prior to the transition to the adult stage. This simple observation suggests a crucial role of the CSN in controlling many aspects of plant life.

In the last 20 years, research carried out in *Arabidopsis* has actively contributed to identify the plant developmental pathways regulated by the CSN, and its involvement in a plethora of plant processes well beyond photomorphogenesis has emerged. Some of these processes are described below and summarized in Table 2.

The CSN in Photomorphogenesis and Other Light-Regulated Processes

As mentioned earlier, of the nine *COP/DET/FUS* loci identified in the initial genetic screens, six loci correspond to CSN subunits, and the remaining three, *COP1*, *COP10* and *DET1*, are not part of the CSN. Extensive biochemical and genetic studies have now defined that these proteins are part of the two additional complexes: the *C*OP1–*S*UPPRESSOR OF *P*HYA-105 (SPA) complex and the *C*OP10–*D*ET1–*D*DB1 (CDD) complex (Lau and Deng 2012). These two complexes cooperate together to induce ubiquitination and proteasome-mediated degradation of positive regulators of light responses.

COP1 is itself a ubiquitin ligase and interacts with several photomorphogenesis-promoting factors, such as the phytochromes A and B, and the transcription factors HY5 (*E*LONGATED *H*YPOCOTYL FACTOR 5), HYH (*HY*5-*H*OMOLG), LAF1 (*L*ONG *A*FTER *F*AR-RED FACTOR1), and HFR1 (LONG *H*YPOCOTYL IN *F*AR-*R*ED). However, COP1 requires the entire COP1–SPA and the CDD complexes to promote the ubiquitination and degradation of photomorphogenesis-promoting factors in darkness. In the dark, these complexes function therefore as master repressors of photomorphogenesis, by triggering the ubiquitination and degradation of transcriptional factors that activate specific light responses. On the contrary, in the presence of light, activated photoreceptors repress COP1 function and allow the accumulation of the photomorphogenesis-promoting transcription factors, resulting in photomorphogenic development. This differential mechanism of COP1 function relies also on its cellular localization. In fact, upon light exposure, when the photomorphogenesis takes place, COP1 is shuttled from the nucleus to the cytoplasm, where it is not able to promote the degradation of HY5 and the other transcription factors (Osterlund et al. 1999).

The similar photomorphogenic phenotype between *csn* mutants and *cop1* suggests that they might work together to regulate photomorphogenesis. Indeed, the CSN is necessary for COP1 nuclear translocation and CDD complex stabilization. In addition COP1 interacts directly with the CSN1 subunit, and both COP1 and the CSN interact with CUL4 (Lau and Deng 2012).

The CSN in Flower Development

The CSN regulates the SCF complex SCF^{UFO}, which contains the F-box protein UFO (*U*NUSUAL *F*LORAL *O*RGANS) and which is required for the proper regulation of floral meristem identity and for the floral organ development (Stratmann and Gusmaroli 2012). The molecular mechanism is not fully elucidated, but probably the CSN works together with SCF^{UFO} to promote the ubiquitination of negative regulators of the expression of *APETALA3* (*AP3*), a homeotic gene required to specify petal and stamen identities. Accordingly, weak *csn* mutants show defects in floral development, and *AP3* decreased expression, indicating that the CSN, by mediating SCF^{UFO} activity, regulates flower development (Stratmann and Gusmaroli 2012).

The CSN in Hormone Signaling

Some F-box proteins can function both as hormone receptors and substrate recognition subunits. The involvement of the CSN in hormonal pathways was first revealed by the interaction with SCF^{TIR1} in auxin signaling. Since then, its involvement in jasmonic acid and gibberellin signaling through regulation of the corresponding SCF functions has been determined (Table 2). SCF ubiquitin ligases are therefore a central component of hormone perception and signaling, and since their activity is under the control of the CSN, the CSN itself occupies a very crucial role in plant hormone response.

Auxin Signaling

Auxin (indole-3-acetic acid or IAA) regulates many developmental plant processes, including embryogenesis, root and stem elongation, apical dominance, phototropism and gravitropism, and lateral root initiation, by inducing a rapid change in auxin-responsive gene expression (see review "Signaling: Auxin signaling").

The key player in auxin signaling is SCF^{TIR1} that promotes auxin-dependent degradation of negative regulators of auxin response. The F-box protein TIR1 belongs to a small family of related F-box proteins that includes five additional members called *a*uxin signaling *F-b*ox or AFBs (AFB1-5). These F-box proteins function as the auxin receptor, directly interacting with auxin. Auxin binding increases the affinity of SCF^{TIR1} for its substrates, the Aux/IAA repressor proteins, leading to their ubiquitination and degradation. This allows the release of the *a*uxin *r*esponse *f*actors (ARFs), a large family of transcription factors which bind promoters of auxin-responsive genes, and stimulates ARF-dependent transcription. Thus, auxin, by promoting the degradation of Aux/IAA proteins through SCF^{TIR1}, causes the transcription activation of auxin-responsive genes (Santner and Estelle 2010).

CSN physically interacts with SCF^{TIR1}, regulates its activity, and is required for the stabilization of the F-box protein TIR1. This is in agreement with the model that proposes CSN to be necessary to prevent the auto-ubiquitination of CRL substrate recognition subunits. Plants with reduced level of CSN show reduced degradation of Aux/IAA and decreased auxin response, underlining the relevance of this complex in regulating auxin signaling (Santner and Estelle 2010).

Jasmonic Acid Signaling

Jasmonic acid (JA) is an important signaling molecule that mediates plant responses to biotic and abiotic stress, as well as other aspects of plant development such as growth and fertility.

Similarly to the auxin sensor SCF^{TIR1}, the SCF^{COI1} ubiquitin ligase, which contains the F-box protein COI1 (*CO*RONATINE-*I*NSENSITIVE1) functions as the JA receptor. After its interaction with JA, COI1 is able to bind the JAZ (*ja*smonate *Z*IM-domain) family of transcriptional regulators, and SCF^{COI1} promotes their ubiquitination and degradation, via the proteasome pathway. In the absence of the hormone, JAZ proteins actively repress a subfamily of MYB transcription factors, which bind the *cis*-acting elements of JA-response genes. In the presence of JA, JAZ proteins are degraded, allowing the expression of JA-induced genes (Santner and Estelle 2010).

Similarly to SCFTIR1, CSN associates physically with SCFCOI1, regulates its function, and is thus essential for proper JA signaling. In fact, plants with reduced level of CSN are not able to respond properly to JA, and an adequate cellular level of CSN is required for proper JA responses (Stratmann and Gusmaroli 2012).

Gibberellin Signaling

Gibberellins (GAs) regulate diverse growth and developmental processes such as seed germination, stem elongation, leaf expansion, and flower development.

GA responses are negative regulated by a class of proteins, named DELLA proteins, from their conserved DELLA motif, composed by five members. The five DELLA proteins have both redundant and specialized functions, finally resulting in the repression of GA signaling. GA perception is mediated by their receptor GID1 (*G*IBERELLIN-*I*NSENSITIVE *D*WARF1). In the presence of GAs, the GID1 receptor binds the DELLA proteins, recruits the ubiquitin ligase $SCF^{SLEEPY1}$, and promotes the ubiquitination of DELLAs, thus inducing their degradation by the proteasome. SCF^{SLY1}-mediated degradation releases the DELLA-dependent block and thus promotes the activation of GA positive regulators, such as the bHLH transcription factors PIFs (*p*hytochrome-*i*nteracting *f*actor) (Santner and Estelle 2010).

SCF^{SLY1} fails to efficiently degrade RGA (*R*EPRESSOR OF *ga1-3*), a member of DELLA proteins family in *csn* mutants. Furthermore, *csn* hypomorphic mutants show developmental defects, including in germination and hypocotyl elongation, which might be ascribed to an inefficient activity of SCF^{SLY1} and to the resulting accumulation of DELLA proteins. Thus, the repression of GA signaling could be the cause of some of growth defects displayed by *csn* mutants (Stratmann and Gusmaroli 2012).

CSN in Plant Pathogen Response

R-Mediated Resistance

Pathogens, when in contact with their plant host, use effector proteins that disrupt immunity responses and promote successful infection. Plant cells have evolved sophisticated signaling pathways to recognize and respond to pathogen-delivered effectors. One of these pathways relies on disease-resistant proteins (R), which recognize pathogen effectors and induce the so-called R protein-mediated response. This leads to hypersensitive localized cell death response (HR) at the infection site, rapid oxidative burst, and activation of various defense-responsive genes (Craig et al. 2009).

The first report of an involvement of the CSN in pathogen response came from studies in *Nicotiana tabacuum*. Two proteins, RAR1 (REQUIRED FOR Mla12 RESISTANCE 1) and SGT1 (SUPPRESSOR OF G2 ALLELE OF SKP1), have been defined as a point of convergence between ubiquitination and R gene-mediated resistance. In fact, RAR1 and SGT1 associate with an SCF-type ubiquitin ligase and with the CSN, and both associations are required for *R* gene-mediated responses to induce resistance against a variety of pathogens (Craig et al. 2009). This SGT1-associated SCF ligase and CSN might therefore work in concert to target negative regulators of the defense response, which have yet to be identified. Such regulators could act as repressors of selected defense genes (Craig et al. 2009).

JA-Mediated Defense

The CSN also regulates the response to pathogenic insects and fungi, through its already described interaction with the SCF^{COI} ligase, which mediates JA responses. In fact, JA is not only involved in the regulation of plant development (see paragraph "Jasmonic acid signaling") but also in plant defense. JA accumulates in response to wounding at injury sites, where it is perceived by SCF^{COI1} and promotes JAZ proteins degradation, thus resulting in the expression of a group of JA-responsive genes which encode several plant defense proteins involved in resistance to herbivores and necrotrophic fungal pathogens (Stratmann and Gusmaroli 2012).

Salicylate-Mediated Defenses

After pathogen infection, plants may respond by increasing the production of salicylic acid (SA) (Craig et al. 2009). This increase in SA levels causes, in turn, the SA-dependent upregulation of pathogenesis-related (*PR*) genes.

A key component of such a response is NPR1 (*n*onexpressor of *p*athogenesis-*r*elated genes), a transcriptional positive regulator of *PR* genes. In standard conditions, NPR1 is found mainly in an oligomeric form and sequestered in the cytoplasm. Upon pathogen infection, SA accumulates and promotes NPR1 reduction to a monomeric state, resulting in translocation of NPR1 monomer to the nucleus. Within the nucleus, NPR1 co-activates transcription of *PR* and other response genes. After the activation of gene expression, NPR1 is recruited by a CUL3-based ubiquitin ligase, ubiquitinated, and degraded. This degradation, which also requires a proper functioning CSN, is essential to limit the transcriptional activation of *PR* genes, avoiding a constitutive defense response in the absence of infection (Stratmann and Gusmaroli 2012). This mechanism of action is conserved well beyond *Arabidopsis*, since CSN-silenced tomato plants also display upregulation of *PR* genes (Stratmann and Gusmaroli 2012).

The CSN plays, therefore, a double role in plant defense against pathogens. On one hand, it positively regulates JA- and wound-dependent gene expression, and on the other hand, it functions negatively on the SA-dependent *PR* gene expression (Stratmann and Gusmaroli 2012).

Hijacking the CSN as a Strategy for Successful Microbial Pathogenicity

The CSN has also been selected as a target protein by pathogenic microbes. Any protein that plays an important role in plant defense may potentially serve as a target factor to block host defenses. Since many plant pathogen responses and signaling pathways rely on proteasome-dependent degradation of specific proteins, different plant pathogens have used a successful pathogenic strategy to interfere with the host ubiquitin–proteasome system, through induction, inhibition, or modification of ubiquitin-related host enzymes (Alcaide-Loridan and Jupin 2012). Given the central role played by the CSN in the regulation of proteolysis, this complex is indeed a candidate target for pathogens. Geminiviruses, single-strand DNA viruses that infect a wide range of plant species, interfere with the activity of the CSN as part of their strategy to infect *Arabidopsis*. This virus encode a protein, named C2, that acts both as transcription factor necessary for the expression of viral genes and also as a pathogenicity factor that suppresses host defenses. In *Arabidopsis* C2 directly interacts with the CSN5 subunit, and this association specifically inhibits the deneddylation activity of the CSN on CUL1, without interfering with the CSN or SCF assembly. Thus, C2–CSN interaction impairs the deneddylation of SCF complexes. Accordingly, C2 overexpression leads to downregulation of JA-dependent genes, suggesting that the activity of SCF^{COI1}, the ubiquitin ligase responsible for the jasmonic acid response and involved in plant defense, is affected by the inhibition of the CSN caused by C2 protein (Stratmann and Gusmaroli 2012).

Table 3 Summary of reported physiological phenotypes in mouse CSN mutants

Gene	Phenotypes	References
Csn2	Early embryonic lethal; abnormal elevation of cyclic E in *csn2*$^{-/-}$ embryos	Lykke-Andersen et al. 2003
Csn3	Early embryonic lethal	Yan et al. 2003
Csn5	Early embryonic lethal; cellular senescence; acts as a regulator of p27; apoptosis	Tomoda et al. 2004
Csn6	Early embryonic lethal	Zhao et al. 2011
Csn8	Early embryonic lethal; T-cell specific deletion exhibit lack of T-cell activation; blocked cell cycle re-entry from G0 to G1-S transition; aberrant gene-expression profile	Menon et al. 2007

The CSN and Other Cellular Processes in Plants and Other Organisms

In addition to the well-studied role of the CSN in several aspects of organism development and environmental responses, the CSN is also involved in the regulation of cellular pathways, such as cell cycle, DNA repair, and regulation of gene expression. Although most of the studies on these topics were performed in yeast and mammals (Table 3), some findings suggest that the CSN plays similar functions in plants as well.

Cell Cycle

The CSN is involved in cell cycle progression, and defects in cell proliferation are one of the main features of *csn* mutants in almost all organisms (Wei et al. 2008). CSN-mediated protein degradation ensures cell cycle progression through the timely and ordered elimination of both positive and negative regulators. In mammalian cells, the CSN regulates multiple points of the cell cycle. The CSN regulates G1 phase progression by regulating the activity of the ubiquitin ligase SCF^{Skp2} and the stability of its substrates, such as p27, a *c*yclin-*d*ependent *k*inase (CDK) inhibitor (Kato and Yoneda-Kato 2009). In T-cells, the CSN is necessary for cell cycle reentry from G0 phase. The CSN is required for proper Rb phosphorylation and for activation of G1 cyclins and cyclin-dependent kinases in response to stimulation (Kato and Yoneda-Kato 2009). Finally, the CSN regulates the G2/M transition, since G2 arrest has been reported in *csn* mutants in *Arabidopsis* (Stratmann and Gusmaroli 2012).

In addition, the CSN seems also to be involved in the cell decision to enter or exit cell cycle. *csn8* knockout T-cells show defects in the entry of cell cycle from the quiescent stage (Wei et al. 2008). On the other hand, CSN5 depletion induces cellular senescence, defined as the process by which cells cease to proliferate and eternally withdraw from the cell cycle (Kato and Yoneda-Kato 2009).

CSN5 and CSN6 are overexpressed in several types of tumors, and the CSN5 isopeptidase activity, located in the MPN + domain, is critical for cancer formation. It is still not known if the MPN domain, contained in CSN6, is also involved in tumorigenesis. It remains also to be seen if PCI-containing subunits have a role in cancer; however, it is possible that the role of the CSN in tumors may be assigned only to certain subunits or to a CSN subcomplex and not to the holocomplex (Lee et al. 2011).

DNA Repair

DNA is exposed to exogenous and endogenous agents, such as chemicals, reactive oxygen species, UV, or ionizing radiation, that induce DNA damage. The DNA damage response (DDR) is a complex signal transduction pathway, evolved in the eukaryotes to counteract DNA damage. The outcome of DDR consists in a delay or arrest in the cell cycle, senescence, and/or apoptosis, depending on the context (Hannss and Dubiel 2011).

The role of the CSN in DNA damage response and DNA repair has been established mainly with studies performed in yeast and mammals. The CSN often participates in the DNA repair through CUL4-based CRLs (Wei et al. 2008).

Different types of CUL4-based CRLs recruit and ubiquitinate a large group of specific substrates involved in DNA repair, including components of the nucleotide excision repair machinery and various histone proteins (Hannss and Dubiel 2011). The CSN, together with CUL4-based CRL, is therefore necessary for the timely degradation of factors involved in DNA damage response, and this function of the CSN seems to be shared between plants and animals (Stratmann and Gusmaroli 2012). Indeed, the G2 arrest observed in *Arabidopsis csn* mutants is likely linked to an activation of the DNA damage response pathway (Stratmann and Gusmaroli 2012).

Gene Expression

Many of the substrates of CSN-regulated CRLs are transcription factors. This would suggest that the CSN, by promoting the ubiquitination and the degradation of these factors, might be indirectly capable of regulating gene expression. Indeed, *csn* mutants, both from plants and animals, exhibit aberrant gene expression. For example, *Arabidopsis csn* mutants have aberrant expression of light and hormone-responsive genes (Wei et al. 2008; Chamovitz 2009). *Drosophila csn* mutants show altered temporal gene expression patterns during early development, exhibiting a large set of genes that are derepressed than the wild type, while others genes are repressed. This indicated that the CSN functions not only as transcriptional repressor but also as transcriptional activator (Chamovitz 2009).

It cannot be completely ruled out that the CSN might have a more direct role on the regulation of gene expression by binding DNA directly. Indeed, the CSN can associate with chromatin in vivo and structural studies have also shown that the PCI domain, which is found in six subunits of the CSN, harbors a C-terminal "winged helix" subdomain that is similar to the winged-helix proteins which bind nucleic acid. However, although a CSN–chromatin interaction has been shown, a direct binding to the nucleic acid has not been reported so far (Chamovitz 2009).

Evolutionary Considerations

The CSN has been highly conserved during evolution, in terms of both structure and function. The evolutionary origin of the CSN is still a matter of debate, but the fact that its structure and function are very similar to that of the lid subcomplex of the proteasome (see paragraph "A common molecular structure for the CSN, the proteasome lid, and the eIF3 complexes") suggests that these two complexes might have derived from a common ancestor, following gene duplication events (Serino and Pick 2013). An interesting feature, which comes from the comparison of CSN composition of different organisms, is that smaller but still functional CSN complexes exist. Smaller CSN complexes, composed of fewer subunits, are found in many unicellular fungi as well as in *C. elegans* (Table 1). This raises the question as to whether the missing subunits can be replaced by similar proteins and also suggests that a certain degree of promiscuity might occur between subunits of related complexes, such as the CSN and the lid (Serino and Pick 2013). It is still an open question whether this is a peculiarity of yeast or whether this promiscuity can also occur in other organisms.

Gene duplication events, such as those that led to the formation of the CSN and the lid, seem to be keeping occurring. In *Arabidopsis*, a second round of gene duplication has given rise to the duplication of the two CSN MPN subunits, CSN5 and CSN6, which are encoded by two duplicated genes, named *CSN5a* and *CSN5b* and *CSN6a* and *CSN5b*. *CSN5a* and *CSN5b* are expressed at different levels and their corresponding single mutants have partially overlapping but unique phenotypes, whereas the double mutants have a classic *cop/det/fus* lethal phenotype. This would suggest that while some roles of the duplicated genes are redundant, others are unique. Indeed, CSN5a and CSN5b form two distinct CSN complexes. Little is known about the function of these two alternative CSN complexes, but it cannot be excluded that the two distinct CSN isoforms may be redundant but also acquired a sub-functionalization. On the contrary, some CSN subunits seem to be shared among PCI complexes. This is the case for the eIF3 subunit CIF-1 from *C. elegans*, which moonlights as CSN7, or for the protein Rpn5 from *S. cerevisiae*, which is both a lid and a CSN subunit (Table 1) (Serino and Pick 2013).

Future Directions

Many aspects of the biochemical, cellular, and biological function of the CSN are still unresolved. For example, the CSN can regulate potentially all CRLs; however, only a subset of CRLs seems to be affected by the CSN function. In fact, it is not clear if the CSN always participates in CRL cycles or not; similarly, it is not clear why the CSN favors only some CRLs and not others. Moreover, even if the CSN prevents the auto-ubiquitination of substrate receptors, this autocatalytic mechanism still occurs. It would be very interesting to know which mechanisms are responsible for the CRL choice between these two very different fates – self-destruction or renovation.

In addition, single CSN subunits seem to have evolved specific roles, and CSN mini-complexes have been reported. In the light of the recent advances that underline also a non-catalytic role of the CSN in the regulation of CRLs, future studies should address the physiological and molecular effects of the CSN subcomplexes.

References

Alcaide-Loridan C, Jupin I. Ubiquitin and plant viruses, let's play together! Plant Physiol. 2012;160(1):72–82.

Chamovitz DA. Revisiting the COP9 signalosome as a transcriptional regulator. EMBO Rep. 2009;10(4):352–8.

Craig A, Ewan R, Mesmar J, Gudipati V, Sadanandom A. E3 ubiquitin ligases and plant innate immunity. J Exp Bot. 2009;60(4):1123–32.

Deng XW, Dubiel W, Wei N, Hofmann K, Mundt K. Unified nomenclature for the COP9 signalosome and its subunits: an essential regulator of development. Trends Genet. 2000;16 (7):289.

Emberley ED, Mosadeghi R, Deshaies RJ. Deconjugation of Nedd8 from Cul1 is directly regulated by Skp1-F-box and substrate, and the COP9 signalosome inhibits deneddylated SCF by a noncatalytic mechanism. J Biol Chem. 2012;287(35):29679–89.

Enchev RI, Scott DC, Da Fonseca PC, Schreiber A, Monda JK, Schulman BA, Peter M, Morris EP. Structural basis for a reciprocal regulation between SCF and CSN. Cell Rep. 2012;2 (3):616–27.

Hannss R, Dubiel W. COP9 signalosome function in the DDR. FEBS Lett. 2011;585(18):2845–52.

Hua Z, Vierstra RD. The cullin-RING ubiquitin-protein ligases. Annu Rev Plant Biol. 2011;62:299–334.

Kato JY, Yoneda-Kato N. Mammalian COP9 signalosome. Genes Cells. 2009;14(11):1209–25.

Lau OS, Deng XW. The photomorphogenic repressors COP1 and DET1: 20 years later. Trend Plant Sci. 2012;17(10):584–93.

Lee MH, Zhao R, Phan L, Yeung SC. Roles of COP9 signalosome in cancer. Cell Cycle. 2011;10 (18):3057–66.

Nezames CD, Deng XW. The COP9 signalosome: its regulation of cullin-based E3 ubiquitin ligases and role in photomorphogenesis. Plant Physiol. 2012;160(1):38–46.

Osterlund MT, Ang LH, Deng XW. The role of COP1 in repression of *Arabidopsis* photomorphogenic development. Trends Cell Biol. 1999;9(3):113–8.

Pick E, Hofmann K, Glickman MH. PCI complexes: beyond the proteasome, CSN, and eIF3 Troika. Mol Cell. 2009;35(3):260–4.

Pierce NW, Lee JE, Liu X, Sweredoski MJ, Graham RL, Larimore EA, Rome M, Zheng N, Clurman BE, Hess S, Shan SO, Deshaies RJ. Cand1 promotes assembly of new SCF complexes through dynamic exchange of F box proteins. Cell. 2013;153(1):206–15.

Santner A, Estelle M. The ubiquitin-proteasome system regulates plant hormone signaling. Plant J. 2010;61(6):1029–40.

Serino G, Pick E. Duplication and familial promiscuity within the proteasome lid and COP9 signalosome kin complexes. Plant Sci. 2013;203–204:89–97.

Stratmann JW, Gusmaroli G. Many jobs for one good cop – the COP9 signalosome guards development and defense. Plant Sci. 2012;185–186:50–64.

Wei N, Deng XW. The role of the COP/DET/FUS genes in light control of *Arabidopsis* seedling development. Plant Physiol. 1996;112(3):871–8.

Wei N, Serino G, Deng XW. The COP9 signalosome: more than a protease. Trend Biochem Sci. 2008;33(12):592–600.

Further Reading

Chen H, Shen Y, Tang X, Yu L, Wang J, Guo L, Zhang Y, Zhang H, Feng S, Strickland E, Zheng N, Deng XW. Arabidopsis CULLIN4 Forms an E3 ubiquitin ligase with RBX1 and the CDD complex in mediating light control of development. Plant Cell. 2006;18(8):1991–2004.

Dohmann EM, Nill C, Schwechheimer C. DELLA proteins restrain germination and elongation growth in Arabidopsis thaliana COP9 signalosome mutants. Eur J Cell Biol. 2010;89(2–3):163–8.

Feng S, Ma L, Wang X, Xie D, Dinesh-Kumar SP, Wei N, Deng XW. The COP9 signalosome interacts physically with SCF COI1 and modulates jasmonate responses. Plant Cell. 2003;15 (5):1083–94.

Gfeller A, Liechti R, Farmer EE. *Arabidopsis* jasmonate signaling pathway. Sci Signal. 2010;3 (109):cm4.

Hind SR, Pulliam SE, Veronese P, Shantharaj D, Nazir A, Jacobs NS, Stratmann JW. The COP9 signalosome controls jasmonic acid synthesis and plant responses to herbivory and pathogens. Plant J. 2011;65(3):480–91.

Hotton SK, Callis J. Regulation of cullin RING ligases. Annu Rev Plant Biol. 2008;59:467–89.

Kotiguda GG, Weinberg D, Dessau M, Salvi C, Serino G, Chamovitz DA, Hirsch JA. The organization of a CSN5-containing subcomplex of the COP9 signalosome. J Biol Chem. 2012;287(50):42031–41.

Liu Y, Schiff M, Serino G, Deng XW, Dinesh-Kumar SP. Role of SCF ubiquitin-ligase and the COP9 signalosome in the N gene-mediated resistance response to tobacco mosaic virus. Plant Cell. 2002;14(7):1483–96.

Lykke-Andersen K, Schaefer L, Menon S, Deng XW, Miller JB, Wei N. Disruption of the COP9 signalosome Csn2 subunit in mice causes deficient cell proliferation, accumulation of p53 and cyclin E, and early embryonic death. Mol Cell Biol. 2003;23(19):6790–7.

Menon S, Chi H, Zhang H, Deng XW, Flavell RA, Wei N. COP9 signalosome subunit 8 is essential for peripheral T cell homeostasis and antigen receptor-induced entry into the cell cycle from quiescence. Nat Immunol. 2007;8(11):1236–45.

Schwechheimer C, Isono E. The COP9 signalosome and its role in plant development. Eur J Cell Biol. 2010;89(2–3):157–62.

Schwechheimer C, Serino G, Callis J, Crosby WL, Lyapina S, Deshaies RJ, Gray WM, Estelle M, Deng XW. Interactions of the COP9 signalosome with the E3 ubiquitin ligase SCFTIRI in mediating auxin response. Science. 2001;292(5520):1379–82.

Sharon M, Mao H, Boeri EE, Stephens E, Zheng N, Robinson CV. Symmetrical modularity of the COP9 signalosome complex suggests its multifunctionality. Structure. 2009;17(1):31–40.

Spoel SH, Mou Z, Tada Y, Spivey NW, Genschik P, Dong X. Proteasome-mediated turnover of the transcription coactivator NPR1 plays dual roles in regulating plant immunity. Cell. 2009;137 (5):860–72.

Tomoda K, Yoneda-Kato N, Fukumoto A, Yamanaka S, Kato JY. Multiple functions of Jab1 are required for early embryonic development and growth potential in mice. J Biol Chem. 2004;279(41):43013–8.

Vierstra RD. The ubiquitin-26S proteasome system at the nexus of plant biology. Nat Rev Mol Cell Biol. 2009;10(6):385–97

Wager A, Browse J. Social network: JAZ protein interactions expand our knowledge of jasmonate signaling. Front Plant Sci. 2012;3:41.

Wang F, Deng XW. Plant ubiquitin-proteasome pathway and its role in gibberellin signaling. Cell Res. 2011;21(9):1286–94.

Wang X, Feng S, Nakayama N, Crosby WL, Irish V, Deng XW, Wei N. The COP9 signalosome interacts with SCF UFO and participates in *Arabidopsis* flower development. Plant Cell. 2003;15(5):1071–82.

Wei N, Deng XW. The COP9 signalosome. Annu Rev Cell Dev Biol. 2003;19:261–86.

Yan J, Walz K, Nakamura H, Carattini-Rivera S, Zhao Q, Vogel H, Wei N, Justice MJ, Bradley A, Lupski JR. COP9 signalosome subunit 3 is essential for maintenance of cell proliferation in the mouse embryonic epiblast. Mol Cell Biol. 2003;23(19):6798–808.

Zhao R, Yeung SC, Chen J, Iwakuma T, Su CH, Chen B, Qu C, Zhang F, Chen YT, Lin YL, Lee DF, Jin F, Zhu R, Shaikenov T, Sarbassov D, Sahin A, Wang H, Lai CC, Tsai FJ, Lozano G, Lee MH. Subunit 6 of the COP9 signalosome promotes tumorigenesis in mice through stabilization of MDM2 and is upregulated in human cancers. J Clin Invest. 2011;121(3):851–65.

Plant Circadian Network

12

Jordi Malapeira, Reyes Benlloch, Rossana Henriques, and Paloma Mas

Contents

J. Malapeira • R. Benlloch • R. Henriques • P. Mas (✉)
Molecular Genetics, Centre for Research in Agricultural Genomics (CRAG), Barcelona, Spain

Consortium CSIC-IRTA-UAB-UB, Center for Research in Agricultural Genomics (CRAG), Parc de Recerca UAB, Barcelona, Spain
e-mail: jordi.malapeira@cragenomica.es; reyes.benlloch@cragenomica.es; rossana.henriques@cragenomica.es; paloma.mas@cragenomica.es

S.H. Howell (ed.), *Molecular Biology, The Plant Sciences 2*,
DOI 10.1007/978-1-4614-7570-5_6

Abstract

- The circadian clock generates biological rhythms with a period of 24 h.
- Circadian clock function is essential for plant physiology and development and provides an adaptive advantage.
- The circadian clock is synchronized by changes in environmental signals.
- Pre- and posttranscriptional mechanisms regulate the expression and activity of the circadian oscillator.
- The expression of many genes and key metabolic and developmental processes are controlled by the circadian clock.

Introduction

The circadian clock is a cellular mechanism present in almost all organisms examined to date. This timing device is able to perceive environmental changes as an indication of passing time and use this information to generate rhythms in multiple biological processes. It was proposed that the 24-h rhythms generated by the circadian clock provide an adaptive advantage by allowing the anticipation of the environmental changes and by synchronizing the biological activities to the most appropriate times during the day or night. The mechanisms responsible for generating and maintaining the rhythms are complex and require the orchestrated function of many players. For optimal growth and survival, plants rely on a sophisticated network of perception and responses to the fluctuating environment. The circadian clock is placed at the center of this network, connecting multiple input and output signals essential at all stages during the plant life cycle.

Despite the pervasive influence of time in all aspects of life, it is rather challenging to precisely define "time." The Merriam-Webster dictionary defines time as "a non-spatial continuum that is measured in terms of events which succeed one another from past through present to future." A more lyrical quote by Jorge Luis Borges states that "time is the substance I am made of. Time is a river which sweeps me along, but I am the river; it is a tiger which destroys me, but I am the tiger; it is a fire which consumes me, but I am the fire." Regardless whether time is considered as a nonspatial continuum or as the substance we are made of, it is obvious that time rules our lives and consequently devices (watches, atomic clocks) have been created that allow us to measure time and organize our lives around it. However, nature had long anticipated human clocks by creating timing devices in our cells that not only measure time but also modulate the timing of our physiology and development.

The timing devices that generate biological rhythms with a period of 24 h in close synchronization with the day–night cycle are known as circadian clocks. The different functional modules that are required for clock function include the following: (1) they must be synchronized every day by the changes in environmental conditions, mostly by the diurnal variations in light and temperature, (2) clock

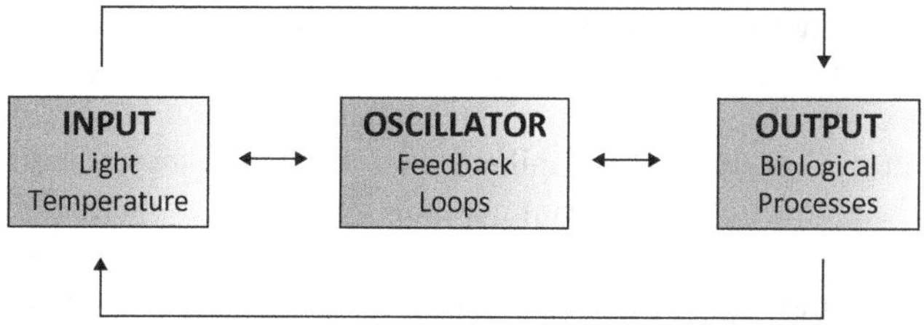

Fig. 1 Schematic representation depicting a simplified version of the main functional modules comprising the circadian system. *Arrows* indicate the relationship between the different modules

components at the so-called central oscillator are reciprocally regulated by feedback loops to generate the rhythms in their own expression, and (3) the rhythms are translated into multiple output biological processes that are controlled by the clock (Fig. 1).

The classical view of the circadian system as input–oscillator–output is clearly oversimplified, and it is now known that the circadian clockwork relies on a much more complex network of components and activities interconnected with other relevant biological pathways (Harmer 2009). Nevertheless, this classical view is still conceptually pertinent, and therefore, the current review has been structured around the three main functional modules (input, oscillator, and output) (Más and Yanovsky 2009). Overall, the review touches upon the inner workings of the circadian clockwork using current knowledge mainly (but not exclusively) from studies of the small plant *Arabidopsis thaliana*.

Input Signals to the Clock

Circadian Entrainment

The circadian clock is synchronized every day by the environmental changes that occur during the day and night cycle, particularly at the dawn/dusk transitions. The environmental cues that synchronize the clock, mostly changes in light and temperature, are called *zeitgebers* (from German: time givers). Based on their importance for clock function, considerable research efforts have been focused in the past decade on the molecular mechanisms responsible for clock entrainment by these signals. A wealth of information is currently available for clock synchronization by light, particularly in the model system *A. thaliana*. More recent studies have also provided insightful clues about the role of temperature in clock resetting and on the clock's ability for compensating (or buffering) temperature changes, in a property known as temperature compensation (McWatters and Devlin 2011).

Circadian oscillations are characterized by periodical changes in biological activities that vary between maximum (peaks) and minimum values (troughs) along a 24-h period. The waveform of a circadian oscillation can be described by a number of parameters, including the period, the amplitude, and the phase of the oscillation. The period corresponds to the duration of one complete rhythmic cycle; the phase describes the state of a rhythm relative to another reference rhythm (e.g., the day–night cycle), while the amplitude corresponds to the difference between mean value and maximum or minimum of a sinusoidal oscillation. Different methods can be used to study how environmental signals entrain the plant clock. One of these methods relies on measuring the variation in circadian parameters upon plant exposure to different stimuli. Using this approach, early studies focused on the characterization of circadian-controlled processes, such as cotyledon movement or expression of circadian-regulated genes, such as *CHLOROPHYLL A-/B-BINDING PROTEIN 2* (*CAB2*) or *COLD-CIRCADIAN RHYTHM-RNA BINDING2* (*CCR2*), in response to different light regimes. The generation of plants expressing the firefly *LUCIFERASE* (*LUC*) fused to the promoters of clock-controlled genes (*CAB2::LUC, CCR2::LUC*) allowed researchers to precisely follow the circadian rhythmic expression of the different output genes.

It is now well accepted that in diurnal organisms, the length of the circadian period under constant light conditions is inversely proportional to the light intensity, a relation that is known as the Aschoff's rule. Following this rule, circadian oscillations display a shorter period under constant (parametric) high light intensities (or fluences) compared to low intensities. Measurements of period length at different light fluences results in the so-called fluence–response curves (FRCs), which show the effects of light quality (wavelength) and quantity (intensity) on circadian period length. The use of monochromatic light has shown that red, blue, and ultraviolet-B (UV-B) light can entrain the circadian clock. Plant light perception is mediated by a set of photoreceptors, each one specifically sensing different qualities of light. In section "Molecular Components Involved in Clock Entrainment by Light: Photoreceptors," the nature of the main photoreceptors involved in circadian entrainment and the molecular mechanism underlying regulation of clock resetting by light is discussed.

The effect of a particular stimulus on clock function can be also assessed by studying the variations in circadian phase in response to discrete pulses of the stimulus (nonparametric) delivered at different times of the day. In diurnal circadian systems, light pulses delivered in the middle of the day (light period) had little or no effect in the circadian phase, whereas light pulses delivered around dusk or dawn cause significant phase advances or delays, respectively. These effects are quantified in the so-called phase–response curves (PRCs), in which phase advances or delays are represented as a function of the time when the stimulus is delivered. PRC analysis in response to light or temperature pulses revealed that the clock controls its own sensitivity to input signals during the day/night cycle. This property is known as "gating" and the current knowledge on the molecular mechanisms that explain the gating response is summarized in section "Properties of Circadian Function: Gating Light Input to the Clock." Although light is the main signal entraining the plant circadian clock, much effort has been devoted in the last years to understand the contribution of temperature changes to the clock's resetting. A summary of our current knowledge on temperature and the crosstalk between

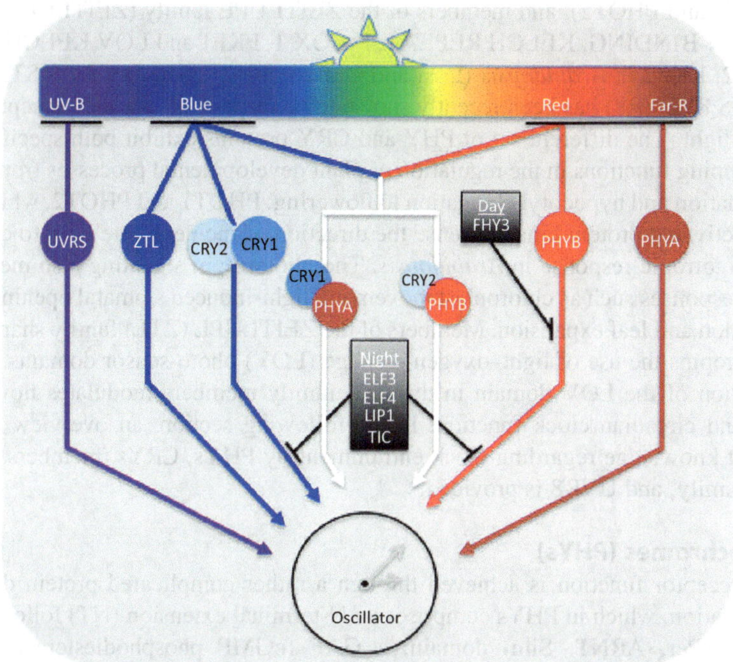

Fig. 2 Schematic representation of photoreceptor involvement in light input to the clock. Photoreceptors perceive the different wavelengths from ultraviolet-B (*UVR8, violet arrow*), blue (*ZTL* and *CRYs, blue arrow*), red (*PHYB, red arrow*), and far-red (*PHYA, dark red arrow*). Integration of *red-* and *blue-light* inputs is achieved by photoreceptor interaction (*white arrows*). FHY3 gates specifically *red-light* input during the day (*black box*), while a set of gating factors regulate light input during the night (*ELF3, ELF4, LIP1*, and *TIC*)

light and temperature in the entrainment of the circadian clock is discussed in sections "Entrainment by Temperature" and "Crosstalk Between Light and Temperature Entrainment," respectively. In addition to light and temperature, the plant clock can be also entrained by other signals including imbibition, hormones, and metabolites. Many metabolic and hormone pathways can be placed both as input and output signals to the clock, which complicates the analysis of their precise function within the circadian system.

Molecular Components Involved in Clock Entrainment by Light: Photoreceptors

Perception and responses to light is achieved in higher plants by a number of photoreceptors that can sense light from UV-B to the near infrared (Fig. 2; Millar 2004). The red and far-red (FR) light are mostly sensed by the phytochrome family (PHYA to PHYE in *Arabidopsis*), while UV-A/blue light is perceived by three classes of photoreceptors: cryptochromes (CRY1, CRY2, and CRY3), phototropins

(PHOT1 and PHOT2), and members of the ZEITLUPE family (ZEITLUPE, ZTL, FLAVIN-BINDING, KELCH REPEAT, F-BOX 1, FKF1 and LOV KELCH PROTEIN 2, LKP2 in *Arabidopsis* (Losi and Gärtner 2012)). The UV RESISTANCE LOCUS 8 (UVR8) has been recently shown to be responsible for the perception of UV-B light. The different set of PHY and CRY proteins exhibit both specific and overlapping functions in the regulation of plant developmental processes from seed germination and hypocotyl elongation to flowering. PHOT1 and PHOT2, which are light-activated protein kinases, sense the direction of incident blue light to control the phototropic response in *Arabidopsis*. The phototropin signaling also mediates other responses such as chloroplast movement, light-induced stomatal opening, and cotyledon and leaf expansion. Members of the ZEITLUPE (ZTL) family share with phototropins the use of light–oxygen–voltage (LOV) photo-sensor domains. Light activation of the LOV domain in the ZTL family members modulates flowering time and circadian clock function. In the following section, an overview of the current knowledge regarding clock entrainment by PHYs, CRYs, members of the ZTL family, and UVR8 is provided.

Phytochromes (PHYs)

Photoreceptor function is achieved through a rather complicated protein domain organization, which in PHYs comprises an N-terminal extension (NT) followed by a PAS (Per, ARNT, Sim) domain, a GAF (cGMP phosphodiesterase/adenyl cyclase/FhlA) domain, a PHY domain, and a complex C-terminus domain composed of two PAS domains and a histidine-kinase-related domain (HKRD). PHYs use phytochromobilin (PΦB) as a chromophore, which is covalently bound to a cysteine residue in the GAF domain. PHYs photoreversibly switch between the so-called Pr and Pfr conformers upon isomerization of a double bond between two rings of the tetrapyrrole. It is proposed that each one of these domains have a relevant function in the final outcome of the photoreceptor activity. In relation to the circadian clock, the use of FRCs comparing circadian period at different light qualities and quantities in wild-type (WT) and *phy* mutant plants was essential for elucidating a role for PHYA and PHYB in circadian entrainment (Chen et al. 2004). The studies showed that PHYA is a crucial factor entraining the clock at low fluences of red and blue light, while PHYB entrains the clock at higher red-light fluences. PHYD and PHYE are partially redundant with PHYB entraining the clock in high-fluence red light. Furthermore, the direct interaction of PHYB and CRY2 was shown to be essential for transmitting together red-light (through PHYB) and blue-light (through CRY2) environmental information to synchronize the oscillator (Fig. 2). The interaction of PHYA with CRY1 might also explain the role of PHYA in clock synchronization by blue light, although the mechanistic details have not been fully demonstrated. Overall, and despite the importance of PHYs for red-light-dependent circadian synchronization, robust circadian rhythms are maintained in the quintuple *phy* mutant (*phyA*/*phyB*/*phyC*/*phyD*/*phyE*), which suggests that other photoreceptors might also contribute to clock synchronization in the absence of PHYs.

Cryptochromes (CRYs)

CRY proteins have two distinct domains: an N-terminal domain of about 500 amino acids (aa) with homology to type II DNA photolyases (photolyase homologous region, PHR) and a C-terminal extension or CCE (for cryptochrome C-terminal extension). The CCE domain is an intrinsically unstructured domain that acts as the effector domain. The light-sensing properties are provided by two chromophores: a flavin adenine dinucleotide (FAD) and a pterin (Chaves et al. 2011). Similar to PHYs, the different domains of CRYs appear to play relevant functions for the activity of the proteins. Both CRY1 and CRY2 are important for the entrainment of the clock under blue light. *cry1* mutant shows an increased period length in plants grown under high and low fluences of blue light, but not at intermediate fluences. The effect of the *cry2* mutation is evident under low intensities of blue light and, in combination with *cry1*, increases the severity of the phenotype at all blue-light fluences. These results suggest that CRY1 and CRY2 have partial redundant roles in the entrainment of the clock (Chen et al. 2004). As mentioned above, the interaction of CRYs and PHYs seems to be important for transmitting the red- and blue-light information to the clock. Consistent with this notion, the interaction of CRY2 and PHYB is key for synchronization under low-fluence blue- and red-light conditions. The fact that the double mutant *cry1/cry2* is still able to maintain rhythms suggests that CRYs are not integral parts of the oscillator as it has been shown for the mammal cryptochromes, mCRY1 and mCRY2.

ZEITLUPE Family

In addition to phototropins, *Arabidopsis* has two other types of proteins containing LOV domains. The first type of proteins is known as PAS/LOV proteins (PLP) (or LOV/LOV proteins, LLP) and contains two LOV domains. However, the physiological relevance of the PLP family remains unknown. The other family, known as the ZEITLUPE family, comprises three proteins (ZTL, FKF1, and LKP2) containing a single LOV domain responsible for flavin binding and photoreception, the F-box domain participating in Skp/Cullin1/F-box type E3 ligase complex or SCF complex, and a carboxy-terminal Kelch repeat involved in protein–protein interactions.

Genetic screenings for mutants with altered period length identified mutations at the *ZTL* locus that lengthened the free-running period of circadian gene expression and altered the photoperiodic regulation of flowering time. The effects of ZTL within the clock are mediated through proteasomal degradation of a member of the pseudo-response regulator (PRR) family, the key clock protein TOC1 (TIMING OF CAB EXPRESSION 1) (see section "Molecular Components of the Circadian Oscillator"). ZTL directly interacts with TOC1 and the interaction is essential for modulating TOC1 protein stability, which in turn is responsible for maintaining stable circadian period by the clock. PRR5, another member of the PRR family is also targeted by ZTL for degradation through the proteasome. Phenotypic characterization of higher-order mutants with other members of the ZTL family showed that LKP2 and FKF1 act redundantly with ZTL in the degradation of both PRR5 and TOC1. The ZTL protein family also directly interacts with the clock component GIGANTEA (GI), a member of the evening loop of the oscillator (see section "Molecular Structure and Mechanisms at the Core of the *Arabidopsis* Circadian Oscillator"). The ZTL–GI interaction is enhanced by blue light and results in the stabilization of ZTL, which also appears to require the function of the molecular chaperone HEAT SHOCK PROTEIN 90 (HSP90), a central element involved in the maturation of several regulatory proteins. During the night, the affinity of ZTL for GI decreases allowing the interaction of ZTL with TOC1. These studies highlight the importance of posttranslational regulation and the complexity of the network controlling circadian period by the clock (further discussed in section "Molecular Structure and Mechanisms at the Core of the *Arabidopsis* Circadian Oscillator").

UV RESISTANCE LOCUS 8 (UVR8)

UV-B light is a harmful component of sunlight, damaging DNA, protein, and other cellular molecules. However, at lower fluences, UV-B light can also function as an environmental signal to regulate the production of flavonoids and to modulate plant development. The *UV RESISTANCE LOCUS 8* (*UVR8*) encodes a specific UV-B photoreceptor that is required for virtually all physiological UV-B responses. UVR8 shows sequence similarity with the human guanine nucleotide exchange factor REGULATOR OF CHROMOSOME CONDENSATION 1 (RCC1), although the two proteins have probably diverged functionally. UVR8 contains no external chromophore but its light-sensing mechanism appears to rely on the presence of key tryptophan residues. In the dark, UVR8 protein is dimeric, but upon exposure to UV-B light, the protein undergoes a conformational change and becomes monomeric. Monomeric UVR8 is the active isoform responsible for triggering the light-induced responses. Analysis of circadian period under continuous white light supplemented with low-fluence UV-B light showed that UV-B is able to entrain the circadian clock. In WT plants, UV-B light leads to a shortening of the circadian period compared to white light conditions (without UV-B). The shortening is absent in the *uvr8* mutant, suggesting that the UVR8 photoreceptor is needed for circadian entrainment by UV-B light. The entrainment of the clock in response to UV-B light appears to be mediated by transcriptional regulation of essential clock genes.

Molecular Components and Signaling Pathways Contributing to Photoreceptor Circadian Function

The basic leucine-zipper transcription factor LONG HYPOCOTYL 5 (HY5) provides a direct link between photoreceptors and the control of the oscillator genes. HY5 acts as a positive regulator of photomorphogenesis by directly affecting the expression of light-induced genes. The CONSTITUTIVE PHOTOMORPHOGENESIS PROTEIN 1 (COP1) is an E3 ubiquitin ligase that targets HY5 protein for degradation. The interaction between photoreceptors and COP1 results in the inhibition of COP1 function that consequently leads to an increase in HY5 protein accumulation. The function of HY5 downstream of both PHYs and CRYs might involve the binding of HY5 to the promoters of several oscillator genes. However, the molecular and mechanistic details of HY5 function in the clock are not fully known (Chen et al. 2004).

Other components of the PHYB signaling cascade that might influence clock function include the SENSITIVITY TO RED LIGHT REDUCED 1 (SSR1), the PSEUDO-RESPONSE REGULATOR 7 (PRR7), the EARLY FLOWERING 3 (ELF3), and the ARABIDOPSIS RESPONSE REGULATOR 4 (ARR4). SRR1 was initially identified by characterization of mutant plants defective on hypocotyl elongation under red light. The SSR1 protein contains no known domains, and its expression is acutely induced by red light. The precise function of SSR1 is unknown, although it is a broadly conserved protein among eukaryotes. In addition to the hypocotyl elongation defects, loss-of-function *ssr1* mutant shows a short-period circadian phenotype, along with a general reduction in the amplitude of the oscillations of clock genes when exposed to continuous light conditions. In WT plants, red-light pulses produce an acute induction of the *CAB2* expression. This acute induction is enhanced in the *ssr1* mutant, which suggests a circadian role for SSR1, plausibly through the PHYB signaling cascade. PRR7 has an essential role in different aspects of circadian oscillator function. PRR7 is not only part of the morning loop of the oscillator (sections "Molecular Components of the Circadian Oscillator" and "Molecular Structure and Mechanisms at the Core of the *Arabidopsis* Circadian Oscillator"), but it is also involved in clock resetting by temperature and in temperature compensation (sections "Entrainment by Temperature" and "Crosstalk Between Light and Temperature Entrainment"). Genetic studies showed that the defects of hypocotyl elongation in *prr7* mutants are enhanced in *phyA* mutants, suggesting that PRR7 functions downstream of both PHYA and PHYB in the transduction of light input information. Consistently, the red-light induction of the essential clock components *CCA1* and *LHY* is reduced in the *prr7* mutant.

ELF3 protein abundance oscillates under the control of the circadian clock, reaching a maximum just before the dark period. *elf3* displays phenotypes correlated with the PHYB signaling cascade, including long hypocotyl under continuous red light and arrhythmia of clock outputs (*CAB2::LUC* and *CCR2::LUC*) under constant light conditions. The ELF3 protein is nuclear and contains several differentiated regions including a proline-rich region, an acidic domain, and a glutamine-/threonine-rich region. These regions are often associated with transcriptional regulators. ELF3 physically interacts with PHYB and this interaction has been proposed to be important for the control of circadian period by ELF3. However, ELF3 overexpression (*ELF3*-ox) causes a long-period phenotype of *CAB2* oscillations under red-light but also under all blue-light fluences. This suggests that ELF3 might act as a general repressor of light input to the circadian oscillator. ELF3 has lately been proposed to play a role as part of the core oscillator as a member of the so-called evening complex (EC, consult sections "Molecular Components of the Circadian Oscillator" and "Molecular Structure and Mechanisms at the Core of the *Arabidopsis* Circadian Oscillator"). The role of ELF3 in gating the light input to the clock is also discussed in section "Properties of Circadian Function: Gating Light Input to the Clock."

ARR4 belongs to the *Arabidopsis Response Regulator* family of proteins involved in cytokinin signaling. Type-A ARRs have a receiver domain with short N- and C-terminal extensions, and their expression is induced by exogenous cytokinin; type-B ARRs have longer C-terminal extensions and their expression is not affected by exogenous cytokinins. The double mutation in *arr3/arr4*, corresponding to two genes belonging to type-A ARRs, exhibits long period of circadian gene expression and cotyledon movement when entrained under both photocycles and thermocycles. The alteration of the period is evident in a number of different light conditions including continuous red and blue light and darkness. ARR4 physically interacts with PHYB, which might explain its light-dependent phenotypes. However, other components and pathways are more likely responsible for the circadian defects observed under dark conditions.

Properties of Circadian Function: Gating Light Input to the Clock

As previously mentioned, the sensitivity of the clock towards environmental stimuli is not constant during the day–night cycle. In plants, a light pulse delivered close to dawn or dusk causes evident phase shifts, while the same pulse in the middle of the day barely affects the phase of the clock. The differential sensitivity of the clock depending on the time of day is a very interesting clock property known as "gating." The gating response plausibly protects clock progression against short-term variations of the environmental conditions that should not reset the clock (e.g., the lightning caused by a storm in the middle of the night). Gating is not limited to light signals, as the clock gates many other responses, including the hormone-dependent induction of clock gene expression.

The mechanisms responsible for gating light signals are not understood in detail. One likely mechanism involves the circadian expression and regulation of the photoreceptors and their downstream factors, which may render plants more or less sensitive to particular wavelengths at different times of day. Consistently, expression of both *PHYs* and *CRYs* is regulated in a circadian fashion. *CRY1* is highly expressed in the morning, while *CRY2* shows a peak of expression just before dusk. *PHYs* also show different peaks of expression during the day; *PHYA* has a double peak with one maximum after dawn and another before dusk; *PHYB*, *PHYD*, and *PHYE* peak in the morning and *PHYC* just before dusk. It is noteworthy that photoreceptor peak of expression mostly occurs around dawn or dusk, at times when plants are more sensitive to entrainment cues. However, regulation of photoreceptor expression and activity do not fully explain the gating ability of the circadian clock. The characterization of PRCs for different environmental stimuli applied at different times of day ("gating experiments") has led to the identification of additional components involved in the gating response. The identified factors include components with a relevant role in the clock gating response during the day such as FAR-RED ELONGATED HYPOCOTYL 3 (FHY3) and during the night

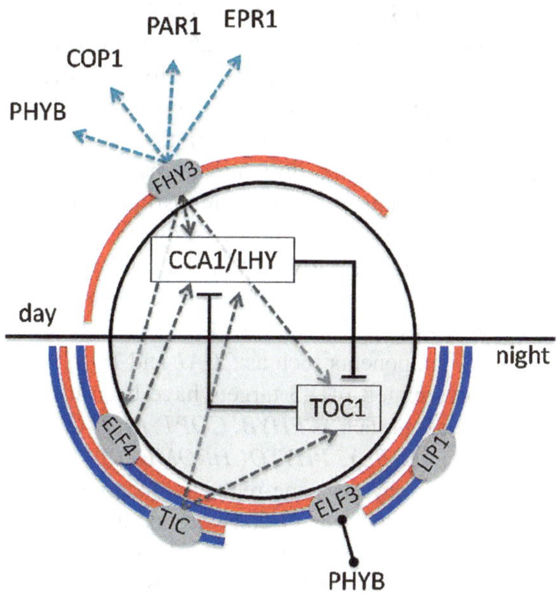

Fig. 3 Factors controlling circadian gating of light input. Different gating factors (*grey ovals*) modulate light input to the clock during subjective day or night. These factors can gate specifically *red light* or *red* and *blue light* (*red/blue semicircles*). The length of the semicircles indicates the time of day at which those factors are active. *Grey arrows* indicate contribution to the regulation of core clock components. *Blue arrows* indicate contribution to the regulation of other factors possibly implicated in the gating mechanism. *Blunt arrow* indicates direct protein–protein interaction

such as ELF3, EARLY FLOWERING 4 (ELF4), LIGHT INSENSITIVE PERIOD 1 (LIP1), and TIME FOR COFFEE (TIC) (Fig. 3). Each one of these factors belongs to different protein families and most likely contributes to clock gating with different mechanisms and signaling pathways.

Gating Responses During the Day

Gating through the early part of the day partially relies on the function of FHY3. FHY3 is a member of the FAR family of transcription factors, which comprises 14 members in *Arabidopsis*. The FHY3 protein contains an N-terminal zinc-finger domain, a nuclear localization signal, a central core transposase domain, and a C-terminal SWIM motif (from SWI2/SNF and MuDR transposases). Loss-of-function *fhy3* mutants display defects in PHYA-controlled processes such as hypocotyl growth inhibition under FR light or anthocyanin accumulation. FHY3 and the related protein FAR-RED-IMPAIRED RESPONSE 1 (FAR1) are key elements of the PHYA signal transduction by regulating PHYA stability. PHYA regulation involves posttranslational modifications such as phosphorylation and ubiquitination. Different forms of phosphorylated photoreceptor are present in the cellular PHYA pool. On one side, hyperphosphorylated PHYA isoforms interact preferentially with the COP1/SUPPRESSOR OF PHYA 1 (SPA1) complex, which targets the photoreceptor for degradation by the 26S proteasome. On the other side, hypophosphorylated PHYA forms interact specifically with FHY3. A mechanism has been proposed by which FHY3 would protect phosphorylated PHYA from COP1/SPA1-mediated proteolysis. In *Arabidopsis*, the circadian clock gates hypocotyl growth to the late night. In the *fhy3* mutant, hypocotyl growth rhythms in seedlings grown under continuous red light were disrupted, as it was the circadian *CAB2::LUC* oscillation. Notably, *fhy3* mutant seedlings exhibit altered circadian phase advances mainly during the early part of the subjective day. The observed effects were specific to red light, as no difference between the WT and *fhy3* was detected at other wavelengths. The absence of FHY3 activity affects the rhythmic expression of key clock components such as *CCA1* and *TOC1*.

Other light- and clock-related FHY3 targets have been identified. The targets include important components such as *PHYB*, *COP1*, *PHYTOCHROME RAPIDLY REGULATED 1* (*PAR1*), *EARLY PHYTOCHROME RESPONSIVE 1* (*EPR1*), *CCA1*, and *ELF4*. ELF4 is a phytochrome-regulated factor essential for the correct progression of the circadian clock under constant conditions. Detailed functional analysis of the *ELF4* promoter showed that FHY3, FAR1, and HY5 act as constant positive regulators, while CCA1 and LHY periodically repress *ELF4* by decreasing FHY3 DNA-binding activity. This leads to a coordinated regulation that is necessary for robust oscillation of *ELF4* and for clock function.

Gating Responses During the Night

As previously mentioned, *elf3* loss-of-function mutants cause arrhythmia under constant light conditions, which initially placed ELF3 as part of a pathway that modulates light input to the clock. In gating experiments, the *elf3* mutant show larger phase shifts than wild type when red- or blue-light pulses are applied. The phase shift differences were more evident during the early and middle part of the subjective night. Compared to the wild-type response, the *ELF3*-ox plants attenuate the phase shifts caused by the light pulses. The mechanism by which ELF3 buffers the response to light is not fully understood although the direct interaction of ELF3 with PHYB could partially mediate the gating function. In addition to its gating function, recent studies have revealed other roles for ELF3 in clock progression, mainly as part of the so-called evening complex (EC). The EC is composed of ELF3, and two other clock components, ELF4 and LUX ARRHYTHMO/PHYTOCLOCK (LUX/PCL). The EC was shown to regulate the circadian modulation of hypocotyl growth and the expression of the key component *PRR9*, most likely by direct binding to its promoter (section "Molecular Structure and Mechanisms at the Core of the *Arabidopsis* Circadian Oscillator").

ELF4 has been also associated with gating responses during the night. In *elf4* mutants, the oscillations dampen to arrhythmia mainly at the end of the first subjective day, a time at which the clock is highly sensitive to light inputs. Accordingly, *elf4* is partially a gating mutant, as it is hypersensitive to red-light pulses, especially during the subjective night. The ability of a particular genotype to reset the clock can be evaluated by inverting the light/dark cycles and scoring the time needed for circadian resetting to the new environmental signals. This kind of experiments revealed that *elf4* mutant resets the clock much faster than wild type. The loss of gating during the subjective night leads to an increased expression of *CCA1* and *LHY* in the inverted cycles, a response that is gated by the clock in the wild-type plants. The increased expression of the morning genes might explain the faster clock resetting in the *elf4* mutant. Therefore, ELF4 contributes to the gating response during the night, but it is also needed for robust circadian oscillations and maintenance of clock function. *ELF4* expression is dependent on PHYB, providing a link between photoreception and clock resetting. ELF4 functionally shares many similarities with ELF3, as *elf4* mutant shows a long hypocotyl phenotype when grown under continuous red light, along with defects in gating of light input signals. Genetic studies have proven that ELF3 and ELF4 function in the same pathway controlling light input into the clock, being ELF4 downstream of ELF3.

Another factor involved in gating the light input to the clock is LIP1, a GTPase most likely acting at the same time as ELF3 and ELF4, i.e., during the first part of the subjective night. The *lip1* mutants show a circadian period length that is insensitive to changes in light intensity. Red light pulses induce larger phase shifts in the *lip1* mutant than in the wild type during the subjective night. The mutant also displays enhanced photomorphogenic responses to both red and blue light. Interestingly, nuclear localization of LIP1 is essential for its role gating the light input to the clock but not for its additional function in development. This suggests the existence of distinct mechanisms for LIP1 function in clock entrainment and photomorphogenesis. LIP1 represents the first small GTPase implicated in the circadian system of plants. Recent studies have shown that LIP1 might also have additional roles in the control of endoreplication and tolerance to salt stress.

The gating role of ELF3, ELF4, and LIP1 during the first part of the subjective night is followed by the gating function of additional players within the clock system. One of these players, TIC, is a nuclear regulator important for the modulation of the light input during mid to late subjective night. In loss-of-function *tic* mutants, light pulses activate *CAB* expression regardless the time of the day, revealing a defect in gating the light input. The amplitude of the oscillation of core clock components is decreased in *tic* mutants compared to WT. The reduced amplitude in clock gene expression might be responsible for the low amplitude of output rhythms, as is the case for *CAB* oscillation. Recently, it has been shown that TIC is also involved in plant iron homeostasis, as the iron-responsive genes are regulated both by TIC and the circadian oscillator (see section "Iron and Copper Homeostasis"). However, it is not clear the connection between the control of iron homeostasis by TIC and its function gating the light input to the clock. TIC also acts as a negative regulator of the hormone jasmonic acid (JA) and links the circadian clock with the JA responses and pathogen infection (see section "Circadian Regulation of Hormone Signaling"). It is a common feature of gating factors to have additional roles maybe acting in an independent fashion from their gating function.

Less information is known about gating the UV-B light input into the clock. PRC analysis of *CCR2::LUC* in response to short pulses of UV-B shows that this treatment indeed produces phase changes. The response is gated by the clock, with a maximum phase delay when pulses are applied during the early subjective night. The gating mechanisms appear to be partially shared between visible and UV-B light. Indeed, visible and also UV-B light pulses acutely induce clock genes such as *CCA1*, *PRR9*, *GI*, and *ELF4*. The clock gates this induction, so that the effects of light exposure are higher around their respective peaks of expression. In addition to their similarities, there are also differences between the visible and the UV-B light-gating mechanisms. For instance, although *HY5* expression is induced by red and UV-B light treatments, the response to red light is gated by the clock, while *HY5* induction by UV-B light is not. Moreover, HY5 activity is not needed for the gated response to UV-B light. Surprisingly, two nonrhythmic genes were found to have a gated response to UV-B pulses with a maximum response at the middle of the subjective day. Further studies are needed to reveal the molecular mechanisms underlying gating of UV-B light.

Entrainment by Temperature

The mechanistic insights of clock entrainment by temperature have been much more elusive. Overall, it seems that thermocycles are weaker synchronizers than photocycles (McClung and Davis 2010). Early studies using *Kalanchoë* plants showed that small temperature changes (0.5 °C) could entrain CO_2 assimilation. In *Arabidopsis*, the rhythms in the expression of essential oscillator components can be phased by thermocycles in a similar fashion than by photocycles. However, we are still far from understanding how circadian clocks perceive and respond to temperature changes. A partial answer to these questions came from studies showing that two components of the oscillator, PRR7 and PRR9, are essential for temperature entrainment of the *Arabidopsis* circadian clock. The study used *prr7*/*prr9* double mutant plants and showed that in these plants the clock does not entrain to thermocycles. However, the function of PRR7 and PRR9 is also important for clock entrainment to photocycles and for circadian rhythmicity in the dark. Therefore, PRR7 and PRR9 might be important for the integration of light and temperature input signaling to the oscillator.

In addition to PRR7 and PRR9, other clock factors are also important for temperature entrainment. A recent study has shown that *TOC1::LUC* oscillations in *elf3* mutant seedlings entrained under light/dark cycles were arrhythmic upon release to constant darkness. Furthermore, the same *TOC1::LUC* arrhythmic phenotype was obtained with entrainment to thermocycles followed by warm or cool constant temperature. This indicates that the *elf3* arrhythmic phenotype is not exclusively dependent on light input and suggests that in addition to its role in the control of light input signals, ELF3 has a more general function as a central component of the oscillator. In order to understand the role of ELF3 in the control of thermic input signals, clock gene expression was analyzed in response to a short warm pulse delivered at discrete time points, at midday or midnight, in WT and *elf3* mutant seedlings. Sensitivity to warm temperatures was higher at night, when expression of *PRR7* and *PRR9* are strongly induced in WT seedlings. Consequently, ELF3 was proposed to have a role gating the expression of these temperature-responsive morning genes.

Crosstalk Between Light and Temperature Entrainment

A recent study has focused on the interaction between light and temperature effects on the clock. The authors analyzed changes in circadian period length of clock gene expression in wild-type and several photoreceptor mutants, including *cry1, cry2, phya, phyb*, and the double mutant *cry1/cry2*. Following light/dark entrainment, seedlings were placed under different regimes of light and temperature (red, blue, or combined red/blue; 12 °C, 17 °C, or 27 °C), and changes in the circadian period were measured. The study revealed a very strong interaction between light quality and temperature regulation. The most striking feature was the arrhythmia of the double mutant *cry1/cry2* at 27 °C and under blue-light constant conditions. Blue light seems to be essential to maintain robust oscillations under these conditions, and the effect is mediated by CRYs. The results place CRYs as an integrating hub for light and temperature input signals in a common regulatory pathway.

The comparison of an *Arabidopsis* mapping population under thermocycles and photocycles has also shown that quantitative variation in response to thermal and photic cues have partially separate genetic bases. It was found a period-shortening effect from thermal entrainment as compared to the photic entrainment, which further reinforces the notion of differential pathways for input responses within the oscillator. This notion is consistent with early studies reporting the existence of different oscillators with variable sensitivity to temperature and light, which might be important for responses to the changing seasons through the integration of both temperature and photoperiodic cues.

Another clock component that appears to be involved in both light and temperature clock responses is ELF4. The *elf4* mutant plants can be entrained to warm/cold cycles, but the rhythmic oscillations become very weak after 1 day under constant temperature. This suggests that *elf4* mutant can sense the thermic changes but functional ELF4 is needed to sustain the circadian oscillation under constant conditions. Therefore, ELF4 contributes to both light and temperature entrainment, although its contribution to each pathway remains to be fully dissected.

Properties of Circadian Function: Clock Temperature Compensation

Changes in temperature have different effects on clock function, and how the clock is able to separate these effects is still a matter of debate. In addition to being entrained by daily rhythms in temperature, the clock is also characterized by a remarkable property, known as temperature compensation (McClung and Davis 2010). Essentially, and to be valuable as a timing device, the circadian system is able to run at the same pace regardless of the variations in temperature. This way the clock does not run faster at higher temperatures or slower at lower temperatures, being able to maintain a period close to 24 h under changing temperatures. The capacity of buffering the variations in temperature (within a physiological range) is in clear contrast to what is happening in many biochemical reactions (McClung and Davis 2010).

An initial study exploring natural variation in *Arabidopsis* identified FLOWERING LOCUS C (FLC) and GI as factors contributing to clock temperature compensation. The *FLC* gene encodes a transcriptional repressor belonging to the MADS-box family of transcription factors (MADS, from the first members of the family MCM1, AGAMOUS, DEFICIENS, and SRF). FLC represses flowering by inhibiting the expression of a key central flower component. Besides its central role in flowering time, FLC was arguably proposed to be involved in clock compensation at high temperatures. However, the role of FLC in temperature compensation is controversial, and the mechanism underlying this regulation remains unknown. In the case of GI, characterization of changing temperature effects on the free-running period in loss-of-function *gi* mutants led to the notion that GI might buffer the clock at different temperatures by influencing *LHY* and *CCA1* expression. In wild-type plants, *LHY* expression decreases at high temperatures (27 °C), while *CCA1* expression increases at lower temperatures (12 °C). Those changes are not so evident in *gi* mutants, revealing a possible role for GI in the regulation of the temperature effects in the expression of clock components. The authors suggested that GI is important not only for clock temperature compensation but also for maintaining a precise oscillation at higher and lower temperatures.

PRR7 and PRR9 have been also shown to play an important role in clock temperature compensation in addition to their previously described function in clock entrainment by light (section "Molecular Components Involved in Clock Entrainment by Light: Photoreceptors"). In contrast to wild-type plants, in which the circadian period remains relatively constant and close to 24 h at different temperatures, the circadian period of the double mutant *prr7/prr9* plants increases at higher temperatures (overcompensation phenotype). This result suggests that PRR7 and PRR9 might play an important role in clock temperature compensation. The expression of some central clock components such as *CCA1* and *LHY* is increased in the *prr7/prr9* double mutant, while others such as *TOC1* or *GI* remain unaltered. These results suggest that the function of PRR7 and PRR9 in clock temperature compensation might be mediated through changes in the expression and/or activity of circadian genes expressed in the morning (see sections "Molecular Components of the Circadian Oscillator" and "Molecular Structure and Mechanisms at the Core of the *Arabidopsis* Circadian Oscillator").

The mechanistic insights of clock temperature compensation were unravelled in a recent study. The study shows that protein kinase CK2 (formerly CASEIN KINASE 2) and clock protein phosphorylation play a key role in the regulation of temperature compensation (Fig. 4). CK2 is a heterotetrameric protein formed by two catalytic α-subunits and two regulatory β-subunits ($\alpha 2\beta 2$). Dephosphorylated isoforms of CCA1 bind to the promoters of the oscillator genes including *TOC1*, *LUX*, *PRR7*, and *PRR9*, while the increased activity of CK2 by overexpression of *CKB4* (*CKB4*-ox), a regulatory subunit of CK2, decreases CCA1 binding to these promoters. The study therefore reveals a mechanism by which phosphorylation of a clock component regulates its transcriptional activity. Furthermore, misregulation of CK2 activity by *CKB4*-ox or by CK2 inhibitors leads to severe alterations of clock temperature compensation. When CK2 activity is decreased, plants show an

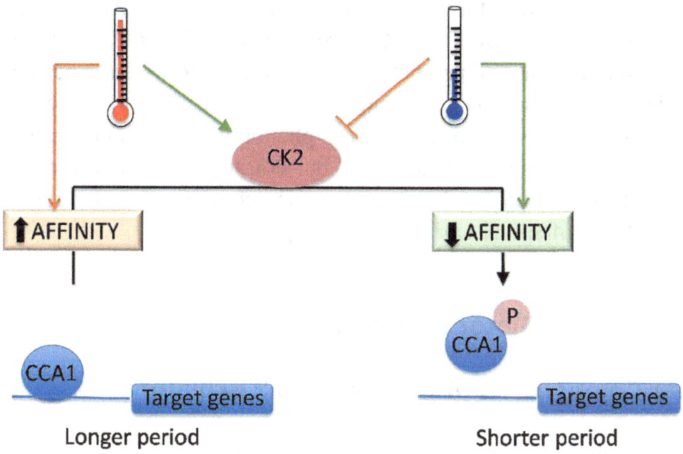

Fig. 4 A role for CK2 in the regulation of temperature compensation. Temperature and phosphorylation of CCA1 (P) by CK2 modulate binding of CCA1 to the promoters of its target genes. High temperature increases CCA1 activity (*orange arrow*) but also increases CCA1 phosphorylation by CK2 (*green arrow*). Opposite, low temperature decreases CK2 activity (*orange arrow*) compensating for a lower CCA1 binding activity (*green arrow*). CK2 counterbalances CCA1 activity at different temperatures maintaining a stable period length over a wide range of temperatures

overcompensation phenotype (i.e., long period at higher temperature). Conversely, when CK2 activity is increased, plants showed an undercompensation phenotype (i.e., shorter period at higher temperatures). The temperature-dependent changes in CK2 activity are also correlated with variations of CCA1 binding to the promoters of its target genes at the different temperatures. Accordingly, higher temperatures increased the affinity of CCA1 for its target promoters *TOC1*, *LUX*, *PRR7*, and *PRR9*, as revealed by comparing CCA1-binding ability by means of chromatin immunoprecipitation (ChIP) techniques in seedlings grown at 12 or 27 °C. Therefore, the study describes a mechanism by which the CCA1-binding activity is precisely counterbalanced by phosphorylation of CK2. This balancing mechanism contributes to maintain the circadian period stable at different temperatures, and thus explaining the clock temperature compensation.

Alternative splicing has been recently identified as an important mechanism regulating temperature compensation. Many clock genes including *CCA1*, *LHY*, *PRRs*, and *TOC1* present alternative splice variants that produce aberrant nonfunctional mRNAs (section "Role of mRNA Processing at the Core of the *Arabidopsis* Circadian Clock"). Controlling the balance between full-length functional mRNA production and nonfunctional variants constitutes a mode of controlling clock gene expression and function. A detailed analysis of splicing variants of clock mRNAs showed that many alternative splicing events are temperature sensitive. For example, splice variants have been detected for *CCA1* and *LHY*, which retain an intron located just after the Myb domain coding region in both transcripts. However, those variants were increased for *LHY* transcripts in plants exposed to cold temperatures, whereas they were reduced in *CCA1* under the same conditions. In the case of *PRR7* and *PRR9*, the SKIP protein (SKI-INTERACTING PROTEIN) is the responsible for the balance between functional mRNA and aberrant splicing variants. The loss-of-function *skip* mutants accumulate aberrant splicing forms of *PRR7* and *PRR9* and display a long-period phenotype under constant light conditions. This phenotype is evident at 17 °C but not at 27 °C. Therefore, alternative splicing might contribute to clock temperature compensation by regulating the relative abundance of splice variants of clock genes.

The Circadian Oscillator in *Arabidopsis thaliana*

Molecular Components of the Circadian Oscillator

The first circadian clock components were identified in *Arabidopsis* through genetic screenings. The idea was to search for mutants with altered circadian expression of the clock output gene *CAB2*. Seeds harboring the firefly *LUCIFERASE* (*LUC*) gene expressed as a reporter under the control of the *CAB2* promoter were mutagenized and seedlings analyzed for changes in circadian period, phase, or amplitude of the *CAB2::LUC* activity. The studies identified *TOC1*/*PRR1* as a circadian oscillator gene. The *toc1* mutation shows a consistent short-period phenotype for multiple clock outputs. Moreover, *TOC1* overexpression causes arrhythmia, confirming that TOC1 plays a key role at the core of the clock (Table 1).

Four additional members of the TOC1 family, known as PRRs (PSEUDO-RESPONSE REGULATORS) were also found to be important for proper circadian clock function. The PRR family includes PRR3, PRR5, PRR7, and PRR9 together with the previously described TOC1 or PRR1. These proteins contain two highly homologous domains: the pseudo-receiver (PR) domain at the N-terminal region and the CCT (CONSTANS, CONSTANS LIKE, and TOC1) domain at the C-terminal region. The PR domain is proposed to be required for protein–protein interactions, while the CCT domain allows DNA binding. The region between the PR and CCT domains of PRR5, PRR7, and PRR9 contains a transcriptional repressive motif. PRR expression is circadian regulated, and the expression of each gene peaks sequentially from the morning to the night in the following order: PRR9, PRR7, PRR5, PRR3, and TOC1/PRR1. The *prr* loss-of-function mutants alter the circadian rhythms in various ways. On one hand, *prr7* or *prr9* single mutants show a long-period phenotype, while the increased period lengthening in *prr7*/*prr9* double mutants suggests a possible redundant role of PRR7 and PRR9 in the control of the circadian period. On the other hand, loss of function of *prr5* or *prr3* causes period shortening, whereas the *prr5*/*prr7*/*prr9* loss-of-function triple mutant leads to arrhythmia. The circadian function of some of the PRRs is also closely related to temperature (consult sections "Entrainment by Temperature," "Crosstalk Between Light and Temperature Entrainment," and "Role of mRNA Processing at the Core of the *Arabidopsis* Circadian Clock"). Taken together, these studies suggest that PRRs are crucial regulators of the circadian oscillator.

Table 1 List of some relevant clock-related genes

Gene	Locus ID	Domain	Loss-of-function circadian phenotype	Overexpression circadian phenotype
CCA1	At2g46830	MYB domain	Short period	Arrhythmic
LHY	At1g01060	MYB domain	Short period	Arrhythmic
RVE8/ LCL5	At3g09600	MYB domain	Long period	Short period
PRR9	At2g46790	Pseudo-receiver	Long period	Short period
PRR7	At5g02810	Pseudo-receiver	Long period	Long period
PRR5	At5g24470	Pseudo-receiver	Short period	Long period
PRR3	At5g60100	Pseudo-receiver	Short period	Long period
TOC1/ PRR1	At5g61380	Pseudo-receiver	Short period	Arrhythmic
NOX/ BOA	At5g59570	GARB domain	Wild type	Long period
LUX	At3g46640	GARB domain	Arrhythmic	Arrhythmic
ELF3	At2g25930	Unknown	Arrhythmic	Long period
ELF4	At2g40080	Unknown	Arrhythmic	Long period
GI	At1g22770	Unknown	Short period	Short period

CCA1 and LHY are closely related single MYB transcription factors that are circadian regulated with a peak of expression early in the morning. These transcription factors show DNA-binding specificity for a DNA sequence (AAAATATCT) known as the evening element (EE). CCA1 was initially identified as a binding factor to the *LHCB1* (*LIGHT-HARVESTING CHLOROPHYLL A-/B-BINDING PROTEIN 1*) promoter. Overexpression of *CCA1* leads to arrhythmia and the loss of function leads to a short-period phenotype, which reflects the importance of CCA1 in maintaining proper circadian rhythms (Table 1). LHY was found in a screening for late-flowering mutants. Like *cca1* mutants, *lhy* loss-of-function mutant plants show short-period phenotypes. Remarkably, the *cca1/lhy* double mutant is arrhythmic, which shows their important function at the core of the *Arabidopsis* oscillator.

RVE8/LCL5 (REVEILLE8/LHY-CCA1-LIKE5) is an MYB transcription factor with high sequence homology to CCA1 and LHY. RVE8/LCL5 was first identified through mass spectrometry analysis of proteins that bind to the EE motif. *RVE8/LCL5* expression is circadian regulated with a peak of expression in the morning. The *rve8/lcl5* loss-of-function mutants show long-period phenotypes and conversely, overexpression of *RVE8/LCL5* leads to short-period phenotypes (Table 1). Although RVE8/LCL5 MYB sequence is highly similar to CCA1 and

LHY, the RVE8/LCL5 function is quite the opposite. Indeed, while CCA1 and LHY act as transcriptional repressors of oscillator gene expression, RVE8/LCL5 binds to the EE present at the oscillator gene promoters and induces transcriptional activity. The mechanism of RVE8/LCL5 function seems to rely on changes in the pattern of histone H3 acetylation at the promoters of its target genes (consult section "Chromatin Remodeling at the Core of the *Arabidopsis* Oscillator"). Interestingly, the triple mutant *lcl5/lcl1/lcl2* (*rve8/rve4/rve6*) shows an extremely long circadian period that is correlated with a reduction in the expression of evening-expressed clock genes. This data suggest that the two RVE8/LCL5 closest homologs, RVE4/LCL1 and RVE6/LCL2, might also be involved in the activation of the EE-containing genes.

In addition to the morning-expressed genes, other components with a peak of expression during the evening have been shown to be relevant for circadian function. Some of these components, such a LUX/PCL1, ELF3, and ELF4, directly interact to form the so-called evening complex (EC), which functions as a transcriptional repressor to regulate plant growth and circadian function. LUX is a GARB transcription factor that was initially identified in a genetic screening for seedlings with long hypocotyls and altered circadian rhythms. Loss-of-function mutations in *LUX* induce arrhythmia and important defects in flowering time and hypocotyl growth. Recent studies have identified a putative LUX binding site (LBS) and its activity as an evening repressor of circadian gene expression. ELF3 and ELF4 are two unrelated proteins that were initially identified in genetic screenings for mutants with altered flowering time. They were later shown to regulate circadian rhythms with key roles at the core of the clock and within the light input (see sections "Molecular Components Involved in Clock Entrainment by Light: Photoreceptors," "Properties of Circadian Function: Gating Light Input to the Clock," and "Molecular Structure and Mechanisms at the Core of the *Arabidopsis* Circadian Oscillator"). Similar to LUX, these two genes are expressed around dusk, and loss-of-function mutations for the individual genes leads to arrhythmia under continuous light conditions and altered flowering time and hypocotyl. Although ELF3 and ELF4 have no structural homology to known functional domains, it has been proposed that ELF4 might act as an effector protein. ELF3 and ELF4 also form a complex with NOX ("night" in Latin) or BROTHER OF LUX ARRHYTHMO (BOA), a close homolog of LUX. NOX/BOA is in turn closely related to the circadian clock. Plants overexpressing NOX/BOA show a long-period phenotype and increased amplitude of *CCA1* expression. The circadian rhythms of other key clock genes are also altered by overexpression of NOX/BOA, which suggest an important regulatory role at the core of the clock. However, and in contrast to *lux* mutants, NOX/BOA RNAi lines display robust circadian rhythms, which suggest that NOX/BOA is not entirely redundant with LUX (Table 1).

Other genes related to clock function include ZTL, the F-box containing protein, which directly interacts with TOC1 and PRR5 and targets these proteins for degradation. Loss-of-function mutations in *ztl* lead to period lengthening and its overexpression causes arrhythmia. The evening-expressed photoperiodic flowering regulator GI has also been proposed to play an important role in the regulation of circadian rhythms. The *gi* loss-of-function mutants show altered amplitude and period of *LHY* and *CCA1* expression as well as a late-flowering phenotype. As mentioned above (section "ZEITLUPE Family"), GI interacts and stabilizes ZTL by forming a blue-light-dependent protein complex. Additionally, LIGHT-REGULATED WD1 (LWD1) and LWD2 are two clock components required for the modulation of circadian period length and photoperiodic flowering. The *lwd1/lwd2* double mutant has an early flowering phenotype and a short-period length. LWD1 has been demonstrated to act as a transcriptional activator of several circadian clock genes. The clock component CHE (CCA1 HIKING EXPEDITION) is another transcription factor that belongs to the class I TCP (TB1, CYC, PCFs) family. CHE has been reported to specifically bind to the consensus class I TCP-binding site (TBS) (GGNCCCAC) at the *CCA1* promoter and inhibit its expression.

Molecular Structure and Mechanisms at the Core of the *Arabidopsis* Circadian Oscillator

A proposed model of the circadian clock in *Arabidopsis* comprises a complex network of interconnected loops (Nagel and Kay 2012; Carré and Veflingstad 2013). TOC1, CCA1, and LHY are the components of the firstly described transcriptional feedback loop, initially defined as a central loop. This transcriptional network was considered the core of the circadian oscillator since overexpression of any of these genes induces arrhythmia, while loss of function leads to short period of circadian gene expression. CCA1 and LHY, the MYB transcription factors expressed in the morning, inhibit *TOC1* expression by specifically binding to the evening element (EE) motif present at the *TOC1* promoter. Initial studies proposed that TOC1 in turn could activate *CCA1* and *LHY* expression. This function was inferred by the analysis of *toc1* mutant plants, which showed low abundance of *LHY* and *CCA1* expression. However, recent studies have shown a more complicated picture in which TOC1 would function as a repressor rather than an activator of *CCA1* and *LHY*. TOC1 represses not only *CCA1* and *LHY* but also nearly all of the oscillator components, including members of the PRR family (*PRR5*, *PRR7*, and *PRR9*), *GI*, and the components of the EC (*LUX* and *ELF4*) (Fig. 5). Repression might occur through direct binding of TOC1, as ChIP-Seq and ChIP-Q-PCR analyses identified the promoters of the oscillator genes as binding targets of TOC1. The studies also suggest that a double repression (i.e., repressing a repressor) might explain the initial notion of TOC1 being an activator of *CCA1* and *LHY* expression.

A second transcriptional loop includes CCA1, LHY, and PRRs. The other members of the TOC1 family also associate to the promoters of *CCA1* and *LHY* and repress their expression during the day. Binding and repression seem to be sequential and consistent with their pattern of expression. This temporal sequence starts close to dawn with PRR9 and ends with TOC1 around dusk. Regarding CCA1 and LHY activity, light induces their expression, and the accumulated proteins bind

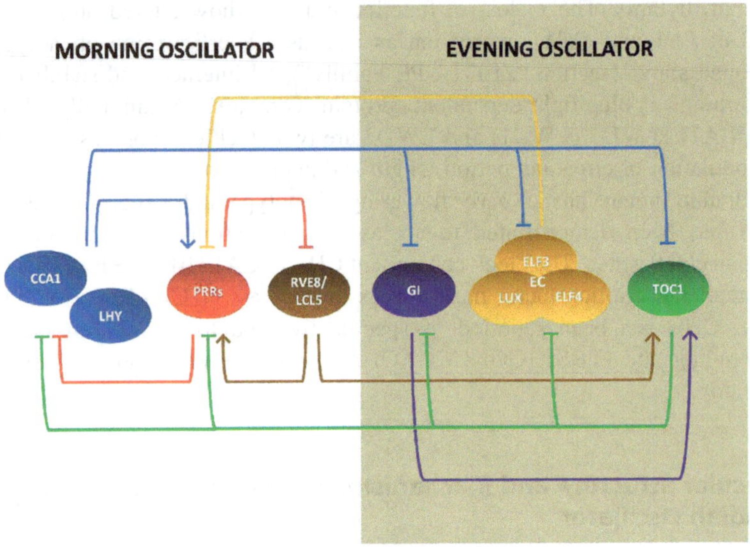

Fig. 5 **Simplified model of the *Arabidopsis* circadian oscillator network.** *Arrows* indicate induction, while perpendicular lines indicate transcriptional repression. In the morning, CCA1 and LHY induce *PRR7* and *PRR9* expression while they inhibit the expression of the evening-expressed genes (*blue line*). The repression of *CCA1* and *LHY* during the day depends on the sequential binding at their promoters of PRR9, PRR7, PRR5, and TOC1 (*red and green lines*). RVE8/LCL5 promotes *PRR5* and *TOC1* expression (*brown line*); in turn, PRR5 down-modulates *RVE8/LCL5* expression (*pink line*). In the evening, GI induces TOC1 accumulation (*purple line*), which represses both the morning-expressed and evening-expressed genes (*green line*). In addition, the EC (*ELF3, ELF4*, and *LUX*) down-modulates *PRR9*, and probably *PRR7*, expression which allows the next cycle transcriptional activation of *CCA1* and *LHY* (*yellow line*)

in turn to the promoters of the evening genes, including *PRR5, TOC1, LUX, ELF4*, and *GI*, to repress their expression. Contrarily, CCA1 and LHY appear to activate the morning-expressed genes *PRR7* and *PRR9*. Further investigations will be necessary to understand at a molecular level how CCA1 and LHY induce *PRR7* and *PRR9* transcription while they transcriptionally repress the evening-expressed oscillator components. The progressive decrease of CCA1 and LHY abundance caused by PRR9 and PRR7 repression early in the day allows the transcription of *PRR5*, which contributes to a further repression of *CCA1* and *LHY*. Close to dusk, the low abundance of CCA1 and LHY facilitates the activation of *TOC1* expression, which controls the timing of the morning- and evening-expressed oscillator genes. Therefore, this network provides a molecular platform that allows the transcriptional inhibition of *CCA1* and *LHY* by the sequential accumulation and repressive activity of PRR9, PRR7, PRR5, and TOC1 at the *CCA1* and *LHY* promoters. An interesting hypothesis that could explain the temporal separation between the PRRs waves of expression is that the late PRRs might repress the previous ones. Consistent with this notion, TOC1 represses *PRR9*, *PRR7*, and *PRR5* expression. Interestingly, PRR5 has been proposed to play a dual role at the core of the circadian oscillator. On one hand, it acts as a transcriptional repressor for the morning genes while it also induces TOC1 activity by promoting its phosphorylation and nuclear accumulation.

The evening loop comprises TOC1, GI, and the members of the EC. TOC1 directly binds to the *GI* promoter in the evening and acts as a transcriptional repressor, while GI was initially proposed to positively regulate *TOC1* transcription. Further characterization has demonstrated that GI binds and stabilizes ZTL (the F-box protein that targets TOC1 and PRR5 for degradation). It has been proposed that GI binding to ZTL might also block access of ZTL substrates to the Kelch interaction region. Therefore, the interaction between GI and ZTL might block TOC1 degradation allowing the accumulation of TOC1 protein. As PRR5 is also targeted for degradation by ZTL, it is possible that GI positively regulates PRR5. The EC (the protein complex composed of ELF3, ELF4, and LUX) is also involved in a transcriptional loop at the core of the clock. The EC indirectly induces *CCA1* and *LHY* expression, while CCA1 and LHY transcriptionally repress *ELF4* and *LUX* by specifically binding to the EE at their promoters. Importantly, the EC was found to repress *PRR9* expression. This downregulation seems to be direct because ELF3 and LUX have been detected at promoter of *PRR9*. A weak interaction has also been observed between ELF3 and *PRR7* promoter, indicating that *PRR7* transcription might be also repressed by the EC. Taken together, these data suggest that the EC may indirectly enhance *CCA1* and *LHY* expression by repressing their transcriptional inhibitor PRR9, and probably PRR7 (Fig. 5). Furthermore, the EC autoregulates its own expression by binding to the promoters of *ELF4* and *LUX* and repressing their expression. This autoregulatory loop might be important for decreasing the EC activity following its induction.

Additional transcriptional loops at the core of the oscillator add new layers of complexity to the *Arabidopsis* circadian network. For instance, the LUX family member NOX/BOA participates in a transcriptional feedback loop with CCA1 and potentially with LHY. NOX/BOA induces *CCA1* transcription most likely by direct binding to the *CCA1* promoter. CCA1, in turn, closes the transcriptional loop by repressing *NOX/BOA* expression. In addition, NOX/BOA seems to induce the expression of *LHY*, *TOC1*, and *GI*, suggesting a more general role as a clock transcriptional activator. Notably, *NOX/BOA* rhythmic expression is also affected in *toc1* and *gi* loss-of-function mutants, suggesting that the function of NOX/BOA at the oscillator also involves TOC1 and GI.

Other clock component participating in feedback regulation is RVE8/LCL5, the morning-expressed MYB transcription factor. RVE8/LCL5 promotes *TOC1* and *PRR5* expression by specifically binding to their promoters and modulating histone H3 acetylation. PRR5, in turn, is responsible for the repression of *RVE8/LCL5* expression. An additional transcriptional loop has been identified between CCA1 and CHE. As mentioned above, CHE binds to the TBS motif at the *CCA1* promoter and inhibits its transcriptional activity. CCA1 and LHY repress in turn *CHE* expression by directly binding to its promoter. Moreover, an interaction between CHE and TOC1 has been described, an interaction that might be important for recruitment of TOC1 to the *CCA1* promoter.

The clock also includes positive feedback loops. One example involves the clock component, LWD1, which can directly bind to the promoter of *PRR9* and induce transcriptional activity, while PRR9 indirectly enhances *LWD1* expression. In addition, LWD1 seems to promote the expression of *CCA1*, *LHY*, *PRR5*, and *TOC1* suggesting that LWD1 might function as a positive regulator of gene expression at the core of the clock. All these studies highlight the complexity of the *Arabidopsis* circadian network far beyond the typical transcriptional feedback loops initially conceived.

Chromatin Remodeling at the Core of the *Arabidopsis* Oscillator

Remodeling activities in chromatin structure play a pivotal role in the regulation of gene expression. The basic unit of chromatin structure is the nucleosome, which includes 146 bp of DNA packed around a histone octamer. Changes in chromatin structure, like histone posttranscriptional modifications, are a key step for controlling the degree of compactness of nucleosomes, which ultimately modulate the transcriptional status. Several covalent modifications decorate the N-terminal tails of histones including, among others, acetylation, methylation, and ubiquitination. Broadly speaking, histone hyper-acetylation leads to an open chromatin structure and induces transcriptional activation, whereas hypo-acetylation induces repression by compacting the chromatin. Histone acetyltransferases (HATs) are the enzymes responsible for the addition of the acetyl group from a donor acetyl-CoA, while histone deacetylases (HDACs) remove the acetylation. Histone methylation, on the other hand, can act as a positive or a negative regulator of transcription depending on the methylated residue and the degree of methylation. It has been proposed that histone methylation might function by modulating the binding of chromatin remodeling factors. Histones can be methylated by histone lysine methyltransferases (HKMTs) at lysine residues or by protein arginine methyltransferases (PRMTs) at arginine residues. Methylation can be removed by histone demethylases (HDMs). Histone monoubiquitination, which mainly occurs on histone H2A and H2B, also contributes to the regulation of gene expression. In *Arabidopsis*, histone H2A monoubiquitination (H2Aub1) is associated with gene silencing. In contrast, histone H2B monoubiquitination (H2Bub1) induces transcriptional activation.

The first evidence showing the importance of histone modifications in the regulation of the plant circadian clock was described at the promoter of *TOC1*. Initial studies showed a strong correlation between H3 acetylation (H3ac) at the *TOC1* promoter and its circadian expression and photoperiodic regulation. Characterization of histone modifications at the promoters of other circadian oscillator genes shows that both H3ac and H3K4 trimethylation (H3K4me3) are also closely correlated with the circadian expression of nearly all oscillator genes. Notably, and despite the fact that both histone marks are essential for activating oscillator gene expression, H3K4me3 oscillates with a delayed phase compared to the phase of H3ac, suggesting that it may contribute through a different mechanism to the transcriptional induction of the circadian genes. Indeed, H3ac might ensure that oscillator gene expression reaches its circadian peak at the proper time, while H3K4me3 facilitates the function of H3ac by blocking clock repressor binding and maintaining the proper time window for activation. This notion was corroborated in studies in which H3K4me3 accumulation was reduced by pharmacological inhibition or by mutation in the enzyme responsible for methylation (see below). Under these conditions, clock repressor binding and oscillator gene repression were enhanced. Therefore, if H3K4me3 is not properly regulated, clock repressors might function much earlier than they should, which would lead to an advanced phase of repression abolishing the circadian peak of expression. On the other hand, H3K4me2 abundance at the promoters of circadian clock genes negatively correlates with their expression. As trimethylation impedes repressor binding, it is tempting to speculate that dimethylation might allow clock repressor association, thus inhibiting gene transcription.

Several studies have started to uncover the molecular mechanisms that regulate histone modifications at the promoters of the circadian clock genes. For example, it is known that CCA1 binding to the *TOC1* promoter at dawn correlates with a hypoacetylated state of histone and with *TOC1* transcriptional repression. On the other hand, RVE8/LCL5 plays the opposite role, i.e., induces *TOC1* transcription by binding to its promoter and promoting histone acetylation. The timing and the antagonistic function of these two transcription factors might be important in the regulation of the pattern of histone acetylation at the *TOC1* promoter. Although there are some clues about the clock effectors regulating chromatin changes, the actual HDACs and the HATs responsible for histone acetylation/deacetylation remain still unknown.

In contrast, the molecular components responsible for the circadian regulation of H3K4me3 at the promoters of the oscillator genes are starting to emerge. For instance, SDG2/ATXR3 (SET DOMAIN GROUP 2/ARABIDOPSIS TRITHORAX-RELATED 3) was identified as a major histone methyltransferase at the core of the clock. Plants misexpressing *SDG2/ATXR3* show a reduced pattern of H3K4me3 at the promoters of the clock genes. The reduced pattern clearly correlated with increased clock repressor binding and reduced circadian expression. These results not only place SDG2/ATXR3 close to the oscillator but also confirm the key function of H3K4me3 controlling the timing of clock repressor binding and hence the circadian peak of expression. Regarding the components responsible for histone demethylation, it is possible that the histone demethylase, JMJD5/JMJD30 (JUMONJI DOMAIN CONTAINING 5/30), could play an important role. *JMJD5/30* expression is circadian regulated with a peak of expression in the evening. CCA1 and LHY strongly repress *JMJD5/30* expression by directly binding to its promoter. The *jmjd5/30* loss-of-function mutation leads to a short-period circadian phenotype for gene expression. Similar short-period phenotypes have been described in mammalian cells deficient in the human JMJD5/30 ortholog, which can be rescued by the *Arabidopsis* JMJD5/30. Despite these interesting findings, the involvement of JMJD5/30 in histone demethylation at the core of the clock remains to be fully demonstrated.

Recently, H2Bub1 (monoubiquitinated H2B) has been shown to contribute to the regulation of some oscillator genes. H2Bub1 is a dynamic mark required for transcriptional activation, it is necessary for transcription elongation and remains stable during down-modulation. The heterodimeric HUB1/HUB2 E3 ubiquitin ligase is the enzyme responsible for H2Bub1. Remarkably, *hub1* mutant plants, with reduced H2Bub, show a decreased induction of several genes, including the oscillator components *CCA1* and *ELF4*. In addition, *hub1-1* mutant shows a reduced pattern of both H2Bub and H3K4me3 at the coding region, suggesting a link between these marks. A similarly mechanism has been described in yeast. H2Bub1 regulates the binding of HKMTs, such as COMPASS, which induces methylation of H3K4 and thus promoting transcription elongation.

Role of mRNA Processing at the Core of the *Arabidopsis* Circadian Clock

Although transcriptional regulation plays an important role at the core of the *Arabidopsis* oscillator (see section "Molecular Structure and Mechanisms at the Core of the *Arabidopsis* Circadian Oscillator"), mRNA processing is emerging as an essential step in the regulation of the circadian clock genes. The following section is focused on the posttranscriptional processing events that modulate the mRNA at the core of the oscillator. Remarkably, alternative splicing has been also described to be temperature dependent, which directly links clock regulation with changes in temperature (consult section "Crosstalk Between Light and Temperature Entrainment"). Other mRNA processing steps such as mRNA stability and transcript translation also contribute to the regulation of oscillator gene expression and activity (Sanchez et al. 2011; Staiger and Green 2011; Staiger and Köster 2011).

Following transcription in the nucleus, the pre-mRNA is processed in various ways to produce the mature mRNA that will be subsequently translated in the cytoplasm. The mRNA processing includes 5′ capping, splicing, 3′ end cleavage, and polyadenylation. All these steps are critical in the regulation of gene expression. Specifically for splicing, a complex protein machinery known as the spliceosome is responsible for the intron removal that generates the intron-free mature mRNA. Different mature mRNAs can be generated from the same pre-mRNA through the selection of alternative splicing sites. This process, known as alternative splicing, can produce proteins with different functionality or even nonfunctional RNA with premature termination codons (PTCs). The nonfunctional RNAs are degraded through different pathways, including the nonsense-mediated decay (NMD) or by the unproductive splicing and translation (RUST). In *Arabidopsis*, more than 60 % of the intron-containing genes are processed by alternative splicing. This regulation has been reported to play an important role in the control of plant growth and development, as well as in plant responses to environmental conditions and pathogens. Recent reports have also revealed a key role for alternative splicing in the regulation of circadian clock function. Indeed, alternative spliced mRNAs have been identified for most of the oscillator genes, including, among others, *CCA1*, *LHY*, *TOC1*, *PRRs*, and *LUX*. Several studies have also identified factors involved in the alternative splicing of circadian clock genes, for instance, STILP1 (SPLICEOSOMAL TIMEKEEPER LOCUS 1). STILP1 is a member of the STIP family of RNA-binding factors required for spliceosome disassembly. The *stilp1* mutant plants show a long-period phenotype caused by altered accumulation of circadian-associated transcripts, including *CCA1*, *LHY*, *PRR9*, *GI*, and *TOC1*. Another example includes PRMT5/SKB1 (PROTEIN ARGININE METHYLTRANSFERASE 5/SHK1-BINDING PROTEIN 1), which is a type II protein arginine methyltransferase responsible for the arginine symmetric dimethylation (Rsme2) of histones, RNA processing factors, and spliceosomal proteins. The *prmt5* loss-of-function mutant shows a long-period phenotype due to aberrant alternative splicing of *PRR9*. Two splicing isoforms of *PRR9* are observed in wild-type plants, a mature mRNA that encodes the full-length protein and an alternative spliced isoform with eight nucleotides at the end of the exon 2 that encodes a truncated form. In *prmt5* mutant plants, the most abundant form contains the eight nucleotides and intron 3 retention while the mRNA that encodes the full-length protein is strongly reduced. These data suggest that the circadian alteration observed in *prmt5* mutants is, at least in part, caused by splicing defects in *PRR9* mRNA.

Another interesting case of alternative splicing is exemplified by GRP7 and GRP8 (GLYCINE-RICH RNA-binding protein 7 and 8), which are RNA-binding proteins involved in the formation of nonfunctional RNAs through alternative splicing. The proteins bind to their own and each other's pre-mRNA, inducing the generation of nonfunctional alternative spliced isoforms. The aberrant alternative spliced variants are targeted for degradation by the NMD pathway, thus contributing to decreased protein abundance. GRP7 functions as a splicing factor and its overexpression affects about 1.7 % of total transcripts, one third of them being circadian regulated. Among those, genes involved in stress responses were enriched, suggesting that GRP7 might function as an integrator of environmental signals, an alternative splicing, and the circadian clock.

The biological implication and mechanistic insights of alternative splicing within the circadian system have been recently explored in detail. The studies show a close connection between temperature and alternative splicing of clock genes. Indeed, low temperatures lead to a transient accumulation of nonfunctional transcripts of *LHY*, *PRR7*, *PRR3*, and *TOC1*, which are degraded by the NMD pathway. Interestingly, the opposite effect was observed for *PRR9*, as low temperatures induce *PRR9* full-length transcript accumulation. Heat stress treatments followed by RNA-Seq analysis revealed intron retention in several circadian clock genes, including *TIC* and *LUX*. Interestingly, mutations in *SKIP* lengthen the circadian rhythms of clock gene expression. Although the *skip* loss-of-function mutant induces genome-wide alternative splicing defects, the long-period phenotype may be caused by aberrant alternative splicing of *PRR7* and *PRR9*. SKIP also modulates alternative splicing of other circadian clock genes such as *CCA1*, *LHY*, *TOC1*, *ELF3*, and *GI*. The circadian defects of *skip* mutant are temperature sensitive, suggesting its possible role as a molecular component linking temperature and alternative splicing of the oscillator genes (consult section "Properties of Circadian Function: Clock Temperature Compensation").

Temperature-dependent alternative splicing also modulates *CCA1* mRNA, and this is important for proper clock functioning and cold tolerance. Two alternative spliced isoforms of CCA1 have been identified: *CCA1α* (full length) and *CCA1β*, which encodes a protein that lacks the DNA-binding domain due to the exclusion of intron 4. The CCA1β protein dimerizes with CCA1 and LHY, but the resulting heterodimer is nonfunctional and shows reduced DNA-binding affinity. Overexpression of *CCA1β* leads to short period, similar to the circadian phenotype of *lhy/cca1* double mutant plants. In addition, plants lacking a functional CCA1β isoform lengthened the circadian period of gene expression, confirming that CCA1β affects CCA1α function. *CCA1* alternative splicing is also suppressed by cold temperatures, which allows the full-length CCA1α to be fully functional. Under cold conditions, CCA1α activates the transcription of the cold-responsive gene *CBF* (*C-REPEAT/DRE-BINDING FACTOR*), thus contributing to initiate the plant cold response. CCA1α also regulates clock gene expression, which is required for cold acclimation. Altogether, these findings demonstrated the intimate connection between temperature and alternative splicing at the core of the clock.

Modulation of mRNA stability of circadian clock genes is an additional mechanism that contributes to the regulation of the oscillator function. The mRNA degradation rates provide a precise tool for controlling transcript accumulation. Indeed, *CCA1* mRNA stability has been described to be light regulated. Under dark conditions, the *CCA1* transcript is stable, while light enhances its degradation. The regulation is quite complex as light not only promotes *CCA1* degradation but also induces its transcription. This dual regulation might be important for clock synchronization with the external light/dark cycles. It was also suggested that the molecular mechanism responsible for controlling *CCA1* stability could rely on the presence of a downstream (DST) instability determinant, an element previously described to target transcripts for rapid decay. Interestingly, a DST element has been identified in several circadian mRNAs that show changes in the half-life over the circadian cycle. Consistently, analyses of *dst1* mutant plants (with a disrupted DST-mediated decay pathway) show a defective circadian mRNA pattern.

Regulation of translation is another interesting example of mRNA-dependent modulation at the core of the clock. A precise control of the translational rate can lead to changes in protein abundance that do not correlate with the pattern of mRNA accumulation. One of the first descriptions of this form of regulation involves the effect of light in promoting *LHY* translation in the morning when mRNA abundance decreases. It had been suggested that the simultaneous transcriptional decrease and translational increase might play a role in sharpening the LHY protein peak at dawn. This dual regulation resembles the double control of *CCA1* induction and degradation by light described above. Interestingly, a similar mechanism has been identified in the mammalian circadian clock. The mouse RNA-binding protein RBM4 (RNA-BINDING MOTIF PROTEIN 4, also known as mLARK) activates posttranscriptional expression of the mouse clock gene *Period1* (m*Per1*). RBM4 associates with the 3′ UTR of the m*Per1* mRNA, inducing translation without affecting its transcript abundance. Like the previously described LHY translational regulation, RBM4 abundance increases at the end of the day, which induces translation when m*Per1* mRNA decreases. This translational regulation is important for the mouse oscillator function as demonstrated by the observation that *RBM4* knockdown by siRNA shortens the circadian period and its overexpression leads to long circadian rhythms.

Posttranslational Regulation of Clock Proteins: Phosphorylation and Proteasomal Degradation

Phosphorylationis a posttranslational modification with a key role in the control of protein activity and function. Protein kinases catalyze the addition of the phosphate group, while phosphatases are required to dephosphorylate the protein. Reversible phosphorylation has been shown to regulate oscillator components in different circadian systems, such as *Arabidopsis*, *Neurospora*, *Drosophila*, and mammals. In *Arabidopsis*, several circadian clock components are subject to phosphorylation, which is important for regulating among others, protein–protein interactions, protein localization, DNA binding, and protein degradation (Kusakina and Dodd 2012).

CK2 is a Ser/Thr protein kinase closely connected with the circadian clock. As mentioned above the CK2 holoenzyme consists of two catalytic alpha subunits and two regulatory beta subunits that form a heterotetrameric complex (2 alpha/2 beta). Overexpression of the regulatory beta subunits *CKB3* (*casein kinase 2 beta subunit 3*) or *CKB4* (*casein kinase 2 beta subunit 4*) increases CK2 activity and leads to a short-period circadian gene expression and early flowering under short-day conditions. The circadian function of CK2 might be mediated by phosphorylation of CCA1 and LHY. Indeed, overexpression of a mutated form of *CCA1* that cannot be phosphorylated by CK2 does not alter circadian clock function, while overexpression of wild-type *CCA1* leads to arrhythmia. The unphosphorylable form of CCA1 also shows reduced dimerization, demonstrating that CCA1 phosphorylation might be required for protein–protein dimerization. It was also proposed that phosphorylation by CK2 might modulate CCA1 protein degradation. This conclusion was inferred by studies using *cka1/a2/a3* (*CK2 α-SUBUNIT 1/2/3*) triple mutant plants, which lack three CK2 alpha subunits. In the triple mutant, the decreased phosphorylation of CCA1 correlates with an increased protein accumulation. The CK2-dependent phosphorylation of CCA1 also has a key role in regulating the DNA-binding properties of CCA1. As previously mentioned (section "Crosstalk Between Light and Temperature Entrainment"), increased CK2 activity has been shown to prevent CCA1 binding to the promoters of the circadian clock genes: *TOC1*, *LUX*, *PRR7*, and *PRR9*. Importantly, the regulation of the CCA1-binding affinity to the promoters by CK2 is temperature dependent. As described before, high temperatures induce CCA1 binding, which is precisely antagonized by a CK2 increase in activity. These observations bring to light an interesting mechanism of clock temperature compensation through the CK2 temperature-dependent modulation of CCA1 transcriptional activity. Taken together, these studies reveal that CCA1 phosphorylation driven by CK2 is necessary for correct circadian oscillator function and temperature compensation.

CK2 is also involved in the modulation of other circadian systems. For instance, in *Neurospora*, CK2 regulates the stability of the central clock component FRQ (FREQUENCY) by modulating its phosphorylation. Indeed, CK2-mediated phosphorylation targets FRQ for degradation, which is important in the control of circadian period length. CK2 is also involved in a temperature compensation mechanism at the *Neurospora* oscillator. These results highlight the existence of a conserved molecular component (CK2) with divergent mechanisms (transcriptional activity versus protein degradation) for temperature compensation in *Neurospora* and *Arabidopsis*.

Phosphorylation is also important in the regulation of the PRR family of proteins. All the members of the family are phosphorylated, and most of them show increased phosphorylation with a peak just before the proteins are degraded. Regarding the kinases responsible for PRR phosphorylation, PRR3 has been demonstrated to be a substrate of the serine/threonine protein kinase WNK1 (WITH NO LYSINE KINASE1). Notably, *WNK1* expression is circadian regulated, and its peak accumulation precisely coincides with the *PRR3* peak of expression. Two-hybrid experiments showed that WNK1 can interact with PRR5 but not with PRR7 or PRR9, opening the possibility that PRR5 might also be phosphorylated by WNK1.

Additional studies have shown that phosphorylation of PRR5 and TOC1 promotes their interaction with ZTL, which targets the proteins for degradation through the proteasome pathway. Phosphorylation of TOC1 also appears to be enhanced by the interaction of TOC1 with PRR5, interaction that presumable occurs through the N-terminal domain of both proteins. PRR5 is required for TOC1 nuclear import, which suggests that the PRR5-mediated phosphorylation might contribute to the transport of TOC1 to the nucleus. Phosphorylation also affects the interaction of TOC1 and PRR3. The interaction might prevent TOC1 degradation because PRR3 and ZTL compete for binding to the same TOC1 N-terminal domain. However, the PRR3-mediated stabilization of TOC1 seems to be restricted to the vasculature because PRR3 is only expressed in this particular tissue. Collectively, the studies demonstrate that phosphorylation is necessary for the modulation of PRR interaction, stability, subcellular localization, and presumably for their activity.

Circadian Clock Outputs

In the previous sections, the most important studies related to clock entrainment by light and temperature, as well as the functioning of the central oscillator that generates and maintains the daily rhythms, have been described. In this section, the findings pertaining to clock outputs will be discussed. In a traditional view of the circadian clock, outputs are the biological processes regulated by the clock. However, recent findings are showing that several outputs can also regulate the clock and thus function as inputs.

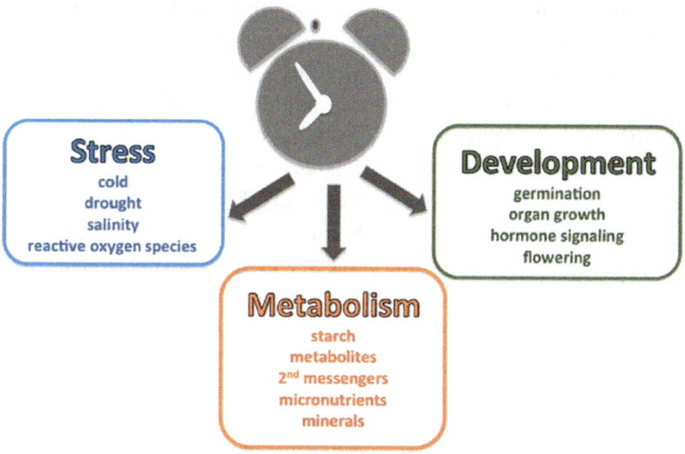

Fig. 6 **The circadian clock regulates different outputs.** Those related to development, metabolism, and stress responses are addressed in this review

In anticipation to the daily environmental changes, the circadian clock adjusts different processes of plant life to the 24-h cycles. The importance of such adjustment was shown in whole-genome transcriptional analysis. Even though there is a variation in the final exact percentage, it is now well accepted that at least 30 % of the *Arabidopsis* genes are regulated by the clock (Doherty and Kay 2010). Moreover, a proper running clock favors increased fitness and biomass. Although most of clock research has dealt with the study of inputs and the molecular mechanisms that constitute the central oscillator, recent findings have uncovered a vast number of circadian-regulated processes. The most recent literature on circadian outputs is addressed by focusing on three major processes: development, abiotic stress responses, and metabolism (Fig. 6).

Circadian Control of Plant Development

Circadian Regulation of Germination

Arabidopsis seeds are maintained in a dormant state until the proper stimuli (whether temperature, nitrate, water) are present to trigger germination. For instance, during imbibition water penetrates the seed coat, the dry tissues inside swell, and the seed coat eventually breaks down, setting in motion the different steps of the germination process. Germination also depends on the ratio between two phytohormones: GA (Gibberellic Acid, which promotes germination) and ABA (Abscisic Acid, which inhibits germination). Besides their opposite function in germination control, GA and ABA also have different roles in the regulation of *Arabidopsis* development. GA promotes growth and cell elongation, whereas ABA is involved in responses to various biotic and abiotic stresses and regulation of stomata opening. The circadian clock is connected with both GA and ABA signaling pathways, an indication that the circadian clock can modulate germination rates (de Montaigu et al. 2010). Indeed, *cca1/lhy* double mutant plants show a reduction in dormancy and increased germination that did not vary with changes in temperature. On the other hand, several *gi* alleles show high dormancy as well as reduced sensitivity to cold treatments. Two other components of the clock, *LUX* and *ZTL*, also seem to regulate germination. The *lux* mutants show lower germination at lower temperatures, whereas *ztl-3* mutants display increased dormancy and a weaker response to cold, similar to *gi* mutants. Although in dry seeds the circadian clock is arrested in an evening-like stage (with high levels of *GI*, *LUX*, *PRR7*, *TOC1*, and *PRR9* and lower of *LHY* and *CCA1*), in imbibed seeds, the transcripts for several clock genes accumulate, particularly *CCA1*, while *PRR*s and *GI* are expressed albeit with reduced amplitude. Further experiments, using inhibitors of ABA and GA treatments revealed that the germination phenotypes observed in *cca1/lhy* and *gi* mutants are due to defects in metabolism and/or response to both hormones. In fact, transcriptional profiling of dormant and nondormant seeds showed that GA biosynthetic genes such as *GA3OX* (*GIBBERELLIC ACID 3-OXIDASE*) accumulate in *cca1/lhy* mutants but were almost undetectable in *gi-11* mutants. *CYP707A2 (cytochrome P450 707A2)*, a cytochrome P450 involved in ABA breakdown, was also misregulated in *gi-11* mutants. In addition, the *ABI3* (*ABSCISIC ACID INSENSITIVE 3*) gene that acts as a germination repressor is a circadian output regulated by *LHY*, *CCA1*, and *GI*. *ABI3* expression is reduced in *cca1/lhy* mutants, which might explain the reduced dormancy observed in these lines. The phenotype might be also explained by the interaction of ABI3 with TOC1.

Imbibition and red light can act together to induce genes from the ABA pathway such as *CYP707A2*. Imbibition alone is also able to induce *CCA1*, suggesting some feedback from this process and the clock. Triple *cca1/lhy/gi* mutants revealed a dual function for GI in germination; it promotes germination in an epistatic way to *cca1/lhy*, but also plays a role in inhibiting germination in an ABA-mediated fashion. This function would be additive to CCA1 and LHY. Analyses of other circadian genes for altered germination responses under ABA or GA control showed that changes in TOC1 transcript oscillation or protein abundance result in defects in hormone signaling in seeds. The importance of the clock in regulating GA and ABA signaling pathways is strengthened by findings showing that more than 40 % of ABA- and GA-regulated genes are under circadian control. Furthermore, ABA-inhibited and GA-activated genes are mostly expressed around dusk, whereas the opposite, ABA-activated and GA-inhibited genes peak closer to dawn. These results highlight the importance of the circadian clock at the initial stages of plant development, which continues during hypocotyl elongation, leaf and root growth, and also flowering.

Circadian Regulation of Hypocotyl Growth

After germination, the young seedling will elongate its embryonic stem (hypocotyl) in order to reach for light. This hypocotyl growth relies mostly on cell expansion events, and it displays a diurnal rhythm, peaking mostly at the end of the night. This specific pattern of growth is due to the coordination of two processes: an internal mechanism (the circadian clock) and external signals (light). These two mechanisms converge on the regulation of several members of the family of basic helix–loop–helix (bHLH) transcription factors, called PIFs (PHYTOCHROME-INTERACTING FACTORS) (Kinmonth-Schultz et al. 2013). Microarray analysis of pif mutants and overexpressing lines indicate that PIFs induce hypocotyl growth by increasing the expression of several genes involved in cell elongation. Light and the clock account for the posttranslational and transcriptional regulation of PIFs, respectively. When seedlings are exposed to light, the red-light circadian photoreceptor PHYB enters the nucleus where it binds to and promotes the degradation of PIFs by the 26S proteasome machinery. On the other hand, the circadian clock regulates PIF transcription by a recently identified mechanism. This comprises the assembly of a protein complex, the Evening Complex (EC) that includes ELF3, ELF4, and LUX (consult section "Molecular Components of the Circadian Oscillator"). The EC binds to and represses the promoters of *PIF4* and *PIF5* until the end of the night, when this repression is released and their transcription occurs. The double regulation by light and the clock permits PIF proteins to accumulate during the end of the dark period, at a time when they activate the expression of genes containing a particular motif or signature element (the G-box) in their promoters (Kinmonth-Schultz et al. 2013). The identification of the EC as a repressor of PIFs constitutes the "missing link" in the process that regulates PIF accumulation at the end of the night. In addition, the EC transcripts and proteins are under clock control (consult also sections "Molecular Components of the Circadian Oscillator" and "Molecular Structure and Mechanisms at the Core of the *Arabidopsis* Circadian Oscillator"), and this leads to their accumulation at dusk. Therefore, the coordinate action of light and the clock regulates *PIF* transcription and protein stability, allowing their accumulation at the exact time to promote the diurnal growth response (Fig. 7).

A detailed analysis of hypocotyl growth dynamics has shown that sucrose promotes growth by extending the number of days in which the hypocotyl is elongating. This response seems to be gated by the clock, as sucrose's effect on elongation is stronger around dawn. The use of *pif* mutants and PIF overexpressing lines has shown that sucrose promotes cell elongation via the PIFs. In fact, overexpression of *PIF5* phenocopies sucrose effects, while sucrose treatments result in higher level of PIF proteins without affecting *PIF* transcription. These findings suggest that sucrose probably affects the stability of PIF proteins, in a posttranslational regulatory mechanism similar to light. In this case, PHYB could be a molecular candidate triggering the degradation of PIFs, since it targets them for degradation by the proteasome. However, the analysis of *phyB* mutants revealed a response to sucrose similar to that in wild type. This would indicate that sucrose affects PIF stability in a PHYB-independent manner, by a yet unknown mechanism.

Circadian Regulation of Leaf and Root Growth and Architecture

Leaves also display a diurnal pattern of growth under continuous light similar to hypocotyls, which is another indication that the clock regulates plant development

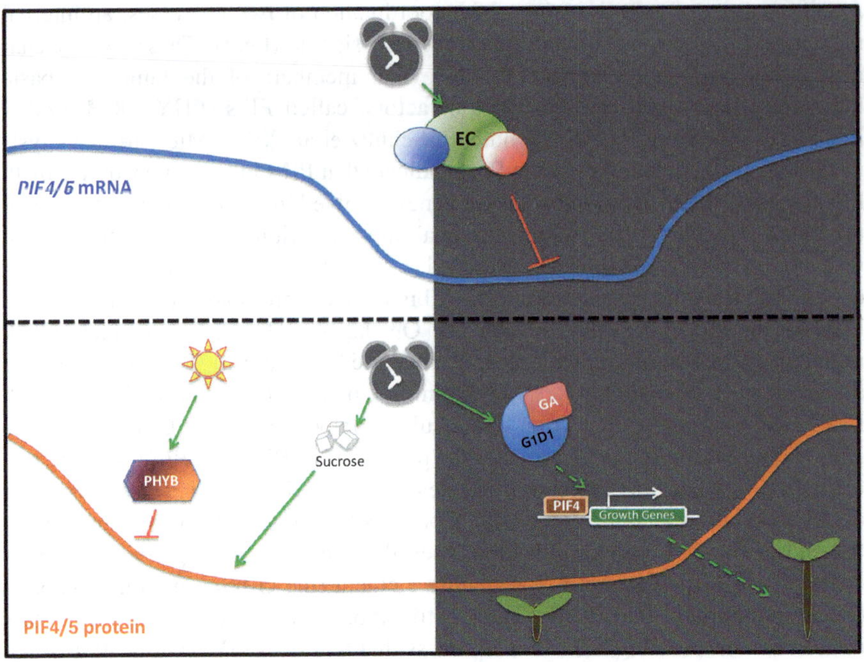

Fig. 7 The coordinated action of light, the clock, and GA regulates PIF4 and PIF5 at different levels. Light, through PHYB, promotes PIF4 and PIF5 protein degradation by the 26S proteasome. The clock regulates both sucrose availability that stabilizes both proteins and the assembly of the EC (evening complex) that represses *PIF4* and *PIF5* transcription at the beginning of the night period. At the end of the night, PIF proteins accumulate. At this time, the GA/GID1 complex is also assembled and promotes the degradation of DELLA proteins, which are repressors of PIF4 transcriptional activity. The dashed line represents this positive regulation of GA/GID1 in PIF4 activity

beyond the initial stages of their life cycle. To fully address the role of the circadian clock in the regulation of the plant growth, detailed analyses of the temporal growth patterns of different organs (root and shoot) of wild-type, *CCA1* overexpressing (*CCA1-ox*) lines, and *prr5*/*prr7*/*prr9* (*prr579*) mutants were performed. These lines showed smaller rosettes with a somewhat lower relative growth rate (RGR) as well as shorter roots also with lower RGR. Their lateral roots were also shorter and grew in a more horizontal direction. These findings highlight a possible function for the clock in controlling root architecture, especially in the formation of lateral roots and their angle of bending. In fact, a proper running clock is required to prevent a severe growth repression in both roots and shoots during the night.

Growth during the night also relies on the adequate use of starch reserves, a process that is under circadian regulation (for further details, see also section "Circadian Regulation of Starch Metabolism," below). It is possible that seedlings with a nonfunctional clock would be unable to mobilize their starch reserves each night. This defect would alter their nocturnal growth pattern leading to an overall loss of biomass, which is characteristic of *CCA1-ox* and *prr579* plants. Consistently, when starch content was assessed at dusk and dawn in sink and source leaves from wild-type and circadian mutants, it was evident that wild-type seedlings were able to mobilize all their reserves. The circadian mutants displayed a lower starch turnover, an effect that could also be a consequence of reduced storage of carbohydrate in these plants. This would indicate that the clock could affect development by also regulating the allocation of carbon supplies. Although there are discrepancies among studies concerning starch levels at the end of the night in several circadian mutants, there is a wide consensus on the role of the circadian clock in promoting nocturnal growth by regulating the allocation of carbohydrates during the day and controlling the rate of starch degradation during the night (Stitt and Zeeman 2012).

The circadian function controlling starch was also confirmed by a different study focused on high-throughput analysis of root growth kinetics. In this report, the rate of primary root elongation was determined under different photoperiods and free-running conditions. Roots were shown to elongate preferably at dawn. Analyses of mutants impaired on the generation/usage of starch further highlighted the connection between a normal diurnal pattern of root elongation and adequate allocation of carbohydrate during the day. The study also suggests that the circadian clock regulates root growth by two independent mechanisms: one requiring *CCA1* and *LHY* and the other relying on *ELF3*. *CCA1* and *LHY* would be involved in promoting or adjusting root growth at night by setting the adequate rate of starch degradation and to maintain the required amount of carbohydrate supply during this period. However, neither *CCA1* nor *LHY* seem to be directly involved in the mechanism integrating the clock with the starch degradation pathway. The regulators for this process are yet to be identified. On the other hand, *ELF3* participates in a different, starch-independent mechanism that represses root growth during the day and promotes it at night. Possibly, *ELF3* expressed in roots interacts with some metabolic signal deriving from shoots. These two complementary loops would account for the regulation of root growth according to the carbohydrate supply and thus avoiding periods of carbohydrate starvation that are detrimental for plant development.

Circadian Regulation of Hormone Signaling

Regulation of plant growth by the clock might also rely on the circadian control of hormone pathways. Indeed, genome-wide analyses have shown that the clock regulates the expression of many genes involved in hormone signaling pathways. In fact, PIFs were proposed as the molecular connectors linking the circadian clock with GA signaling. Besides PIFs, the clock also affects GA signaling by regulating the expression of *GA20ox1* (*GIBBERELLIN 20 OXIDASE 1*), encoding a GA synthetic enzyme, and *GID1* (*GIBBERELLIN INSENSITIVE DWARF 1*) that encodes a GA receptor (Kinmonth-Schultz et al. 2013). The circadian regulation accounts for the higher sensitivity of plants to GA during the night. The interaction between GA and GID1 initiates a signaling cascade that culminates with the proteasomal degradation of DELLA proteins around dawn. DELLAs are repressors of PIF4 and PIF5 transcriptional activity, and their degradation around dawn coincides with the time of accumulation of PIF proteins and the promotion of rhythmic daily growth (de Montaigu et al. 2010). Interestingly, although GA signaling is a clock output, it also mediates the expression of clock-regulated genes involved in stress responses and cell wall modifications. This finding highlights a common pattern that emerges with the study of clock-regulated pathways, i. e., clock outputs can also function as clock inputs, thus creating a feedback mechanism that might account for the high robustness of the circadian-regulated responses.

Similar to GA, signaling of the plant hormone ABA is also tightly connected with the circadian clock. Indeed, a recent study has described a novel feedback loop in which the oscillator component TOC1 is able to bind and repress the circadian expression of *ABAR/CHLH/GUN5* (*ABA RELATED/H SUBUNIT OF THE MAGNESIUM-PROTOPORPHYRIN IX CHELATASE/GENOMES UNCOUPLED 5*), an ABA-related gene. In turn, ABAR is required for the gated induction of TOC1 by ABA. Consistent with this loop, plants misexpressing *TOC1* display characteristic ABA-induced phenotypes that mostly occur under stress conditions. These phenotypes will be addressed in more detail in the next section (consult section "Drought and Salinity Stress Responses Under Clock Control").

The circadian clock is also involved in the regulation of auxin signaling. Genome-wide transcriptional analyses have shown that the circadian clock regulates several auxin-induced genes. This finding provided the molecular mechanism that underlies 70 years old observations describing that plant responsiveness to auxin treatments varied accordingly to the time of the day. The circadian clock seems to control the expression of several genes encoding proteins that participate in all the steps of the auxin signaling pathway, from biosynthesis to transport and the regulation of free or conjugated auxin. Several transcriptional regulators that act as activators (ARFs) or repressors (Aux/IAA) of auxin signaling are circadian regulated, further confirming the hypothesis that the clock can modulate the plant responses to auxin. Interestingly, the majority of genes that are both clock and auxin regulated share a similar phase of expression, peaking within a 4-h window in the middle of the subjective day. In addition, the clock gates the plant responsiveness to endogenous and exogenous auxin levels. The gating effect is visible in growth responses that are auxin dependent, suggesting that the circadian clock is upstream of the auxin signal transduction pathway.

Jasmonic acid (JA) signaling is also connected with the circadian clock. JA is involved in several physiological processes with a relevant role in triggering plant defenses against necrotrophic pathogens and wounding caused by feeding insects. Several types of evidence have connected the JA signaling pathway to the circadian clock: (1) most of the JA-responsive genes that are clock regulated have a morning phase of expression, (2) several JA biosynthesis genes also display a 24-h rhythmic expression, and (3) JA abundance also seems to accumulate with a rhythmic pattern, higher in the middle of the day and lower at night. The existence of a circadian-regulated response could allow plants to predict the timing when pathogen infection and insect attack are most likely to occur and set in motion the necessary time-specific defense responses to counteract these threats. The connection between the circadian clock and JA-mediated pathogen defense seems to depend, at least in part, on a component of the evening oscillator, TIC (TIME FOR COFFEE).

Clock-Controlled Regulation of Flowering Time

Plants perceive changes in photoperiod in their leaves, and the information is conveyed by a mobile signal (or florigen) to the shoot apical meristem (SAM) where floral transition occurs. Circadian regulation of photoperiodic flowering ensures that plants flower under the most favorable conditions (Song et al. 2010; Kinmonth-Schultz et al. 2013). This is achieved by the modulation of the amount of florigen, which is partly encoded by *FT* (*FLOWERING LOCUS T*). *Arabidopsis* are facultative long-day plants, i.e., flower earlier under long-day than under short-day conditions. Consistent with this photoperiodic regulation, *FT* transcripts accumulate at a precise time of day and in the proper season. *FT* is transcribed in the leaf vascular tissue from where the FT protein migrates to the SAM. Here, FT activates the transcription of several genes that will promote the development of the flower meristem. *FT* accumulation under long days depends on the transcriptional and posttranscriptional regulation of its activator *CO* (*CONSTANS*). Under long days, *CO* transcripts accumulate at the end of the afternoon, when the repression by CDFs (CYCLING DOF FACTORs) is attenuated. CDFs are transcriptional regulators that bind directly and repress the *CO* promoter. Similar to *CO*, *CDF*s are circadian-regulated genes that accumulate preferably in the morning and decrease during the day. *CDF* transcription is repressed by the central oscillator components PRR9, PRR7, and PRR5. In addition, CDF proteins are also subject to posttranslational regulation by clock components. This is achieved by the blue-light dependent complex consisting of FKF1 and GI. Under long days, the FKF1–GI complex is stabilized due to a circadian-dependent coincidence mechanism that allows both FKF1 and GI proteins to accumulate at the end of the afternoon. The FKF1–GI complex targets CDFs for degradation by the proteasome, degradation that allows the transcription of *CO*. As mentioned above, the circadian clock also participates in the posttranscriptional regulation of *CO*. Indeed, CO protein accumulation in the afternoon under long days depends on the concerted action of light and the clock. This is achieved by the accumulation of FKF1 (which also has a function as a blue-light photoreceptor; consult section "ZEITLUPE Family") and allows CO stabilization at the end of the day in a blue-light dependent manner. Therefore, through FKF1, the clock regulates CO's accumulation, either by targeting its repressors to degradation or by directly forming heterodimers that prevent its degradation by the proteasome. Although the clock is important in promoting CO accumulation at the end of long days, CO is also tightly regulated by several components of the light signaling pathway such as PHYB, CRY2, SPA1, and COP1. Based on the particular focus of this section on circadian outputs, this regulatory mechanism will not be discussed in detail.

Circadian Regulation of Abiotic Stress

Cold Responses Under Circadian Control

Plants from temperate climates are able to survive freezing temperatures by triggering a mechanism known as cold acclimation. Exposure to low, nonfreezing temperatures sets in motion cellular responses that ultimately result in freezing tolerance. At the center of the cold-responsive mechanism is the CBF pathway formed by CBF1 (C-REPEAT BINDING FACTOR 1 or drought-responsive element-binding factor 1B, DREB1B), CBF2 (DREB1C), and CBF3 (DREB1A) that belong to the AP2/ERF (APETALA 2/ETHYLENE-RESPONSIVE ELEMENT-BINDING FACTOR) family of transcription factors. Upon cold induction, CBFs bind to promoters containing specific elements (known as *CRT/DRE, C-REPEAT ELEMENT/DEHYDRATION-RESPONSIVE ELEMENT*) and induce the expression of approximately 100 cold-regulated (*COR*) genes that constitute the CBF regulon. This cold-induced transcriptome confers freezing tolerance due to the modulation of several metabolic processes such as lipid biosynthesis, synthesis of secondary metabolites, and cryoprotective substances.

The circadian clock seems to have a pervasive role in the modulation of cold responses. Transcriptional analysis of the cold-induced transcriptome has revealed that the outcome of the experiments profoundly varies depending on the time of day. Indeed, the vast majority of variation obtained in different datasets of cold-responsive transcriptomes is due to diurnal- and/or circadian-regulated genes. The clock seems to gate the expression of several cold-induced transcription factors. This is a widespread event with higher accumulation of cold-induced regulators at dawn. Interestingly, cold treatments also affect the expression of clock components, leading to reduced circadian amplitude under LD cycles and even arrhythmia under free-running conditions.

The expression of the *CBF* genes themselves is circadian regulated, and their cold induction is higher at Zeitgeber Time 4 (ZT4) than at ZT16, an additional indication of gating by the clock. The mechanism underlying this regulation has been recently shown. The *CBF1*, *CBF2*, and *CBF3* genes are organized in tandem in a single *locus*, and their promoters contain several EE, suggesting regulation by CCA1 and LHY. CCA1 and LHY are indeed able to bind to the promoters of the *CBFs*. As expected, *cca1-11/lhy-21* double mutants show a reduction in *CBF1* and *CBF3* induction upon cold treatment. Confirming the role of CCA1 and LHY in the control of freezing tolerance, cold induction of the downstream CBF targets *COR15A*, *COR47*, and *COR78* was impaired in *cca1-11/lhy-21* mutant plants, which might be responsible for the reduced freezing tolerance observed in these plants. The circadian regulation of cold responses is extended to other components of the central oscillator, which further suggest that gating the cold response is critical for adequate environmental adaptation.

PIF7, a bHLH transcription factor of the PIF family, was also shown to be a transcriptional regulator of the *CBF2* promoter. PIF7 binds to a G-box motif in the *CBF2* promoter that is required for its temperature and circadian regulation. In addition to *CBF2*, PIF7 might also bind and repress *CBF1*, and possibly *CBF3* although the latter with lower affinity. Interestingly, *PIF7* expression is regulated by the clock but not by temperature. This finding suggests the possibility that PIF7 could participate in the circadian regulation of *CBF2* in a temperature-independent manner. In agreement with this hypothesis, PIF7 was shown to interact with TOC1 in nuclei of mesophyll protoplasts. Furthermore, the PIF7–TOC1 heterodimers enhance the repression of the *CBF2* promoter. Interestingly, TOC1 interacts with CBFs, which also promote resistance to drought and high salinity. Similar to the interaction of PIF7 with TOC1, the interaction of PIF7 with PIF4 might increase the repression of the *CBF2* promoter. It is possible that all the different mechanisms converge on the circadian regulation of *CBF*s in a temperature-independent manner, in order to provide plants with a way of preventing non-proper induction of *CBF*s (e.g., under nonstressful conditions).

Microarray analysis of end-of-day responses (a response that occurs at the subjective late afternoon, from circadian time (CT) CT8 to CT12) confirmed the importance of circadian regulation in cold responses. A comparison of the *prr579* triple mutants with wild-type seedlings showed a significant enrichment in cold-responsive genes, including *CBF1*. The accumulation of *CBF*s probably contributes to the increased freezing tolerance of the *prr579* mutants. Overexpression of CBFs promotes tolerance to other stresses, which is consistent with the increased resistance to high salinity and drought of *prr579* mutant plants. As mentioned above, exposure to low temperatures induces the biosynthesis of cryoprotective compounds. This phenotype was also reported in *prr579* mutants, which accumulated higher levels of raffinose (important for desiccation tolerance in seed development) and L-proline. The mutants were also affected in the ABA biosynthetic pathway and were shown to accumulate high levels of the hormone ABA during the day.

PRR7 and PRR9 act as repressors of the *CBF* pathway, while ChIP-seq (immunoprecipitation of chromatin followed by deep sequencing) analysis of PRR5 also identified the *CBF*s as targets of PRR5. Combining this data with *CBF* microarray expression data showed that their repression matches with the time of PRR5 accumulation. These results reinforce the idea that PRR5 might also act as a direct repressor of *CBF*s. Taken together, the findings suggest that PRR5, PRR7, and PRR9 regulate cold responses at different levels, by controlling the CBF regulon, possibly due to direct repression of the *CBF*'s promoters, and/or by regulating the expression of genes involved in cold-responsive metabolic pathways. In section "Circadian Outputs in Metabolism," the role of the clock in the regulation of metabolic processes is further addressed.

Drought and Salinity Stress Responses Under Clock Control

Genome-wide transcriptome and ChIP-seq analyses uncovered a whole new landscape of circadian-regulated genes and thus have clarified the role of the clock in several biological processes. For instance, transcriptional profiling of wild-type, *TOC1* overexpressing, and *toc1–2* mutant plants has revealed a significant number of genes related to the hormone ABA signaling pathway (consult also section "Circadian Regulation of Hormone Signaling"). Among many other processes, ABA regulates the plant responses to drought conditions. Consistently, *TOC1* overexpressing plants are less tolerant to dehydration due to impaired stomatal function and poor regulation of leaf transpiration rates. Oppositely, *TOC1*-RNAi and *toc1–2* mutant lines show increased tolerance to drought stress. These plants have a more efficient closure of stomata and reduced water-loss rates, an indication that TOC1 mediates ABA-induced responses and drought stress. *TOC1* microarray data has also shown misexpression of *ABAR/CHLH/GUN5*, a gene closely related to the ABA pathway and arguably defined as a putative ABA receptor. *ABAR* expression is regulated by the circadian clock, with a peak of expression around dawn. TOC1 directly regulates *ABAR* expression, by binding to the *ABAR* promoter with a maximum enrichment at the times of lowest *ABAR* transcription. Gene expression analysis with TOC1 misexpressing plants confirmed the repressive function of TOC1 and the opposite correlation between *TOC1* accumulation and drought tolerance. Moreover, *TOC1* is itself induced by ABA, and this induction is gated by the clock (i.e., only occurs during the day) and requires the function of ABAR. This reciprocal regulation defines a novel feedback loop connecting the circadian clock and ABA signaling through TOC1 and ABAR. A modeling approach including the ABA and TOC1 results has recently determined the relationship between the expression of oscillator components and the control of downstream physiological events. The model incorporates several parameters such as the *TOC1–ABAR* feedback loop, the induction of *TOC1* by ABA, and the widespread repressive role of TOC1 in the central oscillator (consult section "Molecular Structure and Mechanisms at the Core of the *Arabidopsis* Circadian Oscillator"). In this model, mathematical simulations of *TOC1* accumulation under high ABA concentrations lead to repression of TOC1 targets and longer circadian period, which fully agree with the experimental data. The model also addresses the gating of ABA responses by incorporating stomata aperture, a well-characterized circadian output. Interestingly, the model predicts that changes in *TOC1* abundance could also affect the diurnal gating of stomata responses to ABA, i.e., the specific increase in sensitivity towards ABA at dusk and its reduction at dawn.

The considerable overlap between circadian microarray datasets and ABA-dependent transcriptional profiling has also been addressed in other studies, which emphasized the importance of the clock in setting the adequate responses to a wide range of stresses (e.g., drought, osmotic, salinity). For instance, a relationship between drought responses and time of day has been shown with the transcriptional analysis of responses to mild drought conditions (4 days of withholding water) at four different times under LD cycles (ZT0, ZT6, ZT12, ZT18). The analysis showed different stress- and hormone-dependent responses depending on the time when drought samples were collected. Subsequent data analysis identified a set of genes that were up- or downregulated independently of the time of sampling, therefore constituting the "core drought response." When the effect of drought in the transcription of these genes was assessed over the entire diurnal period, it was found that at ZT12, the effects were much stronger, an indication of circadian regulation of these responses. Comparative analysis of the ZT12 drought-responsive transcriptome identified higher similarities with drought/rehydration, osmotic, and salt stress transcriptomes. In addition, the ZT12 drought-responsive transcriptome also shared similarities with the ABA-treatment transcriptome, further reinforcing the critical role of ABA in circadian-regulated abiotic stress responses, especially those at the end of the day.

Most of these findings highlight the importance of the circadian clock in the regulation of stress responses in coordination with the ABA signaling pathway. Other clock components have been recently shown to have a role in the regulation of stress responses. For instance, GI has an important function in the circadian response to salinity stress (see also section "Clock-Controlled Regulation of Flowering Time"). The *gi* mutants have increased salt tolerance mostly due to higher activity of SOS1 (SALT OVERLY SENSITIVE 1), a Na^+/H^+ antiporter protein. This regulation seems to depend on protein–protein interactions. Under low salt conditions, GI binds to the SOS2 kinase (SALT OVERLY SENSITIVE 2) preventing the phosphorylation and activation of SOS1. Upon salt stress, GI is degraded by the 26S proteasome, which allows the interaction of SOS2 with the Ca^{2+}-dependent protein SOS3 (SALT OVERLY SENSITIVE 3) to occur. The SOS2/SOS3 complex then phosphorylates and activates SOS1, thus allowing the export of sodium ions and consequently salt tolerance. Therefore, GI not only participates in the regulation of salt tolerance but also integrates the responses to an environmental stress with a developmental process such as flowering. Under salt stress, GI is degraded, which results in delayed flowering, until more favorable conditions are available. Altogether, these findings highlight the role of the circadian clock in mediating the adequate timing for environmental stress responses in coordination with developmental processes.

A Connection Between Reactive Oxygen Species (ROS) and the Circadian Clock

Plants deal not only with external, environmentally imposed stresses (as described above), but also with internal, metabolic cues such as respiration and photosynthesis. These are key processes in the plant life cycle but also a source of Reactive Oxygen Species (ROS) that, if left uncontrolled, promote cellular damage and ultimately cell death. ROS can also function as second messengers during stress responses. Therefore, plants have developed several scavenging strategies to control the level of ROS in cells. ROS production follows an oscillatory pattern, reaching its maximum at ZT7, right after the peak in light-harvesting capacity. This circadian behavior is also found in the expression of ROS-responsive genes and ROS responses, suggesting that the clock could also regulate the responses to ROS. Analysis of mutants with defects in components of the oscillator, e.g., *CCA1*, *LHY*, *PRR5*, *PRR7*, *PRR9*, *ELF3*, *ELF4*, and *LUX*, showed a common phenotype of hypersensitivity to increasing superoxide levels. Transcriptional profiling of ROS-responsive genes has also revealed that approximately 30 % of transcripts are circadian regulated. This regulation was shown to be dependent on *CCA1* rhythmic oscillation. Some of these genes contain the EE and/or CBS motifs in their promoters while their oxidative stress-dependent induction was gated to the early morning, a time when *CCA1* transcripts accumulate. ChIP analysis confirmed the binding of CCA1 to the promoters of several ROS genes (e.g., *JMJD5*, *MYB 59*,

WRKY11). Similar to other outputs, ROS also affect the expression of clock-related components such as *FKF1*. The mechanism underlying this regulation is not fully understood but possibly involves tissue-specific responses. Other circadian regulators, such as GI, might be also involved in plant responses to ROS, since the *gi* mutants are more tolerant to paraquat-induced oxidative stress.

Stress responses are rather demanding for plants, especially from an energetic point of view, since they require allocation of considerable resources for constant monitoring and activation of the adequate response pathways. The existence of gating mechanisms that can anticipate these threats and trigger the responses at the appropriate time window might provide maximal tolerance with minimal costs. The increasing number of stress responses that are under circadian regulation emphasizes the importance of this time-keeping mechanism in facilitating plant adaptation to a constantly changing environment.

Circadian Outputs in Metabolism

Circadian Regulation of Starch Metabolism

The processes of carbon assimilation, storage, and utilization are regulated with a diurnal pattern. Newly assimilated carbon can be readily available as sucrose, or it can be stored in chloroplasts as starch for later use (Stitt and Zeeman 2012). During the day, starch accumulates linearly fulfilling needs for carbohydrate during the night and suggesting an "anticipation" mechanism. This mechanism is also involved in setting a linear rate for starch degradation at night. During this period, starch is converted to maltose in the chloroplast and exported to the cytosol, where it will be metabolized back to sucrose (Haydon et al. 2013). This dynamic process allows the maintenance of a constant supply of carbon throughout the night, which will be almost completely used up until the next dawn.

Several findings point to a model for starch metabolism that has the shape of a triangle with one of the sides representing the rate of starch accumulation during the day, and the other the rate of degradation at night (Fig. 8). Changes in day length result in adjustments of these rates, represented as steeper or flatter slopes in each side of the triangle. For instance, under shorter days, more photoassimilate is converted into starch, thus shifting the balance towards storage in order to cope with the longer night period. Under these conditions, the rate of degradation will adjust accordingly to the duration of the night, to allow the maintenance of an adequate supply of energy, which is required for growth. If the night is extended, root growth is stopped and there is a reprogramming of the transcriptome to a typical carbon starvation situation. After several days, the rate of starch biosynthesis/degradation adjusts to ensure an adequate supply of carbon and growth resumes. If plants, on the other hand, are exposed to an early night, they will immediately adjust their rate of starch degradation so that carbon starvation does not occur.

Circadian clock mutants such as *cca1/lhy* with shorter circadian periods degrade starch at a faster rate when grown under 24-h cycles and prematurely exhaust their reserves. However, if *cca1/lhy* seedlings are grown on 17-h cycles (matching their

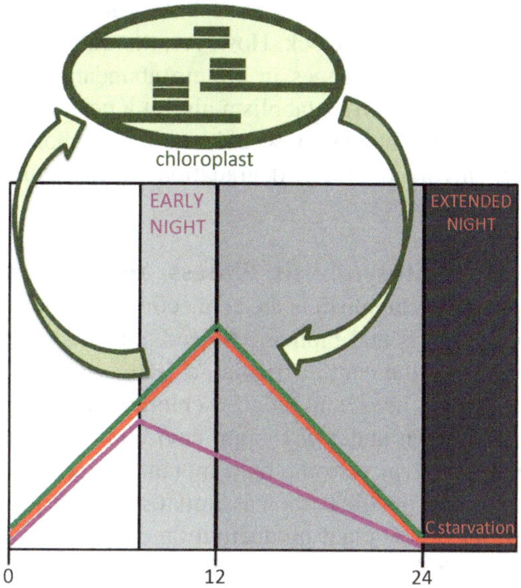

Fig. 8 Schematic representation of starch metabolism. Plants grown under 12 LD cycles store photoassimilated carbon during the light period in the form of starch, which is stored in the chloroplast. During the dark period, starch is degraded in order to maintain nocturnal growth (*green line*). If plants are exposed to an early night, they adjust the rate of starch degradation so that the accumulated reserves last for the entire dark period (*pink line*). Conversely, if the night period is extended, the accumulated starch has been totally exhausted and symptoms of C starvation appear (*red line*)

endogenous circadian period), a proper rate of starch degradation is restored. When seedlings are grown under extended nights, the linear rate of starch degradation is rapidly adjusted to permit maximal growth and overall fitness (see also section "The Clock Controls Photosynthesis, Fitness, and Biomass"). This suggests that mobilization of starch reserves is a critical and limiting factor in plant productivity. Mathematical modeling also suggests the existence of a feedback mechanism between clock and assimilated carbon that allows the adjustment of starch biosynthesis and degradation to changes in photoperiod. Testing the model against changes in photoperiod revealed an accurate fit with the experimental data, showing, for instance, that under shorter days there is a steeper slope for the rate of starch accumulation and a reduction in the slope of the rate of starch degradation at night. The modeling approach also shows that circadian regulation is the best setup to avoid carbon starvation. Together, these findings suggest that circadian clock regulation of starch metabolism is important for increased fitness and productivity.

Despite the considerable progress in the study of this regulation, the molecular mechanisms behind the responses are still not fully understood. Probably, the circadian clock modulates the expression of several components of the starch metabolism pathway, such as MEX1 (MALTOSE EXCESS 1), a major maltose exporter from the chloroplast. In addition, transcripts encoding sugar transporters like *STP1* (*SUGAR TRANSPORTER 1*) as well as a new family of sugar efflux proteins are also regulated by the clock. However, the transcriptional regulation is not followed by concomitant changes in protein abundance, suggesting that the circadian clock modulates starch metabolism also at a posttranslational level. One possible mechanism for this regulation would be the phosphorylation/dephosphorylation of enzymes involved in the degradation of the starch granule in the chloroplast.

The Clock Controls Photosynthesis, Fitness, and Biomass

Closely related to starch metabolism is the connection of the circadian clock to the regulation of photosynthesis and biomass. Indeed, circadian period mutants with defects in components of the central oscillator such as *toc1-1* and *ztl-1* develop poorly, have lower growth rates, and reduced chlorophyll content. Notably, when the mutant plants are grown under day–night environmental cycles that perfectly match their internal circadian periods, both mutants show increased chlorophyll content and higher photosynthetic carbon assimilation. The results suggest that the circadian clock can maximize plant productivity most probably by matching photosynthetic capacity with the proper time of day.

The circadian clock also modulates plant productivity by controlling the morphological characteristics typical of hybrids and allopolyploids, which are routinely described as hybrid vigor. Hybrids and allotetraploids showed epigenetic repression of both *CCA1* and *LHY* and accumulation of *TOC1*, *GI*, as well as genes encoding components of the starch and photosynthetic pathways. These hybrids and allotetraploids have also higher chlorophyll and starch content than their parents. Confirmation of these findings was obtained by analyzing *cca1* single mutants, *cca1/lhy* double mutants, as well as transgenic lines in which *CCA1* transcripts were targeted by RNA interference (RNAi). All these transgenic lines displayed higher chlorophyll and starch content. The results suggest that the clock could participate in the regulation of plant productivity by regulating photosynthetic efficiency and carbon partitioning during the day and the rate of starch degradation during the night.

Metabolite Profiling Under Circadian Control

The role of the circadian clock in the control of a wide variety of metabolic processes has been uncovered by transcriptome and metabolomic (mass spectrometry) analyses of wild-type plants, *prr5/prr7/prr9* (*prr579*) mutants, and *CCA1* overexpressing (*CCA1-ox*) lines. This comprehensive approach revealed that despite their morphological similarities, *CCA1-ox* and *prr579* plants showed very distinct and well-defined transcript and metabolite profiles. In fact, *prr579* mutants have a larger number of changed metabolites that accumulate at higher levels. This phenotype was even more pronounced when comparisons were made between *prr579*, *CCA1*-ox, and mutants affected in flavonoid metabolism. High-resolution time series profiling revealed a significant increase in metabolic intermediates (e.g., citrate, malate, and shikimate) from the tricarboxylic acid cycle (TCA) in *prr579* but not in *CCA1*-ox lines. In addition, *prr579* plants accumulate higher amounts of antioxidant vitamins, such as ascorbate and α-tocopherol as well as osmolytes (e.g., proline, raffinose, galactinol). The accumulation of osmolytes is also related to the increased freezing and drought tolerance of *prr579* plants (see section "Cold Responses Under Circadian Control"). In order to match the metabolic profiling of *prr579* plants with the transcriptional activity of PRRs, transcriptome analyses focused on central metabolism were performed. As expected, *prr579* mutants showed induction of genes involved in the synthesis of osmolytes and the biosynthesis of chlorophyll, ABA, carotenoid, and α-tocopherol. These findings confirm the role of PRRs as negative regulators of these biosynthetic pathways, and they also show their importance in the regulation of ROS responses. The mechanism by which PRR5, PRR7, and PRR9 regulate the TCA cycle is yet to be completely elucidated. These findings highlight the role of PRRs in maintaining central metabolism and uncover an organelle specificity in controlling metabolism: in chloroplasts, by regulating carotenoid and ABA biosynthesis, and in mitochondria, by controlling the TCA cycle.

Second Messengers, Micronutrients, and Mineral Homeostasis as Clock Outputs

Calcium Signaling

Ca^{2+} is a second messenger that participates in several signaling pathways. In response to various stimuli, Ca^{2+} is mobilized from internal and external storage leading to the establishment of acute concentration gradients in $[Ca^{2+}]_{cyt}$ (concentration of cytosolic free Ca^{2+}). Several Ca^{2+}-dependent proteins perceive these changes and induce transcriptional modifications that are ultimately responsible for the cellular responses. $[Ca^{2+}]_{cyt}$ oscillates on a daily basis with a period of 24 h. Under short days, $[Ca^{2+}]_{cyt}$ peaks around dusk, whereas under long days it accumulates around midday. These findings suggest that circadian regulation of $[Ca^{2+}]_{cyt}$ and perception of different photoperiods could be part of a mechanism regulating flowering time (Haydon et al. 2013).

The circadian regulation of $[Ca^{2+}]_{cyt}$ can also account for stomata movements, a well-known circadian output. Stomata movements allow the transpiration and gas exchange on the leaf surface. To achieve maximum water efficiency and photosynthetic capacity, stomata open just before dawn and close at night. Stomata movements depend on the change in turgor pressure of the guard cells which in its turn is due to several processes: (1) the concerted action of several plasma membrane and vacuolar channels (e.g., K^+, Cl^-, NO_3^-, and Ca^{2+}), (2) the accumulation of malate, and (3) the clock. Therefore, by regulating the $[Ca^{2+}]_{cyt}$, the circadian clock could promote the correct match between stomata aperture, the adequate time of day, and environmental conditions. In molecular terms, this could be achieved by modulation of the guard cell signaling network in response to external signals and/or by the direct regulation of proteins involved in the movement of stomata.

The mechanism by which the circadian clock regulates $[Ca^{2+}]_{cyt}$ is not fully understood, but it seems to require a functional *CCA1*, since in *cca1-1* mutants the rhythms of $[Ca^{2+}]_{cyt}$ are almost absent. Analyses of other circadian mutants suggest an interplay between light and the clock in the regulation of $[Ca^{2+}]_{cyt}$ rhythms, as well as the probable involvement of a cell-specific oscillator. Moreover, the circadian clock could regulate the rhythms of $[Ca^{2+}]_{cyt}$ by modulating the transcription of genes encoding proteins involved in Ca^{2+} homeostasis, such as calmodulin-like isoforms and Ca^{2+} channels. The circadian oscillation of $[Ca^{2+}]_{cyt}$ is also correlated with the metabolic status of the cells as it is absent in seedlings grown in media with 3 % sucrose. This finding correlates the energy status with the circadian function, similar to what has been reported in animals. Possibly, this is achieved by the regulation of NAD^+ (nicotinamide adenine dinucleotide) accumulation, a sensor of the energy status in the cell that has also been shown to affect the expression of components of the central oscillator. In fact, the synthesis of cADPR, a product of NAD^+, was shown to oscillate and regulate the fluxes of $[Ca^{2+}]_{cyt}$, suggesting that changes in metabolic state and $[Ca^{2+}]_{cyt}$ could also affect the functioning of the central oscillator.

Iron and Copper Homeostasis

Iron (herein Fe) is an important micronutrient required for photosynthesis and for respiration. Due to its chemical properties, it can be also easily oxidized and thus triggering the production of ROS. Although plants have nutritional requirements for Fe, they also need to control its accumulation to prevent oxidative stress. Several reports have shown that components of the iron homeostasis network are transcriptionally regulated by the clock. Whether under light/dark cycles or in continuous light, *IRT1* (*IRON-REGULATED TRANSPORTER 1*), *bLHL39* (*basic HELIX-LOOP-HELIX 39*), and *FER* (*FERRITIN*) transcripts were shown to oscillate. *IRT1* encodes for the high affinity Fe(II) transporter responsible for Fe uptake from the soil; *bHLH39* is a transcription factor involved in the Fe deficiency response while *FER1* is a major Fe storage protein. Therefore, the clock regulates both the expression of genes involved in the Fe deficiency response and in Fe storage. Further evidence of circadian regulation in Fe homeostasis was provided by the observed lengthening of circadian period of *IRT1* expression in circadian mutants such as *ztl-4* and *prr7/prr9*. Interestingly, and similar to other clock outputs, the status of Fe nutrition seems to affect clock function, since seedlings grown under Fe deficiency have longer periods.

An independent study using a genetic screen for mutants de-repressed in *FER1* under low iron conditions identified several alleles of the circadian-regulated gene *TIC*. The *tic* mutants have short periods and are mostly affected in their evening oscillator. The mutants also show leaf chlorosis, which can be overcome by iron supplementing. In *tic* mutants, *FER1* expression oscillates at higher levels, and this response seems to depend also on light. However, the phenotype seems to be independent of Fe-responsive elements. The *tic* mutant did not affect other members of the Fe homeostasis pathway (e.g., *IRT1*), but it did modulate the expression of components of the ROS-scavenging pathway. This could suggest that TIC regulates *FER1* in connection with its role in regulation of ROS rather than Fe homeostasis. Nevertheless, the central oscillator does affect *FER1* expression since its rhythmic oscillation is abolished in *lhy-21* and *cca1-11* mutants. Future work will uncover the components involved in the mechanism connecting Fe homeostasis and circadian regulation.

In addition to Fe, the circadian clock is also involved in the regulation of copper (Cu) homeostasis. Overexpression of COPT1 (COPPER TRANSPORTER 1) and COPT3 (COPPER TRANSPORTER 3), two plasma membrane proteins involved in Cu transport, results in phenotypes typical of circadian clock mutants. These overexpressing lines develop long hypocotyls under white light and have changes in flowering time, similar to *cca1/lhy* mutants. Interestingly, changes in Cu transport also seem to affect the phase and amplitude of *CCA1* and *LHY* expression, suggesting that Cu could act as an input to the clock, similar to Fe. Collectively, all these studies suggest that metal homeostasis in *Arabidopsis* is an important clock output. The extra layer of regulation might be essential for gating the expression of components of these pathways, to facilitate an efficient use of resources. In parallel, a widespread feedback mechanism could account for the effects that variations in metal availability have on clock function.

Circadian Control of Mineral Assimilation

The circadian clock also regulates the assimilation of mineral nutrients such as N (nitrogen) and S (sulfur). This regulation is achieved by transcriptional control of components involved in the assimilation pathways, such as nitrate, ammonium, and sulfate transporters. Plants assimilate these minerals by a process initiated with the uptake of their inorganic forms from the soil, followed by their subsequent reduction to organic forms that can be readily used. Systems biology studies have revealed the existence of a gene network responsive to organic N, and this network seemed to be under the control of CCA1. In fact, CCA1 was shown to bind to the promoters of several genes encoding proteins required for N assimilation such as GLUTAMINE SYNTHASE (*GLN3.1*) and GLUTAMATE DEHYDROGENASE (*GDH*).

Future Directions

Despite the recent advances in our understanding of the *Arabidopsis* circadian system, we are still far from having a complete picture of the components and mechanisms responsible for the circadian timing in plants. Regarding input pathways, future research most likely will identify the molecular effectors connecting photoreceptor function with the oscillator. Additional resetting cues and the interconnections between the different input pathways will definitively improve our knowledge on how the clock is synchronized every day by the environmental signals. As the majority of the current oscillator components have a repressing function, the identification of the clock components responsible for activation will be essential to understand how the clock temporally coordinates circadian outputs. Future areas of clock research will likely focus on the mechanistic insights of protein–protein interactions, pre- and posttranscriptional and translational regulations, and their integration into global circadian networks. Connections of the oscillator with the different output pathways and their relationships will be also an important area of future research, particularly for its possible application to plants of agronomical interest.

References

Carré I, Veflingstad SR. Emerging design principles in the *Arabidopsis* circadian clock. Semin Cell Dev Biol. 2013;24:393–8.

Chaves I, Pokorny R, Byrdin M, Hoang N, Ritz T, Brettel K, Essen L-O, van der Horst GTJ, Batschauer A, Ahmad M. The cryptochromes: blue light photoreceptors in plants and animals. Annu Rev Plant Biol. 2011;62:335–64.

Chen M, Chory J, Fankhauser C. Light signal transduction in higher plants. Annu Rev Genet. 2004;38:87–117.

de Montaigu A, Tóth R, Coupland G. Plant development goes like clockwork. Trends Genet. 2010;26:296–306.

Doherty CJ, Kay SA. Circadian control of global gene expression patterns. Annu Rev Genet. 2010;44:419–44.

Harmer SL. The circadian system in higher plants. Annu Rev Plant Biol. 2009;60:357–77.

Haydon MJ, Hearn TJ, Bell LJ, Hannah MA, Webb AAR. Metabolic regulation of circadian clocks. Semin Cell Dev Biol. 2013;24:414–21.

Kinmonth-Schultz HA, Golembeski GS, Imaizumi T. Circadian clock-regulated physiological outputs: dynamic responses in nature. Semin Cell Dev Biol. 2013;24:407–13.

Kusakina J, Dodd AN. Phosphorylation in the plant circadian system. Trends Plant Sci. 2012;17:575–83.

Losi A, Gärtner W. The evolution of flavin-binding photoreceptors: an ancient chromophore serving trendy blue-light sensors. Annu Rev Plant Biol. 2012;63:49–72.

Más P, Yanovsky MJ. Time for circadian rhythms: plants get synchronized. Curr Opin Plant Biol. 2009;12:574–9.

McClung CR, Davis SJ. Ambient thermometers in plants: from physiological outputs towards mechanisms of thermal sensing. Curr Biol. 2010;20:R1086–92.

McWatters HG, Devlin PF. Timing in plants – a rhythmic arrangement. FEBS Lett. 2011;585:1474–84.

Millar AJ. Input signals to the plant circadian clock. J Exp Bot. 2004;55:277–83.

Nagel DH, Kay SA. Complexity in the wiring and regulation of plant circadian networks. Curr Biol. 2012;22:R648–57.

Sanchez SE, Petrillo E, Kornblihtt AR, Yanovsky MJ. Alternative splicing at the right time. RNA Biol. 2011;8:954–9.

Song YH, Ito S, Imaizumi T. Similarities in the circadian clock and photoperiodism in plants. Curr Opin Plant Biol. 2010;13:594–603.

Staiger D, Green R. RNA-based regulation in the plant circadian clock. Trends Plant Sci. 2011;16:517–23.

Staiger D, Köster T. Spotlight on post-transcriptional control in the circadian system. Cell Mol Life Sci. 2011;68:71–83.

Stitt M, Zeeman SC. Starch turnover: pathways, regulation and role in growth. Curr Opin Plant Biol. 2012;15:282–92.

Further Reading

Andres F, Coupland G. The genetic basis of flowering responses to seasonal cues. Nat Rev Genet. 2012;13:627–39.

Chow BY, Kay SA. Global approaches for telling time: omics and the *Arabidopsis* circadian clock. Semin Cell Dev Biol. 2013;24:383–92.

Devlin PF, Kay SA. Circadian photoperception. Annu Rev Physiol. 2001;63:677–94.

Graf A, Smith AM. Starch and the clock: the dark side of plant productivity. Trends Plant Sci. 2011;16:169–75.

Henriques R, Mas P. Chromatin remodeling and alternative splicing: pre- and post-transcriptional regulation of the *Arabidopsis* circadian clock. Semin Cell Dev Biol. 2013;24:399–406.
Hotta C, Gardner MJ, Hubbard KE, Baek SJ, Dalchau N, Suhita D, Dodd AN, Webb AAR. Modulation of environmental responses of plants by circadian clocks. Plant Cell Environ. 2007;30:333–49.
McClung CR. The genetics of plant clocks. In: Stuart B, editor. Advances in genetics. Elsevier Academic Press; 2011;74:105–39.
McClung CR, Gutiérrez RA. Network news: prime time for systems biology of the plant circadian clock. Curr Opin Genet Dev. 2010;20:588–98.
Sanchez A, Shin J, Davis SJ. Abiotic stress and the plant circadian clock. Plant Signal Behav. 2011;6:223–31.
Yamashino T. From a repressilator-based circadian clock mechanism to an external coincidence model responsible for photoperiod and temperature control of plant architecture in *Arabidopsis thaliana*. Biosci Biotechnol Biochem. 2013;77:10–6.

Drought Stress Signaling Network

13

Takashi Kuromori, Junya Mizoi, Taishi Umezawa, Kazuko Yamaguchi-Shinozaki, and Kazuo Shinozaki

Contents

T. Kuromori • K. Shinozaki (✉)
Gene Discovery Research Group, RIKEN Center for Sustainable Resource Science, Yokohama, Japan
e-mail: takashi.kuromori@riken.jp; kazuo.shinozaki@riken.jp

J. Mizoi • K. Yamaguchi-Shinozaki
Laboratory of Plant Molecular Physiology, Graduate School of Agricultural and Life Sciences, The University of Tokyo, Tokyo, Japan
e-mail: ajmizoi@mail.ecc.u-tokyo.ac.jp; akys@mail.ecc.u-tokyo.ac.jp

T. Umezawa
Faculty of Agriculture and Graduate, School of Bio-Applications and Systems Engineering, Tokyo University of Agriculture and Technology, Tokyo, Japan
e-mail: taishi@cc.tuat.ac.jp

S.H. Howell (ed.), *Molecular Biology, The Plant Sciences 2*,
DOI 10.1007/978-1-4614-7570-5_7

Abstract

- Stress signaling networks in drought responses are composed of intracellular signaling systems, transcriptional regulatory complexes, and intercellular communication systems.
- The signaling mechanisms underlying changes in gene expression enable plant responses to drought stress.
- Signaling factors and transcription factors are themselves regulated transcriptionally and/or posttranslationally (e.g., phosphorylation or proteolysis) in response to drought stress.
- Abscisic acid (ABA) is a key phytohormone, and ABA signaling is a major part of the drought response regulatory networks. However, ABA-independent pathways are also involved.
- The complexity of and cross talk between ABA-dependent and ABA-independent pathways in drought stress signaling networks have been extensively analyzed at the cellular level, but not at the intercellular level.
- Intercellular signaling in response to water deficit has to be elucidated for a comprehensive understanding of plant responses and adaptation to drought stress.

Introduction

Our society is at present facing various challenges, such as climate change, competition for land use, rising demands for food, etc. Addressing these problems will depend in part upon expanding our knowledge of plant biology, because plants provide us with food and other sustainable resources. Drought is an abiotic stress responsible for some of the greatest crop losses. Water deficiency limits plant yields; therefore, improvement in drought stress tolerance may markedly increase crop yields.

Understanding drought stress responses in plants has been an active area of plant research. Historically, drought stress responses have been observed through numerous physiological experiments in various plants, including crops, vegetables, trees, and horticulture plants. In addition to classical studies, there has been marked progress in molecular biological analyses over the last decade. Particularly, the availability of the *Arabidopsis* genome sequence has had a major impact on all fields

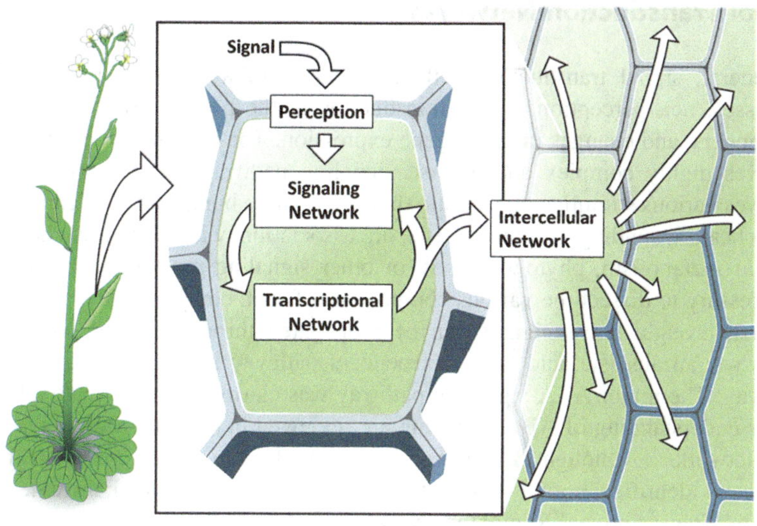

Fig. 1 Overview of signaling flow in the plant drought response. The signaling network is activated after perception of stress, which enhances transcriptional networks to produce molecules required to cope with the stress, and induces signaling to distant tissues to make adjustments at the whole-plant level

of plant research and has resulted in rapid developments in plant molecular biology. The availability of the complete genome sequence has provided much information regarding all genes, including predicted genes, *cis* elements, gene structures, and gene families. In addition to *Arabidopsis*, the complete genome sequences of other plant species have been reported. Comparative analysis between species has resulted in significant progress in research regarding the mechanisms of plant drought response and tolerance in nature. Plant genome resources have been elaborated upon through comprehensive transcriptome, proteome, and metabolome analyses. Integration of data obtained from each -omics hierarchy has provided a more comprehensive picture of drought stress responses (Hirayama and Shinozaki 2010).

When plants respond to drought, stress signals are issued immediately after perception of the change in water status, transcription of stress response genes is induced according to these signals, and further signals are spread systemically throughout the plant (Fig. 1). During drought stress, the phytohormone, abscisic acid (ABA), accumulates and plays important roles in various responses to drought stress, such as stomata closure, induction of stress proteins, and accumulation of various metabolites for the protection of cells from water deficit stress (Umezawa et al. 2010). In addition, molecular analyses have revealed ABA-independent processes in drought stress responses. This entry describes the drought stress response signaling networks, which consist of intracellular signal transduction and transcriptional regulatory and intercellular communication networks.

Signal Transduction Networks

In general, signal transduction pathways consist of a number of well-ordered processes, i.e., perception, signal transduction at the protein level or by second messengers, and changes in target gene expression. The response to drought stress involves highly complex and diverse signaling pathways as this type of stress involves various stimuli, e.g., osmotic shock, oxidative burst, and sometimes strong light, heat, wounding responses, etc., in association with dynamic changes in cellular metabolism, phytohormones, or other signaling molecules. Therefore, it is necessary to dissect the various components of such complex pathways.

First, drought stress signals can be propagated through one or both of two pathways, an abscisic acid (ABA)-dependent pathway and an ABA-independent pathway. The major ABA-signaling pathway was clarified recently, and advances in our understanding of ABA signaling are reviewed in section "Signal Transduction Networks." Although a number of factors involved in other signaling pathways have been identified, not all such factors have yet been determined, as reviewed in section "Transcriptional Regulatory Networks."

The Core ABA-Signaling Pathway

As described above, ABA is a major phytohormone involved in the drought stress response in plants, which has been shown to induce various responses, such as stomatal closure and stress-responsive gene expression. Hence, ABA signaling has been suggested to play an important part in drought stress responses in plants. This section presents a review of early ABA signaling, focusing on three major components: the PYR/PYL/RCAR-type ABA receptor, 2C-type protein phosphatase (PP2C), and SNF1-related protein kinase 2 (SnRK2) (Umezawa et al. 2010).

PP2C: A Classical Regulator in ABA Signaling

As in the case of other phytohormones, genetic studies have made significant contributions to our understanding of the ABA signal transduction pathway. In the 1990s, ABA-insensitive (ABI) *Arabidopsis* mutants were isolated through genetic screening for ABA sensitivity. *ABI1* and *ABI2* were shown to encode homologous 2C-type protein phosphatases (PP2C), suggesting the importance of protein phosphorylation for ABA signaling. ABI1 and ABI2 belong to the group A subfamily of plant PP2Cs; nine members of group A PP2C have been identified in the *Arabidopsis* genome. Initially, ABI1 and ABI2 were identified from strongly ABA-insensitive mutants, *abi1-1* and *abi2-1*. However, subsequent studies demonstrated that disruption of *ABI1* or *ABI2* genes caused ABA hypersensitivity. Furthermore, some other PP2Cs, e.g., AHG1 and AHG3/PP2CA, were identified from ABA-hypersensitive mutants. Therefore, group A PP2Cs were shown to act as negative regulators of ABA signaling. The reason why *abi1-1* and *abi2-2* mutations cause ABA insensitivity remained unclear until 2009 (see below).

A series of genetic studies suggested that group A PP2Cs may be functionally redundant but still play distinctive roles in ABA signaling. For example, *abi1-1* or *abi2-1* shows an ABA-insensitive phenotype in overall ABA responses, such as seed dormancy, germination, stomatal closure, and ABA-responsive gene expression, suggesting that both ABI1/ABI2 play roles in these various plant tissues. However, AHG1 and AHG3 play more specific roles in seeds. Therefore, members of the group A PP2C family act as global regulators of ABA signaling in plants.

ABA Receptors

The evidence outlined above indicated the importance of group A PP2Cs to ABA signaling; however, it was not clear how they were involved. The mechanism was finally clarified by two studies in 2009 that identified soluble ABA receptors, PYR/PYL/RCARs. The groups making these discoveries used different approaches. Cutler's group identified PYR1 by chemical genetics using an ABA agonist, and Grill's group identified RCAR1 by identifying PP2C-interacting proteins. There are 14 members of the PYR/PYL/RCAR family in *Arabidopsis*. These members have been suggested to be not only functionally redundant but also to have partially distinct roles in different plant tissues. Each PYR/PYL/RCAR protein has a hydrophobic pocket that can recognize ABA as a ligand. Binding of ABA alters the conformation of the PYR/PYL/RCAR protein by closing the "gate and latch" structure over the ABA pocket. Such conformational changes enable PYR/PYL/RCAR to interact with group A PP2Cs, resulting in inhibition of their protein phosphatase activity.

SnRK2: A Major Positive Regulator of ABA Signaling

Events downstream of PP2C were clarified in several studies of ABA-activated protein kinases. The SNF1-related protein kinase 2 (SnRK2) family was identified as ABA-activated protein kinases in plants. There are 10 members of the SnRK2 family in *Arabidopsis*; these can be classified into three subclasses. Among them, subclass III SnRK2s, i.e., SRK2E/OST1/SnRK2.6, SRK2D/SnRK2.2, and SRK2I/SnRK2.3, are strongly activated by ABA and play pivotal roles in the plant ABA response. This conclusion was obtained from reverse genetic studies in which a triple-knockout mutant of subclass III SnRK2s (*srk2dei* or *snrk2.2/2.3/2.6*) was established. The mutant exhibited extreme ABA insensitivity; for example, *srk2dei* seeds can germinate even in the presence of 100 μM ABA. In addition, *srk2dei* impaired most ABA responses, e.g., stomatal closure, ABA-responsive gene expression, seed dormancy, etc. These results indicated that subclass III SnRK2s are indispensable for ABA signaling as positive global regulators.

Current Model of the Core ABA-Signaling Pathway

Subclass III SnRK2s are activated in response to ABA and in turn positively regulate ABA signaling. A series of biochemical studies were performed to determine the mechanism underlying ABA-dependent SnRK2 activation. First, in vitro and in vivo assays indicated that subclass III SnRK2s physically interact with group

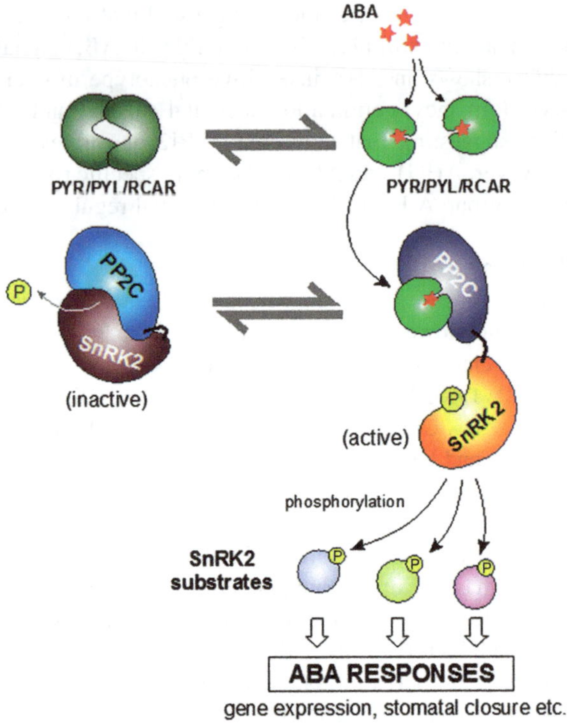

Fig. 2 The core ABA signal transduction pathway in plants. PYR/PYL/RCAR, ABA receptor; PP2C, type 2C protein phosphatase; SnRK2, SNF1-related protein kinase 2

A PP2Cs. Then, an in vitro reconstitution assay demonstrated that group A PP2Cs can dephosphorylate and inactivate subclass III SnRK2s, indicating that group A PP2Cs negatively regulate subclass III SnRK2s.

Subclass III SnRK2s can phosphorylate the AREB/ABF-type bZIP transcription factor, a major transcriptional regulator of ABA-responsive gene expression. The major ABA signal transduction pathway has been determined, as shown in Fig. 2. In the absence of ABA, group A PP2Cs inactivate subclass III SnRK2s by direct dephosphorylation. After elevation of endogenous ABA levels in response to environmental stimuli or developmental cues, ABA is captured by the receptor PYR/PYL/RCARs, and ABA-bound receptors interact with and inhibit group A PP2Cs. Then, subclass III SnRK2s are activated and phosphorylate downstream factors (substrates), such as AREB/ABFs, ion channels, etc., and induce cellular and physiological ABA responses. This model was strongly supported by structural analyses of PYR/PYL/RCAR-PP2C or PP2C-SnRK2 complexes.

Furthermore, a long-standing mystery surrounding the *abi1-1* mutation was also finally resolved. The *abi1-1* mutation is a point mutation, which results from a substitution of Lys to Asp in a PP2C catalytic domain. Biochemical analyses indicated that protein phosphatase activity against SnRK2 was not markedly changed by *abi1-1* mutation, but ABA-bound PYR/PYL/RCAR could not inhibit *abi1-1*-type PP2C. Thus, *abi1-1* mutation leads to continuous inactivation of SnRK2 even in the presence of ABA in plant cells, resulting in ABA insensitivity of *abi1-1* mutant plants.

Some other proteins have been identified as SnRK2 substrates/downstream factors. For example, SnRK2 phosphorylates and regulates slow anion channels, SLAC1, or its homologs, which are essential for stomatal closure. In addition, SnRK2 can phosphorylate the potassium channel KAT1 and the NADPH oxidase AtrbohD, suggesting that a wide range of proteins can be phosphorylated by SnRK2. Recently, it was reported that the potassium transporters KUP6 and KUP8 are involved in osmotic-stress responses, and SnRK2 can interact with and phosphorylate them. Identification of in vivo SnRK2 substrates is important to understand both ABA and osmotic-stress signaling. Phosphoproteomics analysis of SnRK2 substrates through the *srk2dei* mutant has revealed downstream proteins that are regulated by ABA receptor complex (Umezawa et al. 2013).

Other Signaling Pathways

Early Osmotic-Stress Signaling

In the 1990s, osmosensors and osmotic-stress signaling were characterized in yeast cells in which it was shown that *Sln1* encodes a membrane-spanning histidine kinase that regulates a MAPK cascade, the Hog1 pathway, that directly controls osmotic-stress-responsive gene expression. In *Arabidopsis*, AHK1/AtHK1 were identified by their ability to complement Sln1 in yeast and are therefore candidate plant osmosensors. Reverse genetic studies revealed that AHK1/AtHK1 are involved in osmotic-stress signaling. Although the signal transduction mechanism is unknown, response-regulator-like proteins (ARRs) may function downstream of AHK1/AtHK1.

Several protein kinases have been suggested to be involved in osmotic-stress signaling. For example, SnRK2s are also activated in response to osmotic stress. In contrast to ABA signaling, all subclasses of SnRK2 show osmotic-stress-responsive activation. Among them, SRK2C/SnRK2.8, a subclass II SnRK2, is activated by osmotic stress and positively regulates drought tolerance in *Arabidopsis*. In SRK2C-overexpressing plants, a number of drought stress-responsive genes, including the DREB1A/CBF3 transcription factor, were hyperinduced by drought stress. In addition, a reverse genetic study suggested that the subclass II SnRK2s, SRK2C/SnRK2.8 and SRK2F/SnRK2.7, may be the major factors involved in osmotic-stress signaling, although also involved in ABA signaling. Finally, a pivotal role of SnRK2s in osmotic-stress signaling was reported by a recent study in which an *Arabidopsis*-decuple mutant in all ten SnRK2s was developed. The decuple mutant was hypersensitive to osmotic stress, suggesting that SnR2s positively regulates plant tolerance to osmotic stress.

As described above, SnRK2s are activated by both ABA and osmotic stress. Although the mechanism of ABA-dependent activation of SnRK2 has been established, it remains unclear how osmotic-stress signals activate SnRK2s (Kulik et al. 2011). It is likely that osmotic-stress signaling involves a number of mechanisms of SnRK2 activation, because osmotic stress-dependent activation is more rapid than ABA-dependent activation. In addition, subclass III SnRK2s are negatively regulated by group A PP2Cs, as described above, while subclasses I and II SnRK2s show no or only weak interactions, respectively, with PP2C. Therefore, some as-yet-unidentified negative regulators may function upstream of SnRK2 in the osmotic-stress response. A candidate regulator is the calcium-binding protein designated SnRK2-interacting calcium sensor (SCS), which interacts with a tobacco SnRK2, NtOSAK. An in vitro kinase assay demonstrated that SCS inhibits SnRK2 activity in a Ca^{2+}-dependent manner, suggesting that SCS acts as a negative regulator of SnRK2 activation.

Calcium-Dependent Pathway

Among the several second messengers involved in drought stress signaling, Ca^{2+} plays an important role in both ABA-dependent and ABA-independent pathways by affecting different types of protein kinases, including the calcium-dependent protein kinase (CDPK) family. CDPK contains internal EF-hand motif(s), and its activity is directly regulated by Ca^{2+} binding. ABA or osmotic stress induces an intracellular Ca^{2+} spike or oscillation, which likely affects CDPK activity. Several studies have indicated that CDPKs are also involved in ABA or osmotic-stress signaling in plants.

Several CDPK members are involved in typical ABA responses, e.g., stomatal closure or gene expression control. In addition to the SnRK2 family, recent studies highlighted the possible involvement of CDPKs in the core ABA pathway. Interestingly, SnRK2s and CDPKs may share substrates in ABA signaling, e.g., the *Arabidopsis* CDPKs, AtCPK4 and AtCPK11, positively regulate ABA responses and can phosphorylate AREB/ABF transcription factors. As described above, AREB/ABFs are major substrates of SnRK2 in the current ABA-signaling model. Another well-known SnRK2 substrate is SLAC1, a slow anion channel, and its homologs, which play an essential role in the regulation of ion concentration in guard cells for stomatal closure. Recent studies demonstrated that SLAC1 is also phosphorylated by some CDPKs, i.e., AtCPK6, AtCPK21, and AtCPK23. In addition, recent studies revealed that some CDPKs functionally and physically interact with group A PP2Cs, as well as SnRK2s, suggesting that CDPKs may function as an alternative pathway in core ABA signaling.

Another Ca^{2+}-regulated protein kinase family, i.e., the CBL-interacting protein kinase (CIPK)/SnRK3 family, plays significant roles in ABA and osmotic-stress signaling. The protein kinase activity of CIPK is regulated in a Ca^{2+}-dependent manner. In the absence of Ca^{2+}, CIPK activity is inhibited by autoregulation. When intracellular Ca^{2+} is elevated, calcineurin B-like (CBL) proteins bind to Ca^{2+} and then associate with CIPKs to mediate their autoregulation. The CBL-CIPK system participates in various signal transduction pathways, e.g., salt, ABA, or osmotic stress, and has a crucial role in salt tolerance in plants. In the early 2000s, a genetic screen identified *Arabidopsis* mutants, *salt-overly-sensitive 2* (*sos2*) and *sos3*, and *SOS2* and *SOS3* encode CIPK24 and CBL4, respectively. In this case, SOS2 activity is upregulated by the Ca^{2+}-bound form of SOS3, and then SOS2 phosphorylates membrane-bound Na^{+}/H^{+} antiporters to maintain cellular ion homeostasis by excluding Na^{+}. Another example is CBL1/9-CIPK23, which regulates the AKT1 potassium channel. In addition, the CBL-CIPK system has been implicated in ABA and osmotic-stress signaling.

Mitogen-Activated Protein Kinase (MAPK)-Mediated Signaling

The MAPK family is a group of protein kinases that is well conserved in eukaryotes. In general, MAPK activity is regulated by MAPK cascade(s), consisting of MAPK, MAPK kinase (MAPKK), and MAPKK kinase (MAPKKK). MAPKKs activate MAPKs by phosphorylation of the TxY motif, while MAPK phosphatases dephosphorylate and inactivate MAPKs. Plant MAPKs play key roles in various signaling pathways in responses to pathogens, phytohormones (ethylene, jasmonate, etc.), cell differentiation, etc. Several MAPK cascades are involved in ABA or osmotic stress, e.g., AtMPK1, AtMPK2, AtMPK3, and AtMPK6 are activated by ABA or osmotic stress. However, our knowledge of the molecular mechanism of MAPK signaling in ABA or osmotic-stress responses is limited. MAPKs can be activated by ABA or osmotic stress in several ways. For example, plant MAPKs are activated by reactive oxygen species (ROS), and previous studies demonstrated that ABA and osmotic stress cause an oxidative burst in plant cells. In addition, a recent study indicated that MAPK(s) can be regulated directly by calmodulin(s). The in vivo targets of MAPKs should be identified to understand MAPK-mediated ABA or osmotic-stress signaling. SnRK2 has been shown to activate MAPK1 and MAPK2 in response to osmotic stress.

Phospholipids

In addition to Ca^{2+}, phospholipid signaling had been well studied as a second messenger system in the context of ABA or osmotic-stress signaling. The several classes of phospholipid-derived signal messengers, such as phosphatidic acid (PA), diacylglycerol, inositol polyphosphates, etc., are generated by modification of membrane lipids by several types of enzymes, including phospholipases, lipid kinases, and/or phosphatases. One such enzyme, phospholipase D (PLD), is a major family of phospholipases in plants, which hydrolyses phospholipids to generate PA. Recent studies indicated that PA production by PLDs plays a significant role in plant drought responses. Different PLDs and their PAs likely have different functions, e.g., AtPLDα1 and AtPLDδ are involved in the ABA and ROS responses, respectively. PA binds to several target proteins, such as ABI1 (PP2C), phosphoinositide-dependent protein kinase 1 (PDK), and NADPH oxidases. In addition, PA can activate H^+-ATPase or MAPKs. Further studies are required to determine the functional roles of PA or PLDs in drought stress signaling in plants.

Reactive Oxygen Species (ROS)

As described above, ROS may function as signal mediators in ABA or osmotic-stress signaling. There are several sources of ROS generation in plant cells. NADPH oxidase is involved in ROS generation. In *Arabidopsis*, two NADPH oxidases, AtrbohD and AtrbohF, have been implicated in ABA responses in guard cells. In vitro analysis showed that ABA-activated SnRK2 phosphorylates AtrbohF, suggesting that ROS generation may be affected by ABA signaling.

In general, ROS attack proteins, thus inducing functional changes and affecting signal transduction systems. Therefore, it is important to identify which proteins are attacked by ROS; however, our knowledge regarding those attacked in ABA and osmotic-stress signaling is limited. Biochemical studies have shown that ABI1 and ABI2, group A PP2Cs, are sensitive to H_2O_2 at the protein level. However, the functional relationships between such biochemical evidence and in vivo ABA signaling are unclear.

Nitric oxide (NO) is another oxidant that functions as a signaling molecule. The importance of NO was originally identified in the blood vessels of animals, and it was then confirmed to be an endogenous signal in plant cells. For example, nitrate-reductase-mediated NO generation is required for ABA-induced stomatal closure, by affecting ion channels in guard cells. Furthermore, NO stimulates MAPK activation as well as ROS, suggesting its involvement in the responses to abiotic stress, phytohormones, or developmental cues. However, our knowledge of the mechanism of NO-mediated signaling in plants is inadequate. As in the case of ROS, identifying the targets of NO in plant cells is vital.

Proteolysis

In addition to posttranslational modifications and second messengers, proteolysis is a major signal transduction regulatory mechanism. Proteolysis affects signal transduction systems by specifically controlling the amounts of signaling factors at the protein level. Such selective proteolysis depends on ubiquitination of target proteins by E3 ubiquitin ligases. Several types of E3 ligases exist in plants, a number of which are implicated in ABA signaling, e.g., RING E3 ligases, such as KEEP ON GOING (KEG), RHA2a/b, SDIR1, AIP2, etc.; or components of CUL4-based E3 ligases, DWA1 and DWA2. It will be important to identify such proteolysis pathways to gain insight into feedback regulation in ABA signaling. In addition to ubiquitination, some other modifications, such as SUMOylation, have also been implicated in ABA and osmotic-stress signaling. The SUMO ligase AtSIZ1 plays multiple roles not only in ABA signaling but also in phosphate deficiency and flowering. For both ubiquitination and SUMOylation, target proteins should be identified to gain further understanding of these systems.

Transcriptional Regulatory Networks

Importance of Gene Expression in the Plant Response to Drought

Plants respond continuously to changes in the surrounding environment and optimize their stress tolerance and growth. Many molecular, physiological, and morphological changes occur in response to drought and contribute to the enhancement of stress resistance. Stress-induced changes in the expression levels of many genes are essential to drive these processes. Stress-inducible genes that function in the acquisition of stress tolerance are categorized into the following two groups:

functional and regulatory genes (Yamaguchi-Shinozaki and Shinozaki 2006). The former include genes for proteins that contribute directly to cellular stress tolerance, such as LEA (late embryogenesis abundant) proteins, molecular chaperones, enzymes for detoxification of reactive oxygen species, and those for the biosynthesis of sugars or proline, which are important as osmolytes and/or protectants. The other group of genes includes those for proteins that are involved in signal transduction and gene expression, such as protein kinases, components of ABA signaling, enzymes for lipid signaling, and various transcription factors.

In plants, complex gene expression networks are involved in stress-induced gene regulation (Yamaguchi-Shinozaki and Shinozaki 2006). Recent advances in microarray technology revealed that hundreds to thousands of genes are up- and downregulated in response to changes in the surrounding environment in a stress-specific manner. Transcription factors can induce expression of a specific set of genes, and therefore stress-specific transcription factors play important roles as hubs of transcription networks, which decode inputs of inter- and intracellular signals into specific gene expression patterns.

The plant hormone ABA plays an important role in responses to water deficit, including at the transcriptional level (Yamaguchi-Shinozaki and Shinozaki 2006; Fujita et al. 2011). A significant proportion of genes that are induced by water deficit are also highly induced by exogenous application of ABA. The expression of these genes is significantly decreased in mutants that are deficient in ABA biosynthesis or ABA signal transduction, indicating the importance of ABA-dependent pathways in the transcriptional responses to water deficit. Nevertheless, there are groups of genes that are induced in response to water deficit but are not highly induced by exogenously applied ABA. Furthermore, the expression of these genes is induced even in ABA- or ABA-signaling-deficient mutants. Therefore, ABA-independent pathways, in addition to the ABA-dependent pathways, are involved in plant transcriptional responses to water deficit (Yamaguchi-Shinozaki and Shinozaki 2006). Hereafter, major ABA-dependent and ABA-independent gene expression pathways that respond to water deficit will be introduced (Fig. 3).

AREB/ABF Transcription Factors and Their Target cis-Acting Element ABRE Represent a Major ABA-Dependent Pathway

In addition to its role in the response to water deficit in vegetative tissues, ABA plays other important roles in seed maturation and dormancy (Fujita et al. 2011). In particular, the cellular environment in the late stage of embryogenesis shares similarity with vegetative tissues under conditions of severe drought in terms of cellular dehydration. Many genes that support cellular function during dehydration are commonly induced under these two conditions. Such genes include those encoding late embryogenesis abundant (LEA) proteins, which are considered to protect cells from dehydration. ABA induces a large proportion of these genes and contributes to the development of cellular tolerance to dehydration. Most of these

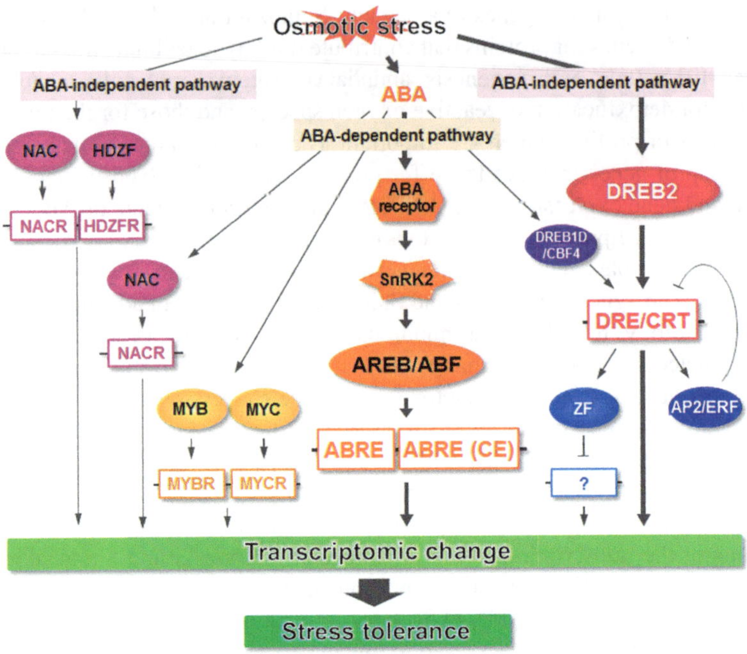

Fig. 3 Major transcriptional regulatory networks of transcription factors and their target *cis*-acting elements involved in osmotic-stress-responsive gene expression in *Arabidopsis thaliana*. Transcription factors and *cis*-acting elements are indicated by ellipses and boxes, respectively. Key components of ABA signaling, the ABA receptor complex and the protein kinase SnRK2, are also shown. Early and late gene expression responses are shown in the *upper* and *lower parts*, respectively. The two major pathways are indicated by *thick gray arrows* and other pathways by *thin gray arrows*

ABA-inducible genes are regulated by the ABA-responsive element (ABRE) in their promoters. The ABRE is an 8-bp sequence (PyACGTGG/TC), with a core ACGT sequence, which is targeted by plant bZIP-type transcription factors. A single copy of an ABRE is not sufficient to induce ABA-responsive gene expression. To function as an active *cis*-acting element, ABRE requires in proximity another copy of ABRE or another specific *cis*-acting element, which is termed a coupling element. Examples of these coupling elements are CE1 and CE3, and DRE/CRT (a major *cis*-acting element in an ABA-independent pathway; refer to the section on "DREB Transcription Factors" below) has been shown to function as an ABRE coupling element. In fact, the promoters of many dehydration-inducible genes have more than two ABRE copies, and combinations of an ABRE and a coupling element are enriched in the promoters of ABA-inducible genes, which support the importance of ABREs in ABA-dependent gene expression (Maruyama et al. 2012).

The major transcription factors that activate gene expression through ABREs in response to water deficit in the vegetative tissues are the ABRE-binding proteins/factors (AREBs/ABFs), which belong to the bZIP transcription factors (Fujita et al. 2011). *AREB/ABF* family genes (*AREB1/ABF2*, *AREB2/ABF4*, *ABF1*, and *ABF3*) are expressed mainly in the vegetative tissues under conditions of abiotic stress. Three members, *AREB1/ABF2*, *AREB2/ABF4*, and *ABF3*, are highly induced in response to ABA and osmotic-stress conditions, such as high salinity and dehydration. A triple AREB/ABF mutant exhibits reduced resistance to drought and decreased sensitivity to exogenous ABA compared with wild type, single mutants, or double mutants. Transcriptome analysis of the triple mutant under osmotic-stress conditions showed that the levels of expression of many osmotic-stress-inducible genes, including those for many LEA proteins, protein phosphatase 2Cs, and transcription factors, were reduced. Thus, these three AREBs/ABFs have been shown to be central transcription factors that cooperatively function in ABA-dependent transcriptional activation through their ABREs under these abiotic stress conditions. The function of AREBs/ABFs in ABA-dependent transcriptional activation in response to abiotic stress has also been established in rice.

In addition to their stress inducibility, the activity of AREBs/ABFs is posttranslationally regulated through phosphorylation (Fujita et al. 2011). The phosphorylation of AREBs/ABFs is necessary for their full activation and is catalyzed by SnRK2s that have been activated by ABA. The importance of the posttranslational activation of AREBs/ABFs was suggested by the observation that overexpression of *AREB1/ABF2* activated downstream gene expression only when an artificial active form of this gene was used. Thus, in the major ABA-dependent transcriptional network, an increase in the cellular ABA concentration in response to water deficit is recognized by the ABA receptor complexes, and then the receptor complex in turn activates the SnRK2-AREB/ABF pathway to induce expression of genes that are under the control of the *cis*-acting element ABRE.

DREB Transcription Factors and Their Target cis-Acting Element DRE/CRT Represent a Major ABA-Independent Pathway

In *Arabidopsis*, the *RD29A/COR78/LTI78* gene is induced by dehydration and low temperature. This gene is inducible by exogenously applied ABA. The analysis of its promoter, together with expression analyses in mutants defective in ABA biosynthesis or signaling, revealed that the dehydration inducibility of this gene is regulated by both ABA-independent and ABA-dependent pathways through different *cis*-acting elements (Yamaguchi-Shinozaki and Shinozaki 2006). Promoter fragments of *RD29A* that respond to dehydration independent of ABA contain a *cis*-acting element, i.e., the dehydration-responsive element (DRE; TACCGACAT). Similar *cis*-acting elements, named C-repeat (CRT) and low-temperature-responsive element (LTRE), were found in low-temperature-inducible genes. These sequences share a common core sequence, A/GCCGAC, which is referred to as the DRE/CRT core sequence. The DRE/CRT core sequence is found in the promoters of many dehydration-inducible genes and represents a major ABA-independent pathway. Two major transcription factors that recognize DRE/CRT and activate downstream gene transcription are DRE-binding protein 1/CRT-binding factor (DREB1/CBF) and DREB2. Both DREB1/CBF and DREB2 belong to the plant-specific APETALA2/ethylene-responsive element-binding factor (AP2/ERF) type transcription factors, and each of DREB1/CBF and DREB2 forms a conserved subgroup. *Arabidopsis* has six and eight *DREB1/CBF* type and *DREB2*-type genes, respectively. Among them, three DREB1/CBF-type transcription factor genes, *DREB1A/CBF3*, *DREB1B/CBF1*, and *DREB1C/CBF2* are highly induced by low temperature and act as major transcription factors that activate gene transcription through DRE/CRT. In contrast, two DREB2-type transcription factor genes, *DREB2A* and *DREB2B*, are highly induced in response to osmotic-stress conditions, such as dehydration and high salinity and are considered to be involved in DRE-mediated gene transcription in response to water deficit.

Transcriptional regulation is critical for the activity of DREB1/CBF, whereas additional regulation at the posttranslational level plays important roles in activating DREB2A-mediated gene transcription (Qin et al. 2011). The DREB2A protein has a negative regulatory domain (NRD), and removal of the NRD from DREB2A converts the protein into its constitutively active form (DREB2A CA). The DREB2A protein is a target of selective proteolysis by the 26S proteasome, and the DREB2A CA protein accumulates more stably in the nucleus than the wild-type protein, suggesting that stability control is a posttranslational regulatory mechanism of DREB2A. In contrast to wild-type *DREB2A*, overexpression of *DREB2A CA* induces the expression of many dehydration-inducible genes and improves drought resistance in transgenic *Arabidopsis* plants. The downstream genes of DREB2A include not only functional genes, such as those encoding LEA proteins or sugar metabolism enzymes, but also those for regulatory genes, such as transcription factors. It is interesting that some of these transcription factors are members of a group of AP2/ERF-type repressors with an EAR-type repression motif that bind to DRE/CRT, suggesting the existence of a negative feedback loop downstream of DREB2A. The transcriptomic and metabolomic patterns in DREB2A CA-expressing plants are similar to those in drought-stressed wild-type *Arabidopsis* plants. Moreover, DREB2A CA-transgenic *Arabidopsis* plants exhibit dwarf morphology similar to stressed plants, which is a phenomenon that often occurs in transgenic plants overexpressing an active positive regulator of stress tolerance. Thus, DREB2A is an important regulator of DRE/CRT-mediated gene induction in the ABA-independent pathway in response to water deficit.

DREB2A is also induced in response to heat shock, and DREB2A CA can induce heat-responsive genes and enhance heat-shock tolerance in transgenic plants (Mizoi et al. 2012). It is interesting that DRE/CRT is involved in gene expression not only during dehydration but also under conditions of low and high temperature. There are overlaps between dehydration-inducible genes and low- or high-temperature-inducible genes. These overlaps in the downstream genes are reasonable because both low and high temperatures are closely related to dehydration. For example, at freezing temperatures, ice formation in apoplastic spaces causes severe cellular dehydration, and protective molecules, such as LEA proteins and sugars, are considered to be important for maintaining cellular integrity. Heat stress is associated with cellular dehydration under severe drought conditions with strong solar radiation and/or high temperature. Under these conditions, plants cease cooling their leaves by transpiration, which causes an increase in leaf temperature; the expression of molecular chaperones facilitates maintenance of cellular functions. The cross talk between dehydration and low-temperature signals at DRE/CRT is mediated by the different target specificities of the dehydration-inducible DREB2 and the low-temperature-inducible DREB1/CBF. In contrast, the cross talk between DRE/CRT-regulated gene expression under dehydration and high-temperature conditions is related to an unknown mechanism of stress-specific target selection by DREB2A.

DREB1D/CBF4, a DREB1-type transcription factor, is highly induced by dehydration and exogenous application of ABA. The overexpression of DREB1D/CBF4 results in the expression of DREB1/CBF target genes and improves the drought resistance of transgenic plants, suggesting a role for DREB1D/CBF4 in the ABA-dependent transcriptional response to dehydration (Mizoi et al. 2012).

NAC Transcription Factors Function in Both ABA-Independent and ABA-Dependent Pathways

The *ERD1* gene encoding a ClpA homologue of *Arabidopsis* is induced by dehydration but not by ABA and is therefore under the control of an ABA-independent pathway. Analysis of the *ERD1* promoter revealed the involvement of two separate *cis*-acting elements, an MYC-like sequence (CATGTG) and a 14-bp *rps1* site 1-like sequence, in the induction of this gene. The MYC-like sequence was found to be a target of three NAM, ATAF1, ATAF2, and CUC (NAC) transcription factors, ANAC019, ANAC055, and ANAC072/RD26. NAC transcription factors are plant specific, and these three proteins are included in the stress-responsive NAC (SNAC) group of NAC transcription factors, which is one of the six major groups of NAC transcription factors. Consequently, a zinc-finger homeodomain (ZFHD) transcription factor, ZFHD1, was identified as a transcriptional activator that recognizes the 14-bp *rps1* site1-like sequence. Overexpression of any one of the three NAC transcription factors upregulated the expression of multiple stress-inducible genes and enhanced stress tolerance but was not sufficient to induce *ERD1* expression. Expression of the *ERD1* gene was induced only when both the NAC and ZFHD proteins were overproduced simultaneously in a transgenic plant. Thus, these transcription factors cooperatively activate the transcription of the *ERD1* gene through an ABA-independent pathway (Nakashima et al. 2012).

In addition to their roles in the ABA-independent pathway, NAC transcription factors also regulate gene expression through an ABA-dependent pathway (Nakashima et al. 2012). The three NAC transcription factors are induced by ABA, and the genes upregulated by overexpression of *ANAC072/RD26* include many that are ABA-inducible. In addition, the overexpression of *ANAC072/RD26* enhances ABA sensitivity, whereas transgenic plants in which its activity was repressed were insensitive.

In rice, five SNAC genes, *OsNAC3*, *4*, *5*, *6*, and *SNAC1*, are induced by abiotic stress, such as dehydration, high salinity, and low temperature. Of these, *OsNAC5*, *OsNAC6*, and *SNAC1* are responsive to ABA, with *OsNAC5* and *OsNAC6* showing particularly marked upregulation. The overexpression of *OsNAC5*, *OsNAC6*, or *SNAC1* induces expression of abiotic stress-responsive genes and enhances the drought resistance of transgenic rice. Several stress-inducible *NAC* genes are also induced by jasmonates and/or during senescence in *Arabidopsis* and rice. The targets of these *SNAC* genes include genes involved in biotic stress responses or senescence. Thus, these stress-responsive NAC transcription factors not only function in the transcriptional response to abiotic stress conditions, including water stress, but are likely involved in the cross talk between abiotic and biotic stress responses, as well as between stress responses and senescence (Nakashima et al. 2012).

Other Transcriptional Pathways Involved in Water Stress Responses

In addition to the aforementioned pathways of *cis*-acting elements and transcription factors, many other transcriptional pathways function in water stress responses. The *Arabidopsis RD22* gene is inducible by dehydration, which is dependent on ABA biosynthesis. The induction of *RD22* is regulated by two *cis*-acting elements, MYC and MYB recognition elements. A MYC transcription factor, MYC2, and a MYB transcription factor, MYB2, can bind to these *cis*-acting elements and cooperatively activate the transcription of this gene. Simultaneous expression of MYC2 and MYB2 in transgenic *Arabidopsis* induces several target genes, including not only ABA-inducible genes but also jasmonate-inducible genes. Conversely, a mutation in *MYC2* decreased the expression of target genes, including *RD22*. MYC2 has also been reported to act as a transcription factor that functions in defense responses and light signaling; it is, therefore, considered to be a key regulator of cross talk between abiotic and biotic stress responses and light signaling pathways (Yamaguchi-Shinozaki and Shinozaki 2006).

A group of C_2H_2-zinc-finger transcription factor genes are induced in response to water deficit (Kiełbowicz-Matuk 2012). The promoters of these genes contain stress-related *cis*-acting elements, such as DRE/CRT, ABRE, MYB recognition sequences, and a MYC recognition sequence, and were shown to be expressed downstream of DREB1A/CBF3 or DREB2A. Ectopic overexpression of the C_2H_2-zinc-finger gene, *STZ*, has been shown to increase the drought resistance of transgenic *Arabidopsis*. Thus, these C_2H_2-zinc-finger transcription factors form a cascade in a transcriptional network downstream of another transcription factor.

The involvement of WRKY transcription factors, which target W-box sequences, in ABA-dependent gene expression in response to water deficit has been reported (Rushton et al. 2012). AtWRKY40, a negative regulator of ABA responses, binds to the promoters of multiple stress-inducible transcription factor genes, including *DREB1A/CBF3*, *DREB2A*, and *MYB2*, and represses their expression at low ABA concentrations, whereas an elevated ABA concentration induces relocalization of AtWRKY40 from the nucleus to the plastids, which releases the target genes from repression. Another *WRKY* gene, *ABO3/AtWRKY63*, is the causative gene for the *aba overly sensitive 3* mutant. Although the *abo3* mutant is hypersensitive to ABA in seedling establishment and growth, stomatal closure is less sensitive to ABA, and the *abo3* mutant exhibits lower drought resistance. The induction of *AREB1/ABF2* in response to exogenous ABA was reduced in the *abo3* mutant, and AtWRKY63 was shown to bind to a W-box sequence in the promoter of *AREB1/ABF2* in vitro, suggesting the involvement of AtWRKY63 in ABA-dependent regulation of gene expression.

Nuclear factor Y (NF-Y) is a heterotrimeric transcription factor that is conserved in eukaryotes and which binds to the CCAAT-box (Laloum et al. 2013). In plants, each subunit is encoded by multiple genes, which generates a diversity of trimers. In *Arabidopsis*, *NF-YA5* was shown to be upregulated during drought stress and in response to ABA. A loss-of-function mutant of this gene is hypersensitive to drought. Overexpression of this gene improves drought resistance, which is associated with altered expression of stress-responsive genes. *NF-YB1* and its ortholog in maize, *ZmNF-YB2*, increase drought resistance in transgenic *Arabidopsis* and maize, respectively.

Regulation of Transcriptional Networks by Small Noncoding RNAs

In addition to transcription factors, small noncoding RNAs, which consist of 20–24 nucleotides, represent another type of regulators in the transcriptional networks of plants (Khraiwech et al. 2012). The two major classes of small noncoding RNAs in plants are microRNAs (miRNAs) and small interfering RNAs (siRNAs). miRNAs are transcribed from endogenous *MIR* loci as a precursor containing a hairpin structure and then are processed into a mature form. They recognize the reverse complementary sequence of target mRNAs and negatively regulate the targets by either cleaving or inhibiting translation. Many *MIR* loci and their target genes have been identified and found to be conserved among plants. A large number of miRNAs are upregulated or downregulated in response to drought; their targets include transcription factors and signaling components that can regulate gene expression, development or morphological changes, and thus these miRNAs are believed to be involved in adaptive responses of plants.

siRNAs are generated and function in a similar pathway to miRNAs, but they are derived from long double-stranded RNAs and can also cause transcriptional gene silencing via DNA methylation. The source of the long double-stranded RNAs are variable: endogenously, they are mainly generated from *cis*-antisense transcript pairs (i.e., overlapping transcripts in the opposite directions) or RNA-dependent transcription. An example of siRNA-dependent regulation is the degradation of an mRNA for a proline-catabolizing enzyme in response to salt stress, which helps the accumulation of proline (Khraiwech et al. 2012).

Plant Transcriptional Network Complexity May Be Important for Optimizing Responses to the Ever-Changing Natural Environment

As described above, plants use the complex transcriptional networks, which include a number of *cis*-acting elements and transcription factors that recognize them, to respond to water deficit. Plant transcription factors often form homologous groups, which increases the complexity of the networks. In addition, there is cross talk between these transcriptional networks, i.e., those among networks that respond to other abiotic stress conditions, such as low and high temperatures, as well as to biotic stresses. Moreover, the configuration and components of these networks differ depending on the internal environment, such as between different growth stages and tissues. In nature, plants are exposed to annually and diurnally fluctuating environments. In addition, the surrounding environment varies depending on where a seed germinates and puts down roots. The extremely complex transcriptional network system of plants may have evolved to adapt to the diverse conditions found in natural environments. Although only a part of the cross talk between multiple transcriptional networks in plants has been determined, further research will help us gain a more comprehensive understanding of the responses of plants to environmental stress conditions, including drought.

Intercellular Communication Networks

Stress responses in plants are very different from those in animals. Plants have neither specialized sensory organs nor a nervous system. Animals have nervous systems to react rapidly and avoid adverse conditions. By contrast, plants are sessile and cannot escape from stress conditions, so that they have evolved the capability to sense and respond to various stresses in their changing environment. In addition, animals have nervous systems to rapidly propagate signals from one part of the organism to another. Plants, too, have means of locally or systemically spreading stress signals. However, our knowledge about how plants transmit intercellular and systemic signals is lacking compared to our understanding of intracellular signaling (Jia and Zhang 2008). Nonetheless, intercellular signaling in response to drought is described in this section from three viewpoints: root-to-shoot, vasculature-to-guard-cell, and cell-to-cell signaling (Fig. 4).

Root-to-Shoot Signaling

Stomatal regulation is well recognized as a model system for studies of cellular signal transduction. However, recent studies have suggested that stomatal movements can be used not only as a model system for research on intracellular signal transduction but also for research on plant remote signaling as stomatal behavior may be regulated remotely in response to a variety of environmental stresses.

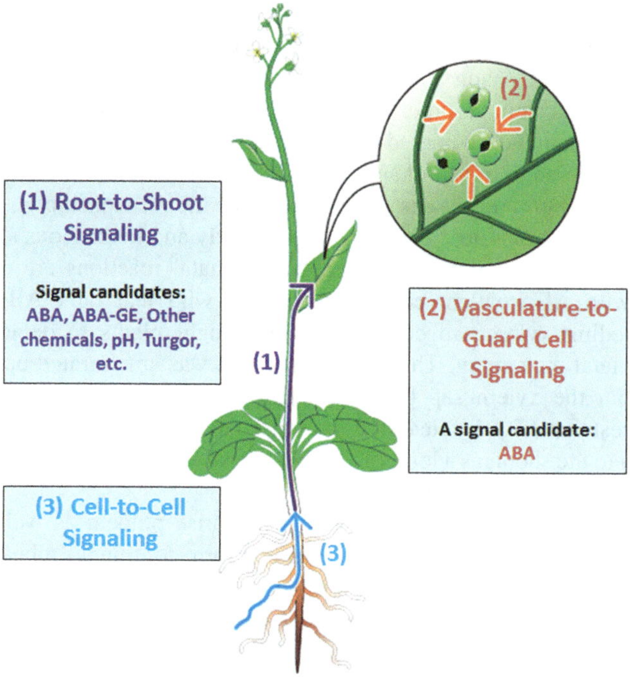

Fig. 4 An overview of regional viewpoints of intercellular signaling in the plant drought response. The intercellular drought response signaling network can be considered in terms of the following regional divisions: root-to-shoot signaling (*purple line*) (*1*), vasculature-to-guard-cell signaling (*brown line*) (*2*), and cell-to-cell signaling (*blue line*) (*3*). Each signal candidate is described in the text

Among the studies concerning stomatal movements in relation to remote signaling, the best examples are those of stomatal movements in relation to root-to-shoot signaling under drought stress (Jia and Zhang 2008). In some plant species, including maize (*Zea mays*) and apple (*Malus domestica*), stomatal closure was shown to occur when part of the root system was exposed to water deficit in a root-split experiment, even though the water status in the leaves remained unchanged. These findings indicated that a root-derived signal was transported to the leaves, inducing stomatal closure (Seo and Koshiba 2011). While the factor involved in root-to-shoot signaling under drought stress has not been elucidated, several candidates have been proposed: abscisic acid (and related molecules), pH, turgor (hydraulic signal), and some other biomolecules.

Abscisic Acid (ABA)

Abscisic acid (ABA) regulates stomatal movements. Therefore, ABA is a good candidate as a root-to-shoot signal in drought responses. Many studies have indicated that soil drying results in an increased ABA level in the roots. Under mild stress conditions when soil drying has just begun, even though the water potential of the leaves is only affected slightly or unaffected, ABA accumulates in root tissues. This increase in ABA level is closely correlated with a decrease in leaf stomatal conductance (Jia and Zhang 2008).

In addition to accumulation of ABA in the roots, many studies have shown that drought stress can induce a marked increase in ABA content in xylem sap. Xylem is generally considered to transport and store water, nutrients, and hormones from the roots to the aboveground tissues. Investigation of the relationship between ABA xylem concentration and leaf conductance in woody and herbaceous species originating from different habitats revealed that stomatal reactions are much better correlated with ABA xylem concentration than with leaf bulk ABA level. In addition, feeding xylem sap collected from drought plants to detached leaves inhibits stomatal movement. This inhibitory effect was ameliorated upon removal of ABA from the xylem sap by ABA affinity column chromatography. These studies suggest that ABA is released from roots to the xylem vessels and transported in the xylem to the shoot, as a long-distance hormonal signal that remotely regulates stomatal movement under drought stress.

The vasculature-based transport systems comprise phloem as well as xylem, because both consist of conductive elements that form continuous tubular columns. ABA has been reported to be generally present at much higher levels in phloem than in xylem tubes. Some of the ABA accumulated in leaves is transported to the roots, and tracer experiments using isotope-labeled ABA have indicated that the movement of ABA from leaves to roots is activated by water deficiency in the roots. If a stem is girdled (tissue stripped off annularly to disrupt phloem), ABA accumulation is restrained to the roots, indicating that ABA translocation from shoot to root is conducted through phloem sap. Based on these studies, ABA recirculation, in which ABA is loaded into the phloem and then transported to the roots where it is deposited in the xylem vessels, has also been discussed (Umezawa et al. 2011).

Although the crucial roles of ABA in root-to-shoot signaling are established, it should be noted that the mechanism underlying long-distance signaling is not completely understood and ABA may not be a universal signal that plays a central role in all cases. There is evidence that the remote regulation of stomatal movement by root-derived ABA is dependent on plant species. For example, root-derived ABA may be a central regulator of stomatal movement in many plant species, including tomato and wheat, but not in other plant species, including maize and sycamore.

ABA Glucosyl Ester (ABA-GE)

ABA is conjugated to glucose to form ABA glucosyl esters (ABA-GEs), another transportable candidate that may function in root-to-shoot signaling under stress conditions. ABA-GE is a reversibly inactive form of ABA and is widespread among plant species. Therefore, ABA-GE is thought to serve as a storage form of ABA. In *Arabidopsis*, this conjugated form is hydrolyzed to the bioactive ABA form by the β-glucosidase AtBG1 localized to the endoplasmic reticulum. Water stress can induce polymerization of AtBG1, thereby activating the enzyme to release ABA, leading to an increase in the active ABA concentration. AtBG1-deficient

Arabidopsis plants exhibit lower ABA levels in leaves and exhibit stress-sensitive phenotypes. While ABA-GE is located in xylem sap, cell vacuoles and likely in the cytosol and cell wall, AtBG1 β-glucosidase is located in the endoplasmic reticulum where it remains during stress responses. Recently, vacuolar-localized AtBG2 has been reported to release active ABA by hydrolyzing ABA-GE. The glucosyl-transferase UGT71B6 that glucosylates ABA into ABA-GE in vitro and in vivo has also been identified. Accordingly, ABA-GE is a candidate long-distance signaling molecule in the vascular system, especially under stress conditions. Nevertheless, it has been argued that the ABA-GE level in roots is too small to contribute significantly to the overall increase in ABA during water stress, and no mechanism for ABA-GE transport has yet been reported (Goodger and Schachtman 2010).

Other Chemical Agents

Various compounds have been proposed as alternatives to ABA as xylem-borne stress signals. The conjugate form of ABA is one such candidate described above, while ABA precursors are other candidates related to ABA biosynthesis. AAO3, which catalyzes the final step of ABA biosynthesis, shows a marked increase in transcript level in response to drought stress within the guard cells, suggesting abscisic aldehyde as a candidate signaling molecule, which is a compound immediately upstream of ABA in the ABA biosynthetic pathway that converts ABA by AAO3. However, the AAO3 protein level does not increase rapidly after dehydration; relatively prolonged drought treatment is necessary for its accumulation. In addition, the transport of ABA precursors has not been investigated.

Cytokinin is a phytohormone synthesized mainly in roots and is an antagonist of ABA. Cytokinin is correlated with stomatal control in some plants. Acetylcholine is another candidate as a long-distance signaling molecule, because it functions as a neurotransmitter in animals and has physiological functions in both plants and animals. Nevertheless, both cytokinin and acetylcholine have been suggested to play roles in the regulation of stomatal movement only under normal conditions, not in drought stress (Jia and Zhang 2008).

Malate is another candidate; this molecule is a xylem sap constituent the level of which has been shown to increase in response to water stress and is involved in the guard-cell signal transduction network. Nevertheless, the xylem sap malate concentration increases significantly only in the later stages of water stress. In addition, the use of malate in a transpiration bioassay had no effect on stomata. Recently, sulfate was suggested as a candidate xylem-borne factor that combines with ABA in the early water stress response (Goodger and Schachtman 2010). However, further investigations are required in detail.

pH

As described above, ABA is one of the most likely candidates for the root-to-shoot signaling molecule. However, as ABA is a weak acid, it can be compartmentalized by pH conditions in plant tissues. ABA compartmentalization is commonly explained by "anion trap theory"; ionized ABA is accumulated in alkaline sections and compartmentally redistributed by the pH change gradient over lipid membranes. Based on this theory, if leaf apoplastic pH is changed, the local ABA level would be altered, causing variations in stomatal status. Thus, even without induction of de novo ABA biosynthesis or ABA transport from roots, stomatal closure can be induced by the compartmental redistribution of ABA in leaf tissues due to pH gradient changes. Therefore, pH may act as a regulator of stomatal movement if a pH change is induced by drought stress. Indeed, it has repeatedly been shown that the pH of the xylem sap increases in response to dry soil conditions in many plant species.

It has been proposed that the soil drying-induced pH increase in xylem would cause the leaf apoplast to become more alkaline, which would contribute to increased sequestration of ABA in the apoplastic space around guard cells, thus promoting stomatal closure. An experiment involving feeding detached leaves of *Commelina communis* artificial xylem sap buffered to different pH values indicated that an increase in pH reduces the transpiration rate. Another study demonstrated that modulation of apoplastic pH regulates stomatal movements in response to root-derived ABA signals. These data suggest that pH, in conjunction with ABA, acts to coordinately regulate stomatal movement under drought stress conditions. However, other studies indicate that pH signaling is not used by all plant species. For example, drought stress may decrease xylem sap pH in some plant species (Jia and Zhang 2008; Seo and Koshiba 2011).

Turgor (Hydraulic)

While it has been demonstrated increasingly that chemical signals are responsible for root-to-shoot signaling in response to soil drying, there is also evidence supporting hydraulic signaling as a root–shoot signal. In some trees, the soil drying-induced reduction in leaf conductance can be progressively reversed by pressurization of the root system. In addition, the leaf conductance returns rapidly to its pre-pressurization levels once depressurized. A more recent experiment in *Arabidopsis* indicated that the stomatal closure induced by root-applied water deficits could be relieved by supplying water directly to leaves. This indicates that a hydraulic signal may be involved in regulation of stomatal behavior. Similar findings have been reported in the woody plant *Betula occidentalis* and in bell pepper (*Capsicum annuum*).

Accordingly, it has been postulated that ABA acts downstream of the hydraulic signal in communicating water stress between roots and shoots, and soil water stress appears to elicit a hydraulic response in the shoot, which precedes ABA signaling and results in stomatal closure. This hypothesis is supported by the finding that a relatively low water potential in *C. communis* epidermis significantly promotes ABA-induced stomatal closure, although it has no direct effect on the stomatal aperture. The same observation was obtained by feeding ABA into field-grown plants over a range of leaf water potentials.

In addition to the above explanations, the functions of the chemical and hydraulic signals appear to depend on plant species and/or stress stage. Hydraulic signaling is more likely in woody plants. If both chemical and hydraulic signals are involved in a specific signaling process, it will be important to determine at what stage which signaling pattern plays a predominant role and at what stage(s) the various signaling patterns control stomatal behavior in a synergistic manner (Jia and Zhang 2008).

Vasculature-to-Guard-Cell Signaling in Leaves

Much experimental data support the involvement of ABA in root-to-shoot signaling under drought stress. However, whether ABA is first synthesized in roots and transported to shoots or is directly produced in leaf vascular tissues as an immediate response to stress remains unclear. Some studies indicate that the vasculature of leaves, but not that of roots, is the primary site of ABA biosynthesis under either control or drought stress conditions. Using antibodies specific for AtNCED3, AtABA2, and AAO3, leaf vascular parenchyma cells have been demonstrated to be the main sites of ABA biosynthesis under drought stress as well as well-watered conditions. Reciprocal grafting between ABA-deficient mutants and wild-type tomato and *Arabidopsis* plants demonstrated that stomatal closure was affected by the leaf (shoot) genotype, but not the root genotype, and therefore ABA biosynthesis in the shoot was necessary and sufficient to mediate stomatal closure of plants water stressed at the roots. Studies of beans and *Arabidopsis* have shown that ABA is produced in greater amounts at an earlier stage in leaves relative to roots in response to water stress, and therefore ABA-induced stomatal closure is not dependent on root-released ABA. Such studies suggest that ABA acting on stomata may be produced mostly in leaves of plants subjected to water stress. As a consequence, signal(s) other than ABA must be produced in roots under water stress and transmitted to leaves, following which ABA is likely transported from its site of synthesis to its site of action, e.g., the stomata, under drought conditions (Seo and Koshiba 2011).

Based on the above findings, it has been proposed that ABA in the vascular tissue is transported to guard cells, leading to its accumulation in stomata. Auxin is another phytohormone that co-ordinates plant development. Recently, the auxin transport system has been shown to include both cellular efflux and influx carriers. In current models, auxin migrates to adjacent cells through polar transport to regulate disproportionate cell growth. Although it is not clear whether ABA is transported in a polar manner, as is the case for auxin, ABA is generally considered to be transported in the transpiration stream. The ABA signal should be transmitted to peripheral guard cells to prevent transpiration as soon as possible, especially under drought stress conditions.

As ABA is a weak acid (pKa 4.7), it exists mostly in the ionized form in the cytosol, where the pH is approximately neutral. From the viewpoint of an anion trap, ionized ABA cannot pass through the lipid bilayer of plasma membranes via a passive diffusion mechanism without specific transporters. Thus, an ABA exporter is necessary for ABA transport from the interior to the exterior of the cell to overcome the anion trap. Similarly, as stress conditions elevate the apoplastic pH, an ABA importer is also necessary for cellular uptake of ionized ABA, e.g., at guard cells, particularly under drought stress conditions (Umezawa et al. 2011).

Although there have been indications of their presence, ABA transporters have been reported only recently. Based on mutant analysis, two ABA transporters, AtABCG25 and AtABCG40, which belong to the ATP-binding cassette transporter family, have been characterized biochemically and shown to play a role in mediating ABA export from inside to outside the cell and import from outside to inside the cell, respectively. The promoter of AtABCG25 is active in the vascular tissue and that of AtABCG40 is active in guard cells, consistent with the hypothesis that the primary site of ABA synthesis is vascular tissues, from which ABA is transported to guard cells. However, mutants defective in each of these transporters do not perfectly match the typical phenotypes of ABA-deficient mutants. This suggests the existence of redundant transporters or a passive transport mechanism mediated by pH gradient (Sreenivasulu et al. 2012).

Recently, another class of ABA-importing transporters (AIT1–4) was identified from the yeast two-hybrid system, which belongs to the low-affinity nitrate transporter family suggested to function as ABA importers at the site of ABA synthesis involved in regulation of the stomatal aperture. These findings strongly suggest active control of ABA transport by cellular carriers. Indeed, this model is consistent with recent reports that some ABA receptors are soluble and localized to the cytosol and nucleus. However, investigation of the ABA transport mechanism is still in the preliminary stages, and so more studies of ABA intercellular regulation are necessary (Boursiac et al. 2013).

Taken together, the above data suggest the following working model; some signals, which have not been fully identified, are transmitted from roots to shoot; drought stress is recognized in leaf vascular tissues; ABA biosynthesis is activated in leaf vascular tissues; and ABA synthesized in the vascular tissues is transported to the guard cells to close stomata. It should be noted that this model is based mostly on experimental evidence in *Arabidopsis*, while plants may show different stomatal closure responses to water stress depending on species and stress conditions.

Cell-to-Cell Signaling

In plants, the apoplast is the extracellular space that allows the moderately free diffusion of molecules, while the symplast is the intracellular space that is separated from the apoplast by the plasma membrane and that consists of the cytoplasm of multiple cells connected by plasmodesmata, which are plasma membrane-enclosed pores that cross the cell wall and contain a central element of endoplasmic reticulum. As described above, under drought conditions the ABA concentration increases in the apoplast, resulting in stomatal closure to prevent transpiration. As the stress hormone ABA is required for fast signaling, ABA may spread rapidly to peripheral cells to cope with environmental changes, especially under stress conditions. The ABA transport system allows rapid transmission of ABA molecules and effective transmission of drought signaling among plant cells. Identification of cellular factors for ABA flux suggests signal distribution by apoplastic diffusion (Umezawa et al. 2011).

In contrast, plants have a unique cell-to-cell network of cytoplasmic connections directly linking plant cells, called the plasmodesmata, which comprise symplastic pathways. Various molecules can use this system to move between cells. Unlike gap junctions in animal systems, the size limits for passage through plasmodesmata vary markedly among tissues. There are two modes of movement between cells via plasmodesmata: selective movement, which is determined by specific sequences within the transported protein, and nonselective movement, which has no such requirements (Busch and Benfey 2010). Although challenging, further research is required to determine whether (or how) ABA and/or other signal molecules are related to symplastic distribution for drought responses.

Generally, the direction of the transpiration stream is from the roots to shoots. Thus, root-derived chemical signals are transmitted within the water stream toward shoots. In roots, as water diffusion into the inner cell layers of the stele is blocked by Casparian strips, water flow into the vasculature requires aquaporins, which are specific transporters of water molecules. Therefore, all molecules, not only water, require specific transporters for diffusion into the vasculature over Casparian strips, suggesting that they enter symplastic pathways at least once.

With regard to crossing of apoplastic and symplastic pathways, it should also be noted that guard cells in shoots are developmentally unique. Most cells in plants are linked symplastically with adjacent cells by plasmodesmata. However, guard cells are symplastically isolated from neighboring cells. This supports the rapid spread on apoplastic arrival of drought signals to guard cells.

As a different type of cell-to-cell signaling, electrical signals are related to reactive oxygen species (ROS), which play roles in cellular signaling in plants as well as various other organisms, from bacteria to mammalian cells. Many reports have suggested that ROS are closely related to abiotic stress signals in plants; for example, ABA-induced stomata closure functions downstream of ROS signals. It has recently been reported that such signals propagate by wounding in *Arabidopsis thaliana*, suggesting a model for extracellular propagation of the ROS wave. Each cell along the path of the wave activates its NADPH oxidase (RbohD) and produces ROS in an autonomous manner, resulting in an autopropagating ROS wave. As a membrane potential could be directly affected by ROS and because the rate of certain electrical signals in plants matches the rate of the ROS wave, it is possible that the generation of a ROS wave affects the formation, amplitude, and/or rate of the electrical signal (Mittler et al. 2011). Further research is needed to address this intriguing possibility, as plants have no neurological capacity.

Future Directions

This article focused on stress signaling networks in drought responses in terms of signal transduction, transcriptional regulation, and intercellular transmission. Drought perception is one of the major outstanding questions in drought signaling (Fig. 1). For example, which types of cells first sense or are sensitive to changes in the drought status, and how do these cells recognize and send drought signals?

Some signal transduction molecules and the related transcriptional factors have been discovered, but their upstream regulators are not known.

Drought stress signaling comprises several complex networks, in which protein modification, proteolysis, and transcriptional and posttranscriptional regulations are involved. Although our knowledge of each signaling factor is increasing, it is still difficult to grasp the whole picture of drought stress signaling. Further investigations are required to interconnect known signaling factors, as well as efforts to discover novel factors. Various approaches, including either mature or emerging technologies, should be utilized in future research. For example, recent in-depth next-generation sequencing analyses identified a number of unknown noncoding RNAs and short ORFs expressed in response to ABA or osmotic stress. With regard to intracellular signal transduction at the protein level, recent advances in proteomics, such as modification-specific proteomics or interactomics, will facilitate further elucidation of drought stress signaling in plants.

Many of the links in intercellular signaling networks are not yet known. Several systemic compounds are candidates for cell-to-cell signaling, and specific transporters for these compounds are under investigation in roots, between roots and shoots, and within shoots. Some ideas for the operation of intercellular signaling networks have been proposed based on model plants, but significant variations in drought tolerance mechanisms among different plant species may exist.

The trade-off between growth ability and stress tolerance is a difficult issue, which will become more important in future studies. Improving stress tolerance through genetic engineering and modifications may have negative effects on plant growth and development. In particular, trade-offs between carbon assimilation and water loss related to drought responses, especially the closing of stomata, should be taken into account. Such studies using transgene expression systems and suitable promoters are at present underway. A simple and easy method of determining water-use efficiency (WUE) is required. Technologies for matching water input with plant requirements at the whole-plant level are essential to improve drought tolerance.

References

Boursiac Y, Léran S, Corratgé-Faillie C, Gojon A, Krouk G, Lacombe B. ABA transport and transporters. Trends Plant Sci. 2013;18:325–33.

Busch W, Benfey PN. Information processing without brains – the power of intercellular regulators in plants. Development. 2010;137:1215–26.

Fujita Y, Fujita M, Shinozaki K, Yamaguchi-Shinozaki K. ABA-mediated transcriptional regulation in response to osmotic stress in plants. J Plant Res. 2011;124:509–25.

Goodger JQD, Schachtman DP. Re-examining the role of ABA as the primary long-distance signal produced by water-stressed roots. Plant Signal Behav. 2010;5:1298–301.

Hirayama T, Shinozaki K. Research on plant abiotic stress responses in the post-genome era: past, present and future. Plant J. 2010;61:1041–52.

Jia W, Zhang J. Stomatal movements and long-distance signaling in plants. Plant Signal Behav. 2008;3:772–7.

Khraiwech B, Zhu JK, Zhu J. Role of miRNAs and siRNAs in biotic and abiotic stress responses of plants. Biochim Biophys Acta. 2012;1819:137–48.

Kiełbowicz-Matuk A. Involvement of plant C_2H_2-type zinc finger transcription factors in stress responses. Plant Sci. 2012;185–6:78–85.

Kulik A, Wawer I, Krzywińska E, Bucholc M, Dobrowolska G. SnRK2 protein kinases–key regulators of plant response to abiotic stresses. OMICS. 2011;15:859–72.

Laloum T, De Mita S, Gamas P, Baudin M, Niebel A. CCAAT-box binding transcription factors in plants: Y so many? Trends Plant Sci. 2013;18:157–66.

Maruyama K, Todaka D, Mizoi J, Yoshida T, Kidokoro S, Matsukura S, Takasaki H, Sakurai T, Yamamoto YY, Yoshiwara K, Kojima M, Sakakibara H, Shinozaki K, Yamaguchi-Shinozaki K. Identification of *cis*-acting promoter elements in cold- and dehydration-induced transcriptional pathways in *Arabidopsis*, rice and soybean. DNA Res. 2012;19:37–49.

Mittler R, Vanderauwera S, Suzuki N, Miller G, Tognetti VB, Vandepoele K, Gollery M, Shulaev V, Van Breusegem F. ROS signaling: the new wave? Trends Plant Sci. 2011;16:300–9.

Mizoi J, Shinozaki K, Yamaguchi-Shinozaki K. AP2/ERF family transcription factors in plant abiotic stress responses. Biochim Biophys Acta. 2012;1819:86–96.

Nakashima K, Takasaki H, Mizoi J, Shinozaki K, Yamaguchi-Shinozaki K. NAC transcription factors in plant abiotic stress responses. Biochim Biophys Acta. 2012;1819:97–103.

Qin F, Shinozaki K, Yamaguchi-Shinozaki K. Achievements and challenges in understanding plant abiotic stress responses and tolerance. Plant Cell Physiol. 2011;52:1569–82.

Rushton DL, Tripathi P, Rabara RC, Lin J, Ringler P, Boken AK, Langum TJ, Smidt L, Boomsma DD, Emme NJ, Chen X, Finer JJ, Shen QJ, Rushton PJ. WRKY transcription factors: key components in abscisic acid signalling. Plant Biotechnol J. 2012;10:2–11.

Seo M, Koshiba T. Transport of ABA from the site of biosynthesis to the site of action. J Plant Res. 2011;124:501–7.

Sreenivasulu N, Harshavardhan VT, Govind G, Seiler C, Kohli A. Contrapuntal role of ABA: does it mediate stress tolerance or plant growth retardation under long-term drought stress? Gene. 2012;506:265–73.

Umezawa T, Nakashima K, Miyakawa T, Kuromori T, Tanokura M, Shinozaki K, Yamaguchi-Shinozaki K. Molecular basis of the core regulatory network in abscisic acid responses: sensing, signaling, and transport. Plant Cell Physiol. 2010;51:1821–39.

Umezawa T, Hirayama T, Kuromori T, Shinozaki K. The regulatory networks of plant responses to abscisic acid. Adv Bot Res. 2011;57:201–48.

Umezawa T, Sugiyama N, Takahashi F, Anderson JC, Ishihama Y, Peck SC, Shinozaki K. Genetics and phosphoproteomics reveal a protein phosphorylation network in the abscisic acid pathway in *Arabidopsis thaliana*. Sci Signal. 2013;6:rs8.

Yamaguchi-Shinozaki K, Shinozaki K. Transcriptional regulatory networks in cellular responses and tolerance to dehydration and cold stresses. Annu Rev Plant Biol. 2006;57:781–803.

Index

S.H. Howell (ed.), *Molecular Biology, The Plant Sciences 2*,
DOI 10.1007/978-1-4614-7570-5

CPSIA information can be obtained at www.ICGtesting.com
Printed in the USA
LVOW01*2108100215

426476LV00001B/191/P

9 781461 475699